NEUROSCIENCE RESEARCH PROGRESS

ENCYCLOPEDIA OF NEUROSCIENCE RESEARCH

VOLUME 1

NEUROSCIENCE RESEARCH PROGRESS

Additional books in this series can be found on Nova's website under the Series tab.

Additional e-books in this series can be found on Nova's website under the e-books tab.

NEUROSCIENCE RESEARCH PROGRESS

ENCYCLOPEDIA OF NEUROSCIENCE RESEARCH

VOLUME 1

EILEEN J. SAMPSON
AND
DONALD R. GLEVINS
EDITORS

Nova Science Publishers, Inc.
New York

For permission to use material from this book please contact us:
Telephone 631-231-7269; Fax 631-231-8175
Web Site: http://www.novapublishers.com

NOTICE TO THE READER

Additional color graphics may be available in the e-book version of this book.

Library of Congress Cataloging-in-Publication Data

Encyclopedia of neuroscience research / editors, Eileen J. Sampson and Donald R. Glevins.
p. cm.
Includes index.
ISBN 978-1-61324-861-4 (hardcover)
1. Neurosciences--Research. I. Sampson, Eileen J. II. Glevins, Donald R.
RC337.E53 2011
616.80072--dc23
2011017613

Published by Nova Science Publishers, Inc. † New York

Contents

VOLUME 1

Preface		**ix**
Chapter I	Prefrontal Morphology, Neurobiology and Clinical Manifestations of Schizophrenia *Tomáš Kašpárek*	**1**
Chapter II	Prefrontal Cholinergic Receptors: Their Role in the Pathology of Schizophrenia *M. Udawela, E. Scarr and B. Dean*	**47**
Chapter III	Participation of the Prefrontal Cortex in the Processing of Sexual and Maternal Incentives *Marisela Hernández González and Miguel Angel Guevara*	**73**
Chapter IV	From Conflict to Problem Solution: The Role of the Medial Prefrontal Cortex in the Learning of Memory-Guided and Context-Adequate Behavioral Strategies for Problem-Solving in Gerbils *Holger Stark*	**115**
Chapter V	The Orbitofrontal Cortex and Emotional Decision-Making: The Neglected Role of Anxiety *Sabine Windmann and Martina Kirsch*	**141**
Chapter VI	PEEF: Premotor Ear-Eye Field. A New Vista of Area 8B *Bon Leopoldo, Marco Lanzilotto and Cristina Lucchetti*	**153**
Chapter VII	Prefrontal Cortex: Its Roles in Cognitive Impairment in Parkinson's Disease Revealed by PET *Qing Wangab, Kelly A. Newella, Peter T. H. Wongd and Ying Luc*	**171**

Chapter VIII	Noradrenergic Actions in Prefrontal Cortex: Relevance to AD/HD *Amy F. T. Arnsten*	**177**
Chapter IX	Prefrontal Cortex: Brodmann and Cajal Revisited *Guy N. Elston and Laurence J. Garey*	**199**
Chapter X	Developmental Characteristics in Category Generation Reflects Differential Prefrontal Cortex Maturation *Julio Cesar Flores Lázaro and Feggy Ostrosky-Solís1*	**215**
Chapter XI	Common Questions and Answers to Deep Brain Stimulation Surgery *Fernando Seijo, Marco Alvarez-Vega1, Beatriz Lozano, Fernando Fernández-González, Elena Santamarta and Antonio Saíz*	**227**
Chapter XII	Deep Brain Stimulation and Cortical Stimulation Methods: A Commentary on Established Applications and Expected Developments *Damianos E. Sakas and Ioannis G. Panourias*	**255**
Chapter XIII	Cortical Stimulation versus Deep Brain Stimulation in Neurological and Psychiatric Disorders: Current State and Future Prospects *Damianos E. Sakas and Ioannis G. Panourias*	**271**
Chapter XIV	Invasive Cortical Stimulation for Parkinson's Disease and Movement Disorders *B. Cioni, A. R. Bentivoglio, C. De Simone, A. Fasano, C. Piano, D. Policicchio, V. Perotti and M. Meglio*	**303**
Chapter XV	Deep Brain Stimulation in Epilepsy: Experimental and Clinical Data *M. Langlois, S. Saillet, B. Feddersen, L. Minotti, L. Vercueil, S Chabardès, O. David, A. Depaulis P. Kahane and C. Deransart*	**319**
Chapter XVI	Psychosurgery of Obsessive-Compulsive Disorder: A New Indication for Deep Brain Stimulation? *Dominique Guehl, Abdelhamid Benazzouz, Bernard Bioulac and Pierre Burbaud,Emmanuel Cuny and Alain Rougier, Jean Tignol and Bruno Aouizerate*	**339**
Chapter XVII	Deep Brain Stimulation in Adult and Pediatric Dystonia *Laura Cif, Simone Hemm, Nathalie Vayssiere and Philippe Coubes*	**353**

VOLUME 2

Chapter XVIII Deep Brain Stimulation of the Subthalamus: Neuropsychological Effects **365**
Rita Moretti, Paola Torre, Rodolfo M. Antonello and Antonio Bava1

Chapter XIX Subthalamic High-Frequency Deep Brain Stimulation Evaluated by Positron Emission Tomography in a Porcine Parkinson Model **381**
Mette S. Nielsen, Flemming Andersen, Paul Cumming, Arne Møller, Albert Gjedde, Jens. C. Sørensen and Carsten R. Bjarkam1

Chapter XX Current and Future Perspectives on Vagus Nerve Stimulation in Treatment-Resistant Depression **395**
Bernardo Dell'Osso, Giulia Camuri, Lucio Oldani and A. Carlo Altamura

Chapter XXI Cognitive Aspects in Idiopathic Epilepsy **407**
Sherifa A. Hamed

Chapter XXII Cognitive Impairment in Children with ADHD: Developing a Novel Standardised Single Case Design Approach to Assessing Stimulant Medication Response **443**
Catherine Mollica, Paul Maruff and Alasdair Vance

Chapter XXIII Novel Therapies for Alzheimer's Disease: Potentially Disease Modifying Drugs **475**
Daniela Galimberti, Chiara Fenoglio and Elio Scarpini

Chapter XXIV Cognitive Interventions to Improve Prefrontal Functions **499**
Yoshiyuki Tachibana, Yuko Akitsuki and Ryuta Kawashima

Chapter XXV Insights from Proteomics into Mild Cognitive Impairment, Likely the Earliest Stage of Alzheimer's Disease **521**
Renã A. Sowell and D. Allan Butterfield

Chapter XXVI Animal Models for Cerebrovascular Impairment and its Relevance in Vascular Dementia **541**
Veronica Lifshitz and Dan Frenkel

Chapter XXVII The Critical Role of Cognitive Function in the Effective Self-administration of Inhaler Therapy **561**
S. C. Allen

Chapter XXVIII Foetal Alcohol Spectrum Disorders: The 21st Century Intellectual Disability **573**
Teresa Whitehurst

Chapter XXIX Where There are no Tests: A Systematic Approach to Test Adaptation **587**
Penny Holding, Amina Abubakar and Patricia Kitsao Wekulo

Chapter XXX Paid Personal Assistance for Older Adults with Cognitive Impairment Living at Home: Current Concerns and Challenges for the Future **599**
Claudio Bilotta, Luigi Bergamaschini, Paola Nicolini and Carlo Vergani

Chapter XXXI Neurotoxicity, Autism, and Cognitive Impairment **609**
Rebecca Cicha, Brett Holfeld and F. R. Ferraro

VOLUME 3

Chapter XXXII Molecular Mechanisms Involved in the Pathogenesis of Huntington Disease **617**
Claudia Perandones, Martín Radrizzani and Federico Eduardo Micheli

Chapter XXXIII Huntingtin Interacting Proteins: Involvement in Diverse Molecular Functions, Biological Processes and Pathways **655**
Nitai P. Bhattacharyya, Moumita Datta, Manisha Banerjee, Srijit Das and Saikat Mukhopadhyay

Chapter XXXIV DNA Repair and Huntington's Disease **677**
Fabio Coppedè

Chapter XXXV Putting Together Evidence and Experience: Best Care in Huntington's Disease **691**
Zinzi Paola1, Jacopini Gioia1, Frontali Marina and Anna Rita Bentivoglio

Chapter XXXVI Oral Health Care for the Individual with Huntington's Disease **705**
Robert Rada

Chapter XXXVII Suicidal Ideation and Behaviour in Huntington's Disease **717**
Tarja-Brita Robins Wahlin

Chapter XXXVIII Making Reproductive Decisions in the Face of a Late-Onset Genetic Disorder, Huntington's Disease: An Evaluation of Naturalistic Decision-Making Initiatives **745**
Claudia Downing

Chapter XXXIX The Control of Adult Neurogenesis by the Microenvironment and How This May be Altered in Huntington's Disease **799**
Wendy Phillips and Roger A. Barker

Index **857**

Preface

This book presents current research in the field of neuroscience. Topics discussed include noradrenergic actions in prefrontal cortex; prefrontal morphology, neurobiology and clinical manifestations of schizophrenia; prefrontal cholinergic receptors; deep brain stimulation and cortical stimulation methods; deep brain stimulation in adult and pediatric dystonia; cognitive aspects in idiopathic epilepsy; cognitive interventions to improve prefrontal functions; molecular mechanisms involved in the pathogenesis of Huntington's disease and DNA repair and Huntington's disease

[*]Chapter 1- Schizophrenia is a heterogeneous condition with highly variable clinical manifestation, time-course, and response to various treatment approaches. During the course of the illness, subjects with schizophrenia experience periods of psychosis with symptoms such as hallucinations, delusions or disorganization of thoughts and behavior. Simultaneously as well as after the psychosis alleviation, they may experience so-called negative symptoms such as emotional withdrawal, apathy, anhedonia, lack of initiative, etc. and a varying degree of cognitive dysfunction, with working memory, verbal learning and memory, sustained attention or executive functions being affected the most. One of the cardinal features of schizophrenia is its typical time window of onset: although there are exceptions, the illness manifests during late adolescence or early adulthood in most cases. The outcome of the illness is inherently variable in the sense of the expression of symptoms and functional impairment. Unfortunately, it is very hard to predict the probable course of the illness in individual patients. The failure of current operationally-based diagnostic approaches to provide homogenous subgroups of patients with similar outcomes and responses to treatment led

to a search for distinct neurobiologically-defined syndromes that may share clinical manifestations. Studies of brain morphology may help in this effort.

Chapter 2- Schizophrenia is a debilitating mental illness, affecting approximately 1% of the population. Importantly, CHRM antagonists have been extensively used to control the extrapyramidal side effects associated with the first generation of antipsychotic drugs but it is now becoming clear that the cholinergic receptors may be a target for drugs to alleviate the symptoms of schizophrenia. The symptoms of schizophrenia can be broadly divided into

[*] Versions of chapters 1-10 were also published in *Prefrontal Cortex: Roles, Interventions and Traumas*, edited by Lorenzo LoGrasso and Giovanni Morretti, published by Nova Science Publishers, Inc. They were submitted for appropriate modifications in an effort to encourage wider dissemination of research.

positive symptoms, such as hallucinations and delusions, negative symptoms, such as social withdrawal and lack of motivation, and cognitive deficits, which can include impairments in attention/information processing, problem-solving, speed of processing, working memory and verbal and visual learning and memory. Whilst the positive symptoms have proven malleable to treatment with antipsychotic drugs, the negative symptoms and cognitive deficits are still essentially resistant to available treatments. This in part could be due to schizophrenia being a syndrome, rather than a distinct disease with a single pathology, with the different diseases within the syndrome showing differential responsiveness to available drugs. Thus, a major challenge is to be able to dissect the syndrome of schizophrenia into its components, to understand the pathologies of these biologically distinct entities and to design effective treatments for each disorder. Perhaps the most pressing need is for treatments for the cognitive deficits associated with schizophrenia as these are now recognised as the most debilitating symptoms associated with the disorder.

Chapter 3- In studying the physiology of sexual and maternal behaviors it is important to gain insight into the early phases of sexual and maternal relations among animals. These phases include detection and perception of the stimuli, gender identification, motivation and arousal. Agmo has defined the term sexual motivation as the process that makes an animal search for sexual contact with another animal. It is well known that sexual motivation is activated when a suitable stimulus –e.g., a mate– is perceived. A similar definition can be applied to the maternal motivation, as the process that makes a mother search for maternal contact with a pup (her own or alien); *i.e.*, the process that promotes nurturing behaviors toward an offspring. This is activated when the suitable stimulus –e.g., a pup– is perceived. Given that the suitable stimulus generally activates approach behavior, it can be suggested that these stimuli may function as a positive incentive.

Chapter 4- The tremendous capability of living creatures to individually acquire and store information ensures their survival by fast and adequate adaptation to their environment. Therefore, learning and memory have multiple facets and involve a complex and distributed network in the brain . Our contribution in the field of cognition will be limited to observations on cortical and prefrontal dopamine as well as cortical theta activity during aversive learning and retrieval. Our hypotheses are derived from correlations between the behavior of learning animals and the aforementioned physiological learning-dependent observables. Nevertheless, we are dealing here with fundamental processes of individual acquisition and storage of information. The discussion will focus on two points: (1) how the learning process is initiated, and (2) the way in which the learner forms a goal for guiding the learning process.

This chapter is composed in the form of a lab report to illustrate that the sequence of experiments and discussions presents a learning process itself driven by steadily arising scientific questions.

Chapter 5- The orbital part of the prefrontal cortex is known to be crucially involved in emotional decision-making. Its function seems to be, first, to associate affective value with behavioral bias (i.e., approach or withdrawal) in complex choice situations, and secondly, to flexibly control and transiently inhibit affective and behavioral impulses. Dysfunction of this region due to damage or immaturity of OFC cells leads to impulsiveness, behavioral perseveration, and reduced sensitivity for risk, as observed in children as well as various patients groups. We review the evidence with a focus on the much less investigated question of whether and how *hyper*activation (or hyperfunctioning of modules and processes) involving OFC leads to the opposite behavioral tendencies, namely risk-aversion, enhanced inhibition, undecidedness and higher risk sensitivity including heightened anxiety. We identify two perspectives with opposing views onto this issue, and report some unpublished studies of populations with children and patients with anxiety disorders using the Iowa gambling task, a task that is presumed to be sensitive to OFC function. We find that, if anything, anxiety has the same beneficial effects onto performance as cortical maturation.

Chapter 6- The prefrontal cortex is the rostral part of the frontal cortex. It is subdivided in many cytoarchitectonic areas with different functions. The prefrontal cortex is more developed in non-human and human primates than in other animals. The anatomical interconnections and physiological data suggest that the prefrontal cortex is involved in high-order brain functions and in the organization of motor behaviour.

In this review we will talk about area 8B, which has remained relatively an unexplored part of the prefrontal cortex. Area 8B, together with area 6 and 8A, based on cytoarchitecture, anatomical connections and electrophysiological data, may be considered as part of the premotor cortex and area 8B may be considered a "premotor ear-eye field" (PEEF).

Chapter 7- Parkinson's disease (PD) is a neurodegenerative disorder, classically characterized by the chronic loss of dopaminergic neurons primarily in substantia nigra pars compacta, which leads to a reduction of dopamine levels in the striatum. The motor symptoms do not become obvious until at least an 80% reduction in striatal dopamine levels. Recent works have demonstrated that cognitive impairment occurs even in the early course of PD which is correlated with dopaminergic dysfunctions in the prefrontal cortex and not necessarily related to motor disorders. This commentary will discuss the current knowledge of the prefrontal cortex and its association with cognitive deficits in PD.

Chapter 8- Basic science has made great strides in understanding the neurochemical influences needed for optimal prefrontal cortical (PFC) regulation of behavior and attention. Although much research has focused on the important role of dopamine (DA), there is increasing appreciation that norepinephrine (NE) has a vital influence as well. NE plays an essential role in arousal and attention, and may orchestrate the interplay of posterior sensory cortices, subcortical structures mediating primitive responses, and PFC regulation of attention and behavior. The current chapter reviews this literature, and outlines its direct relevance to our understanding and treatment of Attention Deficit Hyperactivity Disorder (AD/HD).

Chapter 9- Apparently mostly unbeknown to neuroscientists, Korbinian Brodmann published an unprecedented and unsurpassed set of data almost 100 years ago in which he quantified, among other things, the size of the frontal lobe and "prefrontal cortex" in a variety of primate and non-primate species. Brodmann's definition of prefrontal cortex is clear: granular cortex (that containing an identifiable layer IV) in the frontal lobe. Because this term has become confused in modern studies we use the term granular prefrontal cortex (gPFC) here in accordance with Brodmann's definition. We highlight in italics examples of where the term has been misused. Brodmann's observations of gPFC were derived from his Nissl preparations of the brains of a large number of species, both primate and non-primate (Figure 1). Brodmann's data on the frontal lobe and gPFC (Table 1) reveal several interesting observations which predate modern findings. Moreover, statements made in modern studies are often inconsistent with the Brodmann data. In particular Brodmann's data reveal that the size of the frontal lobe of humans is smaller than that predicted from the non-human primate data (Figure 2A). In addition, his data reveal that the proportion of the frontal lobe occupied by the gPFC varies considerably between species (Figure 2B).

Chapter 10- The ability to categorize is fundamental for cognitive development. The evolution from concrete to abstract thinking goes trough different developmental stages that shape cognitive performance in children.

Young children frequently compare and categorize objects using perceptual features, like for example the shape of an object. Progressively when children learn new words that represent semantic labels, they prefer to use the category such as "animals" or "furniture", rather than a perceptual criteria. While performing comparisons between objects, children systematically introduce cognitive changes and variations in their semantic representations. These changes allow the transformation and construction of experiences into categories, achieved by continuously delineating each category's characteristics and its boundaries, thus creating a conceptual structure for each representation.

Chapter 11- When a new technique is introduced, a certain period of time is required for it to be improved and adjusted. The success of functional stereotactic procedures depends on a variety of factors: patient selection, methodology of choice and localization of the target, and the experience of the neurosurgery team. The learning curve resulting from the neurosurgery team's acquired knowledge is also notable.

Chapter 12- Electricity and some of its various effects on the human body have been known since ancient times. The use of electrical stimulation, however, as a therapeutic tool has been limited. Traditionally, medicine has been applied by administration of pharmaceuticals and surgical interventions. Over the last 50 years, great progress in neurophysiology and neurosurgery made it possible to investigate the therapeutic potential of electrical brain stimulation. During the last two decades, the introduction of deep brain stimulation (DBS) and, to a lesser extent, of chronic electrical cortical stimulation (CS) transformed functional neurosurgery and neuroscience, in general. DBS offers superior clinical efficacy with fewer complications compared to the selective lesional procedures of conventional functional neurosurgery. More importantly, the success of DBS in improving outcome has signified the role of electricity in neurological disorders and the great benefits that can be derived by its modulation, thus creating a new field in medicine this of *selective focal electrotherapy* of the nervous system.

Chapter 13- Therapeutic brain stimulation can be applied: a) in networks which are located in cortical and subcortical layers, b) in deep nuclei groups (relay nodes) or c) in combination. CS and DBS belong to the domain of *operative neuromodulation* and particularly to the field of *neural networks surgery*, defined as the field that studies and applies advancements in neural networks research, digitized stereotactic brain imaging and implantable electrical or electronic devices in order to alter electrically the signal transmission in the nervous system, modulate neural networks and produce therapeutic effects.

Chapter 14- A total of approximately 100 patients with Parkinson's disease have been treated by MI ECS.

1. The first report was that by Canavero and Paolotti, who described the case of a 72 year old woman with advanced PD who showed a substantial improvement of all three cardinal parkinsonian symptoms following unilateral extradural MI ECS (electrode strip parallel to Rolandic fissure). In July 1998, a quadripolar electrode was positioned in the extradural space overlying the hand representation area of MI, contralateral to her worst clinical side. Chronic stimulation was delivered subthreshold for any movement or sensation, at 3V, 180 μs, 25Hz, 3+/0- setting, off during sleep. The clinical improvement was bilateral; she could stand without assistance, climb the stairs and walk for a short distance. The UPDRS III in On-Med

condition decreased from 44 to 23 after 3 months. She had a moderate to severe PD-associated dementia and this aspect also improved. L-dopa was reduced by 80%. Importantly, after 3 months, the patient developed infection and was explanted. However, she lost benefit only gradually over several weeks, a sign of likely neuroplastic changes induced by stimulation. Upon reimplantation, she improved again. These results were confirmed in two further patients operated on in 2001 and 2002. Canavero also reported on a patient suffering multiple system atrophy-associated parkinsonism. Bilateral subthalamic deep brain stimulation was ineffective, but moderately responded to bilateral MI stimulation with the following parameters: 25-40 Hz, 90-180µs, 2-2.5V, bipolar stimulation (3+/0- or viceversa), continuously. Motor symptoms improved for about 9 months, while vegetative symptoms remained improved until death almost 3 years later.

Chapter 15- About 30% of epileptic patients do not respond to antiepileptic drugs, of whom only a minority can benefit from resective surgery. Such a therapeutic option is considered only in patients who suffer from focal seizures with an epileptogenic zone clearly identified and safely removable. Therefore, patients with seizures arising from eloquent cortices, or which are multifocal, bilateral, or generalized, represent a particular challenge to "new" or "alternative" therapies. For these patients, neurostimulation appears with a great potential. Different approaches to neurostimulation in epileptic patients now exist and depend on (i) the brain region which is targeted and (ii) the way the stimulation is applied. The aim of neurostimulation in epilepsy is to reduce the probability of seizure occurrence and/or propagation, either by manipulating remote control systems (vagus nerve stimulation, deep brain stimulation), or by interfering with the epileptogenic zone itself (repetitive transcranial magnetic stimulation, cortical stimulation). In most cases, stimulation is delivered continuously or intermittently according to a scheduled protocol. In particular, new progress in biotechnology and EEG signal analysis now allows stimulation in response to detection of electrographic seizures (e.g., closed-loop stimulation). Here, we review the various experimental and clinical attempts that have been made to control epileptic seizures by the means of deep brain electrical stimulation.

Chapter 16- Obsessive-compulsive disorder (OCD) is a relatively common anxiety disorder that affects 2 to 3% of the general population. According to this estimated lifetime prevalence rate, OCD is the fourth most frequent psychiatric condition following the phobias, substance abuse and major depression. OCD, which is characterized by recurrent intrusive thoughts and repetitive, time-consuming behaviors, is a severely disabling mental illness owing to its intensity, the continuous and unchanging or deteriorative course of its symptoms and the disturbance in psychosocial functioning that they cause..

Chapter 17- In a wide range of medical conditions, a movement disorder can be the only symptom or associated with other neurological deficits. Most of them are of unknown origin. Dystonia, dyskinesia, myoclonus, tremor, atethosis are abnormal movements susceptible to be influenced by high frequency modulation of the basal ganglia. Nevertheless, the control of these symptoms is highly depending on the etiology. Movement disorders (primary dystonia, myoclonus dystonia, tardive dyskinesia), which are not associated to other neurologic deficits are more likely to be controlled by high frequency stimulation.

Chapter 18- The basal ganglia were considered as a site for integrating diverse inputs from the entire cerebral cortex and funneling these influences via the ventrolateral thalamus to the motor cortex. The subthalamic nucleus (STN) has generally been considered as a rely station within frontal-subcortical motor control circuitry. The function of the STN in the current model of the basal ganglia organization was developed from animal model and neurodegenerative diseases with hypo- and hyperkinetic movement disorders: in this model, the STN represents a relay in the so-called "indirect" pathway of the parallel basal ganglia-thalamocortical circuits. STN excitatory efferents reinforce the inhibition of basal ganglia output nuclei, the internal globus pallidus (GPi) and the reticular substantia nigra (SNr) on thalamo-frontal neurons. A "direct" striato-pallidal pathway opposes the STN by normally inhibiting basal ganglia outflow, facilitating thalamo-frontal drive.

Chapter 19- Subthalamic high-frequency deep brain stimulation (STN DBS) has during the last 15 years proven its value in the treatment of Parkinson's disease (PD) complicated by motor fluctuations and levodopa-induced dyskinesias. However, many aspects of subthalamic DBS such as the mechanism of action, potential long-term adverse or neuroprotective effect still remain unelucidated.

Chapter 20- Partial response, lack of response and residual symptoms following antidepressant treatment are common in patients with Major Depression. In fact, it has been estimated that 30 to 45% of depressed patients show either partial response or no response to standard antidepressant treatments. In addition, depressed patients may develop treatment resistant depression (TRD), defined as the lack of response to two antidepressant trials, given in succession, at adequate doses and for an adequate time, in compliant subjects. In patients with TRD, different strategies of intervention have been proposed including pharmacological and psychotherapic augmentation as well as brain stimulation techniques.

[**]Chapter 21- Epilepsy is one of the most common neurological disorders. It nearly affects 1% of the population with the highest incidence during the first year of life. During the last years, the increasing evidence from experimental pathological studies and in vivo imaging, and neuropsychological studies of patients with long-term epilepsy, have substantially increased our awareness about the possibility that briefer recurring focal or generalized seizures may cause a striking progressive cognitive decline which can be more detrimental to an individual's overall life. Determining the frequency of cognitive dysfunction due to epilepsy is difficult to estimate. Community based studies reported that approximately 26.4-30% of children with epilepsy when first diagnosed, have evidence of subnormal global cognitive function or mental retardation with inferior academic achievement. Problems in attention and memory are observed in about 30% of newly diagnosed and untreated epileptic patients with single or several seizures of cryptogenic origin.

Chapter 22- There is limited understanding of the problems associated with repeated neuropsychological assessment in children, including the statistics used to guide decisions about cognitive change. Further, clinicians rarely consider change in cognitive function when evaluating treatment response in individual children with ADHD. This is most likely due to a lack of suitable assessment tasks as well as clinicians' limited awareness of the appropriate statistical techniques for measuring cognitive change in individuals. This chapter outlines a study investigating the application of a statistically principled decision rule to the cognitive and behavioural measures of individual children with ADHD in order to classify a significant, positive response to medication. The data demonstrate an evidence-based approach to clinical decision-making that can be used to evaluate cognitive and behavioural improvement in

[**] Versions of chapters 21-31 were also published in *Cognitive Impairment: Causes, Diagnosis and Treatment*, edited by Melanie L. Landow, published by Nova Science Publishers, Inc. They were submitted for appropriate modifications in an effort to encourage wider dissemination of research.

individual children with ADHD following treatment with stimulant medication. Then, a study investigating a novel "intensive design" method for assessing stimulant medication-related improvement in cognitive function in children with ADHD is presented. The results demonstrate the effectiveness of the medication in improving behavioural symptoms of ADHD, as well as certain features of cognitive function (psychomotor, visual attention and learning). Overall, these findings support the use of a novel "intensive" within-subjects design to examine the short-term effects of stimulant medication on cognitive and behavioural functions in children with ADHD. Further, this design is readily utilised in routine clinical practice.

Chapter 23- Alzheimer's disease (AD) is the most common cause of dementia in the elderly, with a prevalence of 5% after 65 years of age, increasing to about 30% in people aged 85 years or older. It is characterized clinically by progressive cognitive impairment, including impaired judgement, decision-making and orientation, often accompanied, in later stages, by psychobehavioural disturbances as well as language impairment. Mutations in genes encoding for Amyloid Precursors Protein or presenilins 1 and presenilin 2 genes (*APP*, *PSEN1* and *PSEN2,* respectively) account for about 3% of cases, characterized by an early onset (before 65 years of age). So far, 31 different mutations have been described in the *APP* gene in 83 families, together with 175 mutations in *PSEN1* and 14 mutations in *PSEN2*.

Chapter 24- The human prefrontal cortex (PFC) plays major roles in higher cognitive functions necessary for maintaining a healthy social life. Psychological and psychiatric problems are often associated with cognitive impairments associated with PFC. Thus, previous cognitive intervention studies have been conducted to improve the functions associated with PFC. In this article, first we describe the functions associated with PFC and its importance in cognitive intervention studies. Then, we describe recent advancements in cognitive intervention methods, particularly interventions to prevent cognitive decline in healthy older adults and those to enhance their emotional control and resilience in preschool children. We also discuss on transfer effects of previous cognitive intervention, which are often observed. Finally, we discuss on the unresolved issues on the mechanism underlying the effect of cognitive intervention. We consider that a deeper understanding of the effect of cognitive intervention will greatly contribute to human welfare and education for all generations. Further multidisciplinary research will be required to achieve this ultimate goal.

Chapter 25- Mild cognitive impairment (MCI) can be considered as the earliest form of Alzheimer's disease (AD) existing as a transitional state between normal aging and AD. MCI exists in two forms: amnestic MCI and nonamnestic MCI. Amnestic MCI patients are able to perform normal daily living activities and have no signs of dementia; however, they do have cognitive complaints that include bursts of episodic memory loss. In some cases, amnestic MCI patients can develop AD at a rate of ~10 to 15% annually, however in other cases, the patients revert back to normal conditions. Pathologic characteristics of MCI are similar to those of AD. For example, MCI patients have hippocampal, entorhinal cortex (EC), and temporal lobe atrophy based on magnetic resonance imaging studies, synapse loss, neuronal loss, low cerebrospinal fluid (CSF)-resident β amyloid levels, genetic risk factors including preponderance in APOE4 allele, and increased levels of oxidative stress.

Chapter 26- Dementia, derived from the Latin *demens* (without mind), is a clinical syndrome characterized by a cluster of symptoms and indications manifested by difficulties in memory, disturbances in language, psychological and psychiatric changes, and a general impairment in successfully completing everyday tasks. Dementia is a common and growing problem, affecting 5% of the population over the age of 65 and 20% over the age of 80. The number of people living with dementia will almost double every 20 years, to 42 million by 2020 and 81 million by 2040, assuming no changes in mortality and no effective prevention strategies or curative treatments.

Chapter 27- Patients with severe cognitive impairment, for example those with advanced Alzheimer's disease, are clearly not capable of coping with complex aspects of self-care, so arrangements are made to compensate for this disability. Overtly demented patients, if they need for example inhaled drug treatment, have alternative therapies such as nebulisers and direct supervision by attendants. At the other end of the spectrum, elderly patients with intact cognition can almost always manage their own medications, including sequenced devices such as inhalers. Probably the most vulnerable group are those with subtle non-apparent cognitive impairment who are often prescribed devices that they are unable to learn to use effectively. It is that group that will be the main object of this review. Since much of the work in this domain has been conducted on patients' use of inhaler devices and spirometry, the review will largely set out the problem in the context of inhaled respiratory drugs. To set the scene; many elderly patients require maintenance respiratory therapy through delivery of drugs to the airways by means of inhaler devices to achieve optimal management of their obstructive airways disease. This applies to asthmatic patients of all ages, and also confers benefit in an important proportion of people with chronic obstructive pulmonary disease (COPD). A large proportion of elderly subjects with these conditions have a degree of cognitive impairment, including problems with executive function and praxis. However, to achieve an adequate therapeutic effect, there must be sufficient deposition of drugs in the medium and small airways. If the delivery device is patient-activated, this requires a competent inhaler technique, irrespective of the design and relative complexity of the specific inhaler. Even an optimal inhaler technique results in the deposition of only between 15 and 30% of the inhaled dose in the medium and small airways, depending on the design of the inhaler. More importantly, that proportion is reduced significantly by minor inadequacies of technique and drastically reduced to sub-therapeutic proportions for patients with an overtly inadequate technique. Consequently, it can be seen that issues of inhaler technique over ride all other considerations when trying to achieve satisfactory maintenance inhaler therapy in patients with asthma and COPD. In this review it will be explained how cognitive function is the main determining factor for inhaler technique, and it can be argued that the relationship between cognitive function and the effective use of inhalers can be seen as a metaphor for a wide range of self-applied treatments in frail old age. Indeed, in the case of respiratory therapy, while a substantial amount of research has been carried out to compare the relative efficacies of different types of bronchodilator and inhaled corticosteroid medications, the real issue for many patients is the suitability, patient preference and usability of the inhaler itself, and this trend is exaggerated in the presence of cognitive impairment. This review explores the use of inhaler devices in frail elderly patients. Both asthma and COPD have a high prevalence in old age so it is commonplace for family physicians, specialists in respiratory medicine, geriatricians, and specialists in internal medicine to face the dilemma of how best to assess an elderly patient for inhaled maintenance therapy and choose an appropriate means of delivery. This clinical challenge will be used to illustrate the need to account for a patient's cognitive state, including subtle degrees of impairment, when deciding how best to apply patient-managed treatments in old age.

Chapter 28- Care, education and support for children with disabilities and their families is high on the agenda of developing countries. A wide range of government policies and initiatives in the United Kingdom seek to protect and include such children to ensure they are treated as valued and equal members of society with much done to change historic conceptualisations of disability. Ideally, children diagnosed with a disability at birth are surrounded by plethora of services which aim to maximise the potential of early intervention, seeking the best possible quality of life for the child and their family. These opportunities, however, are dependent upon early diagnosis, the provision of local services and a sound body of professional knowledge.

Much is now known about the aetiology, diagnosis and assessment of intellectual and neurodevelopmental disabilities such as Autistic Spectrum Disorder, Down's Syndrome and Fragile X Syndrome with subsequent interventions now well in place. However, whilst medical advances have greatly improved the detection of these conditions, either at birth or as a consequence of later neurodevelopmental assessment, many children affected by maternal prenatal alcohol consumption remain undetected. Even where a full diagnosis can be made knowledge and understanding of the disorder in the United Kingdom, both from a medical and educational perspective, is poor. Families remain therefore bereft of the services and support they need.

Chapter 29- In this chapter we outline a systematic approach to test adaptation developed through more than a decade of research in child development in both urban and rural settings in East Africa.

Chapter 30- Dementia is a leading cause of disability and nursing home placement among older adults living in the community. Nursing home placement is associated with a poorer quality of life for the elderly and with emotional upheaval for caregiving families.

Chapter 31- Within the past decade, a number of models have been proposed to account for the rising incidence (or at least the risen awareness) of developmental disorders, namely autism. The etiology of autism remains unknown but more research of late has focused on the hypothesis that autism is associated with early prenatal exposure to environmental toxins. Currently, though very few studies have been conducted examining possible influences of toxic agents on developmental disorders, neurotoxic exposures are seen to account for 3 to 25% of documented cases. Lathe argued that the large rise in autism spectrum disorder (ASD) cases in recent years is a result of increased exposure to environmental toxins as well as a genetic predisposition that may increase biological vulnerability to said exposure. Fido and Al-Saad also implicated excessive neural trace elements as potentially affecting the development of autism as well as other psychological disorders (e.g., Down's syndrome, Parkinson's and Alzheimer's disease).

[‡]Chapter 32- Huntington's disease (HD) is an autosomal-dominant, progressive neurodegenerative disorder with a distinct phenotype, including chorea, incoordination, cognitive decline, and behavioral difficulties. The underlying genetic defect responsible for the disease is the expansion of a CAG repeat in the gene coding for the HD protein, huntingtin (htt). This CAG repeat is an unstable triplet repeat DNA sequence, and its length inversely correlates with the age at onset of the disease. Expanded CAG repeats have been

---

[‡] Versions of chapters 32-39 were also published in *Huntington's Disease: Etiology and Symptoms, Diagnosis and Treatment*, edited by Thomas J. Visser, published by Nova Science Publishers, Inc. They were submitted for appropriate modifications in an effort to encourage wider dissemination of research.

found in 8 other inherited neurodegenerative diseases. Despite its widespread distribution, mutant htt causes selective neurodegeneration, which occurs mainly and most prominently in the striatum and deeper layers of the cortex.

Chapter 33- Huntington's disease (HD, OMIM ID 143100) is a devastating autosomal dominant progressive neurodegenerative disorder named after George Huntington, who provided a classic systematic account of the conditions in HD. Motor dysfunctions like involuntary purposeless motion, known as choreiform, cognitive impairment and psychiatric disturbances like anxiety, depression, aggression and compulsive behavior are the common symptoms in HD. The genetic basis of HD lies in the expansion of CAG repeats in exon1 of the gene, initially called IT15 (interesting transcript 15), now designated as *huntingtin (HTT)* at chromosome 4p16.3. The *HTT* gene consists of 64 exons and codes for a protein of ~348kDa with unknown function(s). CAG repeat numbers vary from 7-36 among normal individuals. Repeat numbers greater than 36 however give rise to the disease. Appearance of the first symptoms, known as the age at onset, varies normally between 30 to 40 years, although early and late onset is also reported. Clinical features of HD progressively develop with an increase in choreic movements, dementia and other motor deficits like dystonia and rigidity. The disease is terminated in death within 10-20 years after the appearance of the initial symptoms. There is no cure for the disease at present. HD is highly prevalent among Caucasian and less frequent among Japanese, Chinese, Finnish and Indians. It has been proposed that prevalence of the disease is higher in diverse Indian populations than that of in Japan and China, while lower than that of Caucasian populations.

Chapter 34- Huntington's disease (HD) is a progressive neurodegenerative disorder resulting in cognitive impairment, choreiform movements and death which usually occurs 15-20 years after the onset of the symptoms. The disease is also characterized by psychiatric and behavioural disturbances. HD is an autosomal dominant disorder caused by a CAG repeat expansion within exon 1 of the gene encoding for huntingtin (*IT15*) on chromosome 4. In the normal population the number of CAG repeats is maintained below 35, while in individuals affected by HD it ranges from 35 to more than 100, resulting in an expanded polyglutamine segment in the protein. Huntingtin plays a role in protein trafficking, vesicle transport, postsynaptic signalling, transcriptional regulation, and apoptosis. Thus, a loss of function of the normal protein and a toxic gain of function of the mutant huntingtin contribute to the disruption of multiple intracellular pathways, ultimately leading to neurodegeneration. Age at onset (AAO) of the disease is inversely correlated with the CAG repeat length; moreover the length of the expanded polyglutamine segment seems to be related to the rate of clinical progression of neurological symptoms and to the progression of motor impairment, but not to psychiatric symptoms.

Chapter 35- Care for Huntington's Disease (HD) patients has been for long time neglected. The rarity of the disorder (with a prevalence of 5-10/100.000) and the difficulty to have large and homogeneous cohorts of patients for clinical trials were certainly some of the causes, but other factors have to be considered. The progressive cognitive deterioration of these patients, their psychiatric symptoms (including aggressiveness, depression, schizophrenia) and different motor disabilities (be it bradychinesia and rigidity or choreic movements) have for a long time determined their confinement in asylums or nursing homes and justified their exclusion from the interest of clinicians and researchers.

Chapter 36- With the many advances in medical care, the dentist will be quite likely to be asked to treat patients with special health care needs. The dentist may be asked to recommend preventive regimens to maintain good oral health or provide comprehensive treatment of advanced dental disease. The patient with Huntington's disease is but one example of the debilitating effects systemic disease can have upon the oral cavity. There are no direct adverse affects on the oral cavity due to Huntington's disease; however complications associated with the disease increase the risk for dental caries and periodontal disease. As Huntington's disease progress, the person's ability to cooperate will diminish as functional and cognitive abilities decline. This will require the development of realistic treatment plans and easily maintainable dental restorations. Caregivers will need to be involved throughout the process as oral complications are likely throughout this long and difficult disease.

Huntington characterized the disease by a triad of symptoms to include gradual personality changes, dementia and choreiform movements. Other symptoms include dysarthria, disturbances of gait and oculomotor dysfunction. The dementia is characterized by forgetfulness, slowness of thought and altered personality with apathy or depression. The patient can also become moody, irritable and incapable of dealing with the routine details of life. Subtle personality changes often become apparent before any motor symptoms arise. These conditions can all play a role in maintaining a healthy mouth. The dentist is likely to be among the first to notice a deterioration in oral health status.

Chapter 37- Most investigators of Huntington's disease (HD) would agree that George Huntington's (1872) description of the disease was correct when he drew attention to the three peculiarities in this disease; "(1) its hereditary nature; (2) tendency to insanity and suicide; (3) its manifesting itself as a grave disease only in adult life". In this original description George Huntington refers to HD as "sometimes that form of insanity which leads to suicide", suggesting that mental illness and possible suicide is secondary to the disease. Indeed, the potential for suicide in HD has been recognised for more than a century as a serious consequence of HD.

Chapter 38- This chapter is about decision-making in the face of genetic risk. It presents a model of responsibility that encapsulates what families find important when making reproductive decisions in the growing awareness of their risk for a late-onset genetic disorder, Huntington's disease (HD). It will be argued that this model facilitates our understanding of how those living with genetic risk plan their lives in the age of the new genetics. These comprise advances in molecular genetics making it possible to offer genetic testing – both prenatal and predictive – to those identified as at-risk from their family history for an increasing number of disorders. Testing establishes whether mutations, which are likely to result in these disorders, are present in genes. It differs from other forms of medical testing in that any body tissue can be used and it can be performed at any stage of life from conception. Results may be sought to inform personal decision making rather than medical care.

Chapter 39- Neurogenesis is the processes whereby newborn neurons are formed and occurs in mammals in adulthood in specialised areas, known as 'neurogenic niches'. The subventricular zone, and subgranular zone of the dentate gyrus of the hippocampus are such specialised neurogenic niches and newborn neurons formed here contribute to learning and memory, but neurogenesis may occur elsewhere in the brain to a more limited extent. The neurogenic niche is composed of specialised glial cells, basal lamina, ependymal cells, neurotransmitter complement and vasculature. Neurogenesis is a complex process, involving many different cell types, and several stages of neuronal maturation; and each component may be affected by many different microenvironmental perturbations. External perturbations, like seizures and lesions alter the microenvironment and in turn, alter neurogenesis. Chronic disease can also affect neurogenesis and the inherited neurodegenerative condition Huntington's disease may do so through an alteration of the microenvironment. We will use the example of Huntington's disease to explore how changes in the microenvironment might impact on neurogenesis and, thus identify potential therapeutic targets.

In: Encyclopedia of Neuroscience Research
Editors: Eileen J. Sampson and Donald R. Glevins
ISBN 978-1-61324-861-4

Chapter I

Prefrontal Morphology, Neurobiology and Clinical Manifestations of Schizophrenia

Tomáš Kašpárek
Department of Psychiatry,
Masaryk University and Faculty Hospital Brno
Jihlavska 20, 625 00, Brno, Czech Republic

Abstract

Dysfunction of the prefrontal cortex is a hallmark of schizophrenia. This dysfunction can be demonstrated using many approaches, including functional neuroimaging techniques, neuropsychological tests focused on working and verbal memory, problem solving, or verbal fluency. Histopathological studies demonstrate the presence of cortical thickness reduction (layers II and III in particular) that is due to the neuropil loss with reduction of pyramidal neuronal size. These neuropathological changes are reflected in neuroimaging studies (volumetric studies and voxel-based studies) with findings of cortical gray matter shrinkage. All of these changes can be viewed in the context of the prefrontal disconnection that is behind many signs and symptoms of the disease, including dopamine system dysregulation and cognitive abnormalities. We review the present-day evidence for the reduction of the prefrontal cortex and then focus on the links between this abnormality and the functional and clinical features of schizophrenia.

A. Introduction

Schizophrenia is a heterogeneous condition with highly variable clinical manifestation, time-course, and response to various treatment approaches. During the course of the illness, subjects with schizophrenia experience periods of psychosis with symptoms such as hallucinations, delusions or disorganization of thoughts and behavior. Simultaneously as well

* Tel.: +420 53223 2560; Fax: +420 53223 3706; E-mail: tomas.kasparek@centrum.cz

as after the psychosis alleviation, they may experience so-called negative symptoms such as emotional withdrawal, apathy, anhedonia, lack of initiative, etc. and a varying degree of cognitive dysfunction, with working memory, verbal learning and memory, sustained attention or executive functions being affected the most (Nuechterlein et al., 2004). One of the cardinal features of schizophrenia is its typical time window of onset: although there are exceptions, the illness manifests during late adolescence or early adulthood in most cases. The outcome of the illness is inherently variable in the sense of the expression of symptoms and functional impairment. Unfortunately, it is very hard to predict the probable course of the illness in individual patients (Bromet et al., 2005; Fleischaker et al., 2005; Jobe and Harrow, 2005). The failure of current operationally-based diagnostic approaches to provide homogenous subgroups of patients with similar outcomes and responses to treatment led to a search for distinct neurobiologically-defined syndromes that may share clinical manifestations (Braf et al., 2007). Studies of brain morphology may help in this effort.

Schizophrenia was considered a functional condition for a significant period of time. This view stemmed from the inability to find any specific morphological correlates of the disease during the period of the neuropathological era in neuropsychiatry in the late nineteenth and early twentieth century. From the late 1970s when the first CT studies of chronic schizophrenia patients were published (Johnstone et al., 1976), it is more and more obvious that morphological abnormalities play a considerable role in the neurobiology of schizophrenia. Until today, brain morphology is in principle studied by two approaches: histopathology and neuroimaging.

B. Neuropathological Findings in the Prefrontal Cortex in Schizophrenia

Contrary to the expectations in the golden era of neuropathological studies in neuropsychiatry, in the age of Kraepelin and Alzheimer, the structural correlate of schizophrenia remained uncovered for a long time, and the phrase “schizophrenia as a graveyard of neuropathologists” emerged. Despite recent advances, the definite pattern of neuropathological abnormalities in schizophrenia is still missing. Many older studies yielded inconsistent or contradictory findings. As in other areas of schizophrenia research, consistency was complicated by the lack of diagnostic validity and reliability, clinical heterogeneity of the illness, or lack of information from other methodologies for identification of regions of interest in older studies (Powers, 1999).

As recently as the last two decades, with the improvements in both diagnostic and neuropathological methodologies, a much clearer picture has emerged. There seem to be several cortical as well as subcortical morphological changes in schizophrenia. Morphological studies of cortical structures focus mainly on prefrontal cortex, anterior cingulate, and temporal lobe regions such as superior temporal gyrus and medial temporal structures. In general, the changes are relatively inhomogeneous in individual patients, both in type and localization; there is a considerable overlap between the patients and healthy controls with no pathognomic features useful for differential diagnosis (Bogerts, 1999). Furthermore, imaging studies suggest that the neuropathological changes are more widespread than recent neuropathological studies demonstrate – simply because the neuropathological examinations have not yet focused attention on these regions.

Pattern of Neupathological Neocortical Changes

In schizophrenia, a smaller volume of frontal, parietal, and termporal cortices were shown (Pakkenberg et al., 1993; Selemon et al., 2002).

Dorsolateral prefrontal cortex (DLPFC, area 9, 46): Changes of the dorsolateral prefrontal cortex are the most studied and the findings form the most established neuropathological evidence in schizophrenia. In the prefrontal cortex *increased neuronal density* was observed in the dorsolateral prefrontal cortex, - by 17% in area 9 and 21% in area 46 across layers III–VI (Selemon, 1995, 1998), with some data suggesting a similar pattern in the whole frontal lobe or even cortex as a whole (Pakkenberg, 1993). The increased density was observed for pyramidal as well as non-pyramidal neurons (Selemon, 1998). No significant changes of glial density were observed (Selemon et al., 1995; Selemon et al., 1998). The *cortical thickness is reduced,* especially in layers II and III, probably as a consequence of increased neuronal density (Pakkenberg, 1987; Selemon 1998; Woo, 1998). The total *number of neurons* in the frontal cortex seems to be preserved (Pakkenberg, 1993; Thune et al., 2001). Furthermore, the *neuronal size* of pyramidal neurons is reduced, especially in the deep layer III of BA 9, but not in layer V (Rajkowska 1998; Pierri et al., 2001). Reduction of the neuronal size and increased neuronal density are in accord with or are consequences of the *decreased neuropil density* as demonstrated by the findings of reduced presynaptic terminals as inferred from *synaptophysin* levels, which is a protein associated with synaptic vesicles (Glantz and Lewis, 1997) and by lower density of *dendritic spines:* on layer III (Garey, 1998; Glantz and Lewis, 2000) but not on layer V and VI (Kolluri et al., 2005) pyramidal neurons in dorsolateral prefrontal area 46 and on layer V in area 9 (Jones et al., 2002). The magnitude of the reduction was determined to be almost 50 % of the spine density in healthy controls. Changes in the deep cortical layers might be a consequence of exposure to antipsychotics – see below – and thus they should be considered with caution. Reduction of glial cells was also observed (Hof et al., 2003).

Other prefrontal areas: In the *frontopolar cortex* (area 10), increased density of pyramidal neurons with decreased density of interneurons was observed in layer V (Benes et al., 1991), and lower density of dendritic spines (Garey et al., 1998). Lower levels of synaptophysin (Perrone-Bizzozero et al., 1996) were observed. Changes of neuronal cell size or glial density were not seen (Benes et al., 1991). In *Broca´s area* (area 44) not significant changes of cell density were observed in the same subjects in whom the increase was found in the DLPFC (Selemon et al., 2003). *Anterior cingulate* seems to be morphologically distinct from lateral and rostral parts of the frontal cortex. In contrast to findings in the DLPFC, no changes in neuronal density in the cingulate cortex were reported (Benes et al., 1991; Kalus et al., 1997). A decrease in the number of dendrites in layer V in medial prefrontal area 32 was reported (Broadbelt et al., 2002), however, with findings of preserved levels of synaptophysin (Honer et al., 1997; Eastwood et al., 2000). Moreover, an increase of certain GABA-ergic parameters was observed, whereas decrease was observed in dorsolateral prefrontal cortex: Kalus et al. (1997) found an increase of parvalbumin-immunoreactive neurons and similar increase of chandelier axonal cartridges. No changes of glia cells were observed in some studies (Benes et al., 1991; Radewicz et al., 2000), although newer study found significant reduction of glial cells in area 24 (Hof et al., 2003).

Table 1. Neuropathology of dorsolateral prefrontal cortex in schizophrenia

LAYER II, III
Decreased dendritic density, loss of dendritic spines
Decreased neuronal cell bodies of pyramidal and GABA neurons
Increased neuronal density
Reduced cortical thickness
NO GLIOSIS
NO NEURONAL LOSS
LOSS OF DENDRITIC SPINES IN LAYER V???

The findings in all these prefrontal areas are less extensive than the changes in dorsolateral prefrontal cortex suggesting that although not uniquely, DLPFC is affected prominently. It is not known If the changes are present also in other frontal areas since they have not been systematically studied. Overall, the pattern of changes in the prefrontal cortex in schizophrenia subjects suggests abnormalities of cortical connectivity, or, in other words, disease of the cortical microcircuitry.

Changes of Cortical Microcircuitry in Schizophrenia

It is possible to look at the changes in much greater detail: several lines of evidence point to layer III pyramidal neurons, parvalbumin-immunoreactive GABA-ergic interneurons and cortical dopaminergic projections.

From the point of view of layer III pyramidal neuronal inputs, reduction of dendritic spines, which is the hallmark of the neuropathology of schizophrenia, may reflect the impairment of connections from other cortical areas (association fibers), nearby PN of the same layer (intrinsic connections), mediodorsal thalamus (thalamocortical projections) and dopaminergic projections. Dendritic spines of layer III pyramidal neurons are the target of these excitatory inputs (DeFelipe and Farinas, 1992), thus the reduction of the spines seems to reflect a reduction of such inputs. Alternatively, the reduction of the dendritic arbor may result from altered capacity of pyramidal neurons to support it, perhaps due to cytoskeleton regulation abnormalities, and consequently, the number of excitatory synapses is diminished.

As there is a reduction of *mediodorsal thalamus* volume in schizophrenia, a deficit of thalamo-cortical connections may contribute to the overall pattern of the cortical microcircuitry changes. The projections from MD thalamus terminate in deep layer III on basal dendritic spines of layer III PN (Erickson adn Lewis, 2004). The reduction of neurons in mediodorsal nucleus, as well as changes of thalamo-cortical afferent fiber bundles were described (Pakkenberg 1990; Dorph-Petersen et al., 2004). However, as the thalamocortical afferents appear to compose a small proportion (< 10 %) of the total cortical excitatory inputs - as demonstrated in visual cortex of cats (Ahmed et al., 1994) – and lesions of mediodorsal thalamus produce no changes in GAD67 mRNA expression (Volk and Lewis, 2003; see below) it is clear that other inputs should contribute to the pattern of changes.

Cortical architecture, connectivity as well as performance of the dorsolateral prefrontal cortex mature during adolescence, which is the time of typical onset of schizophrenia. During that time the prefrontal cortex starts to augment the activity of other neural circuits during the performance of tasks such as delayed action, working memory paradigms (Alexander and Goldman 1978). This increase of the prefrontal cortex involvement correlates with interrelated changes of *connectivity of layer III pyramidal neurons*, parvalbumin-immunoreactive wide-arbor GABA-ergic interneurons and dopaminergic projections, as emerges from primate studies (Lewis 1997): Layer III pyramidal neurons send axonal associational projections to other cortical regions as well as intrinsic axonal collaterals that run horizontally through gray matter to superficial cortical layers where they terminate on dendritic spines of neighboring pyramidal cells thus giving rise to functionally interconnected clusters of pyramidal neurons (Levitt et al., 1993). During adolescence the dendritic spines of layer III pyramidal neurons are extensively pruned (Bourgeois et al., 1994). Changes in dendritic spine density reflect parallel changes in excitatory inputs formed by intrinsic, and not associational, projections to these neurons (Lund and Holbach, 1991; Woo et al., 1997).

The layer III PN also send collateral connections to the dendrites of *GABA-ergic parvalbumin-immunoreactive (wide arbor and chandelier) interneurons* (Levitt et al., 1993; Lund et al., 1993; Melchitzky et al., 1995). Interestingly, the maturation changes of dendritic spines on pyramidal neurons are accompanied by the changes of PV-interneurons that may be a kind of compensatory mechanism of the changes of excitatory inputs to pyramidal neurons (Lewis 1997). There is a growing evidence of the involvement of GABA dysfunction in schizophrenia (Lewis et al., 2005). In schizophrenia decrease of *GAD67* mRNA encoding GABA synthesizing enzyme glutamic acid decarboxylase in the prefrontal cortex was replicated several times (Akbarian et al., 1995; Volk et al., 2000; Guidotti et al., 2000), with a consequent reduction of the protein levels (Guidotti et al., 2000). GAD67 positive neurons density was decreased in layer III-V, but not VI in DLPFC in schizophrenia (Akbarian et al., 1995; Volk et al., 2000) with no changes of the GAD67 mRNA level in individual neurons (Volk et al., 2000). However, *a proportion of GABA neurons did not express detectable levels of GAD67 mRNA*, together with reduced levels of GABA transporter *GAT1* (Volk et al., 2001). These changes seem to be specific to parvalbumin-immunoreactive cells (Beasley and Reynolds 1997; Hashimoto et al., 2003). *PV levels* were reduced in the layer III and IV neurons in the DLPFC (Hashimoto et al., 2003). There is not a loss of PV positive neurons in DLPFC, rather normal numbers of neurons express lower levels of PV. PV positive neurons are of two types: chandelier and wide-arbor basket cells. *Chandelier cells* project axons to initial axonal segments of pyramidal neurons and form there multiple synaptic connections called *"cartridges"*, carrying GAT1 transporters. The density of these GAT1 immunoreactive cartridges was reduced in middle cortical layers of DLPFC in schizophrenia (Woo et al., 1998). On the postsynaptic side up-regulation of *alpha2 subunit containing GABA-A receptors* was seen (Volk et al., 2002). Similar pattern of synaptic changes is present in the case of *wide-arbor basket cells* (Lewis et al., 2001; Benes et al., 1996). These cells send axons that form synpases on the cell bodies of pyramidal neurons. Possible unifying mechanism of the PV associated pattern of changes was proposed: *BDNF receptor TrkB* dysfunction – expressing regional specificity to PV cells – with consequent changes of GAD67 expression and reduction of GABA synthesis and compensatory decrease of parvalbumin levels leading to higher levels of free calcium, decreased GABA re-uptake through down-regulated GAT1 – both leading to increase GABA synaptic levels - and increase of postsynaptic GABA receptors (Hashimoto et al., 2005; Lewis et al., 2005). Although the changes were not seen for calretinin-immunoreactive cells (Hashimoto et al., 2003), there seem to be changes in other GABA cells, such as somatostatin-positive cells (Morris et al., 2008) since PV interneurons cannot explain the changes in layers I and II. PV interneurons inhibit the somata and initial axonal segments of several pyramidal neurons and they share excitatory inputs from mediodorsal thalamus and other pyramidal neurons with local pyramidal neurons (Melchitzky et al., 1999; Melchitzky et al., 2001; Melchitzky and Lewis, 2003), thus enabling their coordinated and timed activity. Functional consequences of this action are synchronized neuronal oscilations in several discrete frequency bands (Cobb et al., 1995; Tamás et al., 2000; Szabadics et al., 2001; Blatow et al., 2003), such as gamma band oscillations, electrophysiological activity linked to many higher order cognitive operations. Several studies reported abnormalities of gamma band oscillations in schizophrenia (Spencer et al., 2003; Cho et al., 2006). Superficial GABA PV and CB positive interneurons give rise to theta band oscillations (Blatow et al., 2003), another band of interest in schizophrenia.

Furthermore, it was shown that also *dopaminergic projections* to the dorsolateral prefrontal cortex undergo substantial changes during adolescence. They form connections with dendritic spines of pyramidal neurons (Goldman-Rakic et al., 1989; Smiley et al., 1992) as well as with dendrites of PV-interneurons (and not with calretinin-immunoreactive GABA interneurons; Sesack et al., 1995a,b). The maturation changes of DA projections just precede the changes of PN a PV cells and it is speculated that the neuromodulatory effect exerted by dopamine may influence the refinement of cortical architecture (Lewis et al., 1997). In accord with these effects of dopaminergic projections are the findings of dopamine neurotrophic effect: its depletion affects dendritic spine and synapse density (Deutch and Jing, 2003; Deutch, 2005). Amphetamine sensitisation, leading to a chronic dopamine dysregulation, is also linked with dendritic spine reduction on layer III pyramidal neurons in primates (Selemon et al., 2007). Thus, monoamine abnormalities may contribute to the structural abnormalities in schizophrenia. It was also demonstrated that dopaminergic signaling has a key functional role in the prefrontal cortex (Williams and Goldman-Rakic, 1995; Arnsten et al., 1994). Reduction of DA connection may be a correlate of both morphological and functional impairments seen in schizophrenia. Benes suggested a model of neuropathology of schizophrenia where abnormal connectivity of local neuronal circuits is caused by rearrangement of dopaminergic connections, which are supposed to shift from PN to nonpyramidal neurons, possibly as a way of compensation of the loss of their pyramidal targets (Benes et al., 1995; Benes 1997).

These data obtained mostly from animal studies seem to be parallel to the neuropathological changes seen in human subjects with schizophrenia: neuropil reduction of layer III pyramidal neurons, decreased cell size of pyramidal neurons as well as PV-interneurons – see above. In the light of animal as well as human neuropathological data we may view the pathology of schizophrenia as pathology of prefrontal cortical microcircuitry connectivity. Because of the developmental changes are prominent for the intrinsic connections between the clusters of pyramidal and PV-cells, with the tuning connections from dopaminergic centers, the basic pathology concerns local circuitry connectivity. These local circuitry abnormalities of prefrontal cortex may lead to difficulties in coordination of action of large populations of neurons with consequent discoordination or disconnection of neuronal nets subserving top-level information processing – leading to either overall impairment of cortical connectivity or rather specific disconnection of fronto-temporal areas (Friston and Frith, 1995). These ideas were formulated explicitly by Selemon and Goldman-Rakic who view the reduced neuropil as a cardinal feature of schizophrenia: The reduced neuropil hypothesis. In short they state that "disturbances in prefrontal cognitive functioning in schizophrenia may be mediated by a process which involves atrophy of neuronal processes but stops short of actual neuronal loss" (Selemon 1995, Selemon and Goldman-Rakic, 1999a). As shown below, several processes that may selectively afflict the neuropil without the induction of cell death exist. The reduced neuropil hypothesis receives support from an elegant computational simulation: Hoffman and Dobscha (1989) presented a model, where reduction of elements of a neuronal network led to the improvement of performance. However, when a certain threshold was crossed, the network produced outputs even without any inputs, which was interpreted as a correlate of autonomic behavior of cortex as a cause of hallucinations.

Dendritic arbor of layer V and VI PN seems to be relatively spared. These cells are the origin of cortico-striatal and cortico-thalamic projections, which seem to be intact in schizophrenia. Similarly, these cells receive very little inputs from mediodorsal thalamus (Erickson and Lewis, 2004), further supporting the view of layer III specific abnormality in schizophrenia.

Table 2. Cortical neurocircuitry changes in dorsolateral prefrontal cortex in schizophrenia

ABNORMALITIES OF EXCITATORY INPUTS ON DENDRITIC ARBOR OF PYRAMIDAL NEURONS
Reduction of intrinsic and cortico-cortical connections between pyramidal neurons
Reduction of dopaminergic projections
Reduction of thalamocortical projections from mediodorsal thalamus
ABNORMALITIES OF PERISOMATIC AND AXONAL INHIBITION OF PYRAMIDAL NEURONS
Reduction of chandelier cells synaptic cartridges on initial axonal segment of pyramidal neurons
Reduction of wide-arbor cells perisomatic inhibition
Up-reglation of GABA-A alpha 2 subunit postsynatptically on pyramidal neurons
= LOSS OF COORDINATION BETWEEN CORTICAL PYRAMIDAL NEURONS

Regional Specificity of the Changes Seen in Prefrontal Cortex

Are the above described changes specific for the prefrontal cortex? There is evidence of changes in many other areas, where similar as well as distinct neuropathological changes were described: disarray of hippocampal pyramidal neurons, reduction of their size, decrease of synaptic markers in hippocampus (Harrison, 2004); reduction of neurons in the mediodorsal nucleus and pulvinar of the thalamus (Pakkenberg et al., 1990; Popken et al., 2000); reduction of normal cortical asymmetry with some reports suggesting pronounced abnormalities in the left hemisphere (Crow et al., 1989); reduction of dendritic spine density in temporal neocortex (BA 20, 21, 22, and 38; Garey et al., 1993), reduction of dendritic spines, neuronal size of layer III PN and axonal terminals in primary – BA 41 – and association auditory cortex - BA 42 (Sweet et al., 2003; Sweet et al., 2008); or neuropil reduction in occipital cortex (Selemon et al., 1995).

C. Etiology

The origin of structural changes is still elusive as well. However, it is probable that there is not just one unique mechanism responsible for the changes. Intrauterine or neonatal lesions, genetic diathesis, oxidative stress, nutritional factors, infectious agents etc. are all mentioned in the etiopathogenetic frameworks of schizophrenia.

Neurodevelopment

From the first studies in the late 19th century it is known that changes in schizophrenia are not followed by gliosis, suggesting the lack of neurodegenerative processes in schizophrenia. Other features, such as frequent occurrence of cavum septi pellucidi in the adult age, cytoarchitectural changes, or reduction of brain asymmetry point to some abnormality in the development of the brain. There is not just one comprehensive developmental theory of schizophrenia, rather, several of them emerged with some of them being refined, some abandoned (de Haan and Bakker, 2004).

Some theories emphasize the involvement of *early developmental abnormalities,* possibly during the pre/perinatal period (Weinberger 1987; Murray and Lewis 1987; Roberts 1991; Benes 1997). The evidence for this view comes from the findings of absence of brain asymmetry, higher prevalence of schizophrenia in subjects born in winter, with obstetric complications, whose mothers had influenza during the 2nd trimester of pregnancy or first-trimester malnutrition, with some motor and language skills abnormalities or social skills impairments during childhood (for review see Murray 1994; Weinberger 1995, Harrison 1999, Maynard et al., 2001).

Neonatal hippocampal lesions: neonatal ventral hippocampal lesions as a model of the pathophysiology of schizophrenia is one of the most elaborated theories (Lipska et al., 1995; Lipska and Weinberger 1995; Sams-Dodd et al., 1997); it is one of the theories of early developmental lesions that can explain long latency between the lesion and clinical manifestation - early developmental lesions exert adverse influence on the consequent development of the brain leading to abnormal maturation of interconnected prefrontal areas. In particular, after neonatal hippocampal lesions in rats reduction of dendritic spines of layer III PN in the DLPFC (Flores et al., 2005), GAD67 mRNA reduction in the prefrontal cortex (Lipska et al., 2003), decrease of NAA levels in prefrontal cortex (Bertolino et al., 1997), postpubertal vulnerability to stress or dopaminergic challenge (Saunders et al., 1998), behavior in adulthood interpreted as a correlate of schizophrenia phenotype (Lipska et al.,

1993) was observed. Findings in humans such as reduced hippocampal volume correlated with prefrontal activity (Weinberger et al., 1992), and poorer performance in neuropsychological tests focused on prefrontal cortex (Bilder et al., 1995; Szeszko et al., 2002), abnormal EEG coherence between prefrontal cortex and hippocampus (Ford et al., 2002), and correlation between lower level on prefrontal NAA and abnormal mesolimbic dopaminergic activity (Bertolino et al., 1999; Bertolino et al., 2000) are data that conform with the hypothesis. The chain of neuropathological changes begins with neonatal or perinatal hippocampal lesion leading to impairment of development of fronto-temporal connections with consequent morphological and functional changes of prefrontal cortex (which is a correlate of cognitive and negative symptoms of schizophrenia), loss of prefrontal regulation of dopaminergic system with consequent mesolimbic dopaminergic dysregulation, mechanism that lays behind psychosis.

Others argue against the idea that early affections of the brain can have such an effect in the adulthood and emphasize abnormalities of *late development*. The main arguments are the pattern of morphological changes – total brain volume reduction without the changes in the intracranial volume – and several lines of evidence suggesting the progression of the changes during certain phases during the course of schizophrenia.

These features seem to be in accord with abnormalities of synaptic pruning (Feinberg 1982; Keshavan 1994, Lewis 1997, Woods 1998), possibly via abnormalities of synaptic apoptosis (see below) as a way of the refinement of heteromodal cortical mircrocircuits during the late adolescence (Lewis 1997). On the other hand, the presence of (albeit non-specific) symptoms early in the childhood speaks in support of the early developmental theories.

Contemporary *syntheses* understand the importance of the late processes and remain open to the possibility of early developmental lesions: *a two-hit hypothesis* (Bayer et al., 1999; Maynard et al., 2001). In this view the disruption of some early maturation process changes the way the brain develops and causes a vulnerability to the consequent negative effect. Neither the first nor the second hit alone can cause schizophrenia, the key is their combination. *Cell-cell signaling* was proposed as a potential process that might be affected both during the early development and later maturation processes (Maynard et al., 2001).

Another mechanism was proposed in the *"sensitization model"* where the early lesion leads to reduced inhibition of the cortex with subsequent increase of reaction to stressful stimuli occurring during adolescence with neurotoxic and apoptotic consequences (Lieberman at al., 1997). Another view is the concept of schizophrenia as a *progressive developmental disorder* due to a disruption of late developmental processes leading to a progression after the illness onset (Woods, 1998).

Apoptosis

It was argued that in schizophrenia there is (at least transiently during critical periods of the disease) an increased likelihood of a certain apoptotic mechanism, *"synaptic apoptosis"* (Mattson et al., 1998) that could explain the loss of dendritic spines without complete neuronal death (Glantz et al., 2006). Synaptic apoptosis seems to be the mechanism underlying synaptic remodeling and elimination – the key processes in the physiological plasticity of the brain as well as in many pathological states (Gilman and Mattson, 2002;

Garden at al., 2002); in schizophrenia it might be the mechanism of the accelerated synaptic pruning and other late neurodevelopmental changes (Feinberg, 1982; Weinberger, 1987). Apoptotic activity can be triggered by many stimuli, such as oxidative stress, inflamation, excitotoxicity, neurotrophins withdrawal, mitochondrial dysfunction (Sastry and Rao, 2000). On the other hand, neurotrophic activity can reverse or prevent the apoptosis (Guo and Mattson, 2000). In schizophrenia mitochondrial dysfunction, decreased capacity to deal with oxidative stress, glutamatergic abnormalities as well as decrease of neurotrophins was demonstrated, making the involvement of apoptotic mechanisms in the pathogenesis of schizophrenia probable (Glantz et al., 2006). Apoptosis was also implied as a potential mechanism underlying progressive gray matter volume loss around the onset of schizophrenia (Berger et al., 2003; Jarskog et al., 2005).

Genetics of Cortical Microcircuits

It is also apparent that at least some genes linked with schizophrenia play a role in the development, maintaining or plasticity of the brain structure and synapse formation. This is the case for BDNF, DISC1, dysbindin, neuregulin 1 etc. Other genes code proteins involved in receptor formation and signal transduction: COMT, CHRNA7, G72, RSG4 (Perlman et al., 2004; Harrison and Weinberger 2005; Iritani 2007). In accord with the pattern of neuropathological changes it seems that the genetic susceptibility is not expressed on the level of a certain type of the synapse, rather, on the level of a cortical microcircuitry underlying core cognitive functions impaired in schizophrenia. Reduced signal-to-noise ratio, impaired synchronization and other dysfunctions were demonstrated as features of cortical circuit abnormalities in schizophrenia and these might be the consequence of genetic variations of the above mentioned genes. We may view the morphological changes as a consequence of the abnormalities of the synapse formation and activity within these circuitries.

Direct evidence connecting genetic polymorphisms implied in schizophrenia with brain structure emerges. The *BDNF* 11757 G/C polymorphism was significantly associated with frontal gray matter in patients with schizophrenia as well as in healthy controls (Agartz 2006). Similarly, the BDNF Val66Met polymorphism was linked with regional gray matter volume in prefrontal, parietal, temporal as well as occipital areas, where the Val-alele carriers showed larger volume (Ho et al., 2006). Pericentriolar Material 1 (*PCM1*) gene located on chromosome 8p22, involved in the maintenance of centromere integrity and regulation of microtubule cytoskeleton important for synaptogenesis and intracellular transport along axons and dendrites, was identified as susceptible gene for schizophrenia; schizophrenia subjects who were carriers of the susceptible allele demonstrated gray matter reduction in bilateral medial orbitofrontal cortex when contrasted with schizophrenia subjects without susceptible allele (Gurling et al., 2008). In similar manner, schizophrenia-associated haplotypes of *DISC1* were associated with gray matter density reduction in prefrontal areas as well as with cognitive abilities linked with prefrontal functions (Cannon et al., 2005). As stated earlier, not only genes coding proteins involved in the brain development show association with the changes of the brain morphology. For example schizophrenia subjects homozygous for *Val-COMT* (Val158Met polymorphism) differ significantly from the Met-COMT carriers in the gray matter morphology in anterior cingulate, temporal cortex and thalamus (Ohnishi et al., 2006).

Environment

The neuropil abundance and condition is sensitive to many environmental influences. It was demonstrated that neuropil density changes physiologicaly (Wooley et al., 1990), as a correlate of learning associated plasticity (Moser et al., 1994). Many adverse events may change the neuropil density as well. Repeated exposure to *stress* was linked with a substantial reduction of dendritic spines of layer III pyramidal neurons in medial temporal cortex (Radley et al., 2006) or apical dendritic arbor of layer II, III neurons in anterior cingulate, but not in lateral orbitofrontal cortex (Liston et al., 2006) of rats; adverse effect of *alcohol* (Ferrer et al., 1986), *social isolation* (Connor and Diamond, 19982; Glantz and Lewis, 2000), *protein malnutrition* (Brock and Prasad, 1992), or *sensory deprivation* (Bryan and Riesen, 1989; Valverde, 1967) were all shown to lead to neuropil reduction.

D. Brain Imaging Methods in Schizophrenia

Methods: Pros and Cons

The advent of MRI brought sufficient image resolution enabling in vivo morphological assessment. It also allows for clear delineation of gray, white matter and liquor compartments, i.e. segmentation, and their separate analysis. Since gray matter is composed primarily of neuropil and cell bodies, imaging of the gray matter in schizophrenia reflects the reduction of neuropil as described by neuropathological studies. However, the changes are not visible for the "naked eye". It is necessary to process the images, extract the relevant information and describe the changes using statistical apparatus. In general, there are two approaches: *region of interest- and voxel-based methods.*

Which method seems to be the best suitable for the detection of morphological abnormalities in schizophrenia? It is evident that the morphological changes do not respect the macroscopic brain structures; rather they stretch across gyri, sulci and even lobes. The substantially high variability of the macroscopic as well as microscopic structures – such as the course or shape of individual sulci or gyri, Brodmann areas locations and extents – complicates the situation even further. This variability is especially pronounced in the prefrontal cortex and in schizophrenia even more than in healthy subjects. Due to these attributes of the brain morphology it seems that region-of-interest (ROI) based approaches (formerly considered as a gold standard in the field of brain morphology research) are not adequate methods to characterize the pattern of prefrontal morphology in schizophrenia. Among the disadvantages of ROI approaches belong the need to define a macroscopic brain structure to be studied; to label the structure with a sufficient validity and reliability, with inherent trade-off between these parameters, and the necessity to introduce arbitrary borders; the ability to study limited number of structures; the implicit investigation of the structure as a whole.

The *voxel-based methods* (VBM) seem to have several advantages over the ROI-methods. They enable assessment of local features of brain morphology within the whole brain, without the necessity to define a priori the regions of interest. They use the computational resources thus limiting the subjective influence on the results as well as labor intensity. On the other hand, they may have lesser statistical power and not as a last point some still find the digital preprocessing of the images too complicated, elusive and prone to unwanted bias. However, these methods were validated several times, both in terms of direct comparison with ROI-approaches (Giuliani et al., 2005; Saze et al., 2007), or in their ability to detect apparent pathology in the brain, such as mesiotemporal sclerosis in temporal epilepsy (Keller et al., 2004). There are several modification of voxel-based morphometry; generally two approaches can be identified – studies of regional gray matter concentration/density and volume. Both approaches are usually pooled together in systematic reviews; however, it should be beard in mind that they may provide quite different results. It is not clear what approach should be preferred - volume studies are used more frequently nowadays, since they allow better understanding of the units measured and easier interpretation. Keller et al. (2004) found the gray matter volume analysis was able to find more regional changes in temporal lobe epilepsy than did the gray matter concentration approach. On the other hand, Meda et al. (2008) found just the opposite: gray matter concentration analysis yielded much better results in a large sample of schizophrenia subjects than the gray matter volume approach; the reason was that modulation, step necessary to obtain volume information, resulted in higher variability with consequent decrease of statistical power.

In the context of schizophrenia, the VBM studies seem to provide neuropathologically meaningful information about cortical gray matter. It was demonstrated that VBM studying gray matter concentration detect analogues pathology as cortical thickness measurement (Narr et al., 2005). We may view the VBM studies of cortical structures in schizophrenia as a measure of regional cortical thickness and consequently of local abundance of neuropil or structural connectivity.

What follows is a review of the present day evidence concerning the brain morphology in schizophrenia with a focus on the prefrontal cortex and clinical features of the illness.

The Pattern of Morphological Changes

ROI-Based Approaches

There are innumerous volumetric studies in schizophrenia. The advantage of the volumetric studies is they may be relatively easily pooled together for meta-analysis. Such statistical re-assessment is available for several regions. Wright at al., (2000) performed a meta-analysis of most of the regions studied by ROI approaches between the years 1980 and 1998. They pooled studies of whole brain, gray matter, white matter volume, ventricular synstem volume, frontal and temporal lobe, temporal lobe areas: amygdala, hippocampus, superior temporal gyrus, subcortical areas: caudate nucleus, putamen, pallidum, and thalamus. Together they analyzed data from 1588 patients with schizophrenia. They expressed the volume of corresponding structures in healthy controls as 100%, and the volumes in schizophrenia as fractions of that value. Whole brain volume in schizophrenia represented 98% of healthy controls, frontal lobe 95% on both sides, temporal lobe 98% on left and 97% on right side, superior temporal gyrus 97% on both sides, amygdala 91% on both sides, hippocampus 93% on left and 94% on right side, parahippocampal gyrus 89% on left and 92% on right side, lateral ventricles 130% on left and 120% on right side, thalamus 96% on both sides, caudate 101% on left and 99% on right side, putamen 104% on both sides, and pallidum 118% on left and 121% on right side. Studies of both first-episode and chronic schizophrenia subjects, medicated and drug naive patients were included. Thus the pattern of changes may be biased by many confounding factors such as long-term exposure to antipsychotics, progression of the chronic illness, life-style changes as a consequence of severe mental condition, substance abuse etc. To rule out such factors it is advantageous to study first-episode, and if possible antipsychotic-naive schizophrenia patients. From the regions included in the meta-analysis it is also obvious that ROI studies focused mainly on regions where relatively easy delineation is possible, such as hippocampus, superior temporal gyrus or subcortical structures. Frontal regions are studied by a limited number of ROI studies and VBM techniques may be superior in these areas.

VBM Approaches

Honea et al., (2005) published a comprehensive review of 15 voxel-based studies in schizophrenia published through May 2004. Together they included 290 patients and 364 healthy controls. The aim was to specify the overlap of individual studies, which was expressed as a percentage of the studies that found reduction in a given structure. Individual studies found reduction in 50 regions. In more than 50% of studies the reduction was reported for left medial temporal lobe and left superior temporal gyrus. In 50% studies the reduction was seen in the left inferior frontal gyrus, left medial frontal gyrus, right superior frontal gyrus, and left parahippocampal gyrus. However, the changes were reported also in the precentral and postcentral gyrus, in insular, parietal and occipital cortex, in caudate nucleus, thalamus and cerebellum. The analysis included studies of both first-episode and chronic schizophrenia subjects. The authors also analyzed the methodological issues of individual studies and found a significant effect: studies that used smaller smoothing kernel tend to find reductions in more areas and more frequently in small structures such as medial temporal lobe structures or cingulate gyrus.

Similar summary is provided for 5 studies with first episode schizophrenia subjects published till the end of 2005 (Kasparek et al., 2007): all 5 studies found gray matter reduction in anterior cingulate gyrus, 4 in inferior frontal gyrus, 3 in superior temporal gyrus, hippocampus, and cerebellum.

Recent study used activation likelihood estimation (ALE), a statistical approach for meta-analytical analysis of neuroimaging data (Laird et al., 2005), to find a statistically consistent pattern of changes across published VBM studies with first-episode and chronic schizophrenia patients (Ellison-Wright et al., 2008). Data from seven VBM studies in first episode found gray matter reduction in inferior frontal gyrus (BA 47), cingulate gyrus (BA 32), precentral gyrus (BA 6), insula (BA 13), uncus, postcentral gyrus (BA 40), mediodorsal thalamus, caudate nucleus, and cerebellum, as well as gray matter increase in superior frontal gyrus (BA 6, 8), middle frontal gyrus (BA 9), precentral gyrus (BA 4), insula, superior temporal gyrus (BA 22), fussiform gyrus (BA 18), inferior parietal lobule (BA 40), lingual gyrus (BA 17), putamen, and cerebellum relative to healthy controls. Comparison of first-episode studies with studies with chronic schizophrenia patients showed greater decrease in the prefrontal cortex, specifically in the dorsolateral prefrontal cortex and medial frontal gyrus. The extent of decrease in cortical areas was greater in the chronic studies as well.

However, the results of the ALE analysis of VBM studies should be considered with caution. There are several issues that should be bear in mind: 1) no influence of negative studies on ALE results, 2) different methodology of pooled studies: concentration vs. volume,

3) different statistics of pooled studies: corrected, uncorrected, voxel-level, cluster-level etc. All these methodological factors may influence the results of a study and it may be inappropriate to pool studies with different methodologies together. It is especially important in the light of the Ellison-Wright meta-analysis results: many cortical areas were reported to have increased, and more importantly, simultaneously both decreased and increased gray matter. However, only two individual studies reported such substantial increase of gray matter (Salgado-Pineda et al., 2003; Whitford et al., 2006). Whitford et al. (2006) found gray matter volume increases in cortical areas BA 6, 8, 9, 10, 17, 18, 46, and in subcortical claustrum in a sample of first-episode schizophrenia subjects; however, two studies from the same center with the similar methodology and similar study samples did not find them (Farrow et al., 2005; Whitford et al., 2005). Of special interest is the report of increased gray matter in area 9, where a substantial body of evidence shows just the opposite: gray matter volume reduction, with a consistent pattern of findings in the neuropathological literature. Moreover the second study that reported gray matter increase did not find it in the prefrontal cortex; the changes were in areas 13, 22, 40, in the cerebellum, and in subcortical putamen. Thus, we have to understand the results of Ellison-Wright study in a following way: only two from seven studies found gray matter increases, and although the pattern of the two studies did not overlap, the ALE meta-analysis indicates that reported subset of foci of changes is distributed within the brain non-randomly. Apparently, the results of prefrontal gray matter increase lack replication in an independent observation.

However, recently published large scale study, that could not be included in the Ellison-Wright meta-analysis, brought some support for gray matter increase in schizophrenia: it was observed in bilateral orbitofrontal cortex (BA 10, 11), anterior cingulate (BA 24), left precentral gyrus (BA 6), left inferior temporal gyrus, right middle temporal gyrus, bilateral globus pallidus, and right medial cerebellum (Honea et al., 2008). All these findings survived quite stringent correction of multiple comparisons for VBM data: family wise errors with significance set to $p<0.05$ FWE corrected. However, the schizophrenia subjects included in this study were not the first episode cases and were exposed to antipsychotics for a substantial period of time. Of importance, in healthy siblings of the subjects, no gray matter volume increases were found, suggesting that the finding of prefrontal gray matter increases in schizophrenia does not reflect primary neurobiology – or endophenotype - of the illness. Moreover, the other recent study with similarly large samples found no gray matter increase (Meda et al., 2008) and the meaning of the findings of gray matter volume increase remain unclear.

The Meda ´s et al. (2008) study included 200 schizophrenia subjects and 200 healthy controls. It is a study with the biggest and sufficient statistical power from studies published till today to study gray matter morphology with sufficient correction of false positives without critical rise of false negative results. The study identified a reduction of gray matter concentration in many cortical and subcortical areas: in frontal lobe in both medial and lateral parts - in dorsolateral prefrontal cortex (BA 9, 46), orbitofrontal cortex (BA 10, 11, 47), pars opercularis and triangularis of inferior frontal gyrus (BA 44, 45), premotor and motor areas (BA 6, 8), cingulate cortex (BA 23, 24, 25, 32, 42); in insular cortex (BA 13); in temporal lobe in superior temporal gyrus (BA 22, 38, 41, 43), middle temporal gyrus (BA 21), inferior temporal gyrus (BA 19), uncus (BA 28); in parietal lobe in postcentral gyrus (BA 3, 1, 2, 40, 43), inferior parietal lobule (39); in occipital lobe in middle occipital gyrus (BA 19); and in mediodorsal thalamus, pulvinar and mammillary bodies, and caudate head. It is important to note that analysis of the gray matter volume showed similar pattern of changes, however on a lower, non-significant statistical threshold – which was due to the higher variability of volume data. This study also included chronic patients and such large scale and powerful study in first episode schizophrenia subgroup is still missing.

Dynamics of the Changes

Neuroimaging studies also observed progressive reduction of gray matter. Progressive gray matter loss in temporal, inferior frontal and cingulate cortex was seen in high-risk subjects who developed psychosis, i.e. the *conversion to psychosis* is linked with gray matter loss in heteromodal associative cortex (Pantelis et al., 2003; Wood et al., 2008). After the first episode of schizophrenia, *further reduction* of gray matter was demonstrated. Cahn et al. (2002) reported progressive loss of total gray matter, others observed progression in frontal (Farrow et al., 2005; Reig et al., 2008), temporal (Kasai et al., 2003; Farrow et al., 2005), or parietal (Farrow et al., 2005) cortex. No progression was observed in hippocampus (Lieberman et al., 2001). Some VBM studies observed dynamic spreading of the progressive loss of gray matter across several areas including prefrontal (inferior frontal gyrus), cingulate (posterior cingulate) temporal (superior temporal gyrus, parahippocampal gyrus), parietal (precuneus, postcentral gyrus, inferior parietal lobule), and occipital (lingual gyrus) cortices, and cerebellum (Thompson et al., 2001; Whitford et al., 2006). Progressive loss of gray matter in dorsolateral prefrontal cortex, superior temporal gyrus, thalamus, and caudate nucleus was observed also in chronically ill patients (van Haaren et al., 2007). For the interpretation of the findings of progressive gray matter loss a report of simultaneous worsening of functional impairment is of great value (Salisbury et al., 2007), i.e. the progression seen in imaging methods is not an epiphenomenon, rather it reflects certain neurobiological process with functional consequences.

This progressive loss of gray matter was interpreted in a *context of clinical progression*, or deterioration, through the first several years of the illness (Davidson and McGlashan, 1997). If there is a kind of ongoing pathogenetic process, at least in a certain subgroup of patients, the question arises if the treatment may stop or revert the changes and, subsequently, if such prevention of neuropathological progression leads to better outcome of the schizophrenia patients. Some data seem to support this possibility (Lieberman et al., 2005). However, it was speculated that the progression of the morphological changes reflect reversible processes of neuropil remodeling and that medication is just one of many factors that influence this plasticity (Weinberger and McClure, 2002). Several mechanisms that affect the plasticity and may lead to progression were proposed: dopaminergic overstimulation with consequent apoptosis (Ben-Schachar et al., 2004), glutamate excitotoxicity (Deutsch et al., 2001), GABA dysfunction (Farber et al., 2003), impaired apoptotic mechanisms (Luo et al., 1998; Datta et al., 1999), mitochondrial dysfunction (Beal et al., 1993; Ben-Schachar et al., 2004), decreased nitric oxid synthesis (Fiscus, 2002), oxidative stress (Mahadik et al., 2001).

In summary the neuroimaging studies provide quite widespread pattern of gray matter reduction, affecting not only dorsolateral prefrontal cortex, which is according to the neuropathological studies affected the most, and not even only heteromodal associative cortex, but also many primary and secondary sensory areas, as well as motor and premotor areas – see table 3. In some of these areas, fronto-temporal in particular, further gray matter loss was demonstrated. However, it seems that such widespread reduction and further progression is not present in all subjects, or subgroups, and substantial variability exists. How such variability can manifest itself on the clinical level is a topic of following sections of this chapter.

Table 3. Pattern of regional gray matter reduction in schizophrenia: neuroimaging data

FRONTAL LOBE
Motor and premotor cortex
Dorsolateral prefrontal cortex
Orbitofronal cortex
Ventrolateral prefrontal cortex
Medial frontal cortex
INSULA
TEMPORAL LOBE
Heteromodal neocortex
Primary and secondary auditory cortex
Mediotemporal structures (hippocampus, parahippocampal gyrus, amygdala)
PARIETAL LOBE
Somatosensory cortex
Heteromodal cortex
OCCIPITAL LOBE
Primary and higher order visual cortex
SUBCORTICAL REGIONS
Thalamus – mediodorsal nucleus, pulvinar
Caudate head

E. Clinical Correlates of Prefrontal Morphological Changes

Cognitive Functions

As schizophrenia is linked with cognitive dysfunction suggesting the involvement of dorsolateral prefrontal cortex, it is of importance to study if the above described morphological abnormality expresses itself as a functional impairment. Review of volumetric studies (Antonova et al., 2004) demonstrated that total frontal lobe volume is associated with executive functioning, working memory, verbal fluency, and immediate memory in schizophrenia. Dorsolateral prefrontal cortex volume was correlated with executive functions, attention, and verbal and visual memory, left DLPFC was associated with abstraction, flexibility, categorization and non-verbal immediate memory, right DLPFC with sustained attention. Orbitofrontal cortex had a complex relationship with many cognitive domains. There were differences in the pattern of structure/function relationship between patients and healthy subjects which may reflect different variability in performance depending on the relative difficulty of the task (Baare et al., 1999) or different variability of brain morphology features. Another explanation was presented with the note that similar brain volume in schizophrenia patients may not predict the same performance as in healthy controls, possibly due to the abnormalities of brain connectivity.

Several voxel-based morphometric studies tried to refine the spatial pattern of the structure/function relationship and focused on the association between dorsolateral prefrontal gray matter volume and *executive functions*. There are reports demonstrating that this assumption holds true: significant correlation was found with the performance in Wisconsin Card Sorting Task (WCST) or Trial Making Test and dorsolateral prefrontal cortical gray matter (BA 9) (Bonilha et al., 2008), direct comparison of schizophrenia subjects with poor and good ability to solve WCST showed reduction in dorsolateral prefrontal cortex and anterior cingulate in poor realtive to good solvers (Rusch et al., 2007).

However, there are also negative reports: Antonova et al. (2005) performed voxel-based morphometry analysis of associations between gray matter and cognitive functions in 45 schizophrenia patients and 43 healthy subjects. The patients had smaller total gray matter volume by 9%, the whole brain volume by 9,5%, regional gray matter reduction in left inferior frontal gyrus (BA 45, 47), left lingual gyrus (BA 17) and gray matter increase in left precuneus (BA 7) and left putamen. Premorbid intelligence quotient (IQ) and actual performance in IQ tests were associated with total gray matter volume, verbal memory with precuneus in patients. In healthy controls inferior frontal gyrus volume was associated with both verbal and non-verbal memory, whereas in patients no such relationship was found. No association with executive functions as measured by Wisconsin card sorting test or Trial making test was observed. However, in contrast to Bonilha et al. (2008) study - which found quite widespread gray matter reduction, particularly in left supplementary motor area (BA 6), bilateral superior frontal gyrus (BA 6, 8), left middle frontal gyrus (BA 9), right opercular area (BA 44), left angular gyrus (BA 39), left superior temporal gyrus (BA 22) and left cerebellar hemisphere – Antonova et al. did not found changes in the dorsolateral prefrontal cortex in the strict sense (BA 9 and 46). Thus the inconsistency of the results may be a consequnece of neurobiological heterogeneity of the samples studied.

Sustained attention disturbance is another key element of the cognitive dysfunction in schizophrenia (Nuechterlein et al., 2004). Continuous Performance task results correlated significantly with gray matter concentration in left inferior frontal gyrus, left angular, supramarginal and postcentral gyri, as well as in left thalamus in first-episode schizophrenia patients (Salgado-Pineda et al., 2003).

Another key concept in schizophrenia is the impairment in *social cognition* – theory of mind, self-reference or working memory – i.e. abilities also linked with prefrontal cortex (Amodio and Frith, 2006), orbitofrontal and triangular parts in particular. Orbitofrontal cortex is involved in attribution of emotional valence to percieved stimuli: lateral parts were shown to mediate evaluation of punishment value while medial parts were linked with the evaluation of reward (Kringelbach and Rolls, 2004) – both functions seem to be impaired in schizophrenia and related to negative symptoms. Pars triangularis of the inferior frontal gyrus is linked with language skills and formal thought disorder or language disturbances are significant features of schizophrenia (Ceccherini-Nelli and Crow, 2004). It is also of importance that mirror neurons were located within pars triangularis. Activity of mirror neurons reflect observed action and disturbances of their function was also proposed to be a key abnormality in schizophrenia (Arbib and Mundehnk, 2005). Others view schizophrenia as a disorder of social cognition in general: the social brain hypothesis (Burns, 2006). In schizophrenia, gray matter reduction was observed in both medial and lateral orbitofrontal cortex, as well as in pars triangularis, even in first-episode patients (Venkatasubramanian et al., 2008). Yamada et al. (2007) studied the relationship between regional brain morphology and the ability to attribute emotional valence to different stimuli as assessed by Perception of Affect Task and found that the performance was positively correlated with gray matter concentration in medial frontal gyrus (BA 10), where a reduction of gray matter was present in schizophrenia subjects with consequent poor performance.

Symptomathology

As mentioned earlier, a considerable heterogeneity of structural findings persists. It is a question if they may reflect the phenomenological diversity of schizophrenia, i.e. if certain clinical subtypes correlate with distinct neurobiological features. Several studies investigated the effect of various clinical variables on the regional brain morphology.

First, there is a certain effect of the illness itself, i.e. subjects in which the illness lasts longer or who have more frequent psychotic exacerbation or more severe course of the illness have pronounced changes of brain morphology. Longer *duration of prodromal phase* of schizophrenia was associated with smaller gray matter density in anterior cingulate, medial and inferior frontal cortex, and insula (Lapin et al., 2007). Similarly, *duration of untreated psychosis* correlated with gray matter density decreases in temporal areas (BA 21, 37, 39) and occipital areas (BA 18, 19) (Lapin et al., 2006), and longer *duration of illness* was associated with lower gray matter volume in right medial temporal region including hippocampus, parahippocampal gyrus, in bilateral anterior cingulate cortex, and medial cerebellum and more gray matter volume in right globus pallidus (Velakoulis et al., 2002). Smaller lingual gyrus (BA 17) and larger precuneus (BA 7) were linked with greater number of previous psychotic episodes, but no association was found between gray matter in putamen and precuneus with antipsychotic exposure (Antonova et al., 2005). *Poor-outcome patients,* i.e. those who were continuously hospitalized or completely dependent on caregivers, unemployed and expressing severe negative and formal thought disorder, demonstrated significantly smaller gray matter volume in temporal and occipital, but not frontal cortical areas (Mitelman et al., 2003). In another group of chronic schizophrenia patients, whose functioning was classified by Global Assessment of Functioning (GAF), a significant negative correlation between the GAF score and regional gray matter was found in the left inferior frontal gyrus and left inferior parietal lobule (Wilke et al., 2001), i.e. the less gray matter in these regions the poorer outcome.

Second, certain clinical symptoms demonstrate association with specific neurobiologic substrate, reflecting either their specific pathophysiology or neurobiological heterogeneity of the condition. Such specific association is demonstrated for verbal hallucinations, insight and negative symptoms.

Comparison of patients with pronounced history of *verbal hallucinations* with non-hallucinating patients revealed gray matter reduction in left insula and adjacent temporal lobe (Shapleske et al., 2002). Intensity of auditory hallucinations correlated with left inferior frontal and right posterior gyri (García-Martí et al., 2008), in a different study using BPRS items to rate intensity of hallucination, negative correlation with hallucinations was found in the left superior (transverse) temporal gyrus, left thalamus, and left and right cerebellum (Neckelmann et al., 2006).

Insight is a concept with critical importance in schizophrenia. Poor insight is frequent finding in schizophrenia patients (Mintz et al., 2003) and the quality of the understanding of schizophrenia symptoms and the need of treatment seem to be crucial for the long-term outcome of the illness (Drake et al., 2007). Insight is linked with cognitive abilities – executive functions in particular (Laroy et al., 2000; Smith et al., 2000; Drake and Lewis, 2003) and thus it is not surprising that schizophrenia subjects with cognitive impairment may have difficulties to understand the consequences of their situation. The clinical observation of the association between executive dysfunction and the quality of insight suggest the involvement of prefrontal systems linked with perceptual and monitoring systems. Several studies investigated such connection between insight and prefrontal morphology: earlier studies found association with frontal lobe atrophy as defined from CT scans (Laroy et al., 2000), and a negative correlation between the score for the severity of 'lack of insight and judgment' and gray matter concentrations in the left posterior and right anterior cingulate and bilateral inferior temporal regions including the lateral fusiform gyri was found (Ha et al., 2004). However, Bassitt et al., (2007) who investigated the link between regional gray matter and quality of insight rated by a specific scale Unawareness of Mental Illness Scale in fifty schizophrenia subjects did not find any correlation.

Negative symptoms rated by BPRS were inversely correlated with gray matter volume in frontal areas BA 8, 9, 10, and temporal areas BA 20, 21, 22; the range of correlation magnitude was between -0.30 and -0.51 (Hazlett et al., 2008).

Recent large scale (175 schizophrenia patients and 177 healthy subjects) VBM study tried to resolve the neurobiological heterogeneity that is behind distinct clinical syndromes (negative, positive and disorganized) and simultaneously to detect areas of changes that are generally linked with schizophrenia, shared by all clinical subgroups (Koutsouleris et a., 2008).

Changes associated with *negative symptoms* were the most widespread and demonstrated the largest effect size. They were linked to gray matter density reduction in *frontal areas:* medial and lateral parts of superior frontal gyrus, middle frontal gyrus, inferior frontal gyrus (orbital, triangular, and opercular parts), gyrus rectus, and supplementary motor area, precentral gyrus, anterior and middle parts of cingulate gyrus; bilateral *insula; temporal areas:* bilateral superior, middle and inferior temporal gyri, fusiform gyrus, mediotemporal structures - left amygdala, bilateral hippocampus and parahippocampal gyrus; *parietal areas:* supramarginal gyrus, postcentral gyrus; and bilateral cerebellum (VI), left caudate and thalamus. No changes were present in occipital areas.

Positive symptoms were associated with reduction in *frontal areas*: middle, inferior frontal (orbital opercular and triangular parts) gyri, right gyrus rectus, precentral gyrus, rolandic operculum, anterior and middle cingulate gyrus; bilateral *insula; temporal areas*: superior temporal gyrus, left middle temporal gyrus; *parietal areas*: supramarginal gyrus; *occipital areas:* inferior occipital cortex; and bilateral *thalamus.* Although there were some changes in the frontal areas, the reduction in temporal areas and in the thalamus were more pronounced than in negative symptoms. Interestingly, no association was observed with mediotemporal limbic areas or cerebellum.

The disorganized symptoms were associated with reduction in *frontal areas:* inferior frontal gyrus, gyrus rectus, medial orbitofrontal cortex, anterior and middle cingulate gyrus, rolandic operculum; *insula; temporal areas*: temporal pole, superior and middle temporal gyri, fusiform gyrus, and left amygdala; *parietal areas*: supramarginal gyrus. No changes were observed in occipital, cerebellar or subcortical structures.

All three syndromes shared (as demonstrated by conjunction analysis) reduction in orbitofrontal areas, medial and lateral prefrontal cortices, inferior frontal gyrus, anterior cingulate gyrus, bilateral insula, temporal polar area, middle and inferior temporal cortex, rolandic operculum, and supramarginal gyrus. These areas seem to be common neurobiological substrate of schizophrenia, irrespective to the clinical manifestation.

Treatment

There are reports that antipsychotic agents may have some influence on the brain morphology. The evidence comes from animal, postmortem, and neuroimaging studies. Although there is a good evidence for the effect of antipsychotics (classical neuroleptics in most cases) in striatum (Harrison, 1999b), much fewer data are available for cortical areas, which are critical for the understanding how antipsychotics interfere with the patho-physiological processes in schizophrenia. Even in the striatum, the effect of antipsychotics on the pattern of synaptic pathology seems to be qualitatively different from the one associated with schizophrenia (Harrison, 1999b).

Present day evidence suggests that there might be a differential effect of first- and second-generation antispychotics on the *brain plasticity*. First-generation antipsychotics, haloperidol in particular, were demonstrated to be neurotoxic as they decrease levels of neurotrophic factors (Parikh et al., 2004; Pillai et al., 2006) and induce apoptosis (Ukai et al., 2004; Wei et al., 2006). Furthermore, no beneficial effect of first-generation antipsychotics on neurogenesis was observed (Mallberg et al., 2000; Wakade et al., 2002; Halim et al., 2004; Schmitt et al., 2004; Wang et al., 2004). Second-generation antipsychotics seem to be superior to classical neuroleptics in the case of neurogenesis potentiation and their anti-apoptotic properties (Quing et al., 2003; Wei et al., 2003). Olanzapine treatment for 28 days increased hippocampal neurogenesis (Kodama et al., 2004), olanzapine also led to proliferation in prefrontal cortex, which rather than neuronal proliferation represented increase in endothelial or glial cells (Wang et al., 2004). Positive effect was demonstrated also for risperidone (Wakade et al., 2002), paliperidone (Nasrallah and Pixley, 2006), and quetiapine (Luo et al., 2005). However, quite opposite to expectations, no effect on neurogenesis was demonstrated for clozapine (Halim et al., 2004; Schmitt et al., 2004). Olanzapine and risperidone reversed the BDNF and NGF reduction induced by haloperidol in rats (Parikh et al., 2004; Pillai et al., 2006). In humans, NGF levels were reduced in drug-naive patients, which were increased after the treatment with risperidone (Parikh et al., 2003).

The direct observation of the *effect of antipsychotics on the brain morphology* in antipsychotic-treated animals brought rather inconsistent and heterogeneous results. Earlier animal studies did not found changes of the neuronal morphology in the cortex of rats treated with perphenazine (Pakkenberg et al., 1973; Fog et al., 1976), or haloperidol when observed at the light microscopic level (Benes et al., 1985).

Following studies brought significant results: treatment-associated *gray matter volume* reduction was observed in left parietal cortex in macaque monkeys (Dorph-Petersen et al., 2005) after olanzapine or haloperidol treatment for 27 months. Widening of layer V in dorsolateral preronta1 area 46 was observed after six-months of antipsychotic exposure

(Selemon et al., 1999b). No changes of *neuronal number or density* were observed after half-year (Selemon et al., 1999) as well as one year-long treatment with typical or atypical antipsychotics in macaque monkeys (Konopaske et al., 2007). Studies using direct assessment of *neuropil* by electron microscopy showed loss of dendritic spines in layer VI of medial prefrontal cortex in rats treated with haloperidol (Benes et al., 1985; Klintzova et al., 1989). Chronic exposure to haloperidol or clozapine led to increase of axodendritic symmetric, i.e. inhibitory, synapses with the loss of asymmetric, i.e. excitatory, synapses in layer VI of medial prefrontal cortex in rats (Vincent et al., 1991). Haloperidol administration was also linked with increased immunoreactivity for GABA in medial prefrontal cortex of rats (Vincent 1994). These studies indicate that the drugs may induce reorganization of synapses in layer VI, perhaps preferentially the inhibitory ones. Haloperidol treatment-associated changes of cytoskeleton and dendritic spines proteins were observed also in rhesus monkeys: chronic exposure to haloperidol led to decrease of spinophilin, protein associated with dendritic spines, increase of microtubule-associated protein 2 (MAP2) phosphorylation (which reduces its ability to stabilize dendritic microtubules) in many cortical areas – prefrontal, orbital, cingulate, entorhinal – but not in primary visual cortex; no changes of vesicle-associated protein synaptophysin were observed (Lidow et al., 2001). However, human post-mortem studies did not reveal a substantial effect of antipsychotics on the morphology of cortical synapses (Harrison et al., 1999a, Harrison et al., 1999b).

Antipsychotic agents were also linked with *glial cells* changes – although again with inconsistent results. Treatment with haloperidol or olanzapine decreased glial cell numbers in parietal cortex (Konopaske et al., 2007), which was specific to astrocytes (Konopaske et al., 2008), and was accompanied by cortical volume reduction. Previous report of antipsychotic effect on glial cells that used the same methodology as in human studies showed just the opposite: in dorsolateral prefrontal cortex (BA 46) six-months exposure to atypical or typical antipsychotics led to increased glial cell density in layer IV and widening of cortical thickness in layer V (Selemon et al., 1999b). This study, however, used shorter duration of antipsychotic exposure and lower daily doses of antipsychotics than the Konopaske et al. studies. Nevertheless, the results are of importance because they demonstrate the possible correlate of neuroimaging findings of cortical increase in schizophrenia. In this context, predominant distribution of dopamine D2 receptors in layer V might be linked with the increase of layer V thickness (Vincent et al., 1993).

It is apparent that although some antipsychotic-associated changes seem to reflect certain features of brain morphology changes in schizophrenia leaving open the possibility, that at least some of the changes in schizophrenia are a consequence of the antipsychotic exposure, other, such as widening of cortical thickness, go against the pattern seen in schizophrenia. However, there is no evidence suggesting that any antipsychotic agent can reverse the pathological cortical changes in schizophrenia; at least no reports of layer III dendritic spine density increase were published. Such prove is, unfortunately, difficult to obtain, since it is hard to model the schizophrenia-associated neuropathology in animals, and antipsychotic drugs may not exert the same effect in the health, where only adverse effects are observed, as in the disease, where specific antipsychotic effect can prevail.

Neuroimaging studies seem to suggest that antipsychotic agents may have influence on brain morphology. ROI studies demonstrated the effect of classical neuroleptics on basal ganglia (Chakos et al., 1994), as well as some second generation antipsychotics (Massana et al., 2005).

As for *the effect on cortical areas*, little evidence is available. Garver et al. (2005) observed an increase of total gray matter by 20.6 (SD 11.4) cc in patients treated with second generation antipsychotics – risperidone or ziprasidone – for 28 days, with no changes in haloperidol-treated subjects. Lieberman et al. (2005) observed decreased rate of gray matter loss in olanzapin- in contrast to haloperidol-treated first-episode schizophrenia subjects. Molina et al. (2005) studied treatment-naive and chronic treatment-resistant schizophrenia patients for two years of treatment with risperidone (treatment naive subjects) or clozapine (treatment-resistant subjects). Treatment with both agents led to increase in frontal (by 9-12 %), parietal (by 8-11 %) and occipital (by 10-12 %) gray matter volume, only on clozapin total (by 4 %) and temporal (by 2 %) gray matter volume increased.

Regional specificity of the changes was investigated in a few studies using cross-sectional as well as longitudinal design. Dazzan et al. (2005) investigated the potential effect of different antipsychotic agents on brain morphology in a cross-sectional VBM study of first-episode psychosis subjects who were drug-free (n = 22), treated with classical neuroleptics (n = 32) or atypical antipsychotics (n = 30). Majority of the atypical antipsychotic treated subjects was given olanzapine (n = 21). The mean duration of treatment was 9 weeks for atypical antipsychotics and 8 weeks for classical neuroleptics. The first-episode psychosis project included patients with schizophrenia or schizophreniform psychosis, affective psychosis, or other psychotic disorders. The distribution of diagnostic categories was non-random in the three groups: atypical antipsychotic –exposed subjects were in 80% diagnosed with schizophrenia, in 7% with affective psychosis and in 13% with other psychosis. Classical neuroleptics-exposed subjects were in 47% schizophrenia subjects, in 47% diagnosed with affective psychosis, and in 6% with other psychosis. Drug-free subjects were diagnosed with schizophrenia only in 32%, in 41% with affective psychosis and in 27% with other psychosis. Subjects treated with classical neuroleptics had higher gray matter concentration in basal ganglia and clusters of gray matter reduction stretching across insula, inferior frontal gyrus (BA 47), and superior temporal gyrus (BA 22), across paracentral lobule (BA 4, 5), superior and medial frontal gyrus (BA 6, 31), and cingulate gyrus (BA 24), and across precuneus (BA 7) when compared with drug free subjects. Subject exposed to atypical antipsychotics had increased gray matter in thalamus when compared with drug free subjects. Direct comparison of the two treated groups showed gray matter reduction in middle temporal gyrus (BA 21) in classical neuroleptics-exposed patients. Due to the apparent diagnostic heterogeneity between the study groups the results may reflect differences in the neurobiology of different conditions rather than the effect of different kinds of treatment.

In a longitudinal study of 6-week treatment with risperidon in a small group of 15 neuroleptic-naive first-episode psychosis subjects a tiny increase of gray matter in BA 39 (superior and middle temporal gyrus) was observed (Girgis et al., 2006). However, the authors did not correct for multiple comparisons effect in a study with no a priori regional hypothesis, which brings the significant probability of the results being false positives. Another longitudinal assessment of 10 chronic patients treated for twelve weeks with second generation antipsychotic identified no changes in cortical or subcortical areas (McClure et al., 2008). Volumetric study of the effect of first as well second generation antipsychotics on the anterior cingulate during the treatment for 2-3 years yielded interesting results: increased anterior cingulate volume correlated with increased exposure to first generation anti-psychotics, but negative correlation was found for exposure with second-generation agents. The increase of anterior cingulate volume was also correlated with greater improve-ment of positive symptoms (McCormick et al., 2005). These results are important for they demonstrate similar pattern of changes as seen in basal ganglia also for cortical structure. Anterior cingulate is interconnected with basal ganglia, and extrastriate dopamine D2 receptors are expressed here (Olsson et al., 2004); both findings may contribute to the anterior cingulate increase after first-generation antipsychotics. The interpretation of such increase is quite important in the light of the findings of gray matter increase in schizophrenia.

Thus, although there are data suggesting there is certain effect of antipsychotics on the brain morphology, the exact nature of this effect is not known and we cannot conclude that the effect of antipsychotics is beneficial in the sense of the reverse of basic neuropathological pattern of changes as present in schizophrenia. However, this pattern – reduced neuropil without the loss of neurons – leaves a chance open that there might be a treatment that may help remodel local neurocircuits and restore normal function in schizophrenia.

F: Specificity of the Morphological Changes

High-Risk Subjects

The patter of brain morphology changes described above is not entirely specific for schizophrenia subjects. Even subjects at high risk of development of schizophrenia show a discrete pattern of cortical reduction. Job et al. (2003) presented data from the Edinburgh High Risk Study, a project of compelling elaboration that recruited healthy siblings of schizophrenia patients at the time window for development of schizophrenia and followed them prospectively. The high risk subjects (n = 146) had reduced gray matter in the anterior cingulate bilaterally than healthy subjects. First episode patients (n = 34) had further reduced gray matter in the left and right cingulate gyrus, left medial, left and right inferior, and middle frontal gyrus (BA9, BA45), right precentral gyrus, in right hippocampus, amygdala, parahippocampal gyrus, middle temporal gyrus (BA 22, BA39), and in left postcentral gyrus (BA40). Another sample of high risk subjects (n = 40) form Early Detection and Intervention Centre for Mental Crises of Ludwig-Maximilians-University in Germany detected significant clusters of gray matter reduction in as good as the whole frontal cortex (medial, dorsolateral and ventrolateral, orbitofrontal parts, anterior cingulate, motor and supplementary motor cortex), temporal areas in superior, middle temporal gyrus, parietal supramarginal gyrus, postcentral gyrus and Rolandic operculum (Meisenzahl et al., 2008). High risk subjects who later developed psychoses had smaller gray matter volume in posterior cingulate, precuneus, superior parietal gyrus and increased gray matter in left supramarginal gyrus when compared with healthy controls. When compared with first-episode schizophrenia patients, relatively higher gray matter volume was observed in temporal lobe bilaterally and smaller gray matter volume in basal ganglia (Borgwardt et al., 2007).

Relatives

Gray matter changes are present even in healthy siblings of schizophrenia patients who did not develop schizophrenia. This finding is one of the features of intermediate phenotype, or *endophenotype,* reflecting neurobiological substrate of the illness that lies between genotype and clinical manifestation of the illness. In particular, Honea et al. (2008) studied 213 healthy siblings of schizophrenia subjects and found a trend (after stringent thresholding p<0.05 FWE corrected) for a gray matter reduction in medial frontal gyrus (BA 10). Without such stringent significance criteria the siblings shared gray matter reduction also in inferior frontal gyrus, paracingulate sulcus (BA 10), superior temporal gyrus (BA 22) and insula. Furthermore, hippocampal gray matter volume was significantly correlated within sibling pairs. In a smaller study, where the genetic liability to schizophrenia was quantified on a continuous scale, the liability was associated with gray matter volume in dorsolateral and ventrolateral cortex (McIntosh et al., 2006). More elegant approach to study the contribution of genome on the illness neurobiology is to study twins discordant for schizophrenia. In one such study Hulshof Pol et al. (2006) found that twins discordant for schizophrenia share reduction of gray matter density in the left medial orbitofrontal cortex, where differences between the schizophrenia-affected and healthy twins were less pronounced in monozygotic twin pairs than in dizygotic ones.

Schizotypal Disorder

Schizotypal disorder, a condition considered to be a member of schizophrenia-spectrum disorders, may represent the *genetic liability* for, or share some pathophysiological processes with schizophrenia – i.e. the patients with schizotypal disorder may share endophenotypes with schizophrenia subjects. Kawasaki et al. (2004) performed direct comparison of schizophrenia patients (n = 25), schizotypal subjects (n = 25) and healthy controls (n = 50) using VBM: schizotypal subjects demonstrated gray matter reduction in the left inferior frontal gyrus (BA47), insula, superior temporal gyrus (BA22), and medial temporal lobe (BA28). When compared with schizophrenia subjects, the schizotypal subjects had significantly more gray matter in orbitofrontal cortex (BA11). Reduction in inferior frontal gyrus and left medial temporal lobe were reduced in both schizophrenia and schizotypal disorder and therefore may represent the core substrate of vulnerability to schizophrenia. Comparison of only schizotypal women with healthy controls brought similar pattern of gray matter reduction: left inferior frontal gyrus, left superior, and middle temporal gyrus (BA 21, 22), right premotor cortex (BA6), and parietal areas (BA13, BA40) (Koo et al., 2006). Automatic ROI analysis of several Brodman´s areas in schizophrenia (n = 57), schizotypal (n = 79) and healthy (n = 148) subjects found also quite similar pattern of changes: schizophrenia subjects showed gray matter reduction in many fronto-temporal areas (prefrontal area 10; cingulate areas 23, 24, 31; temporal areas 20, 21, 22), schizotypal subjects showed similar pattern of changes but of smaller magnitude (cingulate areas 23, 24, 31; temporal area 21). The difference between schizophrenia and schizotypal subjects was present in frontopolar region BA 10 where schizotypal subjects showed volume increase relative to healthy controls, and in superior temporal gyrus (BA 20, 22), where schizotypal subjects showed similar volume as healthy subjects whereas schizophrenia subjects showed reduction (Hazlett et al., 2008).

Bipolar Affective Disorder

There is a continuing debate on the similarities and differences between schizophrenia and bipolar disorder (Murray et al., 2004). It is then interesting to look at the similarities and differences in the brain morphology. Meta-analysis of volumetric studies with bipolar subjects showed no changes of brain morphology in bipolar patients as a whole, but substantial heterogeneity for left subgenual prefrontal cortex, bilateral amygdala and thalamus was noted, suggesting a possible existence of subgroups of patients with morphological abnormalities (Mc Donald et al., 2004). Bearden et al. (2007) found overall increase of the total gray matter by 6.6%, with regional increase of gray matter density in many cortical areas, notably in the heteromodal associative cortex: by 10 % or more in bilateral superior and middle frontal gyri, left ventrolateral prefrontal cortex, superior temporal cortex, retrosplenial and paracingulate cortex, subgenual coretx, and somatosensory cortex. This study found no area of gray matter density reduction. Moreover, the increase of gray matter density was explained by the effect of lithium treatment; patients not treated with lithium did not differ from healthy subjects. Similar increase of gray matter in the anterior cingulate, ventral prefrontal cortex, fusiform gyrus and primary and supplementary motor cortices in adult bipolar patients on various psychopharmacological treatment was reported (Adler et al., 2005). Direct comparison of first-episode schizophrenia and bipolar subjects showed significantly less gray matter in the left precentral/inferior frontal gyrus - BA 6/45, left anterior cingulate/medial frontal gyrus - BA 32/10, right superior frontal gyrus - BA 10, bilateral middle frontal gyrus – BA 9/46 in schizophrenia and significantly less gray matter in the left inferior/middle temporal gyrus - BA 20/21, right uncus - BA 20, and right anterior superior/middle temporal gyrus - BA 38/21 in the bipolar subjects. However, the bipolar subject showed significant reduction of gray matter in the right inferior frontal/precentral gyrus – BA 44/6, and in several temporal and parietal cortical areas when compared with healthy controls; in these areas the changes were present also in the schizophrenia subjects (Farrow et al., 2005). Gray matter reduction in the right inferior frontal gyrus and the left anterior cingulate was referred also in another sample of bipolar patients in which only minority of patients received lithium (Lyoo et al., 2004). This suggests that although the schizophrenia subjects show greater impairment, notably in the prefrontal cortex, the pattern of changes is not specific to schizophrenia and thus, the information is not useful in differential diagnosis between schizophrenia and bipolar disorder. On the histological level, there seem to be different abnormalities in bipolar disorder than those seen in schizophrenia (Harrison, 2002; Frey et al., 2004): the main positive findings include decreased glial density in subgenual prefrontal cortex – BA24 – and dorsolateral prefrontal cortex – BA 9, a reduction of non-pyramidal neurons in cingulate and hippocampus, and a higher occurrence of hyperintensities in white matter; however, clear cut between affective disorders and schizophrenia cannot be made.

Substance Abuse

As noted several times in the discussion of the findings of possible progression, abuse of psychoactive substances may contribute to the pattern of changes. Gray matter reduction was demonstrated in the alcohol- (Chandraud et al., 2007; Jang et al., 2007), opiate- (Lyoo et al.,

2006), and cocaine-dependent patients (Franklin et al., 2002; Sim et al., 2007), or users of MDMA (Cowan et al., 2003). Alcohol-dependent patients had substantial reduction in prefrontal cortex (BA 9, 10), precentral gyrus (BA 6), cingulate gyrus (BA 32), temporal cortex (BA 19, 21, 22, 37), insula (BA 13), parietal cortex (BA 40), thalamus and cerebellum. Neuropsychological performance correlated significantly with the gray matter in the areas of gray matter reduction (Chandraud et al., 2007; Jang et al., 2007). Opiate-dependent patients showed gray matter concentration reduction in dorsolateral prefrontal cortex (BA 9, 10, 11, 47), insula (BA 13), superior temporal cortex (BA 21, 38), and mediotemporal structures (BA 28) (Lyoo et al., 2006). Cocaine-dependent patients had lower gray matter concentration in prefrontal cortex (BA 10), anterior cingulate, premotor cortex (BA 6, 8), insula, and superior temporal cortex (BA 20, 38), thalamus and cerebellum (Franklin et al., 2002; Sim et al., 2007). MDMA users had decreased gray matter in temporal (BA 21), frontal (BA 45) and cerebellum (Cowan et al., 2003). Of special interest are the findings of gray matter differences between smokers and non-smokers because high proportion of schizophrenia subjects smokes (Hughes et al., 1986). Smokers had significantly lower gray matter volume in the dorsolateral, ventrolateral prefrontal cortex, left cingulate and right cerebellum (Brody et al., 2004). Apparently, abuse of or dependence on psychoactive substances is linked with a very similar pattern of gray matter changes as manifest in schizophrenia, even in the first-episode subjects. They affect areas linked with higher order cognitive functions; however, the functional outcome of the affection depends on many factors: if they affect matured or developing brain, what is the exact nature of the pathology of local circuits caused by the particular pathological process etc.

Quite unique situation is in the case of a link between schizophrenia and *cannabis* use. Still unresolved issue is if cannabis use predisposes for or increase risk of the later schizophrenia development or if it reflects certain pathophysiological features of the illness (Newell et al., 2006) with consequent use of cannabis as a kind of automedication. Advocates may be found for both viewpoints. In this context studies that may shed light on the common pathways of schizophrenia neurobiology and cannabis use are of special interest. Bangalore et al. (2008) studied the link between the cannabis use and regional gray matter in first-episode schizophrenia who abused cannabis (n = 15), who did not (n = 24) and healthy controls (n = 42) without anamnesis of abuse. When compared with schizophrenia subjects without cannabis abuse, schizophrenia subjects abusing cannabis had smaller gray matter density in posterior cingulate, a region rich of CB1 receptors (Newell et al., 2006). Even heavy (more than 5 joints/day) and long term (more than 10 years) cannabis use alone, in subjects without schizophrenia or significant psychosis, was associated with bilateral hippocampal and amygdala volume in a ROI study (Yucel et al., 2008), i.e. in areas linked with pathopysiology of schizophrenia.

Neurotic and Stress Related Disorders

Obsessive-compulsive disorder is another condition considered to be of neurodevelopmental origin. Imaging studies refer a consistent pattern of gray matter in anterior cingulate gyrus (Valente et al., 2005; Carmona et al., 2007; Yoo et al., 2008), with reports of gray matter reduction in inferior and medial frontal gyrus, insula, and superior temporal gyrus (Yoo et al., 2008). However, even conditions considered to be consequences of sole psychological processes, such as *Panic disorder*, were shown to be linked with gray matter deficits. In panic disorder, reduction of gray matter density in the left parahippocampal gyrus (Massana et al., 2003) or bilateral putamen (Yoo et al., 2005) was reported. Similarly, life events, at least unfavorable ones exerting substantial stress, can also contribute to the brain morphology changes. This factor is marked in *Posttraumatic stress disorder* where gray matter reduction in anterior cingulate was reported (Yamasue et al., 2003). Moreover, even some *personality traits* may have neuroanatomical correlate: harm avoidance score correlated significantly with the gray matter in the left amygdala (Iidaka et al., 2006), a limbic structure where Rusch et al. (2003) reported significant reduction of gray matter in borderline personality disorder. Moreover, novelty seeking was associated with gray matter in the left medial frontal gyrus and reward dependence with caudate nucleus (Iidaka et al., 2006).

Histopathological Correlates of Similar Findings

The same pattern of neuroimaging findings does not speak about the same neuropathology. The final judge in the etiology and character of the neuroimaging findings is the histopathology. As demonstrated above, it seems that VBM studies in schizophrenia image the abundance of synaptic connections or neuropil. The histopathological meaning of neuroimaging findings in other psychiatric conditions is, unfortunately, unclear.

Apparently, not all changes in schizophrenia detected using imaging methods have support from neuropathological studies. This is because the neuropathological studies focused mainly on the limbic and prefrontal areas. On the other hand, the findings from neuroimaging studies inspire and guide the neuropathological investigation. Neuroimaging, however, brings information that cannot be achieved by neuropathological studies – such as dynamics of the changes. Both approaches are thus complementary and should be used together when trying to understand quite complex findings of brain morphology changes in schizophrenia and related conditions.

Conclusion

In schizophrenia, a somewhat widespread reduction of cortical gray matter exists. The changes are quite heterogeneous both in terms of intersubject and regional variability, with the most consistent and extensive findings being at hand for the dorsolateral prefrontal cortex. Its dysfunction leads to cognitive symptoms as well as to the dysregulation of the dopaminergic system, which predisposes those affected to psychotic episodes. However, the changes are not limited to the heteromodal associative cortex; changes were observed also in primary sensory or motor cortical areas. These findings substantiate the fact that schizophrenia is not only a disease of higher cognition; on the contrary, many signs and symptoms reflect motor (Kasparek et al., in press) or perceptual (Butler et al., 2008) dysfunction.

Neurobiologically, it corresponds to the loss of dendritic spines of layer III pyramidal neurons and interrelated changes in parvalbumin GABA-ergic cells, thalamocortical and dopaminergic projections. Such a pattern of cortical microcircuitry changes results in abnormal coordination of cortical pyramidal neuron activity. This dyscoordination is expressed on the level of neurophysiological abnormalities, such as reduced theta or gamma band oscillations during cognitive load.

It is apparent that the pattern of neuroimaging findings in schizophrenia is not disease-specific. Thus, the neuroimaging has limited value in the diagnostic or differential diagnostic process. However, in the context of established diagnoses, we are able to interpret the changes on the neuropathological level and we may use the information from neuroimaging methods in a specific and quite useful way, i.e., we may try to link the pattern of changes with clinical features of the illness or we may consider regionally-specific therapeutic approaches, such as rTMS for negative syndrome (Prikryl et al., 2007) or refractory hallucinations (Aleman et al., 2007).

Acknowledgment

Supported by a research project of the Ministry of Education, Youth and Sports of the Czech Republic no. MSM0021622404.

References

Adler, C; Levine, A; DelBello, M; Strakowski, S. Changes in gray matter volume in patients with bipolar disorder. *Biol. Psychiatry*, 2005, 58, 151-157.

Agartz, I; Sedvall, GC; Terenius, L; Kulle, B; Frigessi, A; Hall, H; Jonsson, EG. BDNF gene variant and brain morphology in schizophrenia. *American Journal of medical Genetics Part* B (Neuropsychiatric Genetics), 2006, 141B, 513-523.

Ahmed, B; Anderson, JC; Douglas, RJ; Martin, KAC; Nelson, JC; Polyneurronal innervation of spiny stellate neurons in cat visual cortex. *J. Comp. Neurol.*, 1994, 341, 39-49.

Akbarian, S; Kim, JJ; Potkin, SG; Hagman, JO; Tafazzoli, A; Bunney, WE Jr; Jones EG. Gene expression for glutamic acid decarboxylase is reduced without loss of neurons in prefrontal cortex of schizophrenics. *Arch. Gen. Psychiatry*, 1995, 52, 258-66.

Akil, M; Kolachana, BS; Rothmond, DA; Hyde, TM; Weinberger, DR; Kleinman, JE. Catechol-O-methyltransferase genotype and dopamine regulation in the human brain. *J. Neurosce*, 2003, 23, 2008-2013.

Aleman, A; Sommer, IE; Kahn, RS. Efficacy of slow repetitive transcranial magnetic stimulation in the treatment of resistant auditory hallucinations in schizophrenia: a meta-analysis. *J. Clin. Psychiatry*, 2007, 68, 416-21.

Alexander, GE; Goldman, PS. Functional development of the dorsolateral prefrontal cortex: An analysis utilizing reversible cryogenic depression. Brain Res, 1978, 143, 233-249.

Amodio, DM; Frith, CD. Meeting of minds: the medial frontal cortex and social cognition. *Nat. Rev. Neurosci.*, 2006, 7, 268-77.

Antonova, E; Sharma, T; Morris, R; Kumari, V. The relationship between brain structure and neurocognition in schizophrenia: a selective review. *Schizophr. Res.*, 2004, 70, 117-45.

Antonova, E; Kumari, V; Morris, R; Halari, R; Anilkumar, A; Mehrotra, R; Sharma, T. The relationship of structural alterations to cognitive deficits in schizophrenia: a voxel-based morphometry study. *Biol. Psychiatry*, 2005, 58, 457–467.

Arbib, MA; Mundhenk, TN. Schizophrenia and the mirror system: an essay. *Neuropsychologia*, 2005, 43, 268-80.

Arnsten, AFT; Cai, JX; Murphy, BL; Goldman-Rakic, PS. Dopamine D1 receptor mechanisms in the cognitive performance of young adult and aged monkeys. *Psychopharmacology*, 1994, 116, 143-151.

Baaré, WF; Hulshoff Pol, HE; Hijman, R; Mali, WP; Viergever, MA; Kahn, RS. Volumetric analysis of frontal lobe regions in schizophrenia: relation to cognitive function and symptomatology. *Biol. Psychiatry,* 1999, 45, 1597-605.

Bangalore, SS; Prasad, KMR; Montrose, DM; Goradia, DD; Diwadkar, VA; Keshavan, MS. Cannabis use and brain structural alterations in first episode schizophrenia — a region of interest, voxel based morphometric study. *Schizophrenia Res.*, 2008, 99, 1–6.

Bayer, TA; Falkai, P; Maier, W. Genetic and non-genetic vulnerability factors in schizophrenia: The basis of the "two-hit hypothesis". *J. Psychiatry Res.*, 1999, 33, 543-548.

Beal, MF; Brouillet, E; Jenkins, B. Neurochemical and histological characterization of striatal excitotoxic lesions produced by the mitochondrial toxin 3-nitropropionic acid. *J. Neurosci.*, 1993, 13, 4181-4191.

Bearden, CE; Thompson, PM; Dalwani, M; Hayashi, KM; Lee, AD; Nicoletti, M; Trankhtenbroit, M; Glahn, DC; Brambilla, P; Sassi, RB; Mallinger, AG; Frank, E; Kupfer, D; Soares, JC. Greater cortical gray matter density in lithium-treated patients with bipolar disorder. *Biol. Psychiatry*, 2007, 62, 7-16.

Beasly, CL; Reynolds, GP. Parvalbumin-immunoreactive neurons are reduced in the prefrontal cortex of schizophrenia. *Schizophr. Res.*, 1997, 24, 349-355.

Ben-Schachar, D; Zuk, R; Gazawi, H; Ljuboncic, P. Dopamine toxicity involves mitochondrial complex I inhibition: implications to dopamine-related neuropsychiatric disorders. *Biochem. Pharmacol.*, 2004, 67, 1965-1974.

Benes, FM; Paskevich, PA; Davidson, J; Domesick, VB. Synaptic rearangements in medial prefrontal cortex of haloperidol-treated rats. *Brain Res.*, 1985, 348, 15-20.

Benes; FM; McSparren, J; Bird, ED; SanGiovanni, JP; Vincent, SL. Deficits in small interneurons in prefrontal and cingulate cortices of schizophrenic and schizoaffective patients. *Arch. Gen. Psychiatry*, 1991, 48, 996-1001.

Benes; FM; Todtenkopf, MS; Taylor, JB. A shift in tyrosine hydroxylase-immunoreactive varicosities (TH-IRv) form pyramidal (PN) to nonpyramidal (NP) neurons occurs in layer II of the anterior cingulate cortex of schizophrenics. *Soc. Neurosci. Abst.*, 1995, 21, 259.

Benes, FM; Vincent, SL; Marie, A; Khan, Y. Up-regulation of GABAA receptor binding on neurons of the prefrontal cortex in schizophrenic subjects. *Neuroscience*, 1996, 75, 1021-31.

Benes, FM. The role of stress and dopamine-GABA interactions in the vulnerability for schizophrenia. *J. Psychiatr. Res.*, 1997, 31, 257-275.

Berger, GE; Wood, S; McGorry, PD. Incipient neurovulnerability and neuroprotection in early psychosis. *Psychopharmacol. Bul.l*, 2003, 37, 79-101.

Bertolino, A; Saunders, RC; Mattay, VS; Bachevalier, J; Frank, JA; Weinberger, DR. Altered development of prefrontal neurons in rhesus monkeys with neonatal mesial temporolimbic lesions: a proton magnetic resonance spectroscopic imaging study. *Cereb. Cortex*, 1997, 7, 740-8.

Bertolino, A; Knable, MB; Saunders, RC; Callicott, JH; Kolachana, B; Mattay, VS; Bachevalier, J; Frank, JA; Egan, M; Weinberger, DR. The relationship between dorsolateral prefrontal N-acetylaspartate measures and striatal dopamine activity in schizophrenia. *Biol. Psychiatry*, 1999, 45, 660-7.

Bertolino, A; Breier, A; Callicott, JH; Adler, C; Mattay, VS; Shapiro, M; Frank, JA; Pickar, D; Weinberger, DR. The relationship between dorsolateral prefrontal neuronal N-acetylaspartate and evoked release of striatal dopamine in schizophrenia. *Neuropsychopharmacology*, 2000; 22, 125-32.

Bilder, RM; Bogerts, B; Ashtari, M; Wu, H; Alvir, JM; Jody, D; Reiter, G; Bell, L; Lieberman, JA. Anterior hippocampal volume reductions predict frontal lobe dysfunction in first episode schizophrenia. *Schizophr. Res.*, 1995, 17, 47-58.

Blatow, M; Rozov, A; Katona, I; Hormuzdi, SG; Meyer, AH; Whittington, MA; Caputi, A; Monyer, H. A novel network of multipolar bursting interneurons generates theta frequency oscillations in neocortex. *Neuron*, 2003, 38, 805-17.

Bogerts, B. The neuropathology of schizophrenic diseases: historical aspects and present knowledge. *Eur. Arch. Psychiatry Clin. Neurosci.*, 1999, 249(suppl. 4), IV/2-IV/13.

Bourgeois, JP; Goldman-Rakic, PS; Rakic, P. Synaptogenesis in the prefrontal cortex of rhesus monkeys. *Cereb. Cortex*, 1994, 4, 78-96.

Braff, DL; Freedman, R; Schork, NJ; Gottesman, II. Deconstructing schizophrenia: an overview of the use of endophenotypes in order to understand a complex disorder. *Schizophr. Bull.*, 2007; 33; 21-32.

Broadbelt; K; Byne, W; Jones, LB. Evidence for a decrease in basilar dendrites of pyramidal cells in schizophrenic medial prefrontal cortex. *Schizophr. Res.*, 2002, 58, 75-81.

Brody, AL; Mandelkern, MA; Jarvik, ME; Lee, GS; Smith, EC; Huang, JC; Bota, RG; Bartzokis, G; London, ED. Differences between smokers and nonsmokers in regional gray matter volumes and densities. *Biol. Psychiatry*, 2004, 55, 77–84.

Bonilha, L; Molnar, C; Horner, MD; Anderson, B; Forster, L; George, MS; Nahas, Z. Neurocognitive deficits and prefrontal cortical atrophy in patients with schizophrenia. *Schizophrenia Res.*, 2008; 101; 142-151.

Borgwardt, SJ; McGuire, PK; Aston, J; Berger, G; Dazzan, P; Gschwandtner, U; Pfluger, M; D'Souza, M; Radue, EW; Riecher-Rossler, A. Structural brain abnormalities in individuals with an at-riskmental state who later develop psychosis. *Br. J. Psychiatry*, 2007, 191 (suppl . 51); 69-75.

Brock, JW; Prasad, C. Alterations in dendritic spine density in the rat brain associated with protein malnutrition. *Dev. Brain Res.*, 1992, 66, 266-269.

Bromet, EJ; Naz, B; Fochtmann, LB; Carlson, GA; Tanenberg-Karant, M. Long-term diagnostic stability and outcome in recent first-episode cohort studies of schizophrenia. *Schizophrenia Bulletin,* 2005, 31, 639–649.

Bryan, GK; Riesen, AH. Deprived somatosensory-motor experience in stump-tailed monkey neocortex: dendritic spine density and dendritic branching of layer IIIb pyramidal cells. *J. Comp. Neurol.*, 1989, 286, 208-217.

Burns J. The social brain hypothesis of schizophrenia. *World Psychiatry*, 2006, 5, 77-81.

Butler, PD; Silverstein, SM; Dakin, SC. Visual perception and its impairment in schizophrenia. *Biol. Psychiatry*, 2008, 64, 40-7.

Cahn, W; Hulshoff Pol, HE; Lems, EB; van Haren, NE; Schnack, HG; van der Linden, JA; Schothorst, PF; van Engeland, H; Kahn, RS. Brain volume changes in first-episode schizophrenia: a 1-year follow-up study. *Arch. Gen. Psychiatry*, 2002, 59, 1002-10.

Cannon, TD; Hennah, W; van Erp, TG; Thompson, PM; Lonnqvist, J; Huttunen, M; Gasperoni, T; Tuulio-Henriksson, A; Pirkola, T; Toga, AW; Kaprio, J; Mazziotta, J; Peltonen, L. Association of DISC1/TRAX haplotypes with schizophrenia, reduced prefrontal gray matter, and impaired short- and long-term memory. *Arch. Gen. Psychiatry*, 2005, 62, 1205-13.

Carmona, S; Bassas, N; Rovira, M; Gispert, JD; Soliva, JC; Prado, M; Tomas, J; Bulbena, A; Vilarroya, O. Pediatric OCD structural brain deficits in conflict monitoring circuits: A voxel-based morphometry study. *Neuroscience Letters*, 2007, 421, 218–223.

Ceccherini-Nelli A; Crow TJ. Disintegration of the components of language as the path to a revision of Bleuler's and Schneider's concepts of schizophrenia. Linguistic disturbances compared with first-rank symptoms in acute psychosis. *Br. J. Psychiatry*, 2003, 182, 233-40.

Chakos, MH; Lieberman, JA; Bilder, RM; et al. Increase in caudate nuclei volumes of first-episode schizophrenic patients taking antipsychotic drugs. *Am. J. Psychiatry*, 1994, 151, 1430-1436.

Chanraud, S; Martelli, C; Delain, F; Kostogianni, N; Douaud, G; Aubin, HJ; Reynaud, M; Martinot, JL. Brain morphometry and cognitive performance in detoxified alcohol-dependents with preserved psychosocial functioning. *Neuropsychopharmacology*, 2007, 32, 429–438.

Cho, RY; Konecky, RO; Carter, CS. Impairments in frontal cortical gamma synchrony and cognitive control in schizophrenia. *Proc. Natl. Acad. Sci. USA*, 2006, 103, 19878-83.

Cobb, SR; Buhl, EH; Halasy, K; Paulsen, O; Somogyi, P. Synchronization of neuronal activity in hippocampus by individual GABAergic interneurons. *Nature*, 1995, 378, 75-8.

Connor, JR; Diamond, MC. A comparison of dendritic spine number and type on pyramidal neurons of the visual cortex of old adult rats from social or isolated environments. *J. Comp. Neurol.*, 1982, 210, 99-106.

Cowan, RL; Lyoo, IK; Sunga, SM; Ahna, KH; Kima, MJ; Hwang, J; Hagaa, E; Vimal, RLP; Lukas, SE; Renshaw, PF. Reduced cortical gray matter density in human MDMA (Ecstasy) users: a voxel-based morphometry study. *Drug and Alcohol Dependence*, 2003, 72, 225–235.

Crow, TJ; Ball, J; Bloom, SR, Brown, R; Bruton, CJ; Colter, N; Frith, CD; Johnstone, EC; Owens, DG; Roberts, GW. Schizophrenia as an anomaly of development of cerebral asymmetry. A postmortem study and a proposal concerning the genetic basis of the disease. *Arch. Gen. Psychiatry*, 1989, 46,1145-50.

Datta, S; Brunet, A; Greenberg, M. Cell survival: A play in three AKTs. *Gene Dev.*, 1999, 13, 2905-2927.

Davidson, L; McGlashan, TH. The varied outcomes of schizophrenia. *Can. J. Psychiatry*, 1997, 42, 34-43.

Dazzan, P; Morgan, KD; Orr, K; Hutchinson, G; Chitnis, X; Suckling, J; Fearon, P; McGuire, PK; Mallett, RM; Jones, PB; Leff, J; Murray, RM. Different effects of typical and atypical antipsychotics on grey matter in first episode psychosis: the ÆSOP study. *Neuropsychopharmacology*, 2005, 30, 765–774.

de Haan, L. Bakker, JM. Overview of neuropathological theories of schizophrenia: from degeneration to progressive developmental disorder. *Psychopathology*, 2004, 37, 1-7.

DeFelipe, J; Farinas, I. The pyramidal neuron of the cerebral cortex: morphological and chemical characteristics of the synaptic inputs. *Prog. Neurobiol.*, 1992, 39, 563-607.

Deutch, AY. The effects of dopamine denervation on prefrontal cortical structure and reversal by atypical antipsychotic drugs. *Biol. Psychiatry*, 2005, 57, 82S.

Deutch, AY; Jing, D. Prefrontal cortical dopamine denervation decreases dendritic spine density in pyramidal neurons. *Schizophr. Res.*, 2003, 60, 56.

Deutsch, S; Rosse, R; Schwartz, B; Mastropaolo, J. A revised excitotoxic hypothesis of schizophrenia: therapeutic implications. *Clin. Neuropharmacol.*, 2001, 24, 43-49.

Dorph-Petersen, K-A; Pierri, JN; Sun, Z; Sampson, AR; Lewis, DA. Stereological analysis of the mediodorsal thalamic nucleus in schizophrenia: volume; neuron number; and cell types. *J. Comp. Neurol.*, 2004, 472, 449-462.

Dorph-Petersen, KA; Pierri, JN; Perel, JM; Sun, Z; Sampson, AR; Lewis, DA. The influence of chronic exposure to antipsychotic medications on brain size before and after tissue fixation: a comparison of haloperidol and olanzapine in macaque monkeys. *Neuropsychopharmacology*, 2005, 30, 1649-1661.

Drake, RJ; Lewis, W. Insight and neurocognitive deficits in schizophrenia. *Schizophr. Res.*, 2003, 62, 165–173.

Drake, RJ; Dunn, G; Tarrier, N; Bentall, RP; Haddock, G; Lewis, SW. Insight as a predictor of the outcome of first-episode nonaffective psychosis in a prospective cohort study in England. *J. Clin. Psychiatry*, 2007, 68, 81-6.

Eastwood, SL; Cairns, NJ; Harrison, PJ. Synaptophysin gene expression in schizophrenia. Investigation of synaptic pathology in the cerebral cortex. *Br. J. Psychiatry*, 2000,176, 236-42.

Ellison-Wright, I; Glahn, DC; Laird, AR; Thelen, SM; Bullmore, E. The anatomy of first-episode and chronic schizophrenia: an anatomical likelihood estimation meta-analysis. *Am. J. Psychiatry*, 2008, AiA, 1-9.

Erickson, SE; Lewis, DA. Cortical connections of the lateral mediodorsal thalamus in cynomolgus monkeys. *J. Comp. Neurol.*, 2004, 473, 107-127.

Farber, N; Jiang, X; Dikranian, K; et al. Muscimol prevents NMDA antagonist neurotoxicity by activating GABA-A receptors in several brain regions. *Brain Res.*, 2003, 993, 90-100.

Farrow, TFD; Whitford, TJ; Williams, LM; Gomes, L; Harris, AWF. Diagnosis-related regional gray matter loss over two years in first episode schizophrenia and bipolar disorder. *Biol. Psychiatry*, 2005, 58(9), 713-23.

Feinberg, I. Schizophrenia: caused by a fault in programmed synaptic elimination during adolescence? *J. Psychiatr. Res.*, 1982, 17, 319-334.

Ferrer, I; Fabregues, I; Rairiz, J; et al. Decreased number of dendritic spines on cortical pyramidal spines on cortical pyramidal neurons in human chronic alcoholism. *Neurosci. Lett.*, 1986, 69, 115-9.

Fiscus, RR. Involvement of cyclic GMP and protein kinase G in the regulation of apoptosis and survival in neurol cells. *Neurosignals*, 2002, 11, 175-190.

Fleischhaker, C; Schulz, E; Tepper, K; Martin, M; Hennighausen, K; Remschmidt, H. Long-term course of adolescent schizophrenia. *Schizophrenia Bulletin*, 2005, 31, 769–780.

Flores, G; Alquicer, G; Silva-Goméz, AB; Zaldivar, G; Stewart, J; Quirion, R; Srivastava, LK. Alterations in dendritic morphology of prefrontal cortical and nucleus accumbens neurons in post-pubertal rats after neonatal excitotoxic lesions of the ventral hippocampus. *Neuroscience,* 2005, 133, 463-70.

Fog, R; Pakkenberg, H; Juul, P; Bock, E; Jorgensen, OS; Andersen, J. High-dose treatment of rats with perphenazine enanthate. *Psychopharmacology* (Berl), 1976, 50, 305-7.

Ford, JM; Mathalon, DH; Whitfield, S; Faustman, WO; Roth, WT. Reduced communication between frontal and temporal lobes during talking in schizophrenia. *Biol. Psychiatry,* 2002, 51, 485-92.

Franklin, TR; Acton, PD; Maldjian, JA; Gray, JD; Croft, JR; Dackis, CA; O'Brien, CP; Childress, AR. Decreased gray matter concentration in the insular; orbitofrontal; cingulate; and temporal cortices of cocaine patients. *Biol. Psychiatry*, 2002, 51, 134–142.

Frey, BN; da Fonseca, MMR; Machado-Vieira, R; Soarese, JC; Kapczinski, F. Neuropathological and neurochemical abnormalities in bipolar disorder. *Rev. Bras. Pisquistr*; 2004, 26, 180-8.

Friston, KJ; Frith, CD. Schizophrenia: A disconnection syndrome? *Clin. Neurosci*., 1995, 3, 89-97.

García-Martí, G; Aguilar, EJ; Lull, JJ; Martí-Bonmatí, L; Escartí, MJ; Manjón, JV; Moratal, D; Robles, M; Sanjuán, J. Schizophrenia with auditory hallucinations: a voxel-based morphometry study. *Prog. Neuropsychopharmacol. Biol. Psychiatry*, 2008, 32, 72-80.

Garden, GA; Budd, SL; Tsai, E; Hanson, L; Kaul, M; D´Emilia, DM; Friedlander, RM; Yuan, J; Masliah, E; Lipton, SA. Caspase cascades in human immunodeficiency virus-associated neurodegeneration. *J. Neurosci.,* 2002, 22, 4015-4024.

Garey, LJ; Ong, WY, Patel, PS; Kanani, M; Davis, A; Mortimer, AM; Barnes, TRE; Hirsch, SR. Reduced dendritic spine density on cerebral cortical pyramidal neurons in schizophrenia. *J. Neurol. Neurosurg. Psychiatry*, 1998, 65, 446-453.

Garver, DL; Holcomb, JA; Christensen, JD. Cerebral cortical gray expansion associated with two second-generation antipsychotics. *Biol. Psychiatry*, 2005, 58, 62-6.

Gilman, CP; Mattson, MG. Do apoptotic mechanisms regulate synaptic plasticity and growth-cone motility? *Neuromolecular. Med*., 2002, 2, 197-214.

Girgis, RR; Diwadkar, VA; Nutche, JJ; Sweeney, JA; Keshavan, MS; Hardan, AY. Risperidone in first-episode psychosis: a longitudinal; exploratory voxel-based morphometric study. *Schizophr. Res.,* 2006, 82, 89-94.

Giuliani, NR; Calhoun, VD; Pearlson, GD; Francis, A; Buchanan, RW. Voxel-based morphometry versus region of interest: a comparison of two methods for analyzing gray matter differences in schizophrenia. *Schizophr. Res*., 2005, 74, 135-47.

Glantz, LA; Lewis, DA. Reduction of synaptophysin immunoreactivity in the prefrontal cortex of subjects with schizophrenia. Regional and diagnostic specificity. *Arch. Gen. Psychiatry,* 1997, 54, 943-52.

Glantz, LA; Lewis, DA. Decreased dendritic spine density on prefrontal cortical pyramidal neurons in schizophrenia. *Arch. Gen. Psychiatry*, 2000, 57, 65-73.

Glantz, LA; Gilmore, JH; Lieberman, JA; Jarskog, LF. Apoptotic mechanisms and the synaptic pathology of schizophrenia. *Schizophrenia Res.,* 2006, 81: 4-63.

Goldman-Rakic, PS; Leranth, C; Williams, SM; Mons, N; Geffard, M. Dopamine synaptic complex with pyramidal neurons in primate cerebral cortex. *Proc. Natl. Acad. Sci. USA*, 1989, 86, 9015:9019.

Guidotti, A; Auta, J; Davis, JM; Di-Giorgi-Gerevini, V; Dwivedi, Y; Grayson, DR; Impagnatiello, F; Pandey, G; Pesold, C; Sharma, R; Uzunov, D; Costa, E. Decrease in reelin and glutamic acid decarboxylase67 (GAD67) expression in schizophrenia and bipolar disorder: a postmortem brain study. *Arch. Gen. Psychiatry*, 2000, 57, 1061-9.

Guo, ZH; Mattson, MP. Neurotrophic factors protect cortical synaptic terminals against amyloid and oxidative stress-induced impairment of glucose transport; glutamate transport and mitochondrial function. *Cereb Cortex,* 2000, 10, 50-57.

Gurling, HM; Critchley, H; Datta, SR; McQuillin, A; Blaveri, E; Thirumalai, S; Pimm, J; Krasucki, R; Kalsi, G; Quested, D; Lawrence, J; Bass, N; Choudhury, K; Puri, V; O'Daly, O; Curtis, D; Blackwood, D; Muir, W; Malhotra, AK; Buchanan, RW; Good, CD; Frackowiak, RS; Dolan, RJ. Genetic association and brain morphology studies and the chromosome 8p22 pericentriolar material 1 (PCM1) gene in susceptibility to schizophrenia. *Arch. Gen. Psychiatry*, 2006, 63, 844-54.

Ha, TH; Youn, T; Ha, KS; Rho, KS; Lee, JM; Kim, IY; Kim, SI; Kwon, JS. Gray matter abnormalities in paranoid schizophrenia and their clinical correlations. *Psychiatry Res*, 2004, 132, 251-60.

Halim, ND; Weickert, CS; McClintock, BW; et al. Effects of chronic haloperidol and clozapine treatment on neurogenesis in the adult rat hippocampus. *Neuropsychopharmacology*, 2004, 29, 1063-1069.

Harrison, PJ. Neuropathology of schizophrenia. A critical review of the data and their interpretation. *Brain,* 1999a, 122, 593-624.

Harrison, PJ. The neuropathological effects of antipsychotic drugs. *Schizophr. Res.*; 1999b, 40, 87-99.

Harrison, PJ. The neuropathology of primary mood disorder. *Brain*, 2002, 125, 1428-1449.

Harrison, PJ. The hippocampus in schizophrenia: a review of the neuropathological evidence and its pathophysiological implications. *Psychopharmacology,* 2004, 174, 151-162.

Hashimoto, T; Volk, DW; Eggan, SM; Mirnics, K; Pierri, JN; Sun, Z; Sampson, AR; Lewis, DA. Gene expression deficits in a subclass of GABA neurons in the prefrontal cortex of subjects with schizophrenia. *J. Neurosci.,* 2003, 23, 6315-26.

Hashimoto, T; Bergen, SE; Nguyen, QL; Xu, B; Monteggia, LM; Pierri, JN; Sun, Z; Sampson, AR; Lewis, DA. Relationship of brain-derived neurotrophic factor and its receptor TrkB to altered inhibitory prefrontal circuitry in schizophrenia. J. *Neurosci.*, 2005, 25, 372-83.

Hazlett, EA; Buchsbaum, MS; Haznedar, MM; Newmark, R; Goldstein, KE; Zelmanova, Y; Glanton, CF; Torosjan, Y; New, AS; Lo, JN; Mitropoulou, V; Siever, LJ. Cortical gray and white matter volume in unmedicated schizotypal and schizophrenia patients. Schizophr Res, 2008, 101, 111-23.

Ho, B; Milev, P; O´Leary, DS; Librant, A; Andreasen, NC; Wassink, TH. Cognitive and magnetic resonance imaging brain morphometric correlates of brain-derived neurotrophic factor Val66Met gene polymorphism in patients with schizophrenia and healthy volunteers. *Arch. Gen. Psychiatry*, 2006, 63, 731-740.

Hof, PR; Haroutunian, V; Friedrich Jr, VL; Byne, W; Buitron, C; Perl, DP. Loss and altered spatial distribution of oligodendrocytes in the superior frontal gyrus in schizophrenia. *Biol. Psychiatry*, 2003, 53, 1075-1085.

Hoffman, RE; Dobscha, SK. Cortical pruning and the development of schizophrenia: A computer model. *Schizophr. Bull.*, 1989, 15, 477-490.

Honea, R; Crow, TJ; Passingham, D; Mackay, CE. Regional deficits in brain volume in schizophrenia: a meta-analysis of voxel-based morphometry studies. *Am. J. Psychiatry,* 2005, 162, 2233-2245.

Honea, RA; Meyer-Lindenberg, A; Hobbs, KB; Pezawas, L; Mattay, VS; Egan, MF; Verchinski, B; Passingham, RE; Weinberger, DR; Callicott, JH. Is gray matter volume an intermediate phenotype for schizophrenia? A voxel-based morphometry study of patients with schizophrenia and their healthy siblings. *Biol. Psychiatry*, 2008, 63, 465-74.

Honer, WG; Falkai, P; Young, C; Wang, T; Xie, J; Bonner, J; Hu, L; Boulianne, GL; Luo, Z; Trimble, WS. Cingulate cortex synaptic terminal proteins and neural cell adhesion molecule in schizophrenia. *Neuroscience*, 1997, 78, 99-110.

Hughes, JR; Hatsukami, DK; Mitchell, JE; Dahlgren, IA. Prevalence of smoking among psychiatric outpatients. *Am. J. of Psychiatry*, 1986, 143, 993-997.

Hulshoff Pol, HE; Schnack, HG; Mandl, RC; Brans, RG; van Haren, NE; Baaré, WF; van Oel, CJ; Collins, DL; Evans, AC; Kahn, RS. Gray and white matter density changes in monozygotic and same-sex dizygotic twins discordant for schizophrenia using voxel-based morphometry. *Neuroimage*, 2006, 31, 482-8.

Iidakaa, T; Matsumotob, A; Ozakia, N; Suzukic, T; Iwatac, N; Yamamotod, Y; Okadad, T; Sadato, T. Volume of left amygdala subregion predicted temperamental trait of harm avoidance in female young subjects. A voxel-based morphometry study. *Brain Res*, 2006, 1125, 85–93.

Iritani, S. Neuropathology of schizophrenia: A mini review. *Neuropathology*, 2007, 27, 604-608.

Jang, DP; Namkoong, K; Kimb, JJ; Park, S; Kime, IY; Kime, SI; Kima, YB; Choa, ZH; Lee, E. The relationship between brain morphometry and neuropsychological performance in alcohol dependence. *Neuroscience Letters*, 2007, 428, 21–26.

Jarskog, LF; Glantz, LA; Gilmore, JH; Lieberman, JA. Apoptotic mechanisms in the pathophysiology of schizophrenia. Prog Neuro-psychopharmacol *Biol. Psychiatry*, 2005, 29, 846-858.

Job, DE; Whalley, HC; McConnel, S; Glabus, M; Johnstone, EC; Lawrie, SM. Voxel-based morphometry of grey matter densities in subjects at high risk of schizophrenia. *Schizophrenia Res.*, 2003, 64, 1– 13.

Jobe, TH; Harrow, M: Long-term outcome of patients with schizophrenia: a review. *Can. J. Psychiatry*, 2005, 50, 892-900.

Johnstone, EC; Crow, TJ; Frith, CD; Husband, J; Kree,l L. Cerebral ventricular size and cognitive impairment in chronic schizophrenia. *Lancet*, 1976, 2, 924-6.

Jones, LB; Johnson, N; Byne, W. Alteration in MAP2 immunocytochemistry in area 9 and 32 of schizophrenic prefrontal cortex. *Psychiatry Res.,* 2002, 114, 137-48.

Kalus, P; Senitz, D; Beckmann, H. Altered distribution of parvalbumin-immunoreactive local circuit neurons in the anterior cingulate cortex of schizophrenic patients. *Psychiatry Res.*, 1997, 75, 49-59.

Kasai, K; Shenton, ME; Salisbury, DF; Hirayasu, Y; Onitsuka, T; Spencer, MH; Yurgelun-Todd, DA; Kikinis, R; Jolesz, FA; McCarley, RW. Progressive decrease of left Heschl gyrus and planum temporale gray matter volume in first-episode schizophrenia: a longitudinal magnetic resonance imaging study. *Arch. Gen. Psychiatry*, 2003, 60, 766-75.

Kasparek, T; Prikryl, R; Mikl, M; Schwarz, D; Ceskova, E; Krupa, P. Prefrontal but not temporal grey matter changes in males with first-episode schizophrenia. *Prog. Neuropsychopharmacol. Biol. Psychiatry*, 2007, 31,151–157.

Kasparek, T; Prikryl, R; Schwarz, D; Tronerová, S; Ceskova, E; Mikl, M; Vanicek, J. Movement sequencing abilities and basal ganglia morphology in first-episode schizophrenia. *World Journal of Biological Psychiatry*, 2008 January 30. [Epub ahead of print]

Kawasaki, Y; Suzuki, M; Nohara, S; Hagino, H; Takahashi, T; Matsui, M; Yamashita, I; Chitnis, XA; McGuire, PK; Seto, H; Kurachi, M. Structural brain differences in patients with schizophrenia and schizotypal disorder demonstrated by voxel-based morphometry. *Eur. Arch. Psychiatry Clin. Neurosci.*, 2004, 254, 406–414.

Keshavan, MS; Anderson, S; Pettegrew, JW. Is schizophrenia due to excessive synaptic pruning in the prefrontal cortex? The Feinberg hypothesis revisited. *J. Psychiatr. Res.*, 1994, 28, 239-65.

Keller, SS; Wilke, M; Wieshmann, UC; Sluming, VA; Roberts, N. Comparison of standard and optimized voxel-based morphometry for analysis of brain changes associated with temporal lobe epilepsy. *Neuroimage*, 2004, 23, 860–868.

Klintzova, AJ; Haselhorst, U; Uranova, NA; Schenk, H; Istomin, VV. The effects of haloperidol on synaptic plasticity in rat's medial prefrontal cortex. *J. Hirnforsch*, 1989, 30, 51-7.

Kodama, M; Jujioka, T; Duman, RS. Chronic olanzapine or fluoxetine administration increases all proliferation in hippocampus and prefrontal cortex of adult rat. *Biol. Psychiatry*, 2004, 56, 570-580.

Konopaske, GT; Dorph-Petersen, KA; Pierri, JN; Wu, Q; Sampson, AR; Lewis, DA. Effect of chronic exposure to antipsychotic medication on cell numbers in the parietal cortex of macaque monkeys. *Neuropsychopharmacology*, 2007, 32, 1216-1223.

Kolluri, N; Sun, Z; Sampson, A; Lewis, DA. Lamina-specific reductions in dendritic spine density in the prefrontal cortex of subjects with schizophrenia. *Am. J. Psychiatry*, 2005, 162, 1200-1202.

Konopaske, GT; Dorph-Petersen, KA; Pierri, JN; Wu, Q; Sampson, AR; Lewis, DA. Effect of chronic exposure to antipsychotic medication on cell numbers in the parietal cortex of macaque monkeys. *Neuropsychopharmacology*, 2007, 32, 1216-1223.

Konopaske, GT; Dorph-Petersen, KA; Sweet, RA; Pierri, JN; Wei, Z; Sampson, AR; Lewis, DA. Effect of chronic exposure on astrocyte and oligodendrocyte numbers in macaque monkeys. *Biol. Psychiatry,* 2008, 63, 759-762.

Koo, MS; Dickey, CC; Park, HJ; Kubicki, M; Ji, NY; Bouix, S; Pohl, KM; Levitt, JJ; Nakamura, M; Shenton, ME; McCarley, RW. Smaller neocortical gray matter and larger sulcal cerebrospinal fluid volumes in neuroleptic-naive women with schizotypal personality disorder. *Arch. Gen. Psychiatry*, 2006, 63, 1090-1100.

Koutsouleris, N; Gaser, C; Jäger, M; Bottlender, R; Frodl, T; Holzinger, S; Schmitt, GJ; Zetzsche, T; Burgermeister, B; Scheuerecker, J; Born, C; Reiser, M; Möller, HJ; Meisenzahl, EM. Structural correlates of psychopathological symptom dimensions in schizophrenia: a voxel-based morphometric study. *Neuroimage*, 2008, 39, 1600-12.

Kringelbach, ML; Rolls, ET. The functional neuroanatomy of the human orbitofrontal cortex: evidence from neuroimaging and neuropsychology. *Prog. Neurobiol.,* 2004, 72, 341-72.

Laird, AR; Fox, PM; Price, CJ; Glahn, DC; Uecker, AM; Lancaster, JL; Turkeltaub, PE; Kochunov, P; Fox, PT: ALE meta-analysis: controling the false discovery rate and performing statistical contrasts. *Hum. Brain Mapp*, 2005, 25, 155-164.

Lappin, JM; Morgan, K; Morgan, C; Hutchison, G; Chitnis, X; Suckling, J; Fearon, P; McGuire, PK; Jones, PB; Leff, J; Murray, RM; Dazzan, P. Gray matter abnormalities associated with duration of untreated psychosis. *Schizophr. Res.*, 2006, 83, 145-53.

Lappin, JM; Dazzan, P; Morgan, K; Morgan, C; Chitnis, X; Suckling, J; Fearon, P; Jones, PB; Leff, J; Murray, RM; McGuire, PK. Duration of prodromal phase and severity of volumetric abnormalities in first-episode psychosis. *Br. J. Psychiatry Suppl.*, 2007, 51, s123-7.

Laroi, F; Fannemel, M; Rønneberg, U; Flekkoy, K; Opjordsmoen, S; Dullerud, R; Haakosen, M. Unawareness of illness in chronic schizophrenia and its relationship to structural brain measures and neuropsychological tests. *Psychiat. Res.*, 2000, 100, 49–58.

Lee, DE. Reduced inhibitory capacity in prefrontal cortex of schizophrenics. *Arch. Gen. Psychiatry,* 1995, 52, 267-268.

Levitt, JB; Lewis, DA; Yoshioka, T; Lund, JS. Topography of pyramidal neuron intrinsic connections in macaque monkey prefrontal cortex (areas 9 and 46). *J. Comp. Neurol.*, 1993, 338, 360-367.

Lewis, DA. Development of the prefrontal cortex during adolescence: Insights into vulnerable neural circuits in schizophrenia. *Neuropsychopharmacology*, 1997, 16, 385-398.

Lewis, DA; Cruz, DA; Melchitzky, DS; Pierri, JN. Lamina-specific deficits in parvalbumin-immunoreactive varicosities in the prefrontal cortex of subjects with schizophrenia: evidence for fewer projections from the thalamus. *Am. J. Psychiatry,* 2001, 158, 1411-22.

Lewis, DA; Hashimoto, T; Volk, DW. Cortical inhibitory neurons and schizophrenia. *Nature Reviews Neuroscience,* 2005, 6, 312-324.

Lidow, MS; Song, ZM; Castner, SA; Allen, PB; Greengard, P; Goldman-Rakic, PS. Antipsychotic treatment induces alterations in dendrite- and spine-associated proteins in dopamine-rich areas of the primate cerebral cortex. *Biol. Psychiatry*, 2001, 49, 1-12.

Lieberman, JA; Sheitman, BB; Kinon, BJ. Neurochemical sensitization in the pathophysiology of schizophrenia: Deficits and dysfuntion in neuronal regulation and plasticity. *Neuropsychopharmacol.*, 1997, 17, 205-229.

Lieberman, J; Chakos, M; Wu, H; Alvir, J; Hoffman, E; Robinson, D; Bilder, R. Longitudinal study of brain morphology in first episode schizophrenia. *Biol. Psychiatry*, 2001, 49, 487-99.

Lieberman, JA; Tollefson, GD; Charles, C; Zipursky, R; Sharma, T; Kahn, RS; Keefe, RS; Green, AI; Gure, RE; McEvoy, J; Perkins, D; Hamer, RM; Gu, H; Tohen, M. Antipsychotic drug effects on brain morphology in first-episode psychosis. *Arch. Gen. Psychiatry*, 2005, 62, 361-370.

Lipska, BK; Jaskiw, GE; Weinberger, DR. Postpubertal emergence of hyperresponsiveness to stress and to amphetamine after neonatal excitotoxic hippocampal damage: a potential animal model of schizophrenia. *Neuropsychopharmacology,* 1993, 9, 67-75.

Lipska, BK; Swerdlow, NR; Geyer, MA; Jaskiw, GE; Braff, DL; Weinberger, DR. Neonatal excitotoxic hippocampal demage in rats causes post-pubertal changes of prepulse-inhibition of startle and its disruption by apomorphine. *Psychopharmacology* (Berl.), 1995, 122, 35-43.

Lipska, BK; Weinberger, DR. Genetic variation in vulnerability to the behavioral effects of neonatal hippocampal demage in rats. *Proc. Natl. Acad. Sci. USA*, 1995, 92, 8906-8910.

Lipska, BK; Lerman, DN; Khaing, ZZ; Weickert, CS; Weinberger, DR. Gene expression in dopamine and GABA systems in an animal model of schizophrenia: effects of antipsychotic drugs. *Eur. J. Neurosci.*, 2003, 18, 391-402.

Lund, JS; Holbach, S. Postnatal development of thalamic recipient neurons in monkey striate cortex: I. A comparison of spine acquisition and dendritic growth of layer 4C alpha and beta spiny stellate neurons. *J. Comp. Neurol.*, 1991, 309, 115-128.

Lund, JS; Yoshioka, T; Lewitt, JB. Comparison of intrinsic connectivity in different areas of macaque monkey cerebral cortex. *Cereb Cortex*, 1993, 3, 148-162.

Luo, Y; Umegaki, H; Wang, X; et al. Dopammine induces apoptosis through an oxidation-involved SAPK-JNK activation pathway. *J. Biol. Chem.*, 1998, 273, 3756-3764.

Lyoo, IK; Kim, MJ; Stoll, AL; Demopulos, CM; Parow, AM; Dager ,SR; et al. Frontal lobe gray matter density decreases in bipolar I disorder. *Biol. Psychiatry,* 2004, 55, 648-651.

Lyoo, IK; Pollack, MH; Silveri, MM; Ahn, KH; Diaz, CI; Hwang, J; Kim, SJ; Yurgelun-Todd, DA; Kaufman, MJ; Renshaw, PF. Prefrontal and temporal gray matter density decreases in opiate dependence. *Psychopharmacology*, 2006, 184, 139–144.

Mahadik, S; Evans, D; Lal, H. Oxidative stress and role of antioxidant and omega-3 essential fatty acid supplementation in schizophrenia. *Prog. Neuropharmacol. Biol. Psychiatry,* 2001, 25, 463-493.

Malberg, JE; Eisch, AJ; Nestler, EJ; et al. Chronic antidepressant treatment increases neurogenesis in the adult rat hippocampus. *J. Neurosci.*, 2000, 20, 9104-9110.

Massana, G; Serra-Grabulosa, JM; Salgado-Pineda, P; Gastó, C; Junqué, C; Massana, J; Mercader, JM. Parahippocampal gray matter density in panic disorder: a voxel-based morphometric study. *Am. J. Psychiatry,* 2003, 160, 566–568.

Massana, G; Salgado-Pineda, P; Junqué, C; Pérez, M; Baeza, I; Pons, A; Massana, J; Navarro, V; Blanch, J; Morer, A; Mercader, JM; Bernardo, M. Volume changes in gray matter in first-episode neuroleptic-naive schizophrenic patients treated with risperidone. *J. Clin. Psychopharmacol.*, 2005, 25, 111-7.

Maynard, TM; Sikich, L; Lieberman, JA; LaMantia, AS. Neural development; cell-cell signaling; and the „Two-hit“ hypothesis of schizophrenia. *Schizophrenia Bull.*, 2001, 27, 457-476.

McClure, RK; Carew, K; Greeter, S; Maushauer, E; Steen, G; Weinberger, DR. Absence of regional brain volume change in schizophrenia associated with short-term atypical antipsychotic treatment. *Schizophrenia Res.,* 2008, 98, 29–39.

McCormick, L; Decker, L; Nopoulos, P; Ho, BC; Andreasen, N. Effects of atypical and typical neuroleptics on anterior cingulate volume in schizophrenia. *Schizophr. Res.,* 2005, 80, 73-84.

McDonald, C; Zanelli, J; Rabe-Heketh, S; Ellison-Wright, I; Kalidindi, S; Murray, RM; Kennedy, N. Meta-analysis of magnetic resonance imaging brain morphometry studies in bipolar disorder. Biol Psychiatry, 2004, 56, 411-7.

McIntosh, AM; Job, DE; Moorhead, WJ; Harrison, LK; Whalley, HC; Johnstone, EC; Lawrie, SM. Genetic liability to schizophrenia or bipolar disorder and its relationship to brain structure. *Am. J. Med. Genet. B Neuropsychiatr. Genet.,* 2006, 141B, 76-83.

Meda, SA; Giuliani, NR; Calhoun, VD; Jagannathan, K; Schretlen, DJ; Pulver, A; Cascella, N; Keshavan, M; Kates, W; Buchanan, R; Sharma, T; Pearlson, GD. A large scale

(N=400) investigation of gray matter differences in schizophrenia using optimized voxel-based morphometry. *Schizophr Res.*, 2008, 101, 95-105.

Meisenzahl, EM; Koutsouleris, N; Gaser, C; Bottlender, R; Schmitt, GJE; McGuire, P; Decker, P; Burgermeister, B; Born, C; Reiser, M; Möller, HJ. Structural brain alterations in subjects at high-risk of psychosis: A voxel-based morphometric study. *Schizophr. Res.*, 2008, doi:10.1016/j.schres.2008.02.023.

Melchitzky, DS; Sasack, SR; Pucak, ML; Lewis, DA. Synaptic targets of pyramidal neuron axon collaterals providing horizontal connections in monkey prefrontal cortex. Soc. *Neurosci. Abstr.*, 1995, 21: 409.

Melchitzky, DS; Sesack, SR; Lewis, DA. Parvalbumin-immunoreactive axon terminals in macaque monkey and human prefrontal cortex: laminar; regional; and target specificity of type I and type II synapses. *J. Comp. Neurol.*, 1999, 408, 11-22.

Melchitzky, DS; González-Burgos, G; Barrionuevo, G; Lewis, DA. Synaptic targets of the intrinsic axon collaterals of supragranular pyramidal neurons in monkey prefrontal cortex. *J. Comp. Neurol.,* 2001, 430, 209-21.

Melchitzky, DS; Lewis, DA. Pyramidal neuron local axon terminals in monkey prefrontal cortex: differential targeting of subclasses of GABA neurons. *Cereb. Cortex*; 2003, 13, 452-60.

Mintz, AR; Dobson, KS; Romney, DM. Insight in schizophrenia: a meta-analysis. *Schizophr Res.;* 2003, 61, 75–88.

Mirnics, K; Middleton, FA; Marquez, A; Lewis, DA; Levitt, P. Molecular characterization of schizophrenia viewed by microarray analysis of gene expression in prefrontal cortex. *Neuron.*, 2000, 28, 53-67.

Mitelman, SA; Shihabuddin, L; Brickman, AM; Hazlett, EA; Buchsbaum, MS. MRI assessment of gray and white matter distribution in Brodmann's areas of the cortex in patients with schizophrenia with good and poor outcomes. *Am. J. Psychiatry*, 2003, 160, 2154-68.

Morris, HM; Hashimoto, T; Lewis, DA. Alterations in somatostatin mRNA expression in the dorsolateral prefrontal cortex of subjects with schizophrenia or schizoaffective disorder. *Cereb Cortex*, 2008 Jan 17. [Epub ahead of print]

Moser, M-B; Trommald, M; Andersen, P. An increase in dendritic spine density on hippocampal CA1 pyramidal cells following spatial learning in adult rats suggests the formation of new synapses. *Proc. Natl. Acad. Sci. USA*, 1994, 91, 12673-12675.

Murray, RM; Lewis, SW. Is schizophrenia a neurodevelopmental disorder? *BMJ*, 1987, 295, 681-682.

Murray, RM. Neurodevelopmental schizophrenia: the rediscovery of dementia praecox. *Br. J. Psychiatry*, 1994, 25, 6-12.

Murray, RM; Sham, P; Van Os, J; Zanelli, J; Cannon, M; McDonald, C. A developmental model for similarities and dissimilarities between schizophrenia and bipolar disorder. *Schizophr. Res.*, 2004, 71, 405-16.

Narr, KL; Bilder, RM; Toga, AW; Woods, RP; Rex, DE; Szeszko, PR; Robinson, D; Sevy, S; Gunduz-Bruce, H; Wang, YP; DeLuca, H; Thompson, PM. Mapping cortical thickness and gray matter concentration in first episode schizophrenia. *Cerebral. Cortex*, 2005, 15, 708-719.

Nasrallah, HA; Pixley, S. The atypical antipsychotic paliperidone induces neurogenesis in the rat brain: a controlled study. *Neuropsychopharmacology,* 2006, 31, S126-S127.

Newell, KA; Deng, C; Huang, XF. Increased cannabinoid receptor density in the posterior cingulate cortex in schizophrenia. *Exp. Brain Res*, 2006, 172, 556–560.

Nuechterlein, KH; Barch, DM; Gold, JM; Goldberg, TE; Green, MF; Heaton, RK. Identification of separable cognitive factors in schizophrenia. *Schizophr. Res.*, 2004, 72, 29-39.

Ohnishi, T; Hashimoto, R; Mori, T; Nemoto, K; Moriguchi, Y; Lida, H; Noguchi, H; Nakabayashi, T; Hori, H; Ohmor, M; Tsukue, R; Anami, K; Hirabayashi, N; Harada, S; Arima, K; Satoh, O; Kunugi, H. The association between the Val158Met polymorphism of the catechol-O-methyl transferase gene and morphological abnormalities of the brain in chronic schizophrenia. *Brain,* 2006, 129, 399-410.

Neckelmann, G; Specht, K; Lund, A; Ersland, L; Smievoll, AI; Neckelmann, D; Hugdahl, K. MR morphometry analysis of grey matter volume reduction in schizophrenia: association with hallucinations. Int J Neurosci, 2006, 116, 9-23.

Olsson, H; Halldin, C; Farde, L. Differentiation of extrastriatal dopamine D2 receptor density and affinity in the human brain using PET. *Neuroimage*, 2004, 22, 794-803.

Pakkenberg, H; Fog, R; Nilakantan, B. The long-term effect of perphenazine enanthate on the rat brain. Some metabolic and anatomical observations. *Psychopharmacologia*, 1973, 29, 329-36.

Pakkenberg, B. Post-mortem study of chronic schizophrenic brains. *Br. J. Psychiatry*, 1987, 151, 744-52.

Pakkenberg P. Pronounced reduction of total neuron number in mediodorsal thalamic nucleus and nucleus accumbens in schizophrenics. *Arch. Gen. Psychiatry*, 1990, 47, 1023-1028.

Pakkenberg, B. Total nerve cell number in neocortex in chronic schizophrenics and controls estimated using optical dissectors. *Biol. Psychiatry,* 1993, 34, 768-72.

Pantelis, C; Velakoulis, D; McGorry, PD; Wood, SJ; Suckling, J; Phillips, LJ; Yung, AR; Bullmore, ET; Brewer, W; Soulsby, B; Desmond, P; McGuire, PK. Neuroanatomical abnormalities before and after onset of psychosis: a corss-sectional and longitudinal MRI comparison. *Lancet,* 2003, 361, 281-288.

Parikh, V; Ekams, DR; Khan, MM; et al. NGF in never-medicated first-episode psychotic and medicated chronic schizophrenic patients: possible implications for treatment outcome. *Schizophr Res.* 2003, 60, 117-123.

Parikh, V; Khan, MN; Mahadik, SP. Olanzapine counteracts the reduction of BDNF and TrkB receptors in rat hippocampus produced by haloperidol. *Neurosci. Lett.*, 2004, 356, 135-139.

Perlman, WR; Weickert, CS; Akil, M; Kleinman, JE. Postmortem investigations of the pathophysiology of schizophrenia: the role of susceptibility genes. *J. Psychiatry Neurosci.,* 2004, 29(4), 287-93.

Pierri, JN; Volk, CLE; Auh, S; Sampson, A; Lewis, DA. Decreased somal size of deep layer 3 pyramidal neurons in the prefrontal cortex in subjects with schizophrenia. *Arch. Gen. Psychiatry*, 2001, 58, 466-473.

Pillai, A; Terry, A; Mahadik, SP. Differential effects of long-term treatment with typical and atypical antipsychotics on NGF and BDNF levels in rat striatum and hippocampus. *Schizophr. Res.*, 2006, 82, 95-106.

Popken, GJ; Bunney, WE Jr, Potkin, SG; Jones, EG. Subnucleus-specific loss of neurons in medial thalamus of schizophrenics. *Proc. Natl. Acad. Sci. USA*, 2000, 97, 9276-80.

Powers, RE. The neuropathology of schizophrenia. *Journal of Neuropathology and experimental neurology,* 1999, 58(7), 679-690.

Prikryl, R; Kasparek, T; Skotakova, S; Ustohal, L; Kucerova, H; Ceskova E. Treatment of negative symptoms of schizophrenia using repetitive transcranial magnetic stimulation in a double-blind; randomized controlled study. *Schizophr Res.,* 2007, 95, 151-7.

Quing, H; Xu, H; Wei, Z; Gibson, K; Li, XM. The ability of atypical antipsychotic drugs vs. haloperidol to protect PC12 cells against MPP-induced apoptosis. *Eur. J. Neurosci.*, 2003, 17, 1563-1570.

Radewicz, K; Garey, LJ; Gentleman, SM; Reynolds, R. Increase in HLA-DR immunoreactive microglia in frontal and temporal cortex of chronic schizophrenics. *J. Neuropathol. Exp. Neurol.,* 2000, 59, 137-50.

Radley, JJ; Rocher, AB; Miller, M; Janssen, WG; Liston, C; Hof, PR; McEwen, BS; Morrison, JH. Repeated stress induces dendritic spine loss in the rat medial prefrontal cortex. *Cereb Cortex*, 2006, 16, 313-20.

Rajkowska, G; Selemon, LD; Rakic, PS. Neuronal and glial somal size in prefrontal cortex: a postmortem morphometric study of schizophrenia and Huntington disease. *Arch. Gen. Psychiatry*, 1998, 55, 215-24.

Reig, S; Moreno, C; Moreno, D; Burdalo, M; Janssen, J; Parellada, M; Zabala, A; Desco, M; Arango, C. Progression of brain volume changes in adolescent-onset psychosis. *Schizophr Bull.,* 2008 Jan 24. [Epub ahead of print]

Roberts, GW. Schizophrenia: a neuropathological perspective. *Br. J. Psychiatry*, 1991, 158, 8-17.

Rusch, N; van Elst, LT; Ludaescher, P; Wilke, M; Huppertz, JH; Thiel, T; Schmahl, C; Bohus, M; Lieb, K; Heßlinger, B; Hennig, J; Eberta, D. A voxel-based morphometric MRI study in female patients with borderline personality disorder. *NeuroImage,* 2003, 20, 385–392.

Rusch, N; Spoletini, I; Wilke, M; Bria, P; Di Paola, M; Di Iulio, F; Martinotti, G; Caltagirone, C; Spalletta, G. Prefrontal-thalamic-cerebellar gray matter networks and executive functioning in schizophrenia. *Schizophr. Res.*, 2007, 93, 79-89.

Salgado-Pineda, P; Baeza, I; Perez-Gomez, M; Vendrell, P; Junque, C; Bargallo, N; Bernardo, M. Sustained attention impairment correlates to gray matter decreases in first episode neuroleptic-naive schizophrenic patients. *Neuroimage*, 2003, 19, 365-375.

Salisbury, DF; Kuroki, N; Kasai, K; Shenton, ME; McCarley, RW. Progressive and interrelated functional and structural evidence of post-onset brain reduction in schizophrenia. *Arch. Gen.Psychiatry*, 2007, 64, 521-9.

Sams-Dodd, F; Lipska, BK; Weinberger, DR. Neonatal lesions of the rats ventral hippocampus result in hyperlocomotion and deficits of social behavior in adulthood. *Psychopharmacology* (Berl.), 1997, 132, 303-310.

Sastry, PS; Rao, KS. Apoptosis and the nervous system. *J. Neurochem.*, 2000, 74, 1-20.

Saunders, RC; Kolachana, BS; Bachevalier, J; Weinberger, DR. Neonatal lesions of the medial temporal lobe disrupt prefrontal cortical regulation of striatal dopamine. *Nature,* 1998, 393, 169-71.

Saze, T; Hirao, K; Namiki, C; Fukuyama, H; Hayashi, T; Murai, T. Insular volume reduction in schizophrenia. *Eur. Arch. Psychiatry Clin. Neurosci.*, 2007, 257, 473-9.

Schmitt, A; Weber, S; Jatzko, A; et al. Hippocampal volume and cell proliferation after acute and chronic clozapine or haloperidol treatment. *J. Neural. Transm*, 2004, 111, 91-100.

Selemon, LD; Rajakowska, G; Goldman-Rakic, PS. Abnormally high neuronal density in the schizophrenic cortex. A morphometric analysis of prefrontal area 9 and occipital area 17. *Arch. Gen. Psychiatry,* 1995, 52, 805-18.

Selemon, LD; Rajkowska, G; Goldman-Rakic, PS. Elevated neuronal density in prefrontal area 46 in brains from schizophrenic patients: application of a three dimensional; stereologic counting method. *J. Comp. Neurol.,* 1998, 392, 402-12.

Selemon, LD; Goldman-Rakic, PS. The reduced neuropil hypothesis: a circuit based model of schizophrenia. *Biol. Psychiatry*, 1999a, 45, 17-25.

Selemon, LD; Lidow, MS; Goldman-Rakic, PS. Increased volume and glial density in primate prefrontal cortex associated with chronic antipsychotic drug exposure. *Biol. Psychiatry*, 1999b, 46, 161-172.

Selemon, LD; Lidow MS; Goldman-Rakic, PS. Increased volume and glial density in primate prefrontal cortex associated with chronic antipsychotic drug exposure. *Biol. Psychiatry,* 1999b, 46, 161-172.

Selemon, LD; Kleinman, JE; Herman, MM; Goldman-Rakic, PS. Smaller frontal gray matter volume in postmortem schizophrenic brains. *Am. J. Psychiatry*, 2002, 159, 1983-1991.

Selemon, LD; Mrzljak, J; Kleinman, JE; Herman, MM; Goldman-Rakic, PS. Regional specificity in the neuropathologic substrates of schizophrenia: a morphometric analysis of Broca's area 44 and area 9. *Arch. Gen. Psychiatry*, 2003, 60, 69-77.

Selemon, LD; Begovic, A; Goldman-Rakic, PS; Castner, SA. Amphetamine sensitization alters dendritic morphology in prefrontal cortical pyramidal neurons in the non-human primate. *Neuropsychopharmacology*, 2007, 32, 919-31.

Sesack, SR; Bressler, CN; Lewis, DA: Ultrastructural associations between dopamine terminals and local circuit neurons in the monkey prefrontal cortex: A study of calrctinin-immunoreactive cells. *Neurosci. Lett*, 1995a, 200, 9-12.

Sesack, SR; Snyder, CL; Lewis, DA: Axon terminals immunolabeled for dopamine or tyrosine hydroxylase synapse on GABA-immunoreactive dendrites in rat and monkey cortex. *J. Comp. Neurol.*; 1995b, 363, 264-280.

Shapleske, J; Rossell, SL; Chitnis, XA; Suckling, J; Simmons, A; Bullmore, ET; Woodruff, PW; David, AS. A computational morphometric MRI study of schizophrenia: effects of hallucinations. *Cereb Cortex*, 2002, 12, 1331-41.

Sim, ME; Lyoo, IK; Streeter, CC; Covell, J; Sarid-Segal, O; Ciraulo, DA; Kim, MJ; Kaufman, MJ; Yurgelun-Todd, DA; Renshaw, PF. Cerebellar gray matter volume correlates with duration of cocaine use in cocaine-dependent subjects. *Neuropsychopharmacology*, 2007, 32, 2229–2237.

Smiley, JF; Williams, SM; Szigeti, K; Goldman-Rakic, PS. Light and electron microscopic characterization of dopamine-immunoreactive axons in human cerebral cortex. *J. Comp. Neurol.*, 1992, 321, 325-335.

Smith, TE; Hull, JW; Israel, LM; Willson, DF. Insight; symptoms; and neurocognition in schizophrenia and schizoaffective disorder. *Schizophr. Bull.*, 2000, 26,193–200.

Spencer, KM; Nestor, PG; Niznikiewicz, MA; Salisbury, DF; Shenton, ME; McCarley, RW. Abnormal neural synchrony in schizophrenia. *J. Neurosci.*, 2003, 23, 7407-11.

Stark, AK; Uylings, HB; Sanz-Arigita, E; Pakkengerg, B. Glial cell loss in the anterior cingulate cortex; a subregion of the prefrontal cortex; in subjects with schizophrenia. *Am. J. Psychiatry*, 2004, 161, 882-887.

Sweet, RA; Pierri, JN; Auh, S; Sampson, AR; Lewis ,DA. Reduced pyramidal cell somal volume in auditory association cortex of subjects with schizophrenia. *Neuropsychopharmacology*, 2003, 28, 599-609.

Sweet, RA; Henteleff, RA; Zhang, W; Sampson, AR; Lewis, DA. Reduced dendritic spine density in auditory cortex of subjects with schizophrenia. *Neuropsychopharmacology,* 2008 May 7. [Epub ahead of print]

Szabadics, J; Lorincz, A; Tamás, G. Beta and gamma frequency synchronization by dendritic gabaergic synapses and gap junctions in a network of cortical interneurons. *J. Neurosci*, 2001, 21, 5824-31.

Szeszko, PR; Strous, RD; Goldman, RS; Ashtari, M; Knuth, KH; Lieberman, JA; Bilder, RM. Neuropsychological correlates of hippocampal volumes in patients experiencing a first episode of schizophrenia. *Am. J. Psychiatry*, 2002, 159, 217-26.

Tamás, G; Buhl, EH; Lörincz, A; Somogyi, P. Proximally targeted GABAergic synapses and gap junctions synchronize cortical interneurons. *Nat. Neurosci.*, 2000, 3, 366-71.

Thompson, PM; Vidal, C; Giedd, JN; Gochman, P; Blumenthal, J; Nicolson, R; Toga, AW; Rapoport, JL. Mapping adolescent brain change reveals dynamic wave of accelerated gray matter loss in very early-onset schizophrenia. *Proc. Natl. Acad. Sci. USA,* 2001, 98, 11650-5.

Thune, JJ; Uylings, HB; Pakkenberg, B. No deficit in total number of neurons in the prefrontal cortex in schizophrenics. *J. Psychiatry Res.*, 2001, 35, 15-21.

Ukai, W; Ozawa, H; Tateno, M; et al. Neurotoxic potentiation of haloperidol in comparison with risperidon: implications of AKT-mediated signal changes by haloperidol. *J. Neural. Transm*, 2004, 111, 667-681.

Velakoulis, D; Wood, SJ; Smith, DJ; Soulsby, B; Brewer, W; Leeton, L; Desmond, P; Suckling, J; Bullmore, ET; McGuire, PK; Pantelis, C. Increased duration of illness is associated with reduced volume in right medial temporal/anterior cingulate grey matter in patients with chronic schizophrenia. *Schizophr. Res.*, 2002, 57, 43-9.

Valente, AA; Miguel, EC; Castro, CC; Amaro, E; Duran, FLS; Buchpiguel, CA; Chitnis, X; McGuire, PK; Busatto, GF. Regional gray matter abnormalities in obsessive-compulsive disorder: a voxel-based morphometry study. *Biol. Psychiatry*, 2005, 58, 479–487.

Valverde, F. Apical dendritic spines of the visual cortex and light deprivation in the mouse. *Exp. Brain Res.,* 1967, 3, 337-352.

van Haren, NE; Hulshoff Pol, HE; Schnack, HG; Cahn, W; Mandl, RC; Collins, DL; Evans, AC; Kahn, RS. Focal gray matter changes in schizophrenia across the course of the illness: a 5-year follow-up study. *Neuropsychopharmacology*, 2007, 32, 2057-66.

Venkatasubramanian, G; Jayakumar, PN; Gangadhar, BN; Keshavan, MS. Automated MRI parcellation study of regional volume and thickness of prefrontal cortex (PFC) in antipsychotic-naïve schizophrenia. *Acta Psychiatr. Scand,* 2008, 117, 420-31.

Vincent, SL; McSparren, J; Wang, RY; Benes, FM. Evidence for ultrastructural changes in cortical axodendritic synapses following long-term treatment with haloperidol or clozapine. *Neuropsychopharmacology,* 1991, 5, 147-155.

Vincent, SL; Khan, Y; Benes, FM. Cellular distribution of dopamine D1 and D2 receptors in rat medial prefrontal cortex. *J. Neurosci.,* 1993, 13, 2552-2564.

Vincent, SL; Adamec, E; Sorensen, I; Benes,FM. The effects of chronic haloperidol administration on GABA-immunoreactive axon terminals in rat medial prefrontal cortex. *Synapse,* 1994, 17, 26-35.

Volk, DW; Austin, MC; Pierri, JN; Sampson, AR; Lewis, DA. Decreased glutamic acid decarboxylase67 messenger RNA expression in a subset of prefrontal cortical gamma-aminobutyric acid neurons in subjects with schizophrenia. *Arch. Gen. Psychiatry*, 2000, 57, 237-45.

Volk, D; Austin, M; Pierri, J; Sampson, A; Lewis, D. GABA transporter-1 mRNA in the prefrontal cortex in schizophrenia: decreased expression in a subset of neurons. *Am. J. Psychiatry*, 2001, 158, 256-65.

Volk, DW; Pierri, JN; Fritschy, JM; Auh, S; Sampson, AR; Lewis, DA. Reciprocal alterations in pre- and postsynaptic inhibitory markers at chandelier cell inputs to pyramidal neurons in schizophrenia. *Cereb Cortex,* 2002, 12, 1063-70.

Volk, DW; Lewis, DA. Effects of a mediodorsal thalamus lesion on prefrontal inhibitory circuitry: implications for schizophrenia. *Biol. Psychiatry*; 2003; 53: 385-9.

Wakade, CG; Mahadik, S; Waller, JL; Chiu, FC. Atypical neuroleptics stimulate neurogenesis in adult rat brain. *J. Neurosci. Res.*, 2002, 69, 72-79.

Wang, HD; Dunnavant, FD; Jarman, R; et al. Effects of antipsychotic drugs on neurogenesis in the forebrain of adult rat. *Neuropsychopharmacology*, 2004, 29, 1230-1238.

Wei, Z; Mousseau, DD; Richardson, JS; Dyck, LE; Li, XM. Atypical antipsychotics attenuate neurotoxicity of beta-amyloid 925-35) by modulating Bax and Bcl-X(l/s) expression and localization. *J. Neurosci. Res.*, 2003, 74, 942-947.

Wei, Z; Moussem, DD; Dai, Y; et al. Haloperidol induces apoptosis via the delta-2 receptor system and BCI-XS. *Psychogenomics J.*, 2006, 6, 279-288.

Weinberger, DR. Implications of normal brain development for the pathogenesis of schizophrenia. *Arch. Gen. Psychiatry*, 1987, 44, 660-669.

Weinberger, DR; Berman, KF; Suddath, R; Torrey, EF. Evidence of dysfunction of a prefrontal-limbic network in schizophrenia: a magnetic resonance imaging and regional cerebral blood flow study of discordant monozygotic twins. *Am. J. Psychiatry*, 1992, 149, 890-7.

Weinberger, DR. Schizophrenia. From neuropathology to neurodevelopment. *Lancet,* 1995, 346, 552-557.

Weinberger, DR; McClure, RK. Neurotoxicity; neuroplasticity; and magnetic resonance imaging morphometry: what is happening in the schizophrenic brain? *Arch. Gen. Psychiatry*, 2002, 59, 553-558.

Whitford, TJ; Farrow, TF; Gomes, L; Brennan, J; Harris, AW; Williams, LM. Grey matter deficits and symptom profile in firsts episode schizophrenia. *Psychiatry Res.,* 2005, 139, 229-238.

Whitford, TJ; Grieve, SM; Farrow, TF; Gomes, L; Brennan, J; Harris, AW; Gordon, E; Williams, LM. Progressive grey matter atrophy over the first 2-3 years of illness in first-episode schizophrenia: a tensor-based morphometry study. *Neuroimage*, 2006, 32, 511-519.

Williams, GV; Goldman-Rakic, PS. Modulation of memory fields by dopamine D1 receptors in prefrontal cortex. *Nature,* 1995, 376, 572-575.

Wilke, M; Kaufmann, C; Grabner, A; Pütz, B; Wetter, TC; Auer, DP. Gray matter-changes and correlates of disease severity in schizophrenia: a statistical parametric mapping study. *Neuroimage*, 2001, 13, 814-24.

Woo, TU; Pucak, ML; Kye, CH; Matus, CV; Lewis, DA. Peripubertal refinement of the intrinsic and associational circuitry in monkey prefrontal cortex. *Neuroscience*, 1997, 80,1149-58.

Woo, TU; Whitehead, RE; Melchitzky, DS; Lewis, DA. A subclass of prefrontal gamma-aminobutyric acid axon terminals are selectively altered in schizophrenia. *Proc. Natl. Acad. Sci USA*, 1998, 95, 5341-6.

Wood, SJ; Pantelis, C; Velakoulis, D; Yücel, M; Fornito, A; McGorry, PD. Progressive changes in the development toward schizophrenia: studies in subjects at increased symptomatic risk. *Schizophr Bull*, 2008, 34, 322-9.

Woods, TB. Is schizophrenia a progressive neurodevelopmental disorder? Towards a unitary pathogenetic mechanism. *Am. J. Psychiatry,* 1998, 155, 1661-1670.

Wooley, CS; Gould, E; Frankfurt, M; McEwen, BS. Naturally occuring fluctuation in dendritic spine density on adult hippocampal pyramidal neurons. *J. Neurosci.*, 1990, 10, 4035-4039.

Wright, IC; Rabe-Hesketh, S; Woodruff, PWR; David, AS; Murray, RM; Bullmore, ET. Meta-analysis of regional brain volumes in schizophrenia. *Am. J. Psychiatry*, 2000, 157, 16-25.

Yamada, M; Hirao, K; Namiki, C; Hanakawa, T; Fukuyama, H; Hayashi, T; Murai, T. Social cognition and frontal lobe pathology in schizophrenia: a voxel-based morphometric study. *Neuroimage*, 2007, 35, 292-8.

Yamasue, H; Kasai, K; Iwanami, A; Ohtani, T; Yamada, H; Abe, O; Kuroki, N; Fukuda, R; Tochigi, M; Furukawa, S; Sadamatsu, M; Sasaki, T; Aoki, S; Ohtomo, K; Asukai, N; Kato, N. Voxel-based analysis of MRI reveals anterior cingulate gray-matter volume reduction in posttraumatic stress disorder due to terrorism. *PNAS*, 2003, 100, 9039–9043.

Yoo, HK; Kim, MJ; Kim, SJ; Sung, YH; Sim, ME; Lee, YS; Song, SY; Kee, BS; Lyoo, IK. Putaminal gray matter volume decrease in panic disorder: an optimized voxel-based morphometry study. *European Journal of Neuroscience*, 2005, 22, 2089–2094;

Yoo, SY; Roh, MS; Choi, JS; Kang, DH; Ha, TH; Lee, JM; Kim, IY; Kim, SI; Kwon, JS. Voxel-based morphometry study of gray matter abnormalities in obsessive-compulsive disorder. *J. Korean Med. Sci.,* 2008, 23, 24-30.

Yücel, M; Solowij, N; Respondek, C; Whittle, S; Fornito, A; Pantelis, C; Lubman, DI. Regional Brain Abnormalities Associated With Long-term Heavy Cannabis Use. *Arch. Gen. Psychiatry*, 2008, 65, 694-701.

In: Encyclopedia of Neuroscience Research
Editors: Eileen J. Sampson and Donald R. Glevins
ISBN 978-1-61324-861-4

Chapter II

Prefrontal Cholinergic Receptors: Their Role in the Pathology of Schizophrenia

M. Udawela, E. Scarr and B. Dean
Rebecca L Cooper Research Laboratories,
Mental Health Research Institute of Victoria,
Parkville, VIC, Australia

Abstract

There is a growing body of evidence to suggest that alterations in the cholinergic systems of the central nervous system are important in the pathology of schizophrenia. This data has arisen from studies into preclinical and clinical pharmacology, patient treatment, post-mortem neurochemistry and neuroimaging of patients with the disorder. The neuropharmacology of antipsychotics used to treat the disorder first led scientists to investigate changes in both the neurotransmitters and their receptors that are targeted by these therapeutic agents. These investigations have revealed that some atypical antipsychotics increase acetylcholine release in the prefrontal cortex, possibly contributing to improvements seen in some of the cognitive dysfunction for these drugs. Whilst there are data that show that some current atypical antipsychotic drugs have some impact on negative symptoms and cognitive deficits, none of these drugs are truly effective in treating these symptoms. Thus, in schizophrenia, various components within the cholinergic system are being investigated as potential targets for drugs that might effectively reverse these symptoms that are currently treatment-resistant. In this regard, post-mortem studies have revealed that patients with schizophrenia have low levels of cholinergic muscarinic receptors (CHRMs) in the CNS, and it is now well established that CHRM1 is decreased in the dorsolateral prefrontal cortex from subjects with the disorder; this receptor deficit is believed to play an important role in the cognitive dysfunctions associated with the disease. Recent information from basic research, in particular from studies using $CHRM^{-/-}$ mice, has enabled us to develop hypotheses as to what the functional consequences of deficits in CHRMs might be. Several lines of evidence also suggest a pathophysiological function for nicotinic acetylcholine receptors in schizophrenia. Nicotine administration has been shown to improve cognition in patients with schizophrenia. In the prefrontal cortex this receptor is thought to affect cognition through its activation-dependent effects on dopamine D_1 receptor mediated neurotransmission. Thus, both muscarinic and nicotinic receptors are viable drug targets with respect to cognitive deficits in schizophrenia. Using previous findings, in combination with our own data, we are now beginning to understand the specific changes in the various neurotransmitter receptors in the brains of schizophrenic patients, the functional effects of these changes and the roles the receptors may play in the pathology of the disorder.

Introduction

Schizophrenia is a debilitating mental illness, affecting approximately 1% of the population [1]. Importantly, CHRM antagonists have been extensively used to control the extrapyramidal side effects associated with the first generation of antipsychotic drugs [2] but it is now becoming clear that the cholinergic receptors may be a target for drugs to alleviate the symptoms of schizophrenia. The symptoms of schizophrenia can be broadly divided into positive symptoms, such as hallucinations and delusions, negative symptoms, such as social withdrawal and lack of motivation, and cognitive deficits, which can include impairments in attention/information processing, problem-solving, speed of processing, working memory and verbal and visual learning and memory [3, 4]. Whilst the positive symptoms have proven malleable to treatment with antipsychotic drugs, the negative symptoms and cognitive deficits are still essentially resistant to available treatments [5, 6]. This in part could be due to schizophrenia being a syndrome, rather than a distinct disease with a single pathology, with the different diseases within the syndrome showing differential responsiveness to available drugs. Thus, a major challenge is to be able to dissect the syndrome of schizophrenia into its components, to understand the pathologies of these biologically distinct entities and to design effective treatments for each disorder. Perhaps the most pressing need is for treatments for the cognitive deficits associated with schizophrenia as these are now recognised as the most debilitating symptoms associated with the disorder [7-9].

Cognitive Impairment

Despite the relationship between cognitive deficits and schizophrenia being identified almost a century ago [10], overall treatment of these symptoms remains inadequate [11]. Whilst some atypical antipsychotics offer slight cognitive benefit in patients with schizophrenia, even the patients who benefit from antipsychotic drug treatment continue to exhibit pronounced cognitive impairments [12-16]. There is evidence that dysfunction of the pre-frontal cortex (PFC) is responsible for cognitive impairments in schizophrenia [17, 18]. As stated earlier, cognitive deficits are a better predictor of the functional outcome in patients with schizophrenia compared to other symptoms of the disease [19-21], with these deficits now believed to be the most debilitating symptom for patients with schizophrenia to assimilate back into society. Thus there is a great need for new generation therapeutics, directed towards improving cognition.

Neuroimaging Studies, Cognition and Schizophrenia

Functional neuroimaging studies were first performed on patients with schizophrenia in 1974, using the ^{133}Xenon clearance method [22], and showed that, while whole brain cerebral blood flow was not different compared to healthy volunteers, there was a reduced gradient of frontal to posterior blood flow in patients with schizophrenia, termed 'hypofrontality'. This was replicated in other studies using both ^{133}Xenon [23, 24] and positron emission tomography (PET) [25, 26]. However, hypofrontality was not observed in all investigations [27-30], which may have been due to the lack of behavioural control over the subjects' mental activity. Subsequent studies overcame this by performing cognitive activation tasks during imaging, the most common being the Wisconsin Card Sorting Test, which measures concept formation, cognitive flexibility, feedback processing and working memory [23, 31-33]. These studies revealed a lack of the normal increase in blood flow in the dorsolateral pre-frontal cortex (DLPFC) during cognitive challenge in patients with schizophrenia compared to control subjects. This occurred in patients both on and off medication and did not appear to be due to global cortical dysfunction [34]. These findings are indicative of a potential role for the DLPFC in the cognitive deficits of schizophrenia.

To further characterise the changes in cerebral blood flow seen in schizophrenia to sub regions within the frontal cortex, and to link these to the distinct components of cognitive impairment, investigators later utilised various tasks that could allow differentiation of cognitive components, together with functional magnetic resonance imaging (fMRI). These tasks included working memory tasks such as N-back and Sternberg paradigm, as many of the cognitive deficits observed in schizophrenia have been attributed to working memory abnormalities [18, 34-36], as well as response inhibition tasks such as the Stroop and AX-type continuous performance task. In agreement with the ^{133}Xenon studies, the majority of the fMRI studies demonstrated hypoactivity of the DLPFC in subjects with schizophrenia during working memory tasks compared to matched controls [37-39]. Once more, the results from all the studies were not in agreement; some revealed no difference between control and schizophrenia groups [40], whilst others showed hyperactivity in this region in schizophrenia [41, 42]. More recent studies indicated that the degree of DLPFC activation is linked to performance during various executive and working memory tasks, with good performance associated with relatively greater DLPFC activity and marked behavioural deficits associated with relatively reduced DLPFC signal [39, 41, 43, 44], which could account for inconsistencies seen in other studies.

A recent quantitative meta-analysis by Glahn *et al.* [45] on 12 neuroimaging studies that used the N-back working memory paradigm, where subjects are required to monitor a series of quickly changing stimuli and indicate when the current stimulus is the same as the one presented *n* trials previously [46], confirmed hypoactivity of the DLPFC (Brodmann's Area, BA 9), rostral PFC (BA 11) and right ventrolateral/insular cortex (BA 13), but also revealed abnormally increased activation in the right dorsomedial PFC (BA 9), anterior cingulate (BA 32) and left frontal pole (BA 10) regions in patients with schizophrenia [45]. It was proposed that in these patients adjacent brain regions are being recruited in an effort to compensate for the impaired DLPFC activity during working memory tasks [45, 47].

The notion that changes in functionality of the DLPFC and other areas of the cortex are important in schizophrenia is supported by structural neuroimaging studies. Thus, abnormalities in the rostral medial PFC/orbito-fontal cortex have been observed in some studies [48]. These include a reduction in grey matter [49-51] and increased mean diffusivity [52], as well as reduced neuropil fraction and reduced grey-level index in the rostral PFC (BA 10) [53]. Several cognitive abilities associated with BA 10 are impaired in schizophrenia such as impairments in mentalising, episodic memory, prospective memory, source memory and multitasking, and these are suggested to be associated with particular symptoms [48]. Studies using fMRI have also shown reduced activity in the anterior cingulate gyrus in schizophrenia compared to healthy subjects during a wide range of tasks [54-60]. Although not all imaging studies have reported this finding, reduced activation in the anterior cingulate gyrus appears to be consistent during tasks which reliably activate this region in healthy subjects, such as Stroop; which measures reaction time to determine interference in attention, Go-No-Go; which measures a participants capacity for sustained attention and response control and Verbal Fluency, and during incorrect responding [45, 61].

The attenuated frontal lobe activation observed in patients with schizophrenia can be normalised with antipsychotic treatment. For example, it was shown using PET that haloperidol, a first generation antipsychotic, increased regional cerebral blood flow (rCBF) in part of the right DLPFC (BA 8/9) while reducing rCBF in several other cortical regions: in the ventrolateral frontal cortex (BA 47) on the left, the dorsolateral frontal cortex (BA 44 and 45) on the right, and the occipital cortex (BA 17/18), bilaterally. The second generation antipsychotic, clozapine, increased rCBF in several neocortical regions: the dorsolateral frontal cortex, bilaterally in BA 46 and on the left in BA 6 and 9. Clozapine decreased rCBF in ventrolateral frontal cortex (BA 47), bilaterally, and in parts of the superior frontal cortex (BA 10) and the sensory motor cortex (BA 4) [62]. Substitution of risperidone for typical antipsychotic drugs increased functional activation in the right PFC during working memory task performance [63]. However minimal performance improvement was observed in these cases, which could be due to the small sample sizes, low level of cognitive impairment at study entry or the simplicity of the task [62-64].

The Cholinergic System in Schizophrenia

The dopamine hypothesis of schizophrenia has been held for the last three decades [65, 66] and is based on the observation that stimulation of the dopaminergic system with drugs such as amphetamine often leads to transient psychotic symptoms, while blocking the dopamine D_2 receptor with antagonists leads to reduced positive symptoms in subjects with schizophrenia [66-69]. Dopamine receptor dysfunction has also been linked to cognitive dysfunction in patients with schizophrenia, with a decrease in D_1 receptor-like binding, as measured with PET, in the PFC of patients with schizophrenia, which correlated with cognitive dysfunction [70]. Significantly, chronic blockade of D_2 receptors leads to a decrease in D_1 receptors in the PFC region, along with impairments in working memory, in non-human primates [71]. This raised the possibility that effective treatment for psychoses may contribute to the impaired cognitive functioning seen in subjects with schizophrenia.

Despite a large body of pharmacological evidence supporting the dopamine hypothesis, it cannot account for the whole range of symptoms seen in schizophrenia. It is therefore significant that some of the earliest forms of treatment for schizophrenia involved modulation of the cholinergic system using agents such as atropine and hyoscine to treat positive symptoms [2, 10, 72]. However, with the advent of antipsychotic drugs, interest in these therapeutic agents waned and these treatments were never widely used except in the case of CHRM antagonists, which were used to control the side effects produced by antipsychotic drug treatment [2, 73]. Interest in targeting the CHRMs in the treatment of schizophrenia has seen a resurgence since it became clear that one of the actions of clozapine, the archetypal second generation atypical antipsychotic drug [74], is to antagonise that family of receptors [75-80]. The pharmacology of clozapine was of keen interest because it i) does not cause extrapyramidal side effects or tardive dyskinesia [81, 82], ii) was effective in otherwise treatment-refractory patients [74], iii) can cause some alleviation of negative symptoms [83, 84] and iv) has cognitive enhancing effects [85, 86]. More recently it has been shown that the major metabolite of clozapine, *N*-desmethylclozapine (NDMC), is a potent partial agonist at the muscarinic cholinergic M1 receptor (CHRM1) [87]. Moreover, it has been shown in subjects with schizophrenia that improvement in cognitive deficits after treatment with clozapine correlates with a higher plasma NDMC / clozapine ratio [87]. This suggests that the cognitive improvement associated with clozapine could have been due to NDMC stimulating CHRM1. Other atypical antipsychotic drugs, which have been shown to impact on cognitive deficits compared to typical drugs, significantly increased ACh release in the medial PFC, an effect not observed with typical antipsychotics [88]. These data suggest that all atypical antipsychotic drugs can indirectly target the cholinergic system even though they do not bind to any CHRMs. This could be why some of these drugs may prove partially effective in the treatment of cognitive impairments associated with the disease. However, data from NDMC, and previous studies showing that muscarinic antagonists can worsen neurocognitive impairment [89, 90], would suggest that direct stimulation of CHRMs will be the best approach to treating the cognitive deficits associated with schizophrenia.

Underpinning the notion that targeting cholinergic receptors will be a useful therapeutic approach to treating schizophrenia is a body of evidence indicating that the cholinergic system may be disrupted in schizophrenia [91]. This evidence mainly comes from post-mortem studies, which have revealed decreased nicotinic and muscarinic acetylcholine receptor availability or expression in the cortex and hippocampus of individuals with schizophrenia [92-101]. These are supported by a study showing decreased activity of choline acetyltransferase, which catalyses the synthesis of ACh, measured post-mortem, in correlation with a patient history of poorer cognitive functioning [102]. These findings suggest that the processing of acetylcholine may also be abnormal in subjects with schizophrenia. This concept fits with the first of more recent attempts to treat schizophrenia, which has been to use inhibitors of acetylcholine esterase, an enzyme that rapidly breaks down acetylcholine in the synapse, to indirectly increase levels of acetylcholine. This approach is supported by experiments in rodents where it has been shown that combining atypical antipsychotic drug treatment with cholinesterase inhibitors increased the concentration of ACh in the medial PFC by 2- to 3- fold [88]. This outcome could explain why treatment with galantamine, an acetylcholinesterase inhibitor, produces cognitive enhancement in subjects with schizophrenia [103, 104]. Unfortunately outcomes from subsequent studies using various acetylcholine esterase inhibitors have been inconsistent. Whilst targeting cholinesterase gives a non-focussed approach to improving cognitive deficits in schizophrenia, efforts are also being made to directly target specific muscarinic receptors to improve cognitive impairments [11, 105].

Muscarinic Receptors in Schizophrenia

CHRMs are G-protein coupled receptors found on cholinergic and non-cholinergic cells, where they function as both auto- and hetero- receptors [106-108]. There have been five subtypes identified (CHRM1-5) which are encoded by distinct genes, show different pharmacological properties and are expressed in regionally varying concentrations throughout the CNS [109-113]. As these receptors display a broad range of functions, it has been difficult to determine which CHRM subtypes are involved in the pathology of schizophrenia and / or may be useful therapeutic targets.

Initial studies of CHRM expression in post-mortem CNS looked at [^{3}H]-quinuclidinyl benzilate ([^{3}H]QNB) binding in the frontal cortex from subjects with schizophrenia and showed no significant differences [114]. However [^{3}H]QNB binds to all 5 CHRM subtypes [115], thus there was no change in global CHRM expression in this cortical region. A more comprehensive study measuring [^{3}H]QNB binding in the orbito-frontal and medial frontal cortices from 6 schizophrenia subjects receiving antipsychotic drugs up until death (on-drug), 6 schizophrenia subjects who had not received antipsychotic drugs for at least 40 days prior to death (off-drug) and 10 control subjects showed an increase in receptor number in the orbito-frontal cortex in all schizophrenia subjects compared to controls [116]. The study also revealed that the on-drug group showed a significant increase in receptor number and a significant decrease in affinity in both areas, while the off-drug group showed no significant differences in any binding parameters compared to controls [116]. These results were confirmed in a follow up study, performed on the same subjects, showing increased [^{3}H]QNB binding in the PFC from on drug schizophrenia subjects compared to off drug and control subjects [117]. Therefore it was suggested that differences in CHRM density were a result of long-term antipsychotic use.

Later studies gave a more specific picture of changes in CHRM density in the CNS of people with schizophrenia by utilising pirenzepine, which binds to CHRM1 and to a lesser extent to CHRM4 [118]. Using [^{3}H]pirenzepine, we and others showed decreased binding in the caudate-putamen [119, 120], hippocampus [92] and PFC in schizophrenia [93]. Importantly, we showed that levels of CHRM1 protein and mRNA, but not CHRM4, are decreased in BA 9, indicating the CHRM1 was responsible for the decreased [^{3}H]pirenzepine observed in this region [94]. This picture is made a little complex by our findings that, whilst there was no decrease in CHRM1 or CHRM4 protein or [^{3}H]pirenzepine binding in BA 40, mRNA was decreased for both receptor subtypes, indicating that the pattern of receptor expression observed in BA 9 does not extend to all regions of the PFC. Furthermore, there were no changes in levels of CHRM2 or 3 protein with diagnosis in either BA 9 or BA 40 in the same cohort [121], further indicating that CHRM1 is the receptor predominantly changed in BA 9 in schizophrenia. A change in CHRM1 mRNA was also shown by Mancama *et al.* [98] in BA 6 from subjects with the disorder. Taken together these studies indicate that CHRM1 has important implications in schizophrenia pathology, and thus specifically targeting CHRM1 may prove beneficial in treating cognition with the additional benefit of limited side effects. This hypothesis gains support from data showing that a polymorphism in the *CHRM1* gene in subjects with schizophrenia is associated with higher levels of cognitive impairment [122]. Furthermore, studies in *CHRM1* knockout mice have shown that deletion of this receptor leads to disturbances in learning, memory [123] and circadian rhythms [124], functions that are affected in subjects with schizophrenia [125, 126]. Perhaps the most convincing data supporting this hypothesis comes from the study showing that xanomeline (3-[3-hexyloxy-1,2,5-thiadiazo-4-yl]-1,2,5,6-tetrahydro-1-methylpyridine), a CHRM1/4 selective agonist, has antipsychotic effects and improves cognitive deficits in subjects with schizophrenia [127, 128]. On the other hand, recent data have shown that CHRM4 is more likely the cause of decreased [^{3}H]pirenzepine binding in the hippocampus in schizophrenia, as CHRM4 but not CHRM1 mRNA was decreased in this region [100], where it may be involved in schizophrenia pathology linked to the hippocampus, such as psychosis. Together, these data suggest that decreases in CHRMs in schizophrenia are region and subtype specific, and that different CHRMs may play different roles in the pathology of schizophrenia.

One concern regarding these studies is that changes in CHRM expression may be a result of treatment with antipsychotics, as many of the patients used in these studies had a record of receiving antipsychotic drugs. However, we found no significant correlation between experimental variables and final recorded antipsychotic drug dose [94, 129]. Moreover, the effect of antipsychotics on receptor expression has been examined in rats treated with haloperidol and chlorpromazine; these studies indicated these drugs do not significantly influence either CHRM expression or density [93, 119, 130, 131]. Thus, alterations in CHRMs are likely an entity of the disease rather than an effect of drug use. To date there has been one study performed on medication free subjects that strongly supports the hypothesis that changes in the CHRMs are involved in the pathology of schizophrenia. This study used [^{123}I]iodoquinuclidynyl benzilate, which binds with high affinity to all five CHRM subtypes, and showed decreased binding *in vivo* in cortical and sub-cortical regions in medication free schizophrenia subjects compared to matched controls. The results also showed a significant inverse correlation between level of binding and severity of positive symptoms, suggesting the levels of CHRMs may be associated with symptom levels [99].

Given the numerous roles CHRMs play in a variety of functions, it is imperative to find agents directed at the specific subtypes relevant to the symptoms of interest. Both natural and exogenous ligands of the CHRMs bind to a site that is highly conserved among the receptor subtypes and thus show very low selectivity to the subtypes [132]. Some antagonists, on the other hand, display good selectivity due to their bulky structure, allowing them to interact with amino acids adjacent to the classic (orthosteric) binding site that are less well conserved. However, the smaller orthosteric agonists display poor selectivity as they only interact with a few key amino acids in this site [133]. One approach to increasing ligand selectivity is to use allosteric modulators; an allosteric site can be found on all 5 subtypes of CHRMs [115]. A huge selection of structurally divergent modulators has been described for the different receptor subtypes, including gallamine and alcuronium [134-143]. These can increase or decrease the affinity of agonists and antagonists, including ACh, in a selective manner. Recently, a novel structural class of agonists was identified that are potent and selective for CHRM1, the first to be described being xanomeline [144-147]. These exhibit 'functional selectivity' in that they bind with comparable affinity to all 5 muscarinic subtypes but mainly activate CHRM1 [145]. They are believed to bind to an additional, less conserved 'ectopic' site that is important for receptor activation. However, the precise molecular mechanisms of these agents are still not fully understood [133]. The actions of this class of agonists are promising for treating cognitive deficits, offering a new approach to selectively targeting CHRMs.

In order to determine whether decreases in cortical CHRMs might be related to CHRM1 genotype, or possibly define a subgroup of schizophrenia patients, we performed [H^3]pirenzepine binding in the DLPFC from 80 subjects with schizophrenia and 74 age sex matched control subjects. Interestingly, Kernel density estimation on the binding data revealed two distinct populations among schizophrenia subjects, but not controls, where one group represented 26% of the schizophrenic subjects showing a 74% reduction in mean cortical [H^3]pirenzepine binding compared to control subjects, and were termed the 'muscarinic receptor-deficit schizophrenia' [148]. This segregation showed no correlation with CHRM1 sequence, gender, age, suicide, duration of illness or drug treatment [148]. This is interesting in terms of identifying one of the specific disease entities that make up schizophrenia, and perhaps identifying patients that may respond to some forms of treatments better than others. It would be interesting to see if such patients show altered cognitive behaviours compared to other schizophrenic patients, as indeed findings from previous studies have suggested a link between the level of CHRM expression and severity of symptoms [99].

There is a great deal of evidence demonstrating that the dopaminergic and muscarinic systems interact with each other [149]. For example, *in vivo*, xanomeline increased striatal levels of dopamine metabolites [144]. Studies in rats showed that the cognitive deficits produced by the muscarinic antagonist scopolamine could be reversed by the mixed dopamine D_1-D_2 blocker haloperidol [150]. This was later shown to be due to D_1 and not D_2 blockade [151]. Thus, it is possible that disruption of the interaction between the muscarinic and dopaminergic systems may cause symptoms of schizophrenia. It has been suggested that the balance of dopamine/ACh is of central importance to the pathophysiology of schizophrenia and that increased muscarinic ACh activity may be an attempt to maintain this balance [91]. This model also proposes that this compensatory increase in the muscarinic cholinergic system exerts a damping effect on the positive symptoms while adversely increasing the negative symptoms [91].

Nicotinic Receptors in Schizophrenia

Acetylcholine also acts through another family of receptors, the nicotinic acetylcholine receptors (nAChRs) [152]. The nAChRs are non-selective ligand-gated ion channels, opened by the binding of ACh, but can also be opened by nicotine. nAChRs are present in the CNS and periphery and are made up of pentamers of α, β, γ, δ and ε subunits, each sub-unit being a single gene product [153, 154]. Importantly, only the α_2-α_9 and β_2-β_4 sub-units are expressed in the CNS [154]. The two most predominant CNS nAChRs are $\alpha_4\beta_2$, which is the high affinity receptor, and the receptor made up entirely of α_7 subunits, which is the low affinity receptor [155-158], both of which have been shown to be reduced in schizophrenia [96, 97,

101]. Interestingly, there is an increased rate of smoking in subjects with schizophrenia compared to the general population, up to 90%, and compared to any other psychiatric illness [159]. This has been suggested to reflect self-medication [160-162]; signifying that nAChRs could constitute a drug target for the disorder.

Nicotine affects many aspects of behavior including locomotion, conception, anxiety, learning and memory [163]. Nicotine administration can enhance cognition in patients with schizophrenia, particularly attention [164-172]. Effects of nicotine in schizophrenia, however, do not extend to all areas of cognition. Thus, targeting this system alone would not be an effective form of treatment but it is one that may prove useful in combination with other therapeutics. Another concern with nicotinic agents is the rapid desensitisation observed with repeated administrations of nicotine, which renders nicotine ineffective over time [165, 167, 168, 172-174]. This has prompted research into exploring less toxic partial agonists and allosteric modulators of nAChRs as potential agents for cognitive enhancement, which would be less likely to cause rapid receptor desensitisation. The desensitisation of nicotinic receptors may also explain the ineffectiveness seen in attempts to treat schizophrenia symptoms with anticholinergic agents.

Neurobiological and genetic studies identified the α_7 nicotinic receptor as a potential contributor to the pathology of schizophrenia. Functional polymorphisms in the promoter region of α_7 have shown genetic linkage to the disease [175, 176], while the α_7 receptor was shown to be reduced in the PFC from patients with schizophrenia [97]. Furthermore, rodent studies have shown that α_7 antagonists induce sensory gating deficits (inability to respond appropriately to multiple sensory stimuli) [177] similar to those seen in schizophrenia [178-180]. Sensory gating deficits have been suggested to strongly impact cognitive performance [181-184], and have also been shown to be normalized by cigarette smoking [185]. Hence, drugs directed to the α_7 receptor may alleviate some of the cognitive deficits in schizophrenia. In further support of this, clinical response to clozapine was shown to be higher in schizophrenic subjects who smoke compared to non smokers, and furthermore, treatment with clozapine caused a decrease in smoking behaviours in patients [186, 187], suggesting that nicotinic receptors may also mediate the effects of clozapine. Animal studies have indicated that increased P50 auditory response inhibition by clozapine is mediated by stimulation of α_7 nicotinic receptors [188, 189]. This receptor is, therefore, a highly studied target for the development of drugs to treat cognitive deficits.

A series of α_7 agonists have been developed, one of which is the partial agonist 3-2,4-dimethoxybenzylidene anabaseine (DMZB-A) [190-192]. This compound improved auditory gating deficits in rodents and improved memory in several animal models [193, 194]. In humans, DMZB-A enhanced cognitive functions, particularly attention, in healthy volunteers [195], and in a small proof of concept trial, showed improvement in neurocognition as well as P50 inhibition in subjects with schizophrenia [196]. More recently, the selective α_7 agonist 2-methyl-5-(6-phenyl-pyridazin-3-yl)-octahydro-pyrrolo[3,4-c]pyrrole (A-582941) was shown in rats to activate brain regions involved in working memory and attention; the medial PFC and the ventral/lateral orbitofrontal cortex [197]. The effect was greater in juvenile rats than adult rats, which may be relevant in the treatment of juvenile onset schizophrenia [197]. Although results with α_7 agonists have been promising, the concern of desensitisation of the receptor with long term use remains, thus further development of partial agonists or allosteric modulators is warranted [11, 198]. Galantamine, a cholinesterase inhibitor, which is also an allosteric potentiator of α_7 nicotinic receptors, did not cause α_7 receptor desensitisation, and was effective in enhancing cognition in patients who smoked [199]. This suggests that combining allosteric modulators of the nicotinic receptor with anticholinesterase inhibitors may overcome issues with receptor desensitisation when targeting this receptor.

The $\alpha_4\beta_2$ nicotinic receptor is also likely to be involved in cognition [200], and is also reduced in the hippocampus, cortex and caudate of patients with schizophrenia [101]. These receptors are more sensitive to nicotine than the α_7 receptor, and are thought to account for 90% of the high affinity nicotine binding sites in the rat brain [155]. Agonists of $\alpha_4\beta_2$ have been shown to stimulate the release of dopamine and acetylcholine in the hippocampus and frontal cortex in rats [201]. Furthermore, agonists of this receptor can improve memory in rats and monkeys [201-203]. Thus the $\alpha_4\beta_2$ receptor may also be a potential therapeutic target for treating cognitive symptoms of schizophrenia by several mechanisms. Other nicotinic receptors may also be involved in cognition, however this is difficult to determine given the lack of selective agents to these other receptors.

There is evidence that the role of both the α_7 and $\alpha_4\beta_2$ receptors in cognitive impairments in schizophrenia is mediated through the dopamine receptor system. Unlike haloperidol's reversal effect on CHRM antagonist induced cognitive deficit, this drug potentiated the deficits induced by the nicotinic antagonist mecamylamine [204]. This action was due to dopamine D_2 blockade, since the mecamylamine induced deficit was potentiated by the selective D_2 blocker raclopride, but not the D_1 blocker SCH 23390 [205], and reversed by D_2 but not D_1 agonism [206]. Combined treatment of gallantamine and the neuroleptic drug risperidone showed synergistic improvement of cognitive impairment and increased extracellular concentration of dopamine in the medial PFC in mice. These effects were blocked by a dopamine D_1 receptor antagonist and a nAChR antagonist but not a CHRM antagonist [207], indicating these drugs act independently of the CHRM system, and that when taken in combination may enhance cognition in schizophrenia patients.

It has been suggested that interactions of nicotine with the NMDA glutamate system may play an important role in cognitive functions. This comes from studies showing pre-treatment with the NMDA antagonist MK801 reduced high-affinity nicotinic receptor upregulation induced by repeated nicotine injections in rats, preventing behavioural sensitization [208], and nicotine could reverse MK-801 induced attention deficits [209, 210], while *in vitro* studies have shown that nicotine reduces MK-801 binding to NMDA receptors [211, 212]. These effects were shown to be regionally selective throughout the brain [213].

There is also some evidence of synergistic interactions between the nicotinic and CHRM systems, both at the level of receptor regulation and the functional level [149, 214-216]. In rat brain, administration of scopolamine increased both muscarinic and nicotinic receptors in the frontoperietal cortex [214]. Furthermore, combined scopolamine and mecamylamine treatment acted in a greater than additive fashion in disrupting cognitive function in rats [149]. This action was attenuated by the D_2 agonist LY 171555 but not the D_1 antagonist SCH 23390 [149]. In humans, combined blockade of both receptor types with scopolamine and mecamylamine produced a large impairment in early visual information processing, while selective blockade of the CHRM alone produced only a small impairment and blockade of the nAChR alone showed no impairment [217]. A similar effect was also observed on other cognitive functions, including working memory, visual attention and psychomotor speed

[218]. Thus treatments that can reverse the effects of combined blockade of the muscarinic and nicotinic receptors may be useful in treating cognitive deficits.

Conclusion

Schizophrenia is a complex disease with a broad spectrum of symptoms. It is now believed that several neurotransmitter systems in addition to dopamine are involved, and the disease may be a result of interactions between dopamine and these other neurotransmitter systems. This concept suggests that efficacious treatment of schizophrenia may require a combination of therapies, directed at the various neurotransmitter systems involved, to address the wide range of symptoms associated with the disorder. Studies on the PFC of schizophrenia subjects indicate that the cholinergic system may play an important role in the pathology of schizophrenia, particularly in relation to cognitive functioning, identifying possible new targets for improved therapeutics. Our identification of MRDS suggests that different patients may respond differently to various drugs depending on their biochemical profile, and that identifying which patients would benefit from different types of drugs may be the key to effectively treating this debilitating disorder.

References

[1] Black, Andraeson. *Textbook of Psychiatry*. 2nd ed. Washington, DC: The American Psychiatric Press; 1994.

[2] Tandon, R. Cholinergic aspects of schizophrenia. *Br. J. Psychiatry Suppl.* 1999(37):7-11.

[3] Goldberg, T. E., Gold, J. M., Greenberg, R., Griffin, S., Schulz, S. C., Pickar, D., et al. Contrasts between patients with affective disorders and patients with schizophrenia on a neuropsychological test battery. *Am. J. Psychiatry* 1993;150(9):1355-62.

[4] Nuechterlein, K. H., Barch, D. M., Gold, J. M., Goldberg, T. E., Green, M. F., Heaton, R. K. Identification of separable cognitive factors in schizophrenia. *Schizophr. Res.* 2004;72(1):29-39.

[5] Tandon, R., Belmaker, R. H., Gattaz, W. F., Lopez-Ibor, J. J., Jr., Okasha, A., Singh, B., et al. World Psychiatric Association Pharmacopsychiatry Section statement on comparative effectiveness of antipsychotics in the treatment of schizophrenia. *Schizophr Res.* 2008;100(1-3):20-38.

[6] Heinrichs, R. W. The primacy of cognition in schizophrenia. *Am. Psychol.* 2005;60(3):229-42.

[7] Peuskens, J., Demily, C., Thibaut, F. Treatment of cognitive dysfunction in schizophrenia. *Clin. Ther.* 2005;27 Suppl A:S25-37.

[8] Prouteau, A., Verdoux, H., Briand, C., Lesage, A., Lalonde, P., Nicole, L., et al. Cognitive predictors of psychosocial functioning outcome in schizophrenia: a follow-up study of subjects participating in a rehabilitation program. *Schizophr. Res.* 2005;77(2-3):343-53.

[9] Chen, E. Y., Hui, C. L., Dunn, E. L., Miao, M. Y., Yeung, W. S., Wong, C. K., et al. A prospective 3-year longitudinal study of cognitive predictors of relapse in first-episode schizophrenic patients. *Schizophr. Res.* 2005;77(1):99-104.

[10] Kraepelin, E. Dementia Praecox and Paraphrenia (1919). 1st ed. New York: Thoemmes Continuum; 1971.

[11] Gray, J. A., Roth, B. L. Molecular targets for treating cognitive dysfunction in schizophrenia. *Schizophr. Bull.* 2007;33(5):1100-19.

[12] Bilder, R. M., Goldman, R. S., Volavka, J., Czobor, P., Hoptman, M., Sheitman, B., et al. Neurocognitive effects of clozapine, olanzapine, risperidone, and haloperidol in patients with chronic schizophrenia or schizoaffective disorder. *Am. J. Psychiatry* 2002;159(6):1018-28.

[13] Harvey, P. D., Keefe, R. S. Studies of cognitive change in patients with schizophrenia following novel antipsychotic treatment. *Am. J. Psychiatry* 2001;158(2):176-84.

[14] Purdon, S. E., Jones, B. D., Stip, E., Labelle, A., Addington, D., David, S. R., et al. Neuropsychological change in early phase schizophrenia during 12 months of treatment with olanzapine, risperidone, or haloperidol. The Canadian Collaborative Group for research in schizophrenia. *Arch. Gen. Psychiatry* 2000;57(3):249-58.

[15] Keefe, R. S., Silva, S. G., Perkins, D. O., Lieberman, J. A. The effects of atypical antipsychotic drugs on neurocognitive impairment in schizophrenia: a review and meta-analysis. *Schizophr. Bull.* 1999;25(2):201-22.

[16] Woodward, N. D., Purdon, S. E., Meltzer, H. Y., Zald, D. H. A meta-analysis of neuropsychological change to clozapine, olanzapine, quetiapine, and risperidone in schizophrenia. *Int. J. Neuropsychopharmacol* 2005;8(3):457-72.

[17] Weinberger, D. R. A connectionist approach to the prefrontal cortex. *J. Neuropsychiatry Clin. Neurosci.* 1993;5(3):241-53.

[18] Goldman-Rakic, P. S. Working memory dysfunction in schizophrenia. *J. Neuropsychiatry Clin. Neurosci.* 1994;6(4):348-57.

[19] Meltzer, H. Y. Dimensions of outcome with clozapine. *Br. J. Psychiatry Suppl.* 1992(17):46-53.

[20] Green, M. F. What are the functional consequences of neurocognitive deficits in schizophrenia? *Am. J. Psychiatry* 1996;153(3):321-30.

[21] Green, M. F., Braff, D. L. Translating the basic and clinical cognitive neuroscience of schizophrenia to drug development and clinical trials of antipsychotic medications. *Biol Psychiatry* 2001;49(4):374-84.

[22] Ingvar, D. H., Franzen, G. Distribution of cerebral activity in chronic schizophrenia. *Lancet* 1974;2(7895):1484-6.

[23] Berman, K. F., Zec, R. F., Weinberger, D. R. Physiologic dysfunction of dorsolateral prefrontal cortex in schizophrenia. II. Role of neuroleptic treatment, attention, and mental effort. *Arch. Gen. Psychiatry* 1986;43(2):126-35.

[24] Mathew, R. J., Wilson, W. H., Tant, S. R., Robinson, L., Prakash, R. Abnormal resting regional cerebral blood flow patterns and their correlates in schizophrenia. *Arch. Gen. Psychiatry* 1988;45(6):542-9.

[25] Buchsbaum, M. S., Ingvar, D. H., Kessler, R., Waters, R. N., Cappelletti, J., van Kammen, D. P., et al. Cerebral glucography with positron tomography. Use in normal subjects and in patients with schizophrenia. *Arch. Gen. Psychiatry* 1982;39(3):251-9.

[26] Volkow, N. D., Wolf, A. P., Van Gelder, P., Brodie, J. D., Overall, J. E., Cancro, R., et al. Phenomenological correlates of metabolic activity in 18 patients with chronic schizophrenia. *Am. J. Psychiatry* 1987;144(2):151-8.

[27] Catafau, A. M., Parellada, E., Lomena, F. J., Bernardo, M., Pavia, J., Ros, D., et al. Prefrontal and temporal blood flow in schizophrenia: resting and activation technetium-99m-HMPAO SPECT patterns in young neuroleptic-naive patients with acute disease. *J. Nucl. Med.* 1994;35(6):935-41.

[28] Ebmeier, K. P., Lawrie, S. M., Blackwood, D. H., Johnstone, E. C., Goodwin, G. M. Hypofrontality revisited: a high resolution single photon emission computed tomography study in schizophrenia. *J. Neurol. Neurosurg. Psychiatry* 1995;58(4):452-6.

[29] Gur, R. E., Resnick, S. M., Gur, R. C., Alavi, A., Caroff, S., Kushner, M., et al. Regional brain function in schizophrenia. II. Repeated evaluation with positron emission tomography. *Arch. Gen. Psychiatry* 1987;44(2):126-9.

[30] Gur, R. E., Resnick, S. M., Alavi, A., Gur, R. C., Caroff, S., Dann, R., et al. Regional brain function in schizophrenia. I. A positron emission tomography study. *Arch Gen Psychiatry* 1987;44(2):119-25.

[31] Berman, K. F., Illowsky, B. P., Weinberger, D. R. Physiological dysfunction of dorsolateral prefrontal cortex in schizophrenia. IV. Further evidence for regional and behavioral specificity. *Arch. Gen. Psychiatry* 1988;45(7):616-22.

[32] Weinberger, D. R., Berman, K. F., Illowsky, B. P. Physiological dysfunction of dorsolateral prefrontal cortex in schizophrenia. III. A new cohort and evidence for a monoaminergic mechanism. *Arch. Gen. Psychiatry* 1988;45(7):609-15.

[33] Weinberger, D. R., Berman, K. F., Zec, R. F. Physiologic dysfunction of dorsolateral prefrontal cortex in schizophrenia. I. Regional cerebral blood flow evidence. *Arch. Gen. Psychiatry* 1986;43(2):114-24.

[34] Ragland, J. D., Yoon, J., Minzenberg, M. J., Carter, C. S. Neuroimaging of cognitive disability in schizophrenia: search for a pathophysiological mechanism. *Int. Rev. Psychiatry* 2007;19(4):417-27.

[35] Cohen, J. D., Braver, T. S., O'Reilly, R. C. A computational approach to prefrontal cortex, cognitive control and schizophrenia: recent developments and current challenges. *Philos Trans R Soc. Lond B Biol. Sci.* 1996;351(1346):1515-27.

[36] Silver, H., Feldman, P., Bilker, W., Gur, R. C. Working memory deficit as a core neuropsychological dysfunction in schizophrenia. *Am. J. Psychiatry* 2003;160(10):1809-16.

[37] Barch, D. M., Carter, C. S., Braver, T. S., Sabb, F. W., MacDonald, A., 3rd, Noll, D. C., et al. Selective deficits in prefrontal cortex function in medication-naive patients with schizophrenia. *Arch. Gen. Psychiatry* 2001;58(3):280-8.

[38] Callicott, J. H., Ramsey, N. F., Tallent, K., Bertolino, A., Knable, M. B., Coppola, R., et al. Functional magnetic resonance imaging brain mapping in psychiatry: methodological issues illustrated in a study of working memory in schizophrenia. *Neuropsychopharmacology* 1998;18(3):186-96.

[39] Perlstein, W. M., Carter, C. S., Noll, D. C., Cohen, J. D. Relation of prefrontal cortex dysfunction to working memory and symptoms in schizophrenia. *Am. J. Psychiatry* 2001;158(7):1105-13.

[40] Honey, G. D., Bullmore, E. T., Sharma, T. De-coupling of cognitive performance and cerebral functional response during working memory in schizophrenia. *Schizophr. Res.* 2002;53(1-2):45-56.

[41] Callicott, J. H., Bertolino, A., Mattay, V. S., Langheim, F. J., Duyn, J., Coppola, R., et al. Physiological dysfunction of the dorsolateral prefrontal cortex in schizophrenia revisited. *Cereb Cortex* 2000;10(11):1078-92.

[42] Manoach, D. S., Gollub, R. L., Benson, E. S., Searl, M. M., Goff, D. C., Halpern, E., et al. Schizophrenic subjects show aberrant fMRI activation of dorsolateral prefrontal cortex and basal ganglia during working memory performance. *Biol. Psychiatry* 2000;48(2):99-109.

[43] Ragland, J. D., Gur, R. C., Glahn, D. C., Censits, D. M., Smith, R. J., Lazarev, M. G., et al. Frontotemporal cerebral blood flow change during executive and declarative memory tasks in schizophrenia: a positron emission tomography study. *Neuropsychology* 1998;12(3):399-413.

[44] Ramsey, N. F., Koning, H. A., Welles, P., Cahn, W., van der Linden, J. A., Kahn, R. S. Excessive recruitment of neural systems subserving logical reasoning in schizophrenia. *Brain* 2002;125(Pt 8):1793-807.

[45] Glahn, D. C., Ragland, J. D., Abramoff, A., Barrett, J., Laird, A. R., Bearden, C. E., et al. Beyond hypofrontality: a quantitative meta-analysis of functional neuroimaging studies of working memory in schizophrenia. *Hum. Brain Mapp.* 2005;25(1):60-9.

[46] Gevins, A., Cutillo, B. Spatiotemporal dynamics of component processes in human working memory. *Electroencephalogr. Clin. Neurophysiol.* 1993;87(3):128-43.

[47] Tan, H. Y., Sust, S., Buckholtz, J. W., Mattay, V. S., Meyer-Lindenberg, A., Egan, M. F., et al. Dysfunctional prefrontal regional specialization and compensation in schizophrenia. *Am. J. Psychiatry* 2006;163(11):1969-77.

[48] Dumontheil, I., Burgess, P. W., Blakemore, S. J. Development of rostral prefrontal cortex and cognitive and behavioural disorders. *Dev. Med Child Neurol* 2008;50(3):168-81.

[49] Narr, K. L., Toga, A. W., Szeszko, P., Thompson, P. M., Woods, R. P., Robinson, D., et al. Cortical thinning in cingulate and occipital cortices in first episode schizophrenia. *Biol. Psychiatry* 2005;58(1):32-40.

[50] Yamada, M., Hirao, K., Namiki, C., Hanakawa, T., Fukuyama, H., Hayashi, T., et al. Social cognition and frontal lobe pathology in schizophrenia: a voxel-based morphometric study. *Neuroimage* 2007;35(1):292-8.

[51] Riffkin, J., Yucel, M., Maruff, P., Wood, S. J., Soulsby, B., Olver, J., et al. A manual and automated MRI study of anterior cingulate and orbito-frontal cortices, and caudate nucleus in obsessive-compulsive disorder: comparison with healthy controls and patients with schizophrenia. *Psychiatry Res.* 2005;138(2):99-113.

[52] Rose, S. E., Chalk, J. B., Janke, A. L., Strudwick, M. W., Windus, L. C., Hannah, D. E., et al. Evidence of altered prefrontal-thalamic circuitry in schizophrenia: an optimized diffusion MRI study. *Neuroimage* 2006;32(1):16-22.

[53] Black, J. E., Kodish, I. M., Grossman, A. W., Klintsova, A. Y., Orlovskaya, D., Vostrikov, V., et al. Pathology of layer V pyramidal neurons in the prefrontal cortex of patients with schizophrenia. *Am. J. Psychiatry* 2004;161(4):742-4.

[54] Boksman, K., Theberge, J., Williamson, P., Drost, D. J., Malla, A., Densmore, M., et al. A 4.0-T fMRI study of brain connectivity during word fluency in first-episode schizophrenia. *Schizophr. Res.* 2005;75(2-3):247-63.

[55] Laurens, K. R., Kiehl, K. A., Ngan, E. T., Liddle, P. F. Attention orienting dysfunction during salient novel stimulus processing in schizophrenia. *Schizophr. Res.* 2005;75(2-3):159-71.

[56] Holcomb, H. H., Lahti, A. C., Medoff, D. R., Weiler, M., Dannals, R. F., Tamminga, C. A. Brain activation patterns in schizophrenic and comparison volunteers during a matched-performance auditory recognition task. *Am. J. Psychiatry* 2000;157(10):1634-45.

[57] Rubia, K., Russell, T., Bullmore, E. T., Soni, W., Brammer, M. J., Simmons, A., et al. An fMRI study of reduced left prefrontal activation in schizophrenia during normal inhibitory function. *Schizophr Res.* 2001;52(1-2):47-55.

[58] Dehaene, S., Artiges, E., Naccache, L., Martelli, C., Viard, A., Schurhoff, F., et al. Conscious and subliminal conflicts in normal subjects and patients with schizophrenia: the role of the anterior cingulate. *Proc. Natl. Acad. Sci. USA* 2003;100(23):13722-7.

[59] Heckers, S., Weiss, A. P., Deckersbach, T., Goff, D. C., Morecraft, R. J., Bush, G. Anterior cingulate cortex activation during cognitive interference in schizophrenia. *Am. J. Psychiatry* 2004;161(4):707-15.

[60] Kerns, J. G., Cohen, J. D., MacDonald, A. W., 3rd, Johnson, M. K., Stenger, V. A., Aizenstein, H., et al. Decreased conflict- and error-related activity in the anterior cingulate cortex in subjects with schizophrenia. *Am. J. Psychiatry* 2005;162(10):1833-9.

[61] Raedler, T. J., Bymaster, F. P., Tandon, R., Copolov, D., Dean, B. Towards a muscarinic hypothesis of schizophrenia. *Mol Psychiatry* 2007;12(3):232-46.

[62] Lahti, A. C., Holcomb, H. H., Weiler, M. A., Medoff, D. R., Tamminga, C. A. Functional effects of antipsychotic drugs: comparing clozapine with haloperidol. *Biol. Psychiatry* 2003;53(7):601-8.

[63] Honey, G. D., Bullmore, E. T., Soni, W., Varatheesan, M., Williams, S. C., Sharma, T. Differences in frontal cortical activation by a working memory task after substitution of risperidone for typical antipsychotic drugs in patients with schizophrenia. *Proc. Natl. Acad. Sci. USA* 1999;96(23):13432-7.

[64] Nahas, Z., George, M. S., Horner, M. D., Markowitz, J. S., Li, X., Lorberbaum, J. P., et al. Augmenting atypical antipsychotics with a cognitive enhancer (donepezil) improves regional brain activity in schizophrenia patients: a pilot double-blind placebo controlled BOLD fMRI study. *Neurocase* 2003;9(3):274-82.

[65] Meltzer, H. Y., Stahl, S. M. The dopamine hypothesis of schizophrenia: a review. *Schizophr. Bull.* 1976;2(1):19-76.

[66] Carlsson, A. Antipsychotic drugs and catecholamine synapses. *J. Psychiatr Res.* 1974;11:57-64.

[67] Seeman, P., Lee, T., Chau-Wong, M., Wong, K. Antipsychotic drug doses and neuroleptic/dopamine receptors. *Nature* 1976;261(5562):717-9.

[68] Kapur, S., Remington, G. Dopamine D(2) receptors and their role in atypical antipsychotic action: still necessary and may even be sufficient. *Biol. Psychiatry* 2001;50(11):873-83.

[69] Creese, I., Burt, D. R., Snyder, S. H. Dopamine receptor binding predicts clinical and pharmacological potencies of antischizophrenic drugs. *Science* 1976;192(4238):481-3.

[70] Okubo, Y., Suhara, T., Suzuki, K., Kobayashi, K., Inoue, O., Terasaki, O., et al. Decreased prefrontal dopamine D1 receptors in schizophrenia revealed by PET. *Nature* 1997;385(6617):634-6.

[71] Castner, S. A., Williams, G. V., Goldman-Rakic, P. S. Reversal of antipsychotic-induced working memory deficits by short-term dopamine D1 receptor stimulation. *Science* 2000;287(5460):2020-2.

[72] Pfeiffer, C. C., Jenney, E. H. The inhibition of the conditioned response and the counteraction of schizophrenia by muscarinic stimulation of the brain. *Ann. N Y Acad Sci.* 1957;66(3):753-64.

[73] Bilder, R. M. Neurocognitive impairment in schizophrenia and how it affects treatment options. *Can J Psychiatry* 1997;42(3):255-64.

[74] Kane, J., Honigfeld, G., Singer, J., Meltzer, H. Clozapine for the treatment-resistant schizophrenic. A double-blind comparison with chlorpromazine. *Arch. Gen. Psychiatry* 1988;45(9):789-96.

[75] Zorn, S. H., Jones, S. B., Ward, K. M., Liston, D. R. Clozapine is a potent and selective muscarinic M4 receptor agonist. *Eur. J. Pharmacol.* 1994;269(3):R1-2.

[76] Zeng, X. P., Le, F., Richelson, E. Muscarinic m4 receptor activation by some atypical antipsychotic drugs. *Eur. J. Pharmacol.* 1997;321(3):349-54.

[77] Bymaster, F. P., Shannon, H. E., Rasmussen, K., DeLapp, N. W., Ward, J. S., Calligaro, D. O., et al. Potential role of muscarinic receptors in schizophrenia. *Life Sci.* 1999;64(6-7):527-34.

[78] Michal, P., Lysikova, M., El-Fakahany, E. E., Tucek, S. Clozapine interaction with the M2 and M4 subtypes of muscarinic receptors. *Eur. J. Pharmacol.* 1999;376(1-2):119-25.

[79] Meltzer, H. Y., Chai, B. L., Thompson, P. A., Yamamoto, B. K. Effect of scopolamine on the efflux of dopamine and its metabolites after clozapine, haloperidol or thioridazine. *J. Pharmacol. Exp. Ther.* 1994;268(3):1452-61.

[80] Olianas, M. C., Maullu, C., Onali, P. Mixed agonist-antagonist properties of clozapine at different human cloned muscarinic receptor subtypes expressed in Chinese hamster ovary cells. *Neuropsychopharmacology* 1999;20(3):263-70.

[81] Gross, H., Hackl, H., Kaltenbaeck. Results of double-blind study of clozapine and thioridazine. In: VII CINP Congress; 1970; Prague; 1970.

[82] Casey, D. E. Clozapine: neuroleptic-induced EPS and tardive dyskinesia. *Psychopharmacology (Berl)* 1989;99 Suppl:S47-53.

[83] Brar, J. S., Chengappa, K. N., Parepally, H., Sandman, A. R., Kreinbrook, S. B., Sheth, S. A., et al. The effects of clozapine on negative symptoms in patients with schizophrenia with minimal positive symptoms. *Ann. Clin. Psychiatry* 1997;9(4):227-34.

[84] Llorca, P. M., Lancon, C., Farisse, J., Scotto, J. C. Clozapine and negative symptoms. An open study. *Prog. Neuropsychopharmacol Biol. Psychiatry* 2000;24(3):373-84.

[85] Hagger, C., Buckley, P., Kenny, J. T., Friedman, L., Ubogy, D., Meltzer, H. Y. Improvement in cognitive functions and psychiatric symptoms in treatment-refractory schizophrenic patients receiving clozapine. *Biol. Psychiatry* 1993;34(10):702-12.

[86] Goldberg, T. E., Weinberger, D. R. The effects of clozapine on neurocognition: an overview. *J. Clin. Psychiatry* 1994;55 Suppl B:88-90.

[87] Weiner, D. M., Meltzer, H. Y., Veinbergs, I., Donohue, E. M., Spalding, T. A., Smith, T. T., et al. The role of M1 muscarinic receptor agonism of N-desmethylclozapine in the unique clinical effects of clozapine. *Psychopharmacology (Berl)* 2004;177(1-2):207-16.

[88] Ichikawa, J., Dai, J., O'Laughlin, I. A., Fowler, W. L., Meltzer, H. Y. Atypical, but not typical, antipsychotic drugs increase cortical acetylcholine release without an effect in the nucleus accumbens or striatum. *Neuropsychopharmacology* 2002;26(3):325-39.

[89] Hagan, J. J., Jansen, J. H., Broekkamp, C. L. Blockade of spatial learning by the M1 muscarinic antagonist pirenzepine. *Psychopharmacology (Berl)* 1987;93(4):470-6.

[90] Bartus, R. T., Johnson, H. R. Short-term memory in the rhesus monkey: disruption from the anti-cholinergic scopolamine. *Pharmacol. Biochem. Behav.* 1976;5(1):39-46.

[91] Tandon, R., Greden, J. F. Cholinergic hyperactivity and negative schizophrenic symptoms. A model of cholinergic/dopaminergic interactions in schizophrenia. *Arch. Gen. Psychiatry* 1989;46(8):745-53.

[92] Crook, J. M., Tomaskovic-Crook, E., Copolov, D. L., Dean, B. Decreased muscarinic receptor binding in subjects with schizophrenia: a study of the human hippocampal formation. *Biol. Psychiatry* 2000;48(5):381-8.

[93] Crook, J. M., Tomaskovic-Crook, E., Copolov, D. L., Dean, B. Low muscarinic receptor binding in prefrontal cortex from subjects with schizophrenia: a study of Brodmann's areas 8, 9, 10, and 46 and the effects of neuroleptic drug treatment. *Am. J. Psychiatry* 2001;158(6):918-25.

[94] Dean, B., McLeod, M., Keriakous, D., McKenzie, J., Scarr, E. Decreased muscarinic1 receptors in the dorsolateral prefrontal cortex of subjects with schizophrenia. *Mol. Psychiatry* 2002;7(10):1083-91.

[95] Deng, C., Huang, X. F. Decreased density of muscarinic receptors in the superior temporal gyrusin schizophrenia. *J. Neurosci. Res.* 2005;81(6):883-90.

[96] Freedman, R., Hall, M., Adler, L. E., Leonard, S. Evidence in postmortem brain tissue for decreased numbers of hippocampal nicotinic receptors in schizophrenia. *Biol. Psychiatry* 1995;38(1):22-33.

[97] Guan, Z. Z., Zhang, X., Blennow, K., Nordberg, A. Decreased protein level of nicotinic receptor alpha7 subunit in the frontal cortex from schizophrenic brain. *Neuroreport* 1999;10(8):1779-82.

[98] Mancama, D., Arranz, M. J., Landau, S., Kerwin, R. Reduced expression of the muscarinic 1 receptor cortical subtype in schizophrenia. *Am. J. Med. Genet B Neuropsychiatr Genet.* 2003;119B(1):2-6.

[99] Raedler, T. J., Knable, M. B., Jones, D. W., Urbina, R. A., Gorey, J. G., Lee, K. S., et al. In vivo determination of muscarinic acetylcholine receptor availability in schizophrenia. *Am. J. Psychiatry* 2003;160(1):118-27.

[100] Scarr, E., Sundram, S., Keriakous, D., Dean, B. Altered hippocampal muscarinic M4, but not M1, receptor expression from subjects with schizophrenia. *Biol. Psychiatry* 2007;61(10):1161-70.

[101] Breese, C. R., Lee, M. J., Adams, C. E., Sullivan, B., Logel, J., Gillen, K. M., et al. Abnormal regulation of high affinity nicotinic receptors in subjects with schizophrenia. *Neuropsychopharmacology* 2000;23(4):351-64.

[102] Powchik, P., Davidson, M., Haroutunian, V., Gabriel, S. M., Purohit, D. P., Perl, D. P., et al. Postmortem studies in schizophrenia. *Schizophr Bull.* 1998;24(3):325-41.

[103] Lee, S. W., Lee, J. G., Lee, B. J., Kim, Y. H. A 12-week, double-blind, placebo-controlled trial of galantamine adjunctive treatment to conventional antipsychotics for the cognitive impairments in chronic schizophrenia. *Int. Clin. Psychopharmacol* 2007;22(2):63-8.

[104] Schubert, M. H., Young, K. A., Hicks, P. B. Galantamine improves cognition in schizophrenic patients stabilized on risperidone. *Biol. Psychiatry* 2006;60(6):530-3.

[105] Buchanan, R. W., Freedman, R., Javitt, D. C., Abi-Dargham, A., Lieberman, J. A. Recent advances in the development of novel pharmacological agents for the treatment of cognitive impairments in schizophrenia. *Schizophr Bull* 2007;33(5):1120-30.

[106] Raiteri, M., Leardi, R., Marchi, M. Heterogeneity of presynaptic muscarinic receptors regulating neurotransmitter release in the rat brain. *J. Pharmacol. Exp. Ther.* 1984;228(1):209-14.

[107] Wamsley, J. K., Zarbin, M. A., Kuhar, M. J. Distribution of muscarinic cholinergic high and low affinity agonist binding sites: a light microscopic autoradiographic study. *Brain Res. Bull.* 1984;12(3):233-43.

[108] Vizi, E. S., Kobayashi, O., Torocsik, A., Kinjo, M., Nagashima, H., Manabe, N., et al. Heterogeneity of presynaptic muscarinic receptors involved in modulation of transmitter release. *Neuroscience* 1989;31(1):259-67.

[109] Bonner, T. I., Buckley, N. J., Young, A. C., Brann, M. R. Identification of a family of muscarinic acetylcholine receptor genes. *Science* 1987;237(4814):527-32.

[110] Bonner, T. I., Young, A. C., Brann, M. R., Buckley, N. J. Cloning and expression of the human and rat m5 muscarinic acetylcholine receptor genes. *Neuron* 1988;1(5):403-10.

[111] Bonner, T. I. New subtypes of muscarinic acetylcholine receptors. *Trends Pharmacol. Sci.* 1989;Suppl:11-5.

[112] Bonner, T. I. The molecular basis of muscarinic receptor diversity. *Trends Neurosci* 1989;12(4):148-51.

[113] Peralta, E. G., Ashkenazi, A., Winslow, J. W., Smith, D. H., Ramachandran, J., Capon, D. J. Distinct primary structures, ligand-binding properties and tissue-specific expression of four human muscarinic acetylcholine receptors. *Embo. J.* 1987;6(13):3923-9.

[114] Bennett, J. P., Jr., Enna, S. J., Bylund, D. B., Gillin, J. C., Wyatt, R. J., Snyder, S. H. Neurotransmitter receptors in frontal cortex of schizophrenics. *Arch Gen Psychiatry* 1979;36(9):927-34.

[115] Ellis, J., Huyler, J., Brann, M. R. Allosteric regulation of cloned m1-m5 muscarinic receptor subtypes. *Biochem. Pharmacol.* 1991;42(10):1927-32.

[116] Watanabe, S., Nishikawa, T., Takashima, M., Toru, M. Increased muscarinic cholinergic receptors in prefrontal cortices of medicated schizophrenics. *Life Sci.* 1983;33(22):2187-96.

[117] Toru, M., Watanabe, S., Shibuya, H., Nishikawa, T., Noda, K., Mitsushio, H., et al. Neurotransmitters, receptors and neuropeptides in post-mortem brains of chronic schizophrenic patients. *Acta Psychiatr. Scand.* 1988;78(2):121-37.

[118] Levey, A. I., Kitt, C. A., Simonds, W. F., Price, D. L., Brann, M. R. Identification and localization of muscarinic acetylcholine receptor proteins in brain with subtype-specific antibodies. *J. Neurosci.* 1991;11(10):3218-26.

[119] Dean, B., Crook, J. M., Opeskin, K., Hill, C., Keks, N., Copolov, D. L. The density of muscarinic M1 receptors is decreased in the caudate-putamen of subjects with schizophrenia. *Mol. Psychiatry* 1996;1(1):54-8.

[120] Crook, J. M., Dean, B., Pavey, G., Copolov, D. The binding of [3H]AF-DX 384 is reduced in the caudate-putamen of subjects with schizophrenia. *Life Sci.* 1999;64(19):1761-71.

[121] Scarr, E., Keriakous, D., Crossland, N., Dean, B. No change in cortical muscarinic M2, M3 receptors or [35S]GTPgammaS binding in schizophrenia. *Life Sci.* 2006;78(11):1231-7.

[122] Liao, D. L., Hong, C. J., Chen, H. M., Chen, Y. E., Lee, S. M., Chang, C. Y., et al. Association of muscarinic m1 receptor genetic polymorphisms with psychiatric symptoms and cognitive function in schizophrenic patients. *Neuropsychobiology* 2003;48(2):72-6.

[123] Hamilton, S. E., Loose, M. D., Qi, M., Levey, A. I., Hille, B., McKnight, G. S., et al. Disruption of m1 receptor gene ablates muscarinic receptor-dependent current regulation and seizure activity in mice. *Proc. Natl. Acad. Sci. USA* 1997;94 (24):133116.

[124] Gillette, M. U., Buchanan, G. F., Artinian, L., Hamilton, S. E., Nathanson, N. M., Liu, C. Role of the M1 receptor in regulating circadian rhythms. *Life Sci.* 2001;68(22-23):2467-72.

[125] Heinrichs, R. W., Zakzanis, K. K. Neurocognitive deficit in schizophrenia: a quantitative review of the evidence. *Neuropsychology* 1998;12(3):426-45.

[126] Keshavan, M. S., Reynolds, C. F., 3rd, Miewald, M. J., Montrose, D. M., Sweeney, J. A., Vasko, R. C., Jr., et al. Delta sleep deficits in schizophrenia: evidence from automated analyses of sleep data. *Arch. Gen. Psychiatry* 1998;55(5):443-8.

[127] Mirza, N. R., Peters, D., Sparks, R. G. Xanomeline and the antipsychotic potential of muscarinic receptor subtype selective agonists. *CNS Drug Rev.* 2003;9(2):159-86.

[128] Shekhar, A., Potter, W. Z., Lienemann, J., Sundblad, K., Lightfoot, J., Herrera, J. Efficacy of zanomeline, a slelctive muscarinic agonist, in treating schizphrenia: a double-blind, placebo controlled study. In: 40th Annual Meeting ACNP; 2001 December 9-13, 2001; Hawaii; 2001.

[129] Dean, B. Role of muscarinic receptors in the pathology of schizophrenia. *Am J Med Genet B Neuropsychiatr Genet* 2004;126B(1):8-9.

[130] Zavitsanou, K., Nguyen, V. H., Han, M., Huang, X. F. Effects of typical and atypical antipsychotic drugs on rat brain muscarinic receptors. *Neurochem Res.* 2007;32(3):525-32.

[131] Han, M., Newell, K., Zavitsanou, K., Deng, C., Huang, X. F. Effects of antipsychotic medication on muscarinic M1 receptor mRNA expression in the rat brain. *J. Neurosci. Res* 2008;86(2):457-64.

[132] Hulme, E. C., Lu, Z. L., Saldanha, J. W., Bee, M. S. Structure and activation of muscarinic acetylcholine receptors. *Biochem. Soc. Trans* 2003;31(Pt 1):29-34.

[133] Jakubik, J., Michal, P., Machova, E., Dolezal, V. Importance and prospects for design of selective muscarinic agonists. *Physiol. Res.* 2008.

[134] Jager, D., Schmalenbach, C., Prilla, S., Schrobang, J., Kebig, A., Sennwitz, M., et al. Allosteric small molecules unveil a role of an extracellular E2/transmembrane helix 7 junction for G protein-coupled receptor activation. *J. Biol. Chem.* 2007;282(48):34968-76.

[135] Jakubik, J., Bacakova, L., el-Fakahany, E. E., Tucek, S. Subtype selectivity of the positive allosteric action of alcuronium at cloned M1-M5 muscarinic acetylcholine receptors. *J. Pharmacol. Exp. Ther.* 1995;274(3):1077-83.

[136] Jakubik, J., Bacakova, L., El-Fakahany, E. E., Tucek, S. Positive cooperativity of acetylcholine and other agonists with allosteric ligands on muscarinic acetylcholine receptors. *Mol. Pharmacol.* 1997;52(1):172-9.

[137] Lazareno, S., Birdsall, N. J. Detection, quantitation, and verification of allosteric interactions of agents with labeled and unlabeled ligands at G protein-coupled receptors: interactions of strychnine and acetylcholine at muscarinic receptors. *Mol. Pharmacol.* 1995;48(2):362-78.

[138] Lazareno, S., Dolezal, V., Popham, A., Birdsall, N. J. Thiochrome enhances acetylcholine affinity at muscarinic M4 receptors: receptor subtype selectivity via cooperativity rather than affinity. *Mol Pharmacol* 2004;65(1):257-66.

[139] Mohr, K., Trankle, C., Holzgrabe, U. Structure/activity relationships of M2 muscarinic allosteric modulators. *Receptors Channels* 2003;9(4):229-40.

[140] Proska, J., Tucek, S. Mechanisms of steric and cooperative actions of alcuronium on cardiac muscarinic acetylcholine receptors. *Mol. Pharmacol.* 1994;45(4):709-17.

[141] Tucek, S., Jakubik, J., Dolezal, V., el-Fakahany, E. E. Positive effects of allosteric modulators on the binding properties and the function of muscarinic acetylcholine receptors. *J. Physiol Paris* 1998;92(3-4):241-3.

[142] Tucek, S., Musilkova, J., Nedoma, J., Proska, J., Shelkovnikov, S., Vorlicek, J. Positive cooperativity in the binding of alcuronium and N-methylscopolamine to muscarinic acetylcholine receptors. *Mol. Pharmacol.* 1990;38(5):674-80.

[143] Dolezal, V., Tucek, S. The effects of brucine and alcuronium on the inhibition of [3H]acetylcholine release from rat striatum by muscarinic receptor agonists. *Br. J. Pharmacol.* 1998;124(6):1213-8.

[144] Bymaster, F. P., Wong, D. T., Mitch, C. H., Ward, J. S., Calligaro, D. O., Schoepp, D. D., et al. Neurochemical effects of the M1 muscarinic agonist xanomeline (LY246708/NNC11-0232). *J. Pharmacol. Exp. Ther.*1994;269(1):282-9.

[145] Shannon, H. E., Bymaster, F. P., Calligaro, D. O., Greenwood, B., Mitch, C. H., Sawyer, B. D., et al. Xanomeline: a novel muscarinic receptor agonist with functional selectivity for M1 receptors. *J. Pharmacol. Exp. Ther.* 1994;269(1):271-81.

[146] Spalding, T. A., Trotter, C., Skjaerbaek, N., Messier, T. L., Currier, E. A., Burstein, E. S., et al. Discovery of an ectopic activation site on the M(1) muscarinic receptor. *Mol. Pharmacol.* 2002;61(6):1297-302.

[147] Langmead, C. J., Fry, V. A., Forbes, I. T., Branch, C. L., Christopoulos, A., Wood, M. D., et al. Probing the molecular mechanism of interaction between 4-n-butyl-1-[4-(2-methylphenyl)-4-oxo-1-butyl]-piperidine (AC-42) and the muscarinic M(1) receptor: direct pharmacological evidence that AC-42 is an allosteric agonist. *Mol. Pharmacol.* 2006;69(1):236-46.

[148] Scarr, E., Cowie, T. F., Kanellakis, S., Sundram, S., Pantelis, C., Dean, B. Decreased cortical muscarinic receptors define a subgroup of subjects with schizophrenia. *Mol Psychiatry* 2008.

[149] Levin, E. D., Rose, J. E., McGurk, S. R., Butcher, L. L. Characterization of the cognitive effects of combined muscarinic and nicotinic blockade. *Behav. Neural. Biol.* 1990;53(1):103-12.

[150] McGurk, S. R., Levin, E. D., Butcher, L. L. Cholinergic-dopaminergic interactions in radial-arm maze performance. *Behav. Neural. Biol.* 1988;49(2):234-9.

[151] Levin, E. D. Scopolamine interactions with D1 and D2 antagonists on radial-arm maze performance in rats. *Behav. Neural. Biol.* 1988;50(2):240-5.

[152] Itier, V., Bertrand, D. Neuronal nicotinic receptors: from protein structure to function. *FEBS Lett* 2001;504(3):118-25.

[153] Paterson, D., Nordberg, A. Neuronal nicotinic receptors in the human brain. *Prog Neurobiol.* 2000;61(1):75-111.

[154] Le Novere, N., Changeux, J. P. Molecular evolution of the nicotinic acetylcholine receptor: an example of multigene family in excitable cells. *J. Mol. Evol.* 1995;40(2):155-72.

[155] Flores, C. M., Rogers, S. W., Pabreza, L. A., Wolfe, B. B., Kellar, K. J. A subtype of nicotinic cholinergic receptor in rat brain is composed of alpha 4 and beta 2 subunits and is up-regulated by chronic nicotine treatment. *Mol. Pharmacol.* 1992;41(1):31-7.

[156] Hill, J. A., Jr., Zoli, M., Bourgeois, J. P., Changeux, J. P. Immunocytochemical localization of a neuronal nicotinic receptor: the beta 2-subunit. *J. Neurosci.* 1993;13(4):1551-68.

[157] Zoli, M., Le Novere, N., Hill, J. A., Jr., Changeux, J. P. Developmental regulation of nicotinic ACh receptor subunit mRNAs in the rat central and peripheral nervous systems. *J. Neurosci.* 1995;15(3 Pt 1):1912-39.

[158] Wada, E., Wada, K., Boulter, J., Deneris, E., Heinemann, S., Patrick, J., et al. Distribution of alpha 2, alpha 3, alpha 4, and beta 2 neuronal nicotinic receptor subunit mRNAs in the central nervous system: a hybridization histochemical study in the rat. *J. Comp. Neurol.* 1989;284(2):314-35.

[159] Nisell, M., Nomikos, G. G., Svensson, T. H. Nicotine dependence, midbrain dopamine systems and psychiatric disorders. *Pharmacol. Toxicol.* 1995;76(3):157-62.

[160] Glassman, A. H. Cigarette smoking: implications for psychiatric illness. *Am. J. Psychiatry* 1993;150(4):546-53.

[161] Dalack, G. W., Healy, D. J., Meador-Woodruff, J. H. Nicotine dependence in schizophrenia: clinical phenomena and laboratory findings. *Am. J. Psychiatry* 1998;155(11):1490-501.

[162] Kumari, V., Postma, P. Nicotine use in schizophrenia: the self medication hypotheses. *Neurosci. Biobehav. Rev.* 2005;29(6):1021-34.

[163] Decker, M. W., Brioni, J. D., Bannon, A. W., Arneric, S. P. Diversity of neuronal nicotinic acetylcholine receptors: lessons from behavior and implications for CNS therapeutics. *Life Sci.* 1995;56(8):545-70.

[164] Adler, L. E., Olincy, A., Waldo, M., Harris, J. G., Griffith, J., Stevens, K., et al. Schizophrenia, sensory gating, and nicotinic receptors. *Schizophr. Bull.* 1998;24(2):189-202.

[165] Depatie, L., O'Driscoll, G. A., Holahan, A. L., Atkinson, V., Thavundayil, J. X., Kin, N. N., et al. Nicotine and behavioral markers of risk for schizophrenia: a double-blind, placebo-controlled, cross-over study. *Neuropsychopharmacology* 2002;27(6):1056-70.

[166] Avila, M. T., Sherr, J. D., Hong, E., Myers, C. S., Thaker, G. K. Effects of nicotine on leading saccades during smooth pursuit eye movements in smokers and nonsmokers with schizophrenia. *Neuropsychopharmacology* 2003;28(12):2184-91.

[167] Harris, J. G., Kongs, S., Allensworth, D., Martin, L., Tregellas, J., Sullivan, B., et al. Effects of nicotine on cognitive deficits in schizophrenia. *Neuropsychopharmacology* 2004;29(7):1378-85.

[168] Myers, C. S., Robles, O., Kakoyannis, A. N., Sherr, J. D., Avila, M. T., Blaxton, T. A., et al. Nicotine improves delayed recognition in schizophrenic patients. *Psychopharmacology (Berl)* 2004;174(3):334-40.

[169] Sacco, K. A., Termine, A., Seyal, A., Dudas, M. M., Vessicchio, J. C., Krishnan-Sarin, S., et al. Effects of cigarette smoking on spatial working memory and attentional deficits in schizophrenia: involvement of nicotinic receptor mechanisms. *Arch. Gen. Psychiatry* 2005;62(6):649-59.

[170] Baldeweg, T., Wong, D., Stephan, K. E. Nicotinic modulation of human auditory sensory memory: Evidence from mismatch negativity potentials. *Int. J. Psychophysiol.* 2006;59(1):49-58.

[171] George, T. P., Termine, A., Sacco, K. A., Allen, T. M., Reutenauer, E., Vessicchio, J. C., et al. A preliminary study of the effects of cigarette smoking on prepulse inhibition in schizophrenia: involvement of nicotinic receptor mechanisms. *Schizophr. Res.* 2006;87(1-3):307-15.

[172] Smith, R. C., Warner-Cohen, J., Matute, M., Butler, E., Kelly, E., Vaidhyanathaswamy, S., et al. Effects of nicotine nasal spray on cognitive function in schizophrenia. *Neuropsychopharmacology* 2006;31(3):637-43.

[173] Jacobsen, L. K., D'Souza, D. C., Mencl, W. E., Pugh, K. R., Skudlarski, P., Krystal, J. H. Nicotine effects on brain function and functional connectivity in schizophrenia. *Biol. Psychiatry* 2004;55(8):850-8.

[174] George, T. P., Vessicchio, J. C., Termine, A., Sahady, D. M., Head, C. A., Pepper, W. T., et al. Effects of smoking abstinence on visuospatial working memory function in schizophrenia. *Neuropsychopharmacology* 2002;26(1):75-85.

[175] Freedman, R., Coon, H., Myles-Worsley, M., Orr-Urtreger, A., Olincy, A., Davis, A., et al. Linkage of a neurophysiological deficit in schizophrenia to a chromosome 15 locus. *Proc. Natl. Acad. Sci. USA* 1997;94(2):587-92.

[176] Riley, B. P., Makoff, A., Mogudi-Carter, M., Jenkins, T., Williamson, R., Collier, D., et al. Haplotype transmission disequilibrium and evidence for linkage of the CHRNA7 gene region to schizophrenia in Southern African Bantu families. *Am. J. Med. Genet* 2000;96(2):196-201.

[177] Luntz-Leybman, V., Bickford, P. C., Freedman, R. Cholinergic gating of response to auditory stimuli in rat hippocampus. *Brain Res.* 1992;587(1):130-6.

[178] Adler, L. E., Pachtman, E., Franks, R. D., Pecevich, M., Waldo, M. C., Freedman, R. Neurophysiological evidence for a defect in neuronal mechanisms involved in sensory gating in schizophrenia. *Biol. Psychiatry* 1982;17(6):639-54.

[179] Freedman, R., Adler, L. E., Waldo, M. C., Pachtman, E., Franks, R. D. Neurophysiological evidence for a defect in inhibitory pathways in schizophrenia: comparison of medicated and drug-free patients. *Biol Psychiatry* 1983;18(5):537-51.

[180] Siegel, C., Waldo, M., Mizner, G., Adler, L. E., Freedman, R. Deficits in sensory gating in schizophrenic patients and their relatives. Evidence obtained with auditory evoked responses. *Arch. Gen. Psychiatry* 1984;41(6):607-12.

[181] Gottschalk, L. A., Haer, J. L., Bates, D. E. Effect of sensory overload on psychological state. Changes in social alienation-personal disorganization and cognitive-intellectual impairment. *Arch. Gen. Psychiatry* 1972;27(4):451-7.

[182] Braff, D. L. Attention, habituation and information processing in psychiatric disorders. In: Michels R., Brodie H. K., Cooper A. M., editors. Pychiatry. Philadelphia: Lippincott; 1985. p. 1-13.

[183] Braff, D. L., Geyer, M. A. Sensorimotor gating and schizophrenia. Human and animal model studies. *Arch. Gen. Psychiatry* 1990;47(2):181-8.

[184] Geyer, M. A., Braff, D. L. Startle habituation and sensorimotor gating in schizophrenia and related animal models. *Schizophr Bull* 1987;13(4):643-68.

[185] Adler, L. E., Hoffer, L. D., Wiser, A., Freedman, R. Normalization of auditory physiology by cigarette smoking in schizophrenic patients. *Am J Psychiatry* 1993;150(12):1856-61.

[186] McEvoy, J. P., Freudenreich, O., Wilson, W. H. Smoking and therapeutic response to clozapine in patients with schizophrenia. *Biol. Psychiatry* 1999;46(1):125-9.

[187] George, T. P., Sernyak, M. J., Ziedonis, D. M., Woods, S. W. Effects of clozapine on smoking in chronic schizophrenic outpatients. *J. Clin. Psychiatry* 1995;56(8):344-6.

[188] Simosky, J. K., Stevens, K. E., Adler, L. E., Freedman, R. Clozapine improves deficient inhibitory auditory processing in DBA/2 mice, via a nicotinic cholinergic mechanism. *Psychopharmacology (Berl)* 2003;165(4):386-96.

[189] Adler, L. E., Olincy, A., Cawthra, E. M., McRae, K. A., Harris, J. G., Nagamoto, H. T., et al. Varied effects of atypical neuroleptics on P50 auditory gating in schizophrenia patients. *Am. J. Psychiatry* 2004;161(10):1822-8.

[190] Kem, W. R., Mahnir, V. M., Prokai, L., Papke, R. L., Cao, X., LeFrancois, S., et al. Hydroxy metabolites of the Alzheimer's drug candidate 3-[(2,4-dimethoxy)benzylidene]-anabaseine dihydrochloride (GTS-21): their molecular properties, interactions with brain nicotinic receptors, and brain penetration. *Mol. Pharmacol.* 2004;65(1):56-67.

[191] Walker, D. P., Wishka, D. G., Piotrowski, D. W., Jia, S., Reitz, S. C., Yates, K. M., et al. Design, synthesis, structure-activity relationship, and in vivo activity of azabicyclic aryl amides as alpha7 nicotinic acetylcholine receptor agonists. *Bioorg. Med. Chem.* 2006;14(24):8219-48.

[192] de Fiebre, C. M., Meyer, E. M., Henry, J. C., Muraskin, S. I., Kem, W. R., Papke, R. L. Characterization of a series of anabaseine-derived compounds reveals that the 3-(4)-dimethylaminocinnamylidine derivative is a selective agonist at neuronal nicotinic alpha 7/125I-alpha-bungarotoxin receptor subtypes. *Mol. Pharmacol.* 1995;47(1):164-71.

[193] Woodruff-Pak, D. S., Li, Y. T., Kem, W. R. A nicotinic agonist (GTS-21), eyeblink classical conditioning, and nicotinic receptor binding in rabbit brain. *Brain Res.* 1994;645(1-2):309-17.

[194] Stevens, K. E., Kem, W. R., Mahnir, V. M., Freedman, R. Selective alpha7-nicotinic agonists normalize inhibition of auditory response in DBA mice. *Psychopharmacology (Berl)* 1998;136(4):320-7.

[195] Kitagawa, H., Takenouchi, T., Azuma, R., Wesnes, K. A., Kramer, W. G., Clody, D. E., et al. Safety, pharmacokinetics, and effects on cognitive function of multiple doses of GTS-21 in healthy, male volunteers. *Neuropsychopharmacology* 2003;28(3):542-51.

[196] Olincy, A., Harris, J. G., Johnson, L. L., Pender, V., Kongs, S., Allensworth, D., et al. Proof-of-concept trial of an alpha7 nicotinic agonist in schizophrenia. *Arch. Gen. Psychiatry* 2006;63(6):630-8.

[197] Thomsen, M. S., Mikkelsen, J. D., Timmermann, D. B., Peters, D., Hay-Schmidt, A., Martens, H., et al. The selective alpha(7) nicotinic acetylcholine receptor agonist A-582941 activates immediate early genes in limbic regions of the forebrain: Differential effects in the juvenile and adult rat. *Neuroscience* 2008.

[198] Simosky, J. K., Stevens, K. E., Freedman, R. Nicotinic agonists and psychosis. *Curr Drug Targets CNS Neurol Disord* 2002;1(2):149-62.

[199] Allen, T., McEvoy, J. P., Keefe, R. S., D., L. E., Wilson, W. Galantamine as an adjunctive therapy in the treatment of schizphrenia. In: 11th Congress of the International Psychogeriatric Association (IPA); 2003 August 17-22; Chicago; 2003.

[200] Schreiber, R., Dalmus, M., De Vry, J. Effects of alpha 4/beta 2- and alpha 7-nicotine acetylcholine receptor agonists on prepulse inhibition of the acoustic startle response in rats and mice. *Psychopharmacology (Berl)* 2002;159(3):248-57.

[201] Bontempi, B., Whelan, K. T., Risbrough, V. B., Rao, T. S., Buccafusco, J. J., Lloyd, G. K., et al. SIB-1553A, (+/-)-4-[[2-(1-methyl-2-pyrrolidinyl)ethyl]thio]phenol hydrochloride, a subtype-selective ligand for nicotinic acetylcholine receptors with putative cognitive-enhancing properties: effects on working and reference memory performances in aged rodents and nonhuman primates. *J. Pharmacol. Exp. Ther.* 2001;299(1):297-306.

[202] Levin, E. D., McClernon, F. J., Rezvani, A. H. Nicotinic effects on cognitive function: behavioral characterization, pharmacological specification, and anatomic localization. *Psychopharmacology (Berl)* 2006;184(3-4):523-39.

[203] Lloyd, G. K., Menzaghi, F., Bontempi, B., Suto, C., Siegel, R., Akong, M., et al. The potential of subtype-selective neuronal nicotinic acetylcholine receptor agonists as therapeutic agents. *Life Sci.*1998;62(17-18):1601-6.

[204] McGurk, S. R., Levin, E. D., Butcher, L. L. Nicotinic-dopaminergic relationships and radial-arm maze performance in rats. *Behav. Neural Biol.* 1989;52(1):78-86.

[205] McGurk, S. R., Levin, E. D., Butcher, L. L. Radial-arm maze performance in rats is impaired by a combination of nicotinic-cholinergic and D2 dopaminergic antagonist drugs. *Psychopharmacology (Berl)* 1989;99(3):371-3.

[206] Levin, E. D., McGurk, S. R., Rose, J. E., Butcher, L. L. Reversal of a mecamylamine-induced cognitive deficit with the D2 agonist, LY 171555. *Pharmacol Biochem Behav* 1989;33(4):919-22.

[207] Wang, D., Noda, Y., Zhou, Y., Nitta, A., Furukawa, H., Nabeshima, T. Synergistic effect of combined treatment with risperidone and galantamine on phencyclidine-induced impairment of latent visuospatial learning and memory: Role of nAChR activation-dependent increase of dopamine D1 receptor-mediated neurotransmission. *Neuropharmacology* 2007;53(3):379-89.

[208] Shoaib, M., Schindler, C. W., Goldberg, S. R., Pauly, J. R. Behavioural and biochemical adaptations to nicotine in rats: influence of MK801, an NMDA receptor antagonist. *Psychopharmacology (Berl)* 1997;134(2):121-30.

[209] Rezvani, A. H., Levin, E. D. Nicotinic-glutamatergic interactions and attentional performance on an operant visual signal detection task in female rats. *Eur J Pharmacol* 2003;465(1-2):83-90.

[210] Levin, E. D., Bettegowda, C., Weaver, T., Christopher, N. C. Nicotine-dizocilpine interactions and working and reference memory performance of rats in the radial-arm maze. *Pharmacol. Biochem. Behav.* 1998;61(3):335-40.

[211] Aizenman, E., Tang, L. H., Reynolds, I. J. Effects of nicotinic agonists on the NMDA receptor. *Brain Res.* 1991;551(1-2):355-7.

[212] Court, J. A., Piggott, M. A., Perry, E. K. Nicotine reduces the binding of [3H]MK-801 to brain membranes, but not via the stimulation of high-affinity nicotinic receptors. *Brain Res.* 1990;524(2):319-21.

[213] Levin, E. D., Tizabi, Y., Rezvani, A. H., Caldwell, D. P., Petro, A., Getachew, B. Chronic nicotine and dizocilpine effects on regionally specific nicotinic and NMDA glutamate receptor binding. *Brain Res.* 2005;1041(2):132-42.

[214] Vige, X., Briley, M. Scopolamine induces up-regulation of nicotinic receptors in intact brain but not in nucleus basalis lesioned rats. *Neurosci. Lett.* 1988;88(3):319-24.

[215] Leblond, L., Beaufort, C., Delerue, F., Durkin, T. P. Differential roles for nicotinic and muscarinic cholinergic receptors in sustained visuo-spatial attention? A study using a 5-arm maze protocol in mice. *Behav Brain. Res.* 2002;128(1):91-102.

[216] Brown, E. N., Galligan, J. J. Muscarinic receptors couple to modulation of nicotinic ACh receptor desensitization in myenteric neurons. *Am. J. Physiol. Gastrointest. Liver Physiol.* 2003;285(1):G37-44.

[217] Erskine, F. F., Ellis, J. R., Ellis, K. A., Stuber, E., Hogan, K., Miller, V., et al. Evidence for synergistic modulation of early information processing by nicotinic and muscarinic receptors in humans. *Hum. Psychopharmacol.* 2004;19(7):503-9.

[218] Ellis, J. R., Ellis, K. A., Bartholomeusz, C. F., Harrison, B. J., Wesnes, K. A., Erskine, F. F., et al. Muscarinic and nicotinic receptors synergistically modulate working memory and attention in humans. *Int. J. Neuropsychopharmacol.* 2006;9(2):175-89.

In: Encyclopedia of Neuroscience Research
Editors: Eileen J. Sampson and Donald R. Glevins
ISBN 978-1-61324-861-4

Chapter III

Participation of the Prefrontal Cortex in the Processing of Sexual and Maternal Incentives

Marisela Hernández González*[*] *and Miguel Angel Guevara
Instituto de Neurociencias, Universidad de Guadalajara, México
Francisco de Quevedo 180, Col. Arcos-Vallarta
Guadalajara, Jalisco, México

Abstract

Sexual and maternal behaviors constitute motivated behaviors, the adequate manifestation of which requires an appropriate integration of external stimuli and internal states. In both is required the perception of an appropriate stimulus. For a male rat, one appropriate sexual incentive stimulus could be a receptive female, while for a female mother rat the appropriate maternal incentive could be a pup rat. The sensory clues emitted by either the potential sexual partner or the pups must be first detected and then processed in order to be perceived as incentives and as potentially rewarding. The prefrontal cortex (PFC) plays an important role both in mediating the assignation of the incentive value and in information processing. Thus, using electroencephalographic (EEG) recordings, we have characterized the prefrontal functioning during motivational states of sexual and maternal behavior in rats. In a first study, it was observed that the EEG activity of the medial (mPFC) and orbital prefrontal cortices (oPFC) was modified during the performance of male rats in a T-maze under two different conditions: sexually-motivated (with previous intromission and one receptive and one non-receptive female placed in the goal boxes of the lateral arms); and not sexually-motivated (no previous intromission and empty goal boxes). Only the sexually-motivated males showed a higher proportion in the 6-7 Hz band and a lower proportion in the 8-11 Hz band during the awake-quiet state and while walking in the maze stem. In addition, they showed an increased correlation of the 6-7 Hz band while walking in the maze stem and when remaining close to a receptive female. Similar EEG changes were obtained in estrous female rats that were subjected to a sexually-motivated task using the same paradigm of

[*] Tel: (52) 33 38 18 07 40 ext 5860; E-mails: mariselh@cencar.udg.mx; mguevara@cencar.udg.mx

the T-maze. In the oPFC of the sexually-motivated females and males, different EEG patterns were observed, results that suggest a different functionality of the mPFC and oPFC, which may work together in the processing of sexual incentive stimuli. Another study explored whether the mPFC of female rats shows characteristic EEG patterns during the presentation of olfactory stimuli, when (1) associated with pups (nest-bedding); and, (2) not-associated, with pups (female bedding), during the following distinct reproductive states: proestrus-estrus (P-E), diestrus (D) and lactation (L). During the smelling of nest-bedding, only the L rats showed a higher proportion in the 8-11 Hz band. This finding may represent a characteristic processing of the olfactory stimuli related to pups in the mPFC of L rats. Taken together, these data allow suggest that in rats the prefrontal cortex participates in identifying and processing sexual and maternal incentives.

Introduction

In studying the physiology of sexual and maternal behaviors it is important to gain insight into the early phases of sexual and maternal relations among animals. These phases include detection and perception of the stimuli, gender identification, motivation and arousal. Agmo (1999) has defined the term sexual motivation as the process that makes an animal search for sexual contact with another animal. It is well known that sexual motivation is activated when a suitable stimulus –e.g., a mate– is perceived. A similar definition can be applied to the maternal motivation, as the process that makes a mother search for maternal contact with a pup (her own or alien); *i.e.*, the process that promotes nurturing behaviors toward an offspring. This is activated when the suitable stimulus –e.g., a pup– is perceived. Given that the suitable stimulus generally activates approach behavior, it can be suggested that these stimuli may function as a positive incentive.

Although different neural structures have been implicated in the modulation of sexual and maternal performance, to date there are very few studies that have described which brain areas are involved in the search for sexual and maternal contact, including the processing of the sensory stimuli that are emitted from the potential partner or pup and that are fundamental in generating sexual and maternal motivation. The processing of these different stimuli with sexual and maternal meaning involves the activation of different neural systems, as well as integration with hormonal cues. In recent years, substantial evidence has been accumulated indicating the critical role of the PFC in mediating the assignation of the incentive value of the stimuli and it has been suggested that it is in this cortical region that the sensory inputs generated by distant stimuli from the potential sexual partner or pups are processed. Several techniques, such as lesions and electrical or chemical stimulation, as well as hormonal treatments, were the first ones used in attempts to understand the role of the prefrontal cortex in the motivational aspects of these behaviors. Today, more sophisticated techniques have been developed: the expression of immediate early genes, measurements of the turnover or release of neurotransmitters and descriptions of hemodynamic reactions in response to sexually-relevant stimuli have all provided relevant information concerning the role of the prefrontal cortex in the processing of such stimuli. The spontaneous brain electrical activity or electroencephalographic recording technique (EEG) has proven to be a very useful tool for studying brain functioning during specific behaviors and cognitive processes. One of its main advantages is that the EEG has an excellent temporal resolution that makes it possible to obtain recordings of brain electrical activity from milliseconds to hours, days and even months. With this in mind, the aim of the present chapter is: 1) to review the subdivisions and functions of the rat prefrontal cortex; 2) to define sexual and maternal behaviors as motivated behaviors; 3) to clarify the important role of the incentive stimuli (potential sexual partners or pups) in inducing the corresponding motivational states; and, 4) to demonstrate, by means of EEG recordings, the participation of the prefrontal cortex in the processing of incentive stimuli in relation to sexual and maternal behaviors in rats. This chapter compiles several electroencephalographic studies in which the electrical functionality of different prefrontal subregions was characterized in relation to the approach behaviors (activated by the incentive), which has been considered as a behavioral index of sexual or maternal motivation in the rat.

The Rat Prefrontal Cortex, Subdivisions and Functions

The prefrontal cortex (PFC) can be defined as the cortical region where projections from the mediodorsal nucleus of the thalamus (MD) are received (Rose and Woolsey, 1948; Uylings and Van Eden, 1990). It has been shown that the PFC directly projects to the cholinergic forebrain system and to the monoaminergic cell group in the hypothalamus and brainstem. In this way, the PFC is in a unique position to influence the transmitter systems that modulate large parts of the forebrain. Further, it has become clear that the PFC has a special relationship with the basal ganglia; not only does it project via the cortico-striatal system to the basal ganglia, but it also receives, via the thalamus, projections from these structures, an arrangement that is unique to the frontal lobe. In addition to the anterior connections, the PFC has larger and reciprocal connections with other cortical areas, so that it receives inputs predominantly from somatosensory, visual and auditory cortical association areas in the parietal, occipital and temporal lobes (for a review, see Groenewegen and Uylings, 2000). In this way, the PFC is involved in a number of largely parallel, functionally-segregated cortical-subcortical circuits that subserve sensorimotor, cognitive, emotional/motivational behaviors and visceral functions, so that the PFC is thought to be primarily involved in the cognitive and motivational/emotional processes that underlie and guide complex behavior (Passingham, 1993). Due to its special position in the circuitry of the forebrain, the PFC has been implicated in the high processing of all sensory modalities. Given that the PFC is also provided with visceral sensory input, it seems likely that it plays a role in evaluating the internal and external milieu and, based on this information, in selecting the behavioral strategy and modulating the repertoire of the corresponding autonomic adaptation accordingly.

It has been suggested that many functions are mediated by the prefrontal cortex, including the mediation of working memory for a specific domain or for attributing information, especially that involving spatial and visual object information, and also higher order relationships, such as motivation, learning and remembering complex tasks, the temporal order of events, selecting rules and strategies, and planning, monitoring and modifying sequential behavior (Shimamura et al., 1990; Goldman-Rakic, 1996; Fuster, 1997). These varieties of functions are represented in specific subregions of the PFC, so that it has been suggested that it has a functional heterogeneity based mainly on the divisions of the mediodorsal nucleus of the thalamus (MD). The concept of a functional heterogeneity in the PFC was first postulated in the 1960s (see, for example, Mishkin, 1964), when was suggested that the prefrontal subregions are functionally different. Three main regions have been determined: first, the magnocellular, medial part of the MD projects to the orbital (ventral) surface of the prefrontal cortex (which includes areas 13 and 12). This part of the PFC is called the orbitofrontal cortex, and receives information from the ventral or object-processing visual stream as well as taste, olfactory and somatosensory inputs. Second, the parvocellular, lateral part of the MD projects to the dorsolateral PFC. This part of the PFC receives inputs from the parietal cortex and is involved in tasks such as short-term spatial memory operations (Fuster, 1997; Rolls and Treves, 1998). Third, the pars paralamellaris (most lateral) part of the MD projects to the frontal eye field (area 8) in the anterior bank of the arcuate sulcus. The degree of development of these prefrontal subregions varies among species; for example, the orbitofrontal cortex is well developed in primates, including humans, but poorly developed in rodents.

It is evident from anatomical studies that the prefrontal cortex of the rat is also made up of several subregions that are functionally distinct.

In fact, it is clear from lesion studies that there are at least three distinct prefrontal regions, which include: (1) a medial prefrontal region (mPFC) made up of several subregions that are probably functionally distinct (Cg1, Cg2, Cg3 or prelimbic and infralimbic areas); (2) an orbital prefrontal region (oPFC) that is also made up of several subregions that are likely functionally distinct (ventral, lateral, ventrolateral, and medial orbital areas, and the agranular insular areas); and, (3) a region that is likely analogous to the frontal eye fields of primates (Fr2) (Kolb, 1990; Uylings and Van Eden, 1990) (Figure 1).

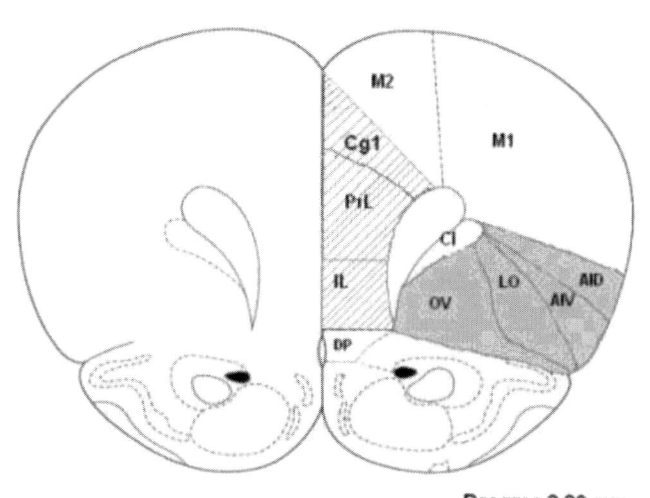

Figure 1. Schematic representation of the rat prefrontal subregions. M1, primary motor cortex; M2 secondary motor cortex; Cg1, cingulated cortex area; PrL, prelimbic cortex; IL, infralimbic cortex; OV, ventral orbital cortex; LO, lateral orbital cortex; AIV, agranular insular cortex ventral; AID, agranular insular cortex dorsal; Cl, claustrum; DP, dorsal peduncular cortex.With shadow is indicated the medial prefrontal subregion, whereas in gray color is indicated the orbital prefrontal subregion. Anterior-posterior coordinates are given with respect to Bregma (Paxinos and Watson, stereotaxic atlas, 1997).

The mPFC is involved in various functions, such as the temporal ordering of sequential events, mnemonic tasks and motive-emotional processes, while the oPFC is mainly involved in the processing of higher levels of somatosensory, visual, auditory, olfactory and gustatory sensory information, participating actively in the performance of high-order olfactory processing. Moreover, it plays a pivotal role in goal-directed behaviors and in guiding behaviors based on reward (for reviews, see Kolb, 1990; Uylings and Van Eden, 1990; Yonemori et al., 2000; Uylings et al., 2003).

The participation of the PFC in reward processing has been demonstrated in various studies. The PFC is part of the mesocorticolimbic dopaminergic system and receives prominent dopaminergic input from the VTA and inputs from other subcortical basal ganglia structures via the mediodorsal thalamus. In turn, it projects back to the VTA and the nucleus accumbens septi (Acc), which are generally considered to be the main components of the brain reward system. Evidence for the involvement of the mPFC in the reward-related mechanism comes primarily from three types of studies: conditioned place preference, intracranial self-stimulation, and self-administration of drugs (Tzschentke, 2000), so that available data strongly suggests that the mPFC forms part of the brain reward circuitry.

Various studies have demonstrated that both sexual (Sheffield et al., 1951, 1954; Denniston, 1954; Kagan, 1955; Beach and Jordan, 1956; Whalen 1961; Whalen et al., 1961; Jowaisas et al., 1971; Hetta and Meyerson, 1978; Everitt and Stacey, 1987; Agmo, 1999; López and Ettenberg, 2001) and maternal behaviors (Fleming and Luebke, 1981; Fleming, 1986; Fleming et al., 1989, 1994; Stern, 1996) are intrinsically rewarding or reinforcing motivated behaviors. Their execution is associated with the activation of several brain regions in the reward circuitry.

The Motivated Behaviors

Biological evolution is characterized by an increase in the integrative capacity of the nervous system accompanied by the expression of more complex behavior patterns. At the same time, however, the basic requirements for life must be met. Thus, during the course of evolution a variety of digestive, respiratory and cardiovascular mechanisms must be coordinated with behavioral responses that assure biological adaptation and survival. The neural integration of visceral responses, which are modulated by the autonomic nervous system and the endocrine system, tends to maintain a relatively constant internal environment, while adaptive behavioral responses involve the somatomotor control system, sensory mechanisms and also sensory-motor interaction. Behavioral responses can be triggered by both internal and external signals and, because they contribute to homeostasis and survival, this is also the case of the primary motivated behaviors: eating, drinking and thermo-regulation. However, there are other motivated behaviors that are also very important for the reproduction and survival of the species though they do not result from a deficit in the organism but, rather, mainly from an adequate hormonal medium and the perception of external stimuli that turn out to be relevant for them. This is the case of sexual and maternal behaviors, which have been considered as secondary motivated behaviors. Sexual behavior in both sexes comprises a complex series of motoric events that ensure the successful transfer of sperm from the male to the female's reproductive tract, whereas maternal behavior comprises another complex series of motoric events designed to ensure the growth and survival of the offspring (one's own or alien young). In both cases of motivated behaviors the nervous system must detect and process a series of sensory stimuli, a task in which, as has been clearly shown, the prefrontal cortex plays a pivotal role.

Before examining sexual and maternal behaviors in detail, however, it will be useful to outline the sequential stages that are common to a variety of adaptive or goal-oriented behaviors, in order to place the problem of the underlying neural mechanism in perspective. For descriptive purposes, such behavioral responses can be divided into initiation, procurement and consummatory phases. Current evidence suggests that the initiation phase may involve: 1) the detection of peripheral and central signals; 2) sensory information from the external environment; and, 3) cognitive factors. The procurement phase is characterized by a general state of behavioral arousal and by searching behavior, which involves locomotor activity, sensory information and previous experience. The final, consummatory, phase relies to a considerable degree on preprogrammed motor responses, that are stereotyped and species-typical (Swanson and Mogenson, 1981). Another useful classification of goal-oriented behaviors that has been used for many years, is one in which motivated behaviors are divided into appetitive and consummatory components (Everitt, 1990; Pryce et al., 1993; Timberlake and Silva, 1995). The appetitive component is made up of those behaviors that bring the organism into contact with an attractive or desired stimulus object or goal, while the consummatory component is composed of the ones that are performed once the goal is achieved. Wallen (1990) has made an analogous distinction by referring to these categories as desire *vs.* ability: appetitive behavior reflects an underlying motivation, appetite, drive or desire to engage in a behavioral interaction with a specific goal or object, while consummatory behavior reflects the ability to perform specific behavioral responses once the goal object is attained.

Thus, the term motivation can refer to a variety of neuronal and physiological factors that initiate, sustain and direct behavior. In the case of sexual and maternal behaviors, voluntary movements are initiated to approach a specific goal (a potential sexual partner or a pup), so that the motivated state impels the individual to perform sexual and/or maternal acts. The sexual and maternal activities of the rat are naturally occurring behavior patterns with a well-organized behavioral sequence and a specific temporal and spatial order, where voluntary goal-directed movements, automatic stereotyped responses and consummatory acts can all be distinguished. The prefrontal cortex is involved in the motivational processing of relevant goals and various studies have demonstrated that lesions in this cortical area alter the sequential organization, sensory processing and mnemonic aspects of motivated behaviors (Kolb, 1984; 1990). The medial prefrontal cortex receives highly processed sensory information from visual, auditory and somatosensory regions (Groenewegen and Uylings, 2000) and, like the ventral tegmental area (VTA), has been implicated in integrating motor action with sensory information (Mogenson et al., 1980; Kolb and Cioe, 2004).

Traditionally, the occurrence of the appetitive component of a species-typical behavior has been taken as evidence of increased motivation: that is, if an organism acts on its environment to obtain a particular stimulus (pups, mate, food), then that stimulus is attractive and thus reflects an underlying increase in motivation.

There are two classes of external stimuli or events: hedonically-potent and hedonically-neutral. The first may be further divided into hedonically-positive and hedonically-negative stimuli or events. Hedonically-positive stimuli or events are associated with approach behavior and may be called positive incentives. Hedonically-negative stimuli –*i.e.*, those associated with withdrawal or escape behaviors– can be called negative incentives. Based on this classification, Agmo (1999) has proposed that the motivation aroused by stimuli with rewarding properties be called incentive motivation. Under this proposition, an appropriate sexual incentive for a male rat would be a sexually-receptive female, for a female receptive rat a sexually active-male, and for a female mother rat, a pup rat. As these suitable stimuli activate approach behaviors it is possible to consider that they may function as positive incentives.

One of the main animal models that have been used to study the neurophysiological and hormonal bases of sexual and maternal behaviors is the rat. Recent research has shown that the rat serves as an excellent animal model of PFC functions (Kesner, 2000; Brown and Bowman, 2002; Uylings et al., 2003), and several studies have demonstrated the differential functioning of the prefrontal subregions in this animal. Taking into account these considerations, the next section discusses a series of experiments that evaluated the possible participation of the PFC in the motivational and performance aspects of sexual and maternal behaviors. However, before examining these different experiments, it is necessary to explain the appetitive and consummatory components of each one of these motivated behaviors in the rat.

Male Rat Sexual Behavior

Two processes can be distinguished in the sexual behavior of the male rat: first, a series of arbitrary responses that lead to the male approaching the female (orientation, smelling, persecution of the female and latency to initiate mating), that together constitute the appetitive component; and, second, a subsequent series of stereotyped motor activities that make up the act of copulation itself (mounting, intromission and ejaculation responses) and constitute the consummatory component of the male rat's sexual behavior (Agmo, 1999; 2002; Pfaff and Agmo, 2002).

Sexual behavior occurs spontaneously when specific external sensory stimuli are present (those emitted by a potential sexual partner); however, to detect a stimuli as sexually attractive an adequate internal state of the subject is necessary (*i.e.*, adequate levels of circulating hormones and neurotransmitters, adequate neural functionality, etc.), that makes it more alert to incentive stimuli from the potential partner. However, simple detection of the sensory stimuli is not sufficient. Also required is an adequate processing of the stimuli that contain sexual meaning that permits the assignation of the incentive value as well as its perception as rewarding. It has been suggested that the processing that entails acquiring the incentive value of the stimuli is mediated by limbic-prefrontal circuits, including mainly the reciprocal projections between the oPFC and the amygdala (Armony et al., 1997; Balkenius and Morén, 2001; Hall et al., 2001). Moreover, Rolls and Treves (1998) have proposed that the oPFC may contribute to the acquisition of the reward value of the stimulus through its connections with the amygdala.

The force that drives the subject to search for sexual interaction with a mate –*i.e.*, the appetitive component or sexual motivation– has been measured under different experimental paradigms. Several studies have confirmed the fact that sex is an efficient incentive (Sheffield et al., 1951, 1954; Denniston, 1954; Kagan, 1955; López and Ettenberg, 2001) and have used sex as a reward in classical operant tasks such as mazes and Skinner boxes (Jowaisas et al., 1971; Everitt and Stacey, 1987). Both, Beach and Jordan (1956) and Whalen (1961) have shown that male rats readily learn to run through a runway when access to a receptive female is used as an incentive, and access to a female can also be used to reinforce other responses; e.g., bar-pressing (Jowaisas et al., 1971; Everitt and Stacey, 1987; Everitt et al., 1987;), or the search for proximity to a female (Hetta and Meyerson, 1978). However, it has also been shown that it is not necessary for the male to work or to perform an operant task in order to gain access to a female and hence demonstrate his sexual motivation. Agmo (2002) has proposed that an adequate and easy measure of the sexual incentive motivation could be approach behavior and the time that a male remains near an inaccessible receptive female. Using this paradigm, the effectiveness of the stimuli emitted by a receptive female in terms of functioning as a positive incentive for male rats has been demonstrated.

Female Rat Sexual Behavior

The female rat's sexual behavior comprises three basic elements: attractivity, proceptivity and receptivity. Attractivity can be defined as a "female's value as a sexual incentive stimulus" and includes both behavioral and non-behavioral components, such as olfactory cues that stimulate the male to engage in sexual behavior with her. Proceptivity includes a number of solicitational behaviors that are followed by increases in the number of sexual mounts by males. These behaviors are exhibited spontaneously by estrous female rats during the normal course of mating and include hopping, darting and ear-wiggling (Beach, 1976; Madlafousek and Hlinak, 1977), a posing or presenting posture (Emery and Moss, 1984a,b), a rapid sequence of approaches toward, orientation to, and withdrawal from, proximity to a sexually-active male (McClintock and Adler, 1978), and the emitting of ultrasonic vocalizations known to stimulate mating by males (White and Barfield, 1989). These behaviors reflect in some way the appetitive aspect of sexual behavior and implied a motivation to mate. Receptivity is indicated by the reflex response of the female to sexual contact with the male and consists in adopting a posture that positions her genitalia to allow penile intromission by the male, dorsiflexion of the spine, elevation of the head, deviation of the tail and extension of the rear legs.

The several solicitational behaviors displayed by the female have been used to ascertain an overall level of sexual readiness as an estimate of the female's motivation to mate, and are considered as appetitive components. On the other hand, the receptivity or lordosis response that results from sexual contact with the male constitutes the consummatory component of the female's sexual behavior.

As in the male's sexual interaction, other behaviors expressed under several experimental conditions have furnished measures of female sexual motivation. Among the different paradigms that have been used to measure female sexual motivation we can mention the following: sexual preference and proximity (Meyerson and Lindstrom, 1973; Clark et al., 1981; Edward and Pfeifle, 1983; Broekman et al., 1988); the female´s willingness to cross an electrified grid to approach a sexually-active male (Meyerson and Lindstrom, 1973); operant tasks such as bar pressing by males (Bermant, 1961; French et al., 1972) and conditioned place preference (Oldenburger et al., 1992; Paredes and Alonso, 1997; Matthews et al., 1997).

Maternal Behavior in the Rat

Parental behavior can be defined as any behavior by a member of a species toward a reproductively-immature conspecific that increases the probability that the recipient will survive to maturity. Since in mammals it is the female that lactates, the mother is the primary caregiver of her offspring; we refer to such behavior as maternal.

Rodents offer good examples of maternal care patterns in species with altricial young (immobile at birth). Since most rodent young are helpless, essentially immobile and incapable of temperature regulation at birth, they are kept in a secluded nest that the mother builds prior to parturition, which also serves to insulate the young in the mother's absence. When the female is in the nest area, she crouches over the young in order to warm and nurse them. The female also licks her young, particularly in their anogenital region, to stimulate urination and defecation. If the nest site is disturbed, or if pups are displaced from it, the rodent mother will engage in transport or retrieval behavior in which she carries pups (one at a time) in her mouth to a new nest site or back to the original one. Another aspect of maternal care in rodents is the occurrence of maternal aggression toward intruders at the nest site, presumably to protect the young. In describing maternal care in rodents, one can therefore refer to pup-directed patterns (retrieving, licking, nursing) and non-pup-directed patterns (nest-building, maternal aggression), all of which contribute to the successful rearing of the offspring. Thus, those behaviors that a female performs in order to gain access to her pups have been classified as appetitive (pup retrieval, pup licking), while nursing (low and high crouching) behaviors are considered as consummatory responses.

Like male and female sexual behaviors, maternal behavior also has an implicit and strong rewarding or reinforcing component. For example, it has been shown that postpartum females and hormone-induced maternal virgins are much more likely than virgin females to retrieve pups under stressful conditions, such as in a T-maze (Stern and Mackinnon., 1976), open-field or in elevated plus maze (Fleming and Luebke, 1981; Bitran et al., 1991). Therefore, a marked reduction in the duration of the freezing reaction in lactating females was found when compared to virgin females who had not been exposed to pups (Hard and Hansen, 1985; Hansen and Ferreira, 1986). Ferreira et al. (1989) examined the punished drinking response in virgin and postpartum females using a test in which licking from a water spout is punished by electric shocks. It was found that lactating females drank more than virgins. Various paradigms, such as place preference (Fleming et al., 1994; Magnusson and Fleming, 1995; Mattson et al., 2001) and pressing a bar in an operant chamber at a high rate in order to obtain access to rat pups as a reward (Lee et al., 2000) have clearly shown that pups can serve as strong reinforcers for postpartum females.

Neural and Hormonal Regulation of Sexual and Maternal Behaviors

It is quite evident that hormones need to act on specific neurons somewhere in the brain or the peripheral nervous system in order to exercise their stimulatory effect on the structures involved in the motivational and execution aspects of sexual and maternal behaviors. The brain needs to be exposed to some hormones in order to react to internal and external stimuli with sexual and/or maternal responses. Thus, it is well known that the testicular and ovarian hormones are necessary for sexual and maternal behaviors in all mammals, including humans. Castration (Bremer, 1958; Rossenblat and Aronson, 1958; Davidson, 1966; Heim and Hursch, 1979; Larsson, 1979; Willie and Beler, 1989), ovariectomy (Boling and Blandau, 1939; Dempsey et al., 1939; Robinson, 1954) and other means of reducing testicular and ovarian hormone concentrations (Bradford and Pawlack, 1993; Reilly et al., 2000) always lead to a reduced motivation or intensity of sexual behavior and altered maternal behavior, at least immediately after parturition (for a review, see Hull et al., 2002; Blaustein and Erskine, 2002; González-Mariscal and Poindron, 2002).

Hormones have no effect unless they are bound to the appropriate receptor. The gonadal hormone receptors are widely distributed in the brain. Among the sites expressing androgen receptors are the medial preoptic area, the medial amygdale, the ventromedial nucleus of the hypothalamus, the bed nucleus of the stria terminalis and the prefrontal cortex. Estrogen receptors are found in the medial preoptic area, the ventromedial nucleus of the hypothalamus, the olfactory bulbs and many other sites (for a review, see Genazzani et al., 1996).

Though the neural structures involved in the regulation of sexual and maternal behaviors are similar, there are certain differences. In the case of *male sexual behavior*, there is evidence of the participation of the olfactory bulb, the amygdala, the bed nucleus of the stria terminalis, the medial preoptic area, several nuclei of the hypothalamus (paraventricular, ventromedial and lateral), the midbrain periaqueductal gray, the nucleus paragigantocellularis of the medulla, among others, with the medial preoptic area playing an essential role in the motivational and performance aspects of that behavior. With respect to *feminine sexual behavior*, the ventromedial nucleus of the hypothalamus, the medial preoptic area, the amygdala, the bed nucleus of the stria terminalis, the ventral tegmental area and the midbrain central gray, among others, have been implicated, with the ventromedial nucleus of the hypothalamus playing a crucial role. In terms of *maternal behavior,* a traditional view has been that in non-primate mammals such behavior is hormonally-regulated by the endocrine events of late pregnancy, parturition and lactation, while the maternal behavior of primates is emancipated from endocrine control and instead is influenced by a variety of social experiences (see Numan and Insel, 2003). In general terms, structures such as the cortex, the trigeminal complex and olfactory circuit, the septum and the bed nucleus of the stria terminalis, the medial preoptic area, the paraventricular nucleus and the habenular complex have all been implicated in the modulation of maternal behavior. In an insightful review, Newman (1999; 2002) has emphasized the large amount of overlap in the neural circuits that regulate a variety of social behavior in mammals. These social behaviors included male and female sexual behavior, offensive aggression and parental behavior, and the neural regions she referred to included the medial preoptic area/bed nucleus of the stria terminalis, the medial amygdala, and the lateral septum. This author speculates that there may be a common central network that regulates a variety of social behaviors, and that the precise behavior that occurs may be determined by the stimuli that have access to this social circuit. In this chapter, we will not describe the specific hormonal and neural regulation of each one of these motivated behaviors, we should not lose sight of the important role that hormones play in the neural substrates involved in the motivational and execution aspects of these reproductive behaviors.

In Which Brain Structures is the Incentive Value of Sexual and Maternal Stimuli Processed?

Although different neural structures have been implicated in the modulation of sexual and maternal performance, there are as yet very few studies that describe which brain areas are involved in the search for sexual and maternal contacts, including the processing of the sensory stimuli emitted by a potential partner or pup, which is fundamental in generating sexual and maternal motivation. The processing of these different stimuli with sexual and maternal meanings involves the activation of different neural systems, as well as integration with hormonal cues.

In males, among the few structures that have been proposed as participants in the processing of incentive stimuli is the medial preoptic area. Studies by Everitt (Everitt and Stacey, 1987; Everitt, 1999) after preoptic lesions showed that while males ceased to copulate with females, they still performed operant responses to gain access to them. Similarly, the pursuit of receptive females, or approaches to inaccessible ones, also ceased (Paredes et al, 1993; Hurtazo et al., 2003), as did the males' interest in odors from receptive females, all of which disappeared after preoptic lesions in male rats and ferrets (Paredes and Baum, 1995; Paredes et al., 1998). Taken together, these data show that reactions to sexual incentives are reduced or abolished by preoptic lesions, thus supporting the idea that this cerebral area is crucial for sexual motivation.

Other brain structures, such as the medial amygdala (ME), have been proposed as one important site for integrating chemosensory and hormonal cues (Wood, 1998). A specific distribution of their functioning has been described, in which the posteromedial amygdalar region (MEpd) is more involved in processing stimuli with sexual meaning (Kollack-Walker and Neuman, 1992, 1995). For example, Maras and Petrulis (2006) demonstrated that MEpd-lesioned male hamsters showed deficits in opposite-sex odor preference that appear to be a result of their decreased attraction to female odors.

Another brain structure that has been involved in the processing of sexual incentive stimuli in males is the prefrontal cortex. For example, Bunnell et al. (1996) showed that the ablation of the cingulated and retrosplenial cortices in hamsters caused an increase in the threshold for reaching sexual arousal. Similarly, Agmo (1995) demonstrated that extensive ablations of the medial and dorsal prefrontal cortex in male rats significantly increased (up to two hours or more) the onset of sexual interaction, despite continued exposure to a receptive female.

In females, the preoptic area is also important in the capacity of distant sexual incentive stimuli to activate approach behaviors. Lesions in this area reduce the intensity of approaches both to an inaccessible male and to a male in a situation where copulatory interactions were possible (Guarraci and Clark, 2006). Moreover, the sexual incentive properties of the odor of a male are abolished by radiofrequency lesions of the medial preoptic area (Xiao et al., 2005). These interesting data show that, in female rats, this area is crucial for the distant sexual incentive properties of a male. Another structure that has been shown to participate in female proceptive and receptive behaviors is the ventromedial nucleus of the hypothalamus (Pfaff and Sakuma, 1979; Floody, 2002); and it has been suggested that it is the combined action of the preoptic area and the ventromedial nucleus that controls all aspects of female sexual behavior, from the approach to a mate to the execution of copulatory reflexes (Agmo, 2007).

Moncho-Bogani et al. (2005) demonstrated that attraction to sexual pheromones and associated odorants in female mice involves the activation of cerebral structures such as the basolateral amygdala. They studied the chemoinvestigatory behavior of female mice towards volatile and non-volatile chemicals contained in male-soiled bedding, in combination with analyses of c-fos expressions induced by such behavior. The cerebral structures were differentially activated by the primary pheromones and secondarily by attractive odorants: the primary attractive pheromone activates the basolateral amygdala and the shell of the nucleus accumbens, but neither the ventral tegmental area nor the orbitofrontal cortex. In contrast, the secondarily attractive male-derived odorants involve activation of a circuit that includes the basolateral amygdala, the prefrontal cortex and the ventral tegmental area.

In maternal behavior: as in male and female sexual behavior, it has been recognized that the medial preoptic area plays a crucial role in the appetitive aspects of maternal behavior and, hence, in the detection and processing of the incentive stimuli emitted by pups. Electrolytic or radiofrequency lesions of this region disrupt retrieving, nest-building and nursing in lactating rats (Numan, 1974; Numan et al, 1977; Jacobson et al., 1980; Gray and Brooks, 1984). Moreover, virgin females (ovariectomized or intact) lesioned in the MPOA do not show maternal behavior, even after many days of exposure to pups (Numan et al., 1977; Miceli et al., 1983; Gray and Brooks, 1984). Therefore, by applying "kindling"-type electrical stimulation to the MPOA, maternal responsiveness to foster pups is promoted in both experienced mothers and virgins (Morgan et al., 1999); so these findings and others (Lee et al., 2000) support the idea that the activity of this brain structure is critical not only for the performance of maternal behavior but also in mediating the reinforcing properties of young pups in postpartum and "sensitized" virgin rats.

Participation of the Prefrontal Cortex in the Processing of Incentive Stimuli with Sexual or Maternal Meaning

The first studies designed to investigate the neural control of sexual behavior in male mammals used lesion techniques, mainly the gross removal of large areas of the brain. Using these techniques, after removal of the neocortex it became clear that the greatest deficits in the copulatory interaction of male rats (Beach, 1940; Larsson, 1962, 1964) and cats (Beach et al., 1955, Beach and Jordan, 1956; Zitrin et al., 1956) were associated with the removal of the frontal cortex. Bunnell et al. (1966) showed in hamsters with ablation of the cingulated and retrosplenial cortices, that the threshold to reach sexual arousal increased; and similar results were reported in rats by Soulairac and Soulairac (1972). These were the first attempts to investigate the possible role of the frontal cortex in sexual behavior. Since then, other studies have been performed to determine the exact role that this cortical region plays in sexual behavior. Studies such as those by Lubar et al. (1973), Michal (1973) and Kolb (1984, 1990) have demonstrated that lesions in this cortical area alter the sequential organization, sensory processing and memory of motivated behaviors, such as sexual interaction. Moreover, prefrontal lesions in male rats increase intromission and ejaculation latencies as well as the duration of the postejaculatory interval, thus rendering them less likely to mate successfully (Fernández-Guasti et al., 1994). As mentioned above, it was not until 1995 that Agmo and collaborators made an interesting study in which extensive ablations of the medial and dorsal prefrontal cortex increased the time required to begin copulatory performance (two hours or more) in males that were in continual exposure to a receptive female. However, as soon as copulation began the sexual behavior of the lesioned animals became completely normal (Agmo et al., 1995). The authors suggested that these data indicated a difficulty for males to identify the female as a sexual incentive.

In the case of incentive stimuli with maternal meaning, there are also several studies that have demonstrated the activation of prefrontal regions in relation to the processing of stimuli emitted by pups in dams. For example, some have described a high activation of the mPFC in human mothers who listen to an infant crying (Lorberbaum et al., 2002), a selective activation of this brain area in dam rats during their first experience with pups (Walsh et al., 1996), and an enhanced c-fos expression in the mPFC of ewes when exposed to lambs (Kendrick et al., 1997). Similar increases in the c-fos mRNA expression have been demonstrated in the mPFC of ewes exposed to odors from their own or strange lambs, although the inactivation of this cortical area only prevents ewes from responding with motor aggression to odor cues from strange lambs and has no effect on the formation of an olfactory recognition memory for their own offspring (Broad et al., 2002). Similarly, an enhanced metabolic activation of the mPFC in response to the presentation of emotionally-relevant maternal nursing calls has also been shown in *Octodon degus* pups (Braun and Poeggel, 2001).

In recent years, substantial evidence has been accumulated indicating the critical role of the PFC in mediating the assignation of the incentive value of the stimuli. For example, it has been shown that many prefrontal neurons encode a representation of olfactory stimuli that is dependent on their association with a reward (Rolls, et al., 2000) and that the medial PFC neurons show different responses to conditioned sensory stimuli (auditory or visual) when associated, or not, with a reward (Takenouchi et al., 1999; Jodo et al., 2000). In more specific studies, it has been suggested that this processing, which implies the acquisition of the incentive value of the stimuli, is mediated by limbic-prefrontal circuits, including mainly the reciprocal projections between the oPFC and the amygdala. These models propose that the oPFC may contribute to acquiring the reward value of the stimulus through its connections with the amygdala (Armony et al., 1997; Rolls and Treves, 1998; Balkenius and Morén, 2001; Hall et al., 2001), where the AMG increases the incentive value of a novel stimulus and the OFC confirms or –in the case of known stimuli– updates this value (Rolls 2000; 2004; Rolls and Treeves, 1998).

Functionality of the Prefrontal Cortex during the Processing of Incentive Stimuli: Electrophysiological Studies

A Brief Introduction to Electrophysiological Techniques

The first studies to investigate the different neuroanatomical areas involved in regulating sexual and maternal behaviors used lesion or electrical stimulation techniques. These techniques tended to be either relatively crude or too invasive. Today, sophisticated techniques have been developed that allow us to determine with greater precision the different cerebral areas that function during the reproductive behaviors that interest us and how they operate. Techniques such as determinations of the expression of immediate early genes and measurements of the turnover or release of neurotransmitters during exposure to incentive stimuli have been used in several studies. Similarly, descriptions of hemodynamic reactions in response to sexually-relevant stimuli have become popular and various imaging studies performed in humans have been published that use, for example, positron emission tomography (Stoléru et al., 1999; Leon-Carrion et al., 2007). Although modern techniques have been developed that allow researchers to obtain instantaneous images from different brain structures with high spatial definition, they have poor temporal definition. Despite its limitation in spatial definition, the electroencephalographic technique (EEG) has an excellent temporal resolution that allows us to obtain recordings of brain electrical activity from milliseconds to hours, days and even months. It is probably due to this advantage that the EEG is still used in numerous laboratories around the world.

The electroencephalogram (EEG) is defined as a mixture of rhythmic sinusoidal-like fluctuations in voltage generated by the brain that, it has been suggested, represent the global activity from the pyramidal cells of the cortex and the activity of neurons in the subcortical structures. Another more specific technique that has been used to study the functioning of brain structures is multiple unit activity (MUA), which is a general index of the net activity within a certain neuronal population. It represents a relatively large and mixed sample of local activity in which different kinds of neurons may be present. The net increase or decrease of MUA may reflect both the involvement of a given brain structure in the neuronal mechanisms of specific behaviors and the predominant response (increase or decrease of firing) of its neuronal components (Buchwald et al., 1973; Olds, 1973; Horio et al, 1986). Finally, the most detailed and specific electrophysiological technique involves the single unit activity recordings of neurons in a particular cerebral region that, it has been suggested, represents the action potentials fired by a certain neuron.

The quantitative analysis of EEG is a useful tool that has allowed researchers to relate changes in brain electrical activity to specific behavioral situations and/or experimental manipulations. Several studies in animals related to sexual and maternal behaviors have been published (Schwartz and Whalen, 1965; Kurtz and Adler, 1973; Holmes and Egan, 1973; Kurtz, 1975; Michael et al., 1977; Lincoln et al., 1980; Bennet et al., 1982; McIntosh et al., 1984; Mead and Vanderwolf, 1992). In contrast, very few studies have been done on humans to correlate the EEG of different cortical areas with the motivational and performance aspects of sexual and maternal behaviors (Mosovich and Tallaferro, 1954; Heath, 1972; Cohen et al., 1976, 1985; Tucker and Dawson, 1984; Graber et al., 1985; Rosen et al., 1986; Cervantes et al., 1992).

Taking into account that behavior results from the simultaneous and coordinated functioning of various cerebral structures, and considering that EEG recordings are a useful tool for investigating the simultaneous functioning of several brain structures in a precise temporal relation with the performance of specific behaviors, several of the studies conducted in our laboratory have been designed to characterize the functioning of the prefrontal cortex and other limbic structures during the motivational states and motor activities involved in the sexual and maternal behaviors of rats. Thus, we have developed different techniques of quantitative analysis to examine EEG recordings, and this has allowed us to obtain information that simple visual observation cannot (Hernández-González et al., 1977a, 1997b; Guevara et al., 2003; Guevara et al., 2005).

The methodology used in all such experiments was similar. Electrodes made of stainless-steel wire (0.2 mm in diameter), insulated with epoxy resin except for a small recording area left exposed at the tip, were implanted in specific cortical and subcortical structures of male or female rats, following the stereotaxic coordinates of Paxinos and Watson atlas (1997). The cable connecting the electrodes implanted in the rat was then connected to a slip-ring. This arrangement allowed the rats free movement and during this time the EEGs from the brain structures were recorded continuously. The cables were connected to the AC preamplifiers of a Grass 7B polygraph (band pass 3-30 Hz) and their outputs plugged into a PCL-812 analog-to-digital converter (Advantech, Co.), which operated as an interface to a microcomputer (PC-compatible). The EEG signals were recorded at a sampling rate of 256 Hz and calibrated with a pulse of 50 microvolts (μV), produced by the preamplifiers and delivered to a PC as a reference to convert the output of the analog-to-digital converter into μV. The capturing of the EEG signals that corresponded to each one of the different conditions was performed using a board unit with eight on/off buttons connected to the digital input lines of the PCL-812. Thus, capturing of 2-sec segments of EEG corresponding to the different behavioral situation began when a specific button on the board unit was pressed; while a different button was pushed to terminate signal input. Absolute power (AP) between 4 and 21 Hz was calculated by means of a Fast Fourier Transformation (FFT) and power densities were obtained for each Hz value. The raw EEG signals were digitally filtered according to the new boundaries using a rectangular window, and relative power (RP) (the proportional contribution of each band expressed as a percentage of total power between 4 and 21 Hz) was calculated for each band of frequencies. It has been suggested that the degree of correlation (r) between two neural regions is indicated by the degree of similarity between the EEG activity at the two sites (López da Silva, 1991), which in turn is the outcome of the functional state of the neuronal networks (Shaw, 1984). Low r reflects higher functional differences and vice versa; higher r reflects a more homogeneous functioning between structures. Thus, the inter- and intra-hemispheric correlation was obtained in the time domain for the same bands by means of Pearson product-moment correlation coefficients between successive amplitude values of EEG segments from the left and right cortical and subcortical regions.

Prefrontal EEG Activity in Relation to Male Sexual Motivation

At the beginning of this chapter, we described that the prefrontal cortex of rats is divided into two main functionally-distinct subregions: the medial (mPFC) and orbital (oPFC) prefrontal cortices (Yonemori, 2000; Kesner, 2000; Uylings et al., 2003). Lesions in these prefrontal areas induce different alterations in the male rat's sexual behavior; however, no one has yet investigated whether or not the subregions of the PFC play different roles in the motivational aspects and, more specifically, in the approach phase during which the detection and processing of the stimuli emitted by the female occur. Hence, to explore the functionality of the mPFC and oPFC in the sexual motivation of the male rat, an experiment was designed (Hernández-González, et al., 2007) to examine the characteristic EEG patterns during approach behavior in a T-maze, and to determine whether these EEG patterns vary through the different behaviors performed by sexually-motivated or non-sexually-motivated male rats.

Brown and colleagues (1974) have reported that every sexual intromission (used as reinforcers) increases sexual arousal considerably until the male reaches ejaculation. Thus, the sexually-motivated condition was generated in the male such that before recording in the T maze he had 1 or 2 intromissions with a receptive female rat and was then immediately placed in the start box, with one receptive and one non-receptive female in the goal-boxes of the T maze (Figure 2). In the non-motivated condition, the male rats were placed in the start box without previous intromission and the goal boxes in the lateral arms were empty.

One group of male rats was bilaterally implanted in the mPFC and the other group in the oPFC. The experiment was conducted on a T-shaped apparatus made of polyurethane-sealed wood that consisted of a brown wood start alley with a start box at one end and two lateral arms with goal boxes at the other. The start box was separated from the start alley by a wooden, guillotine door. At the end of each arm was a small goal box with an opening covered with wire mesh. This meant that the male could hear, see and smell the females, but no copulatory interaction was possible, though some limited physical contact through the wire mesh could occur (Figure 2).

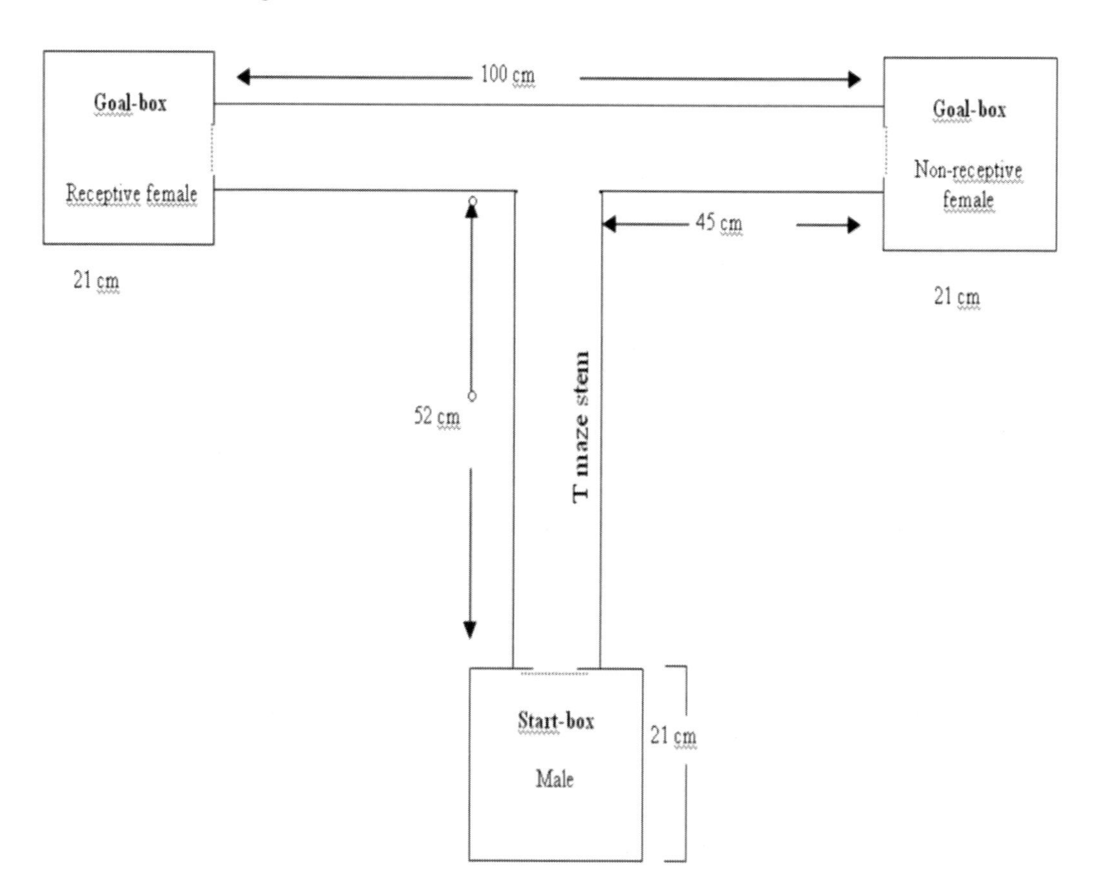

Figure 2. Schematic representation of the wood T-maze used to recording the EEG of male rats during the sexually-motivated (with previous intromission and one receptive and one non-receptive females placed in the goal boxes of the lateral arms); and not sexually-motivated (no previous intromission and empty goal boxes) conditions. For a description of details, see text.

The procedure followed for resolving the T maze in the sexually- and non-sexually motivated task was as follows. Briefly, each rat was placed in the start box for about 5 min, during which time the basal EEG in the awake-quiet condition was recorded. Immediately after this, the guillotine door was raised and the EEG was recorded during the walk through the maze stem and during the time they remained anywhere near the goal boxes of the lateral arms. The EEG from the left and right mPFC and oPFC was recorded simultaneously under different conditions in the T maze, as described previously.

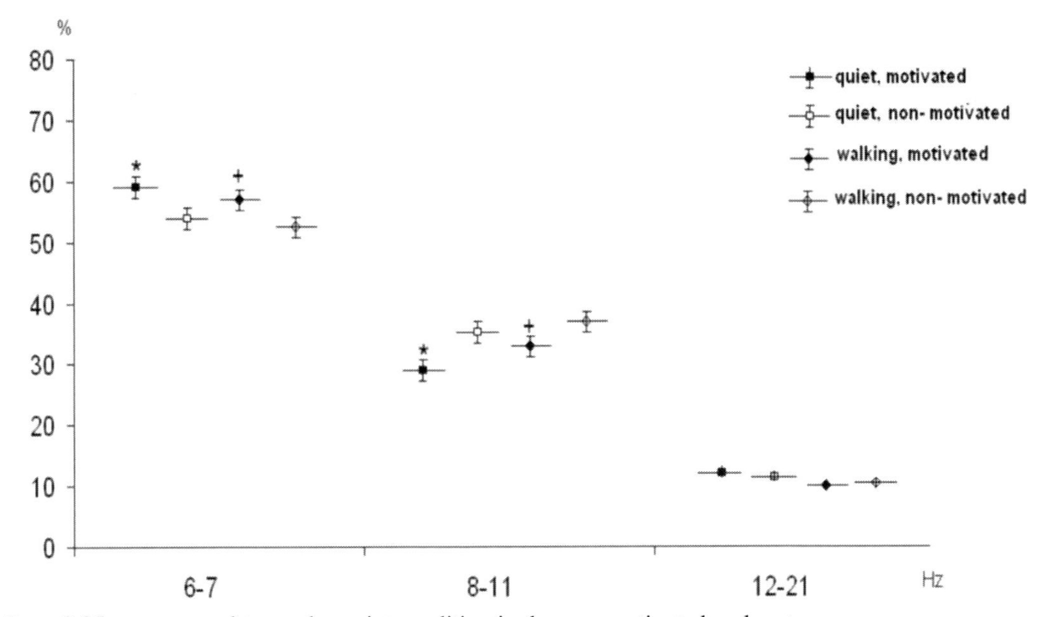

* $p < 0.05$ as compared to awake-quiet condition in the non-motivated male rats.
\+ $p < 0.05$ as compared to walking condition in the non-motivated male rats.

Figure 3, Mean and standard error of Relative power (%) of three frequency bands in the left medial prefrontal cortex (mPFC) during the awake-quiet condition and walking of the sexually and non-sexually motivated male rats. Two way ANOVA for repeated measurements and Tukey tests.

In general terms, the sexually-motivated male rats showed a higher motivation for engaging in sexual behavior compared to the non-sexually motivated ones, as has been demonstrated in other studies (Pfaff and Agmo, 2002; López and Ettenberg, 2001).

An increased relative power (RP) of the low frequencies (6-7 Hz) and a decrease in the 8-11 Hz band was observed in the mPFC during the awake-quiet state in the start box and during the walk in the maze only in the sexually-motivated males (Figure 3). This increase in the RP of the 6-7 Hz band may be associated with the arousal or general activation that characterizes the motivated state, as well as with the orientation of the male's movements, as has been shown in other studies (Gemmel and O´Mara, 1999).

Only the sexually-motivated males presented a higher interprefrontal correlation of the 6-7 Hz band during the walk in the maze stem and during the time they remained near the goal box that contained a receptive female (Figure 4).

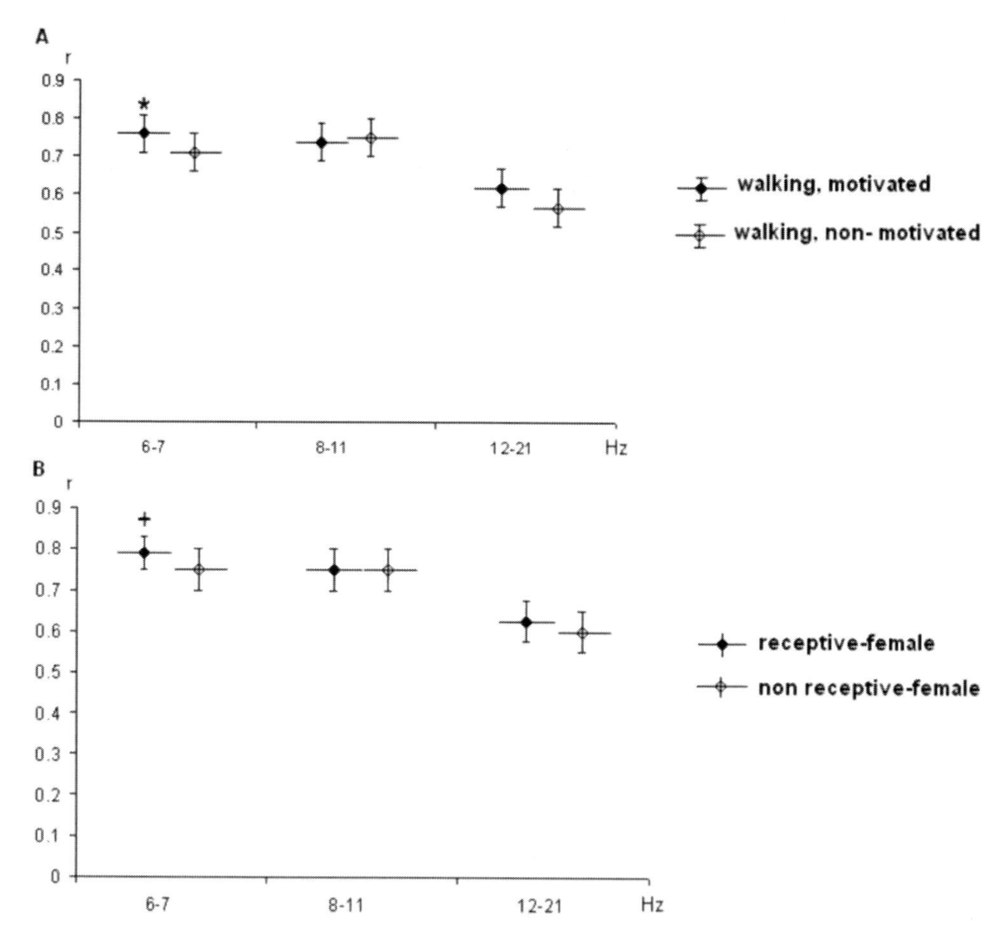

* $p \leq 0.01$ as compared to walking in the non-motivated male rats.

+ $p \leq 0.05$ as compared to sexually motivated rats when remained near to goal boxes with a non-receptive female.

Figure 4. Mean and standard error of Interprefrontal correlation transformed to Fisher´s Z scores of three frequency bands in motivated and non-motivated rats during the walking in the T maze (A) and during the time they remained near to goal boxes with receptive or non-receptive females (B).

In the oPFC, only the RP of the 6-7 Hz band increased whereas the RP of the 8-11 Hz band was decreased as the motivated males walked through the maze stem (Figure 5). Various studies have demonstrated that distal stimuli from a receptive female function primarily to arouse and orient males, so that it is possible that the higher RP values in the 6-7 Hz band in both the mPFC and the oPFC among the sexually-motivated males during the walk in the maze stem may be associated with various processes: the "motivated state" that results from intromission with the female before entering the maze; spatial orientation and selection of the goal; and the processing of the odors and sounds received from the female. It would not be unfounded, then, to suggest that both subregions of the prefrontal cortex manifest a similar functionality during the walk in the maze stem, given the extensive series of anatomical interconnections they possess.

This data suggests that in sexually-motivated males the mPFC is involved in the anticipatory processes –*i.e.*, sensory processing and assigning the incentive value– and in motor execution during the performance of the T maze task, whereas the oPFC is only involved in the motor execution of the T maze (Hernández-González, et al., 2007).

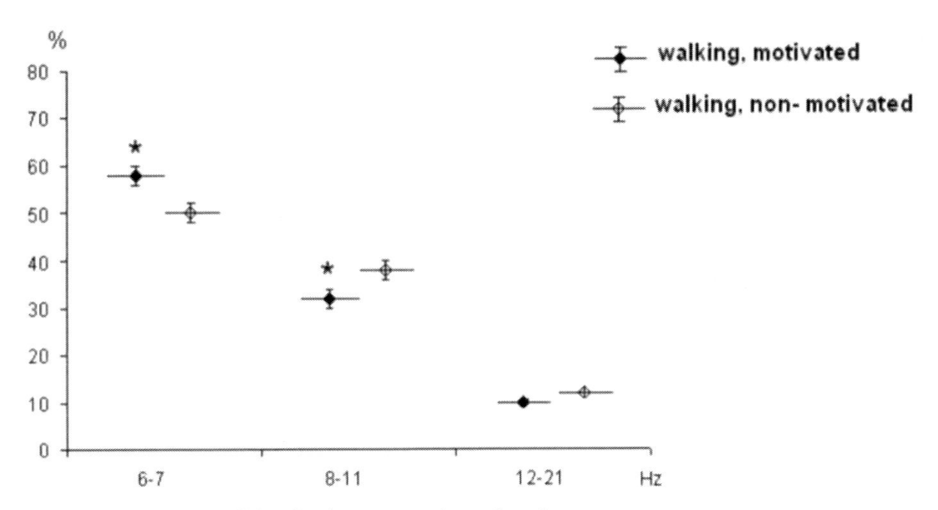

* $p \leq 0.01$ as compared to walking in the non-motivated male rats.

Figure 5, Mean and standard error of Relative power (%) of three frequency bands in the left orbital prefrontal cortex (oPFC) during the walking of the sexually and non-sexually motivated rats. Two way ANOVA for repeated measurements and Tukey tests.

Another study showed that the different frequency bands of the prefrontal EEG were precisely correlated in time with well-defined motor components of the male rat's copulatory activity, as well as with the behaviors related to the motivational or approach phase. The copulatory pelvic movements of the male rats were recorded by means of an accelerometric technique described elsewhere (Contreras and Beyer, 1979; Hernández-González, et al., 1993). A cloth harness was adapted to the rat without causing it any discomfort. At the level of the pelvis, the harness carried a strain gauge transducer (Entran EGB-125-50DC miniature accelerometer) that measures acceleration on a plane. The accelerometer was connected to a DC preamplifier coupled to a Grass 7B polygraph and the output was plugged into a PCL-812 analog-to-digital converter (Advantech, Co.) that operated as an interface to a microcomputer. Electrical signals generated by the accelerometer during pelvic thrusting were continuously recorded at a sampling rate of 128 Hz on the microcomputer for further correlation with EEGs from the prefrontal cortex and the MUA activity from the ventral tegmental area (VTA) and the pedunculopontine nucleus (PPN) (Figure 6).

In the prefrontal cortex, the absolute power (AP) of the different frequency bands (4-16, 18-24 and 26-32 Hz) showed characteristic changes in relation to the execution of pelvic thrusting during mounting, intromission and the ejaculation responses. The approach behaviors, such as orientation and persecution of the female rat that, it has been suggested, are an index of sexual motivation, were correlated with a decreased AP of the fast frequencies and a increases of the low frequencies. Thus, these results confirmed that the electroencephalographic activity of the prefrontal cortex of the male rat is related to the motivation and performance of its sexual behavior (Hernández-González, et al., 1998). The average MUA firing rate in the VTA and PPN increased during pursuit of the female, further augmented during the execution of the pelvic thrusting trains of mounting, intromission and ejaculation responses, and decreased immediately after the pelvic thrusting ended.

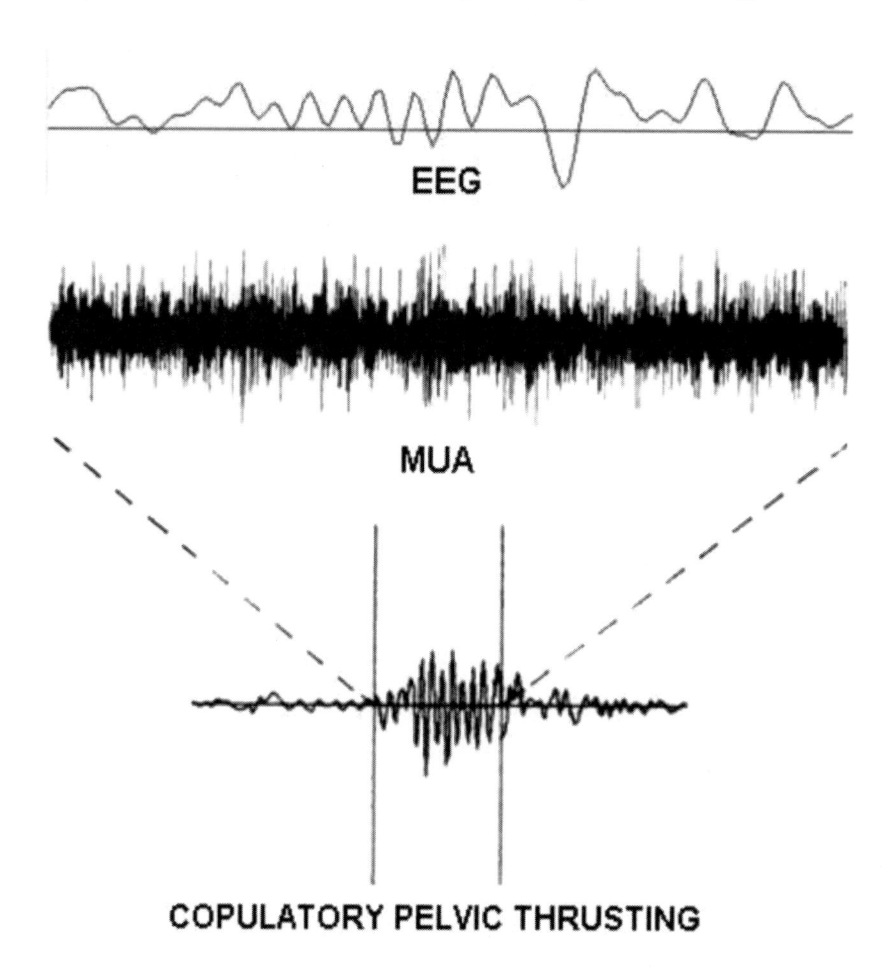

Figure 6. Tracings corresponding to the simultaneous recording of EEG from mPFC and MUA from VTA in relation to the accelerometric recording of pelvic thrusting performed during a 500-msec period corresponding to a ejaculation response.

Similar changes in the MUA firing rate of the medial preoptic area (mPOA) were found by Horio et al. (1986) during both pursuit and pelvic thrusting. It has been suggested that the VTA is involved in the motivational and/or motor aspects of male copulatory behavior (Eibergen and Caggiula, 1973; Hull et al., 1990; Hull et al., 1991), whereas the PPN, located in the caudal mesencephalic tegmentum, is a major component of the mesencephalic locomotor region (MLR) (Grillner and Shik, 1973). Thus, it can be suggested that the VTA and PPN are involved in the motor aspects of the male rat's copulatory activity, though implications in the motivational process could also be proposed.

It is clear, then, that according to the above description of the EEG and MUA studies, the PFC manifests a characteristic functionality in relation to the motivational and motor components of sexual behavior.

Prefrontal EEG Activity in Relation to Female Sexual Motivation

Various reports have shown that the estrous cycle of the rat is a process modulated by hormonal changes that, in turn, are associated with changes of the attraction or sensitivity with which females perceive different sensory stimuli (Pietras and Moulton, 1974; Guevara-Guzman et al., 1997). It is known that during the estrous phase the female rat shows typical receptive and proceptive behaviors associated with the emission of specific vocalizations and odors that together constitute the essential signals with sexual meaning emitted by a receptive female. Moreover, it has been demonstrated that when estrous female rats receive 2-5 intromissions their behavioral proceptivity is increased (Erskine et al., 1989; Rowe and Erskine, 1993; Coopersmith et al., 1996). Thus, in an effort to determine whether the functionality of the oPFC and mPFC prefrontal subregions in female rats is different from that seen in sexually-motivated males, another study was carried out in our laboratory. There, female rats chronically implanted in the oPFC and mPFC were submitted to a sexually-motivated task using the same paradigm of the T maze (Guevara et al., 2004). In this experiment, female rats in the estrous phase received 1 or 2 intromissions in order to increase their sexual proceptivity (and probably sexual motivation to engage in sexual interaction). The procedure for the performance in the T maze test was similar to that used in the previous study, except that in the sexually-motivated test a sexually-active male was placed in one of the goal boxes, whereas in the non-motivated test, the goal boxes remained empty.

The sexually-motivated female rats showed clear behavioral indices of a high motivation to engage in sexual behavior.

During the walk in the T maze stem, the sexually-motivated females showed in the mPFC an increased RP of the low frequencies (4-7 Hz), a decrease in the 8-11 Hz band (Figure 7), and an increased interprefrontal correlation between the left and right mPFC of the 4-7 Hz band; whereas during the awake-quiet state, the mPFC of the sexually-motivated females showed a decreased RP of the low frequencies and an increased RP of the 8-11 Hz band.

These EEG data suggest that in sexually-motivated male and female rats the functionality of the mPFC is similar during the processing of signals that allow the orientation of the subject's movements in the T maze, as well as with the "aroused state" that results from penile or vagino-cervical stimulation (by intromission) before entering the maze. This similar functioning of the mPFC could also be associated with the processing of odors and sounds that are received from the sexual partner. In contrast, during the awake-quiet state, the EEG of the mPFC in sexually-motivated female rats was different from those observed in the sexually-motivated males.

The electroencephalographic functioning of the oPFC in sexually-motivated females was characterized by an increased RP of the low frequencies during the awake-quiet state. A similar EEG pattern in the oPFC of sexually-motivated males was observed during the walk in the maze stem. It is probable that this high proportion of the low frequencies in the oPFC is associated with the general motivated state of the subjects without taking into account the behavior that the male or female rat performs.

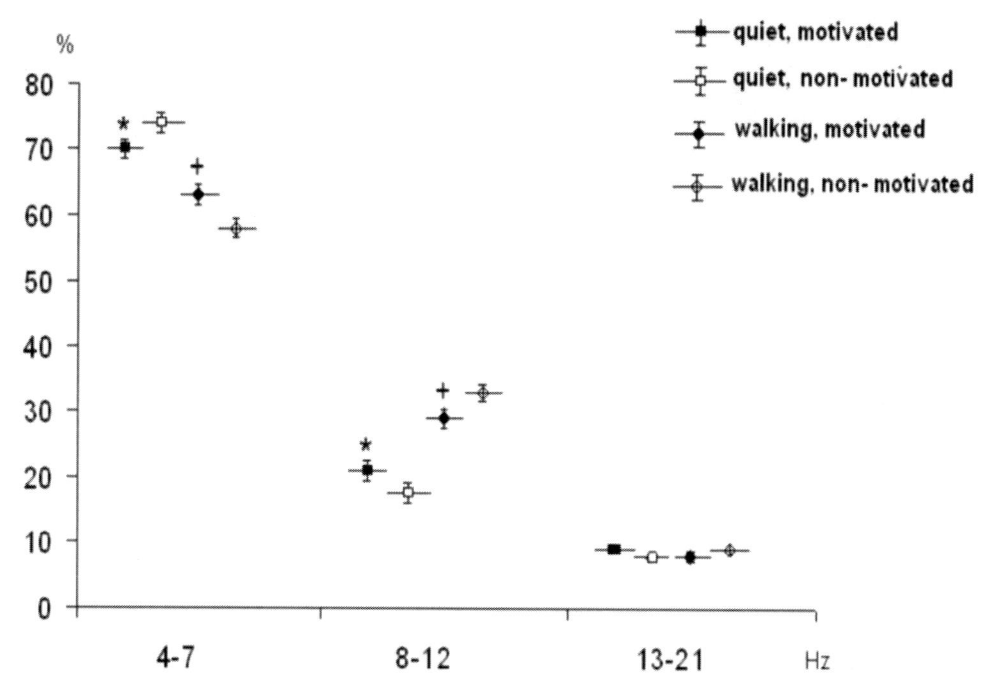

* $p < 0.05$ as compared to awake-quiet condition in the non-motivated female rats.
\+ $p < 0.05$ as compared to walking condition in the non-motivated female rats.

Figure 7. Mean and standard error of Relative power (%) of three frequency bands in the left medial prefrontal cortex (mPFC) during the awake-quiet condition and walking of the sexually and non-sexually motivated female rats. Two way ANOVA for repeated measurements and Tukey tests.

The results of this study concord with the data obtained in males, and show that when rats of both sexes are sexually motivated the functioning of the prefrontal subregions is different compared to that of the prefrontal subregions in non-sexually-motivated subjects. Taken together, these data support the idea that the functioning of the prefrontal subregions varies according to the sexually-motivated state of male and female rats.

Prefrontal EEG Activity in Relation to Maternal Motivation

Maternal behavior is also a motivated behavior, which in rats consists of motorically active behaviors (appetitive behaviors) such as nest repair, pup retrieval, mouthing, licking, and inactive nursing behaviors (consummatory behaviors) that involve a series of postural adjustments and limb immobility that leads to crouching and suckling behaviors (Stern, 1996).

Some lesion and pharmacological studies support the possible involvement of the PFC in maternal behavior. Lesions of the VTA and PFC induce a permanent behavioral syndrome in rats, characterized by locomotor hyperactivity and hypoexploration in an open field (Lemoal et al., 1976; Gaffori and Le Moal, 1979). Lesions of the PFC resulted in alterations such as the disruption of goal-directed behaviors (Kolb, 1984, 1990) and disorganized and persistent retrieving (Stamm, 1955; Slotnick, 1967). A high activation of the mPFC of human mothers while listening to an infant's cries (Lorberbaum et al., 2002), a selective activation of this brain area in dam rats during their first experience with pups (Fleming and Korsmit, 1996; Walsh et al., 1996) and an enhanced c-fos expression in the mPFC of ewes exposed to lambs have all been reported (Kendrik et al., 1997; Broad et al., 2002).

Very few attempts have been made to study brain electrical activity related to maternal behaviors, but the studies that do exist have centered on the hippocampus (Mead and Vanderwolf (1992) or the cortical areas (Lincoln et al., 1980; Cervantes et al., 1992). However, electrical changes in the mPFC in relation to maternal behaviors have not been elucidated; hence, we conducted an experiment to investigate the functioning of the PFC and of the ventral tegmental area (VTA, a subcortical structure that sends the main dopaminergic innervation to PFC) during the motivational (pup retrieval, pup licking) and consummatory (low and high crouching) acts of rat maternal behavior (Hernández-González et al., 2005).

Lactating rats implanted bilaterally in the PFC and VTA were submitted to one or two tests of maternal behavior and EEG recordings were taken between 3 and 10 days post-partum. An EEG baseline was recorded during non-maternal, motorically similar behaviors: 1) walking, recorded during spontaneous horizontal locomotion; 2) forepaw licking, recorded when the female licked her own forepaws; and, 3) awake-quiet condition, recorded when the undisturbed rat, outside the nest, held its head up against gravity with its eyes open and made no movements or only slight vibrissae movements. All baseline recordings were obtained while pups were with their mothers (to eliminate stress in dams due to pup separation). At the start of the 60-min test, the pups were briefly removed and scattered around the opposite side of the nest (to stimulate retrieval) and the EEG was recording during the following maternal behaviors: 1) pup retrieval, recorded when the female picked up a pup in her mouth and carried it to her nest site; 2) pup licking, recorded when the dam picked up a pup with her forepaws and licked the pup's genital region while sitting, but without considering licking or grooming of the head or any other part of the pup's body; 3) low crouching, recorded when the female arched her back over the pups, with the pups underneath, usually with the dam immobile; and, 4) high crouching, recorded when the female positioned herself over the pups, arched her back with all four legs rigidly extended, typically in a splayed position to accommodate the litter mass and allow the pups to suckle. The EEG session was terminated once all behaviors had been performed by the mother or after 1 h.

In both structures, the mPFC and the VTA, a significant increase was obtained in the absolute power of the 8–11 Hz band during pup retrieval compared with basal walking behavior (Figure 8).

During pup licking, the AP of the three frequency bands in the mPFC showed a significant increase compared to forepaw licking, a non-maternal behavior with equivalent postural adjustments and motor elements to those of pup licking. In the VTA, the AP of the 6–7, 8–11 and 12–21 Hz bands also increased during pup licking in relation to forepaw licking (Figure 9).

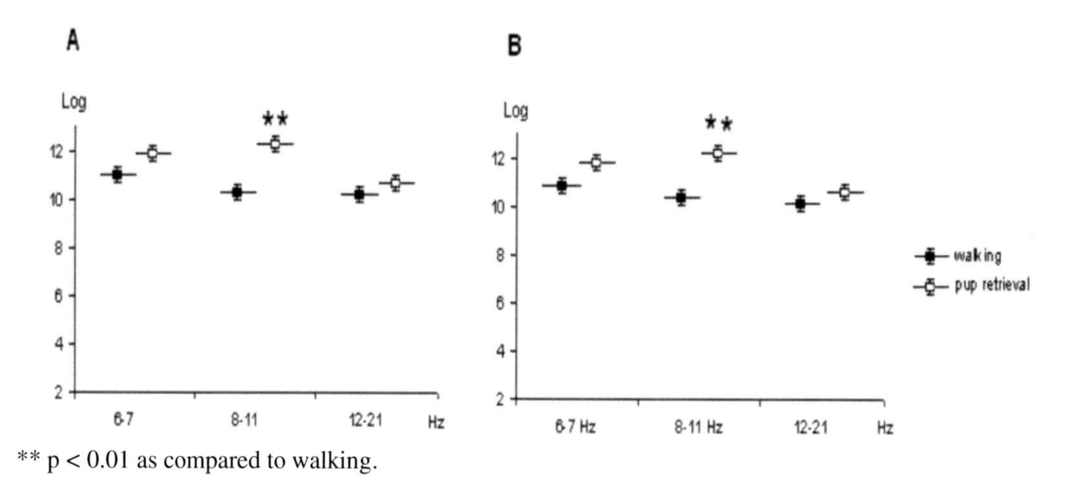

** $p < 0.01$ as compared to walking.

Figure 8. Mean and standard error of absolute power (log) of three frequency bands in medial prefrontal cortex (A) and ventral tegmental area (B) during walking and pup retrieval of dam rats. One-way ANOVA for repeated measurements and Tukey tests.

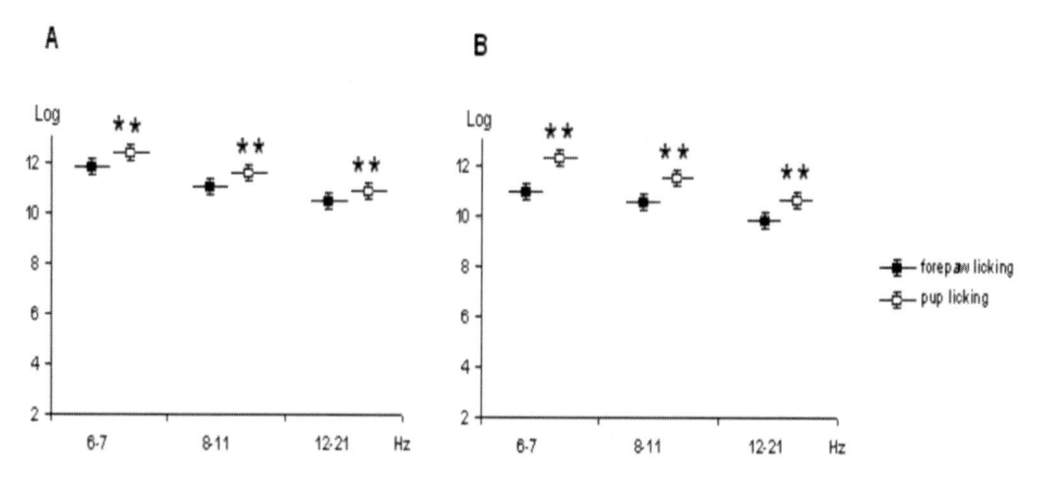

** $p < 0.01$ as compared to forepaw licking.

Figure 9. Mean and standard error of absolute power (log) of three frequency bands in medial prefrontal cortex (A) and ventral tegmental area (B) during forepaw licking and pup licking of dam rats. One-way ANOVA for repeated measurements and Tukey tests.

Various studies have suggested a major role for the medial prefrontal cortex in modulating sequential behaviors and in the spatial orientation of motivated behaviors (Kolb et al., 1983, 1989; Kolb, 1990; Kolb and Cioe, 2004;); thus, one possible explanation is that the increase in the AP of the 8–11 Hz band in mPFC is related to spatial orientation or sequential organization processing performed in this cortical area during pup retrieval acts.

The medial prefrontal cortex receives highly processed sensory information from visual, auditory and somatosensory regions (Groenewegen and Uylings, 2000) and, like the VTA, has been implicated in the integration of motor action with sensory information (Mogenson et al., 1980; Kolb and Cioe, 2004;). Therefore, one possible explanation of the increased AP of the 8–11 Hz band in the mPFC and the VTA during pup retrieval is that this cortical functioning reflects a motivational state and/or sensory processing in these brain structures while the dam is engaged in this activity.

In the left and right mPFC and VTA a clear increment of AP was found in the three bands during pup licking compared to forepaw licking. According to Stern's classification (1996), pup retrieval and pup licking are motorically active behaviors that precede and contribute to the success of dam-initiated nursing and hence form part of the appetitive aspects of maternal behaviors. The VTA has been considered a critical structure in the integration of multimodal stimuli and their transmission to the performance systems (Mogenson et al., 1980; Mirenowicz and Schultz, 1996). Moreover, it has been found that the unit activity of the VTA increases during appetitive behaviors but decreases during consummatory acts (Nishino et al., 1986). Hence, the higher potency of the three bands that appears in the mPFC and the VTA during pup licking is probably associated with the performance of this appetitive behavior and could reflect the particular functioning of these structures in the global neural circuit responsible for pup licking behavior.

An interesting observation based on the present results is that during inactive maternal behaviors –low and high crouching– the mPFC and VTA showed no EEG changes compared to the awake-quiet basal state. Given that these inactive maternal behaviors constitute the consummatory components of maternal care, it is probable that the lack of EEG changes during low and high crouching could be associated with the low participation of the PFC and the VTA in the manifestation of these behaviors, as well as with the consummatory aspects characteristic of the maternal behaviors. Present results demonstrate that both motorically active maternal behaviors are related to a different neural processing in the mPFC and VTA, which could support the suggestion that the mPFC and VTA are involved in processing sensory stimuli to promote adequate motivational and motor responses concerned with maternal behavior.

It is well known that the perception of sensory stimuli, particularly olfactory and somatosensory ones (Pietras and Moulton, 1974; Guevara-Guzman et al., 1997), the willingness of cycling female rats to show maternal behavior or sexual behavior (González and Deis 1986), and the excitability and electrical activity of several brain structures (Corsí-Cabrera et al., 1992; Guevara-Guzman et al., 1997; Contreras et al., 2000) vary through the estrous cycle. Gonadal hormones affect the sensitivity at which the sensory stimuli are perceived and processed by the subjects, and throughout the different reproductive states of the female rat evident changes in hormonal levels appear. Maternal behavior is a motivated behavior that, like sexual interaction, requires an appropriate integration between external stimuli and internal states for its adequate manifestation. There are reports that sensory clues from the pups are important reinforcers for mother rats (Oley and Slotnick., 1970; Stern and Mackinnon, 1976).

Thus, with the aim of investigating the importance of gonadal hormones for the impact of olfactory stimuli on the brain functionality of female rats, in another study we explored whether the PFC and VTA show characteristic EEG patterns during the presentation of olfactory stimuli either associated or not-associated with pups, and whether these EEG patterns vary across the different reproductive states of the female rat (proestrus-estrus, diestrus and lactation).

Virgin female rats were bilaterally implanted in both hemispheres, specifically into the mPFC and the VTA. EEG recordings were obtained in the proestrous-estrous (P-E) and diestrous (D) states (determined through analyses of vaginal smears), and the female rats were then mated with a sexually-active male until they received three ejaculations. During the gestation period, they were housed individually and the first day on which a litter was found was counted as post-partum day 1. The EEG recordings of the lactating female rats (L) were performed during the first 8 days postpartum. The procedure followed for olfactory stimulation was the same as that previously used by Bressler and Baum (1996). Briefly, each rat was placed in the recording cage (with clean sawdust bedding) where it remained at least 5 min before the test. A glass recipient with either nest bedding (obtained from a plexiglas cage where a lactating female rat and its pups had been housed at least 6 h before) or female bedding (obtained from a plexiglas cage where 4-5 virgin female rats, at least two in proestrus-estrus and two in diestrus, had been housed at least 6 h earlier) was placed in a corner of the recording cage, opposite from where the female rat stayed. The EEG from the left and right mPFC and VTA was recorded simultaneously during three different conditions: 1) awake-quiet (no olfactory stimulus present); 2) smelling of nest bedding; and, 3) smelling of female bedding. In all conditions, the EEG was recorded while the undisturbed rat held its eyes open and while the female´s four paws were beneath her, without making any movement with the exception of slight vibrissae movements. The presentation of both olfactory stimuli was randomly counterbalanced and the interval between the presentations of each stimulus was 15 min.

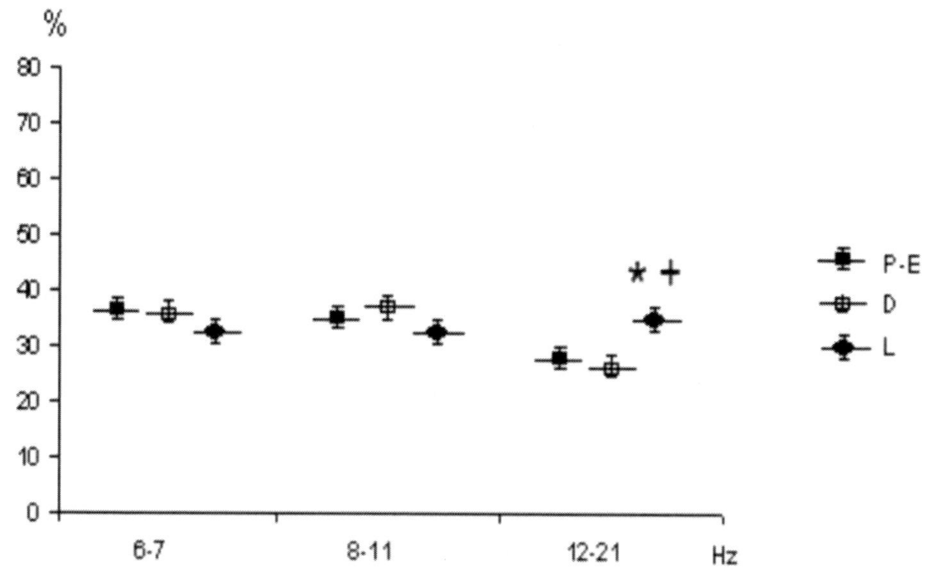

* $p < 0.05$ as compared to P-E rats.
+ $p < 0.05$ as compared to D rats.

Figure 10. Mean and standard error of Relative power (%) of the three frequency bands in the medial prefrontal cortex during the awake-quiet condition of female rats in different reproductive states: Proestrus-estrus (P-E), diestrus (D), and lactation (L). Two way ANOVA for repeated measurements and Tukey tests.

Lactating rats presented the highest frequency of smelling the nest material as compared to P-E and D rats and also remained a longer time smelling the nest bedding compared to P-E rats. When the females were exposed to female bedding, no significant differences were observed in any of these parameters among the different reproductive states, suggesting that the lactating rats showed the highest motivation to approach to the olfactory stimuli associated with pups.

During the awake-quiet condition, which was considered as the basal EEG recording, the lactating rats showed an increased RP of the 12-21 Hz band in the mPFC (Figure 10) and VTA compared to both the P-E and D rats. In L rats, in which estrogen levels are decreased, the EEG activity of mPFC and VTA was characterized by a predominance of fast frequencies that may represent a possible increased arousal state similar to an anxiety state that could result from pup separation (Hard and Hansen, 1985; Ferreira et al., 1989), or could be result from the diminished levels of estrogens as has been reported in other studies (Fernandez-Guasti and Picazzo, 1992; Picazo and Fernández-Guasti, 1993; Guevara-Guzmán et al., 1997; Contreras et al., 2000; Morgan and Paff., 2002). Thus, it is possible that in the present study the increased proportion of fast frequencies in L rats could be associated with a moderate level of anxiety or alertness that presumably optimizes the motivational state of the L rats to search for their pups.

During the smelling of nest bedding, the RP of the 8-11 Hz band in the mPFC of L rats became increased while that of the 6-7 and 12-21 Hz bands decreased as compared to the P-E and D rats (Figure 11).

In the VTA, this phenomenon was observed mainly in the lactating rats. Moreover, the interprefrontal r values of the three frequency bands, mainly those of the 8-11 Hz band, was increased in the D and L rats during the smelling of nest bedding, so that only the olfactory stimuli associated with pups resulted in a higher degree of similarity between the EEG activity of the two cortices (Figure 12).

During the smelling of female bedding, no significant differences were observed between reproductive states. The fact that in both structures of the L rats the slow rhythmical activity showed the main changes in relation to the smelling of nest bedding could be associated with the "relaxed behavior" that L rats showed upon finding pup-associated odors, the higher attraction of the dams for the odor of pups, or both processes. Various L rats showed a behavioral pattern that could be associated with "relaxed behavior"; *i.e.*, they remained on the nest bedding for some time in a sleep-like state. Because these EEG changes characterized by slow rhythmical waves in these structures, is manifested in association to relevant olfactory odors from the nest bedding, which could be associated with a "relaxed state" only in L rats.

Although the mPFC and VTA are not directly implicated in olfactory perception, various studies have suggested an important role for these structures in sensory information processing. For example, it has been found that the mPFC neurons show different responses to conditioned sensory stimuli (auditory or visual) when associated, or not, with a reward (Takenouchi et al., 1999; Jodo et al., 2000), and that the mPFC shows an enhanced metabolic activation in response to the presentation of emotionally-relevant maternal nursing calls (Braun and Poegel., 2001). The VTA has also been shown to present a discriminated response of the neurons in association to rewarded and non-rewarded stimuli (Mirenowicz and Schultz., 1996), and in various studies the VTA has been considered as a critical structure in the integration of multimodal stimuli and their transmission to the performance systems (Mogenson et al., 1980). Thus, it is possible that the higher proportion of slow frequency bands in the mPFC and VTA of L rats in association to smelling of nest bedding could represent a characteristic processing of the olfactory stimuli related to pups, so that, as other studies have suggested (Thompson and Kristal, 1996), these structures may participate in the adequate integration of olfactory stimuli to modulate the motivational and performance aspects of maternal behavior.

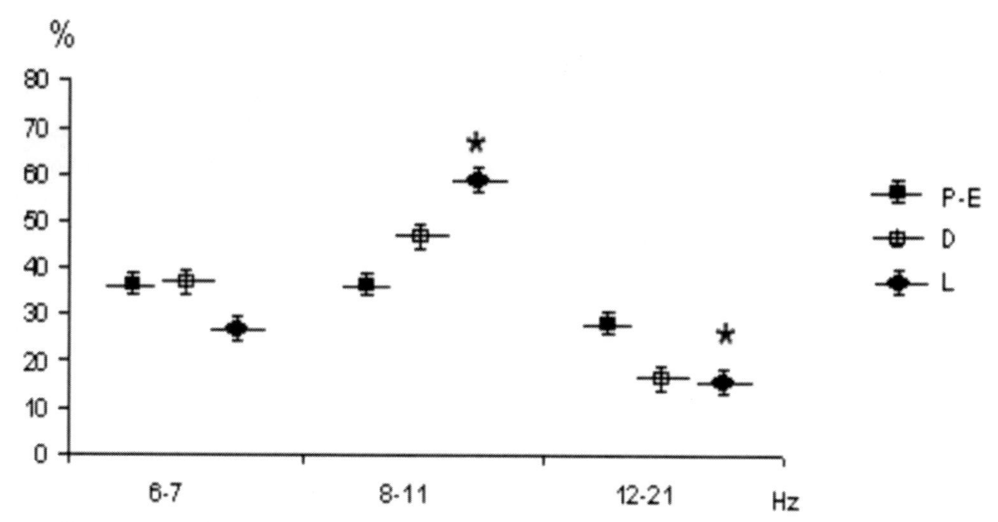

* $p < 0.05$ as compared to P-E and D rats.

Figure 11. Mean and standard error of Relative power (%) of the three frequency bands in the medial prefrontal cortex during the smelling of nest bedding in female rats under different reproductive states: Proestrus-estrus (P-E), diestrus (D), lactation (L). Two way ANOVA for repeated measurements and Tukey tests.

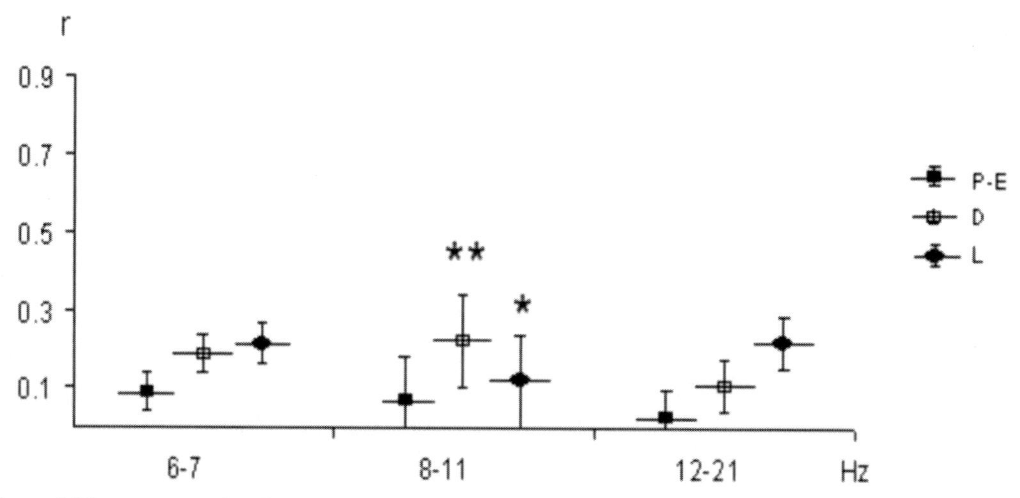

** $p \leq 0.01$ as compared to P-E rats.
+ $p \leq 0.05$ as compared to P-E rats.

Figure 12. Mean and standard error of Interprefrontal correlation transformed to Fisher´s Z scores of the three frequency bands in female rats under different reproductive states: proestrus-estrus (P-E), diestrus (D) and lactation (L) during the smelling of nest bedding.

It has been suggested that distal stimuli from the pups function primarily to arouse and orient the dams (Stern, 1989), so it is possible that the higher r values of the 8-11 Hz band in L rats during the olfaction of nest bedding is associated with various processes: the "relaxed behavior" that L rats showed when they smelled pup-associated odors, the higher attraction of the dams for pup odor, or the processing of olfactory mediated responses (as was proposed by Fleming and Korsmith (1996). It should be noted that all of these events will contribute to reinforcing dams with respect to pups and hence to maternal motivation. Thus, it is possible to suggest that the mPFC and VTA participate in the adequate integration of olfactory stimuli associated with pups in L rats and so modulate the reinforcement and motivational processes that play a crucial role in the expression of maternal behavior.

Conclusion

Since the pioneering work of Beach (1940) on the participation of the cortex in the male rat's sexual behavior, a few studies have been carried out in relation to this topic. One possible reason for the limited interest in studies of the prefrontal cortex and sexual behavior is Beach's proposal that the cerebral cortex is of little importance for such behaviors. However, that view was challenged in a few reports that showed that some cortical areas were indeed important for copulation, while others were less so. More recent studies have found that the selective destruction of the prefrontal cortex has deleterious effects on many aspects of copulatory and maternal behavior, including the processing of incentive stimuli and, therefore, motivational aspects. One of the crucial elements in inducing subjects to engage in a motivated behavior is the adequate processing of the incentive stimuli. The prefrontal cortex is one of the neural substrates that participate in the processing of the stimuli necessary to induce and maintain a motivated state, and this chapter has reviewed several EEG studies of the PFC that show how characteristic changes in the functionality of the mPFC and oPFC, manifested in a predominance of rhythmical slow waves, were observed in association with the presentation of incentive stimuli with sexual or maternal meaning in male and female rats. It seems, therefore, possible to conclude that the PFC does indeed play an important role in the motivated conditions of the subjects. To what degree this different prefrontal functionality could be an index of the processing of the sensory stimuli that induce the motivated state or the general arousal state related to sexual or maternal behaviors is a difficult question to resolve, but taken together, these results constitute a good approximation to knowledge of the neural basis involved in the processing of incentive stimuli and hence, in sexual and maternal motivation.

In the section on the neural bases of sexual and maternal behavior, the pivotal role that the medial preoptic area plays in the three motivated behaviors discussed here was specified. In relation to this aspect Agmo, in his recent book (2007), has suggested an interesting proposition with respect to female sexual behavior "*The sensory input generated by distant stimuli needs to be elaborated somewhere else and given sexual significance before it is directed to the preoptic area. I have suggested that the medial prefrontal cortex may be the site where this stimulus elaboration occurs.*" Considering the anatomical similarities in the neural substrates and the large amount of overlap in the cerebral circuits that regulate male and female sexual behaviors and maternal behavior, it would not be illogical to think that this proposal could be applied to these three motivated behaviors. In this context, Newman (1999) has suggested that it is possible that a common central network exists, and that the exact behavior that occurs may be determined by the stimuli that have access to this neural circuit. Therefore, it is possible that the prefrontal cortex has an important involvement in the processing and assignation of the incentive value of the specific stimuli for sexual and maternal behavior.

Motivated states such as sexual and maternal behaviors are temporal processes that have a beginning and an end, whose intensity varies through the time and –depending on the feasibility of reaching the goal and the consequences of doing so– the willingness of the subjects to engage again in these motivated behaviors may change. In the case of sexual and maternal motivation, this changes during sexual and maternal interaction. It has been suggested that the ordered sequence of events within a specific behavior can be understood as an interplay between reciprocal short-term excitatory and inhibitory influences (Agmo, 1999), so it is probable that this excitatory-inhibitory interplay could be associated with brief periods of high and low motivation that are intercalated in the temporal sequence of the motivated behavior, which may then be associated with specific changes in cerebral functioning. In light of these considerations, the EEG technique has an important advantage in that it allows the researcher to obtain data with a high temporal resolution, a critical point indeed in investigating the functioning of several brain structures along the entire sequence of events from the moment that the animal discovers a potential goal (sexual partner or pup) until the successful performance of the sexual or maternal interaction. Thus, although various reports have described a role for the prefrontal cortex in these motivated behaviors, the use of the EEG technique made it possible to obtain a precise temporal correlation of the PFC functionality with the motivated states of male and female rats, as well as with the performance of sexual and maternal behaviors. Taken together, the EEG data demonstrate that the prefrontal cortex in rats participates in identifying and processing sexual and maternal incentives.

Acknowledgments

The various research projects discussed herein were supported by a grant to M.H.G. from the "Consejo Nacional de Ciencia y Tecnología" (CONACYT), Grant No. 28592N.

References

Ågmo, A. (1999). Sexual motivation – an inquiry into events determining the occurrence of sexual behavior. *Behav Brain Res*, 105,129-150.

Ågmo, A. (2002). Copulation-contingent aversive conditioning and sexual incentive motivation in male rats: evidence for a two-stage process of sexual behavior. *Physiol Behav*, 77,425-435.

Ågmo, A. (2007). Functional and dysfunctional sexual behavior. A synthesis of neuroscience and comparative psychology. Academic Press, San Diego.

Ågmo, A., Villalpando, A., Picker, Z. and Fernández, H. (1995). Lesions of the medial prefrontal cortex and sexual behavior in the male rat. *Brain Res*, 696, 177-186.

Armony, J, Servan-Schreiber, D.J. and LeDoux, J.E. (1997). Computational modeling of emotion, explorations through the anatomy and physiology of fear conditioning. *Psychol. Rev.,* 81, 199-213.

Balkenius, C. and Morén, J. (2001). Emotional Learning, A computational model of the amygdale. *Cybernet Systems,* 32, 611-636.

Beach F.A. (1940). Effects of cortical lesions upon the copulatory behavior of male rats. *J. Comp. Psychol.*, 29,193-245.

Beach F.A., Zitrin A. and Jaynes J. (1955). Neural mediation of mating in male cats. II. Contributions of the frontal cortex. *J. Exp. Zool.*, 130,381-402.

Beach, F.A. and Jordan, L. (1956). Effects of sexual reinforcement upon the performance of male rats in a straight runway. *J. Comp. Physiol. Psychol.*, 49,105-111.

Beach, F.A. (1976). Sexual attractivity, proceptivity, and receptivity in female mammals. *Horm. Behav.,* 7, 105–138.

Bennet, T., Hill P. and French J. (1982). Biotelemetry recording of the electrical activity of the hippocampus and amygdale during sexual behavior in the cat. *Bull Psychom. Soc.*, 20(1),57-60.

Bermant, G. (1961). Response latencies of female rats during sexual intercourse. *Science,* 133, 1771–1773.

Bitran, D., Hilvers, R.J. and Kellogg, C.K. (1991). Ovarian endocrine status modulates the anxiolytic potency of diazepam and the efficacy of γ-aminobutyric acid-benzodiazepine receptor-mediated chloride ion transport. *Behav. Neurosci.,* 105, 653-662.

Blaustein, J.D. and Erskine, M.S. (2002). Feminine sexual behavior: Cellular integration of hormonal and afferent information in the rodent brain. In: In: DW Pfaff., AP Arnold., AM Etgen., SE Fahrbach., and RT Rubin, (eds.) *Hormones, brain and behavior.* Academic Press, Elsevier Science, USA., Vol. 1, pp. 139-214.

Boling, J.L. and Blandau, R.J. (1939). The estrogen-progesterone induction of mating responses in the spayed female rat. *Endocrinology*, 25(582),6.

Bradford, J.M.W. and Pawlaka, A. (1993). Double-blind placebo crossover study of cyproterone-acetate in the treatment of the paraphilias. *Arch. Sex Behav.,* 22, 629-641.

Braun, K. and Poeggel, G. (2001). Recognition of mother´s voice evokes metabolic activation in the medial prefrontal cortex and lateral thalamus of *Octodon degus* pups. *Neuroscience,* 103, 861-864.

Bremer, P. (1958). Asexualization: a follow-up study of 244 cases. Oslo University Press: Oslo Briken, P., Berner, W., Noldus, J., Nika, E. and Michl, U. (2000). Therapie mit dem LHRH-agonisten Leuprolinacetat bei Paraphilien und sexuell aggressiven Impulshandlungen. *Nervenarzt* 71, 380-385.

Bressler, S.C. and Baum, M.J. (1996). Sex comparison of neuronal Fos immunoreactivity in the rat vomeronasal projection circuit after chemosensory stimulation. *Neuroscience*, 71, 1063-1072.

Broad, K.D., Hinton, M.R., Keverne, E.B. and Kendrick, K.M. (2002). Involvement of the medial prefrontal cortex in mediating behavioural responses to odour cues rather than olfactory recognition memory. *Neuroscience,* 114, 715-729.

Broekman, M., de Bruin, M., Smeenk, J., Slob, A.K., and Vander Schoot, P. (1988). Partner preference behavior of estrous female rats affected by castration of tethered male incentives. *Horm. Behav.*, 22, 324–337.

Brown, V.J. and Bowman, E.M. (2002). Rodent models of prefrontal cortical function. *Trends Neurosci.*, 25, 340-343.

Buchwald, J.S., Holstein, S.B. and Weber, D.S. (1973). Multiple unit recording technique, interpretation and experimental application. In RF Thompson., MM Patterson, (eds). *Bioelectric recording techniques*. Vol. 1A. Academic Press, New York. pp. 201-242.

Bunnel, B.N., Friel, J. and Flesher, C.K. (1966). Effects of median cortical lesions on the sexual behaviour of the male hamster. *J. Comp. Physiol. Psychol.,* 61, 492-495.

Cervantes, M., Ruelas, R. and Alcalá, V. (1992). EEG signs of "relaxation behavior" during breast feeding in a nursing woman. *Arch. Med. Res.,* 23, 123-127.

Clark, A.S., Pfeifle, J.K., and Edwards, D.A. (1981). Ventromedial hypothalamic damage and sexual proceptivity in female rats. *Physiol. Behav.,* 27, 597–602.

Cohen H., Rosen R. and Goldstein, L. (1976). Electroencephalographic laterality changes during human sexual orgasm *Arch. Sex Behav.,* 5, 189-199.

Cohen, A.S., Rosen, R.C. and Goldstein, L. (1985). EEG hemispheric asymmetry during sexual arousal: Psyshophysiological patterns in responsive, unresponsive, dysfunctional men. *J. Abnorm. Psychol.,* 94, 580-590.

Contreras, C.M., Molina, M., Saavedra, M. and Martínez-Mota, L. (2000). Lateral septal neuronal firing rate increase during proestrus-estrus in the rat. *Physiol. Behav.,* 68, 279-284.

Contreras, J.L. and Beyer, C.A. (1979). Polygraphic analysis of mounting and ejaculation in the New Zeland white rabbit. *Physiol. Behav.*, 23, 939-943.

Coopersmith C., Candura C., and Erskine M.S. (1996). Effects of paced mating and intromissive stimulation on feminine sexual behavior and estrus termination in the cycling rat. *J. Comp. Psychol.*, 110. 176-186.

Corsi-Cabrera, M., Juárez, J., Ponce-de-Leon, M., Ramos, J. and Velásquez, P.N. (1992). EEG activity during estral cycle in the rat. *Electroenceph Clin Neurophysiol*, 83, 265-269.

Davidson, J.M. (1966). Activation of the male rat´s sexual behavior by intracerebral implantation of androgen. *Endocrinology*, 79, 783-794.

Dempsey, E.W., Hertz, R. and Young, W.C. (1939). The experimental induction of oestrus (sexual receptivity) in the normal and ovariectomized guinea pig. *Am. J. Physiol.*, 116, 201-209.

Denniston, R.H. (1954). Quantification and comparation of sex drives under various conditions in terms of learned responses. *J. Comp. Physiol. Psychol.*, 47, 437-440.

Edwards, D.A., and Pfeifle, J.K. (1983). Hormonal control of receptivity, proceptivity and sexual motivation. *Physiol. Behav.,* 30, 437–443.

Emery, D.E., and Moss, R.L. (1984a). Lesions confined to the ventromedial hypothalamus decrease the frequency of coital contacts in female rats. *Horm. Behav.,* 18, 313–329.

Emery, D.E., and Moss, R.L. (1984b). p-Chlorophenylalanine alters pacing of copulation in female rats. *Pharmacol. Biochem. Behav.,* 20, 337–341.

Eibergen, R.D. and Caggiula, A.R. (1973). Ventral midbrain involvement in copulatory behavior of the male rat. *Physiol. Behav.,* 10, 435-441.

Erskine M.S., Kornberg E., and Cherry J.A. (1989). Paced copulation in rats: Effects of intromission frequency and duration on luteal activation and estrus length. *Physiol. Behav.,* 45, 33-39.

Everitt, B.J. and Stacey, P. (1987). Studies of instrumental behavior with sexual reinforcement in male rats (Rattus norvegicus): II. Effects of preoptic area lesions, castration, and testosterone. *J Comp Psychol,* 101,407-419.

Everitt, B.J., Fray, P., Kostarczyk, E., Taylor, S. and Stacey, P. (1987). Studies of instrumental behavior with sexual reinforcement in male rats (Rattus norvegicus): I. Control by brief visual stimuli paired with a receptive female. *J. Comp. Psychol.*, 101, 395-406.

Everitt, B.J. (1990). Sexual motivation: a neural and behavioral analysis of the mechanisms underlying appetitive and copulatory responses of male rats. *Neurosci. Biobehav. Rev.*, 14, 214-232.

Fernandez-Guasti, A. and Picazo, O. (1992). Changes in burying behavior during the estrous cycle: effect of estrogen and progesterone. *Psychoneuroendocrinology,* 17, 681-689.

Fernández-Guasti, A., Omana-Zapata, I., Luján, M. and Condés-Lara, M. (1994). Actions of sciatic nerve ligature on sexual behavior of sexually experienced and inexperienced male rats: effects of frontal pole decortication. *Physiol. Behav.*, 55, 577-581.

Ferreira, A., Hansen, S., Nielsen, M., Archer, T. and Minor, B.G. (1989). Behavior of mother rats in conflict tests sensitive to anti-anxiety agents. *Behav. Neurosci.,* 103, 193-201.

Fleming, A.S. and Luebke, C. (1981). Timidity prevents the nulliparous female from being a good mother. *Physiol. Behav.,* 27, 863-868.

Fleming, A.S. (1986). Psychobiology of rat maternal behavior. How and where hormones act to promote maternal behavior at parturition. *NY Acad. Sci.*, 474, 234-251.

Fleming, A.S., Cheung, U.S., Myhal, N. and Kessler, Z. (1989). Effects of maternal hormones on "timidity" and attraction to pup-related odors in females. *Physiol. Behav.,* 46, 449-453.

Fleming, A.S., Korsmit, M. and Deller, M. (1994). Rat pups are potent reinforcers to the maternal behavior animal: effects of experience, parity, hormones, and dopamine function. *Psychobiology,* 22, 44-53.

Fleming, A.S. and Korsmith, M. (1996). Plasticity in the maternal circuit: Effects of maternal experience on Fos-Lir in hypothalamic, limbic, and cortical structures in the postpartum rat. *Behav. Neurosci.,* 110, 567-582.

Floody, O.R. (2002). Time course of VMN lesion effects on lordosis and proceptive behavior in female hamsters. *Horm. Behav.*, 41, 366-376.

French, D., Fitzpatrick, D., and Law, O.T. (1972). Operant investigation of mating preference in female rats. *J. Comp. Physiol. Psychol.,* 81, 226–232.

Fuster, J.M. (1997).*The prefrontal cortex: Anatomy, physiology, and neuropsychology of the frontal lobe.* 3rd ed. Lipincott-Raven, New York.

Gaffori, O. and Le Moal, M. (1979). Disruption of maternal behavior and appearance of cannibalism after ventral mesencephalic tegmentum lesions. *Physiol. Behav.*, 23, 317-323.

Gemmel, C. and O´Mara, S.M. (1999). Medial prefrontal cortex lesions cause deficits in a variable-goal location task but not in object exploration. *Behav. Neurosci.*, 113, 465-475.

Genazzani, A.R., Peytaglia F. and Purdy R.H. (1996). *The brain: source and target for sex steroid hormones.* The Parthenon Publishing Group Limited, New York.

Goldman-Rakic, P.S. (1996). The prefrontal landscape: Implications of functional architecture for understanding human mentation and the central executive. *Philos Trans. R Soc London*, 351, 1445-1453.

Gonzalez, D.E. and Deis, R.P. (1986). Maternal behavior in cyclic and androgenized female rats. Role of ovarian hormones. *Physiol. Behav.*, 38, 789-793.

González-Mariscal, G. and Poindron, P. Parental care in mammals:immediate internal and sensory factors of control. (2002). In DW Pfaff., AP Arnold., AM Etgen., SE Fahrbach., and RT Rubin (eds). *Hormones, brain and behavior.* Academic Press, Elsevier Science, USA., Vol. 1, pp. 215-298.

Graber B., Rohrbaugh J., Newlin, D., Varner J. and Ellingson R. (1985). EEG during masturbation and ejaculation orgasm *Arch. Sex Behav.*, 14, 491-503.

Gray, P. and Brooks, P.J. (1984). Effect of lesion location within the medial preoptic-anterior hypothalamic continuum on maternal and male sexual behaviors in female rats. *Behav. Neurosci.*, 98, 703-711.

Grillner, S. and Shik, M.I. (1973). On the descending control of the lumbosacral spinal cord from the mesencephalic locomotor region. *Acta Physiol. Scand*, 87, 320-333.

Groenewegen, H.J. and Uylings, H.B.M., (2000). The prefrontal cortex and the integration of sensory, limbic, and autonomic information. In: HBM Uylings., CG Van Eden., JPC de Bruin., MGP Feenstra., CMA Pennartz. (eds). Cognition, Emotion and autonomic responses: The integrative role of the prefrontal cortex and limbic structures. *Progress in Brain Research,* vol 126, Elsevier, Amsterdam., pp. 3-28.

Guarraci, F.A. and Clark, A.S. (2006). Ibotenic acid lesions of the medial preoptic area disrupt the expression of parter preference in sexually receptive female rats. *Brain Res,* 1076, 163-170.

Guevara, M.A., Hernández-González, M. and Ramos-Guevara, J.P. (2004). Especialización subregional de la corteza prefrontal en la rata: diferente actividad durante una tarea sexualmente motivada. In MA Guevara., M Hernández-González., P Durán-Hernández., (eds.) *Aproximaciones al estudio de la corteza prefrontal.* Universidad de Guadalajara, México. pp.151-178.

Guevara M. A., Hernández González M., Zarabozo D. and Corsi Cabrera M. (2003). POTENCOR: A program to calculate power and correlation spectra of EEG signals. *Comp. Meth. Prog. Biomed.*, 72, 241-250.

Guevara, M.A., Ramos, J., Hernández-González, M. and Corsi-Cabrera, M. (2005). FILDIG: a program to filter brain electrical signals in the frequency domain. *Comp. Meth Prog. Biomed.* 80, 93-186.

Guevara-Guzmán, R., Barrera-Mera, B. and Weiss, M.L. (1997). Effect of the estrous cycle on olfactory bulb response to vaginocervical stimulation in the rat: Results from electrophysiology and Fos immunicytochemistry experiments. *Brain Res. Bull.,* 44, 141-149.

Hall, J, Parkinson, J, Connor, T, Dickinson, A, and Everrit B.J. (2001). Involvement of the central nucleus of the amygdala and nucleus accumbens core in mediating pavlovian influences on instrumental behavior. *Eur. J. Neurosci.*, 13, 1984-1192.

Hansen, S. and Ferreira, A. (1986). Food intake, aggression, and fear in the mother rat: control by neural systems concerned with milk ejection and maternal behavior. *Behav. Neurosci.*, 100, 64-70.

Hard, E. and Hansen, S. (1985). Reduced fearfulness in the lactating rat. *Physiol. Behav.*, 35, 641-643.

Heath R. (1972). Pleasure and brain activity in man *The Journal of Nervous and Mental Disease,* 154(1),3-18.

Heim, N. and Hursch, C.J. (1979). Castration of sex offenders: treatment or punishment? A review and critique of recent European literature. *Arch. Sex Behav.,* 8, 281-306.

Hernández-González, M., Guevara, M.A., Oropeza, M.V. and Moralí, G. (1993). Male rat pelvic copulatory movements: computarized analysis of accelerometric data. *Arch. Med. Res.*, 24, 155-160.

Hernández-González, M., Guevara, M.A., Moralí, G. and Cervantes, M. (1997). Computer programs to analyze brain electrical activity during copulatory pelvic thrusting in male rats. *Physiol. Behav.*, 62,701-708.

Hernández-González M., Guevara M. A., Moralí G., and Cervantes M (1997). Subcortical multiple unit activity changes during male sexual behavior of the rat. *Physiol. Behav.*, 61(2), 285-291.

Hernández-González, M., Guevara, M.A., Cervantes, M., Moralí, G. and Corsi-Cabrera, M. (1998). Characteristic frequency bands of the cortico-frontal EEG during the sexual interaction of the male rat as a result of factorial analysis. *J. Physiol. (Paris),* 92, 43-50.

Hernández-González, M., Prieto-Beracoechea, C.A., Navarro-Meza, M. and Guevara, M.A. (2005). Prefrontal and tegmental electrical activity during olfactory stimulation in virgin and lactating rats. *Physiol. Behav.,* 83, 749–758.

Hernández-González M., Prieto-Beracoechea C.A., Arteaga-Silva M. and Guevara M.A. (2007). Different functionality of the medial and orbital prefrontal cortex during a sexually motivated task in rats. *Physiol. Behav.*, 90, 450-458.

Hetta, J. and Meyerson, B.J. (1978). Sexual motivation in the male rat: a methodological study of sex-specific orientation and the effects of gonadal hormones. *Acta Physiol. Scand Suppl.*, 453, 1-67.

Holmes, J. and Egan, K. (1973). Electrical activity of the cat amygdale during sexual behavior. *Physiol. Behav.*, 10, 863-867.

Horio, T., Shimura, T., Hanada, M. and Shikochi, M. (1986). Multiple unit activities recorded from the medial preoptic area during copulatory behavior in freely moving male rats. *Neurosci. Res.,* 3, 311-320.

Hull, E.M., Meisel, R.L. and Sachs, B.D. Male sexual behavior. (2002). In DW Pfaff., AP Arnold., AM Etgen., SE Fahrbach., and RT Rubin (eds). *Hormones, brain and behavior.* Academic Press, Elsevier Science, USA., Vol. 1, pp: 1-138.

Hull, E.M., Bazzet, T.J., Warner, R.K., Eaton, R.C. and Thompson, J.T. (1990). Dopamine receptors in the ventral tegmental area modulate male sexual behavior in rats. *Brain Res.*, 512, 1-6.

Hull, E.M., Weber, M.S., Eaton, R.C., Dua, R., Markowski, V.P., Lumley, L., and Moses, J. (1991). Dopamine receptors in the ventral tegmental area affect motor, but not motivational or reflexive components of copulation in male rats. *Brain Res.*, 554, 72-76.

Hurtazo, H.A., Paredes, R.G. and Ågmo, A. (2003). Inactivation of the medial preoptic area/anterior hypothalamus (MPOA/AH) reduces sexual incentive motivation in male rats. *Soc. Neurosci. Abstr.,* 404,6.

Jacobson, C.D., Terkel, J., Gorski, R.A. and Sawyer, C.H. (1980). Effects of small medial preoptic area lesions on maternal behavior: retrieving and nest building in the rat. *Brain Res.*, 194, 471-478.

Jodo, E., Suzuki, Y. and Kayama, Y. (2000). Selective responsiveness of medial prefrontal cortex neurons to the meaningful stimulus with a low probability of occurrence in rats. *Brain Res.*, 856, 68-74.

Jowaisas, D., Taylor, J., Dewsbury, D.A. and Malagodi, E.F. (1971). Copulatory behavior of male rats under an imposed operant requirement. *Psychonom. Sci.*, 25,287-290.

Kagan, J. (1955). Differential reward value of incomplete and complete sexual behavior. *J. Comp. Physiol. Psychol.*, 48,59-64.

Kendrick, K.M., Da Costa, A.P.C., Broad, K.D., Okhura, S., Guevara, R., Levy, F. and Keverne, E.B. (1997). Neural control of maternal behavior and olfactory recognition of offspring. *Brain Res. Bull.*, 44, 383-395.

Kesner, R.P. (2000). Neural mediation of memory for time: role of the hippocampus and medial prefrontal cortex. *Psychol. Bull Rev.*, 5,585-596.

Kollack-Walker, S. and Neuman, S.W. (1992). Mating behavior induces selective expression of Fos protein within the chemosensory pathways of the male Syrian hamster brain. *Neurosci. Lett*, 143, 223-228.

Kollack-Walker, S. and Newman, S.W. (1995). Mating and agonistic behavior produce different patterns of Fos immunolabeling in the male Syrian hamster brain. *Neuroscience,* 66, 721-736.

Kolb, B., Sutherland, R.J. and Whishaw, I.Q., (1983). A comparison of the contribution of the frontal and parietal association cortex to spatial localization in rats. *Behav. Neurosci.,* 97, 13-27.

Kolb, B. (1984). Functions of the frontal cortex of the rat. A comparative review. *Brain Res. Rev.*, 8, 65-98.

Kolb, B., Buhrmann, K. and McDonald, R., (1989). Dissociation of prefrontal, posterior parietal, and temporal cortical regions to spatial navigation and recognition memory in the rat. *Soc. Neurosci.,* Abstract 15, 607.

Kolb, B. (1990). Prefrontal cortex. In B Kolb., RC Tees. (eds). *The Cerebral Cortex of the Rat.* MA: The MIT Press, Cambridge. pp. 437-58.

Kolb, B. and Cioe, J., (2004). Organization and plasticity of the prefrontal cortex of the rat. In S Otani. (ed). *Prefrontal Cortex: From synaptic plasticity to cognition.* Kluwer Academic Publishers, France. pp. 1-32.

Kurtz, R. and Adler N. (1973). Electrophysiological correlates of copulatory behavior in the male rat *J. Comp Physiol. Psychol.*, 84, 225-239.

Kurtz, R. (1975). Hippocampal and cortical activity during sexual behavior in the female rat *J Comp. Physiol Psychol.*, 89(2),158-169.

Larsson, K. (1962). Mating behaviour in male rats after cerebral cortex ablation. I. Effects of lesions in the dorsolateral and the median cortex. *J. Exp. Zool.,* 151, 167-176.

Larsson, K. (1964). Mating behaviour in male rats after cerebral cortex ablation. II. Effects of lesions oh the frontal lobes compared to lesions un the posterior half of the hemispheres. *J. Exp. Zool.,* 155, 203-214.

Larsson, K. (1979). Features of the neuroendocrine regulation of masculine sexual behavior. In C. Beyer, (ed). *Endocrine control of sexual behavior.* Raven Press, New York. pp. 77-163.

Lee, A., Clancy, S. and Fleming, A.S. (2000). Mother rats bar-press for pups: effects of lesions of the MPOA and limbic sites on maternal behavior and operant responding for pup-reinforcement. *Behav. Brain Res.*, 108, 215-231.

Lemoal, M., Stinus, L. and Galey, D., (1976). Radiofrequency lesion of the ventral mesencephalic tegmentum: Neurological and behavioral considerations. *Exp. Neurol.*, 50, 521-535.

Leon-Carrion, J., Martín-Rodriguez, J.F., Damas-López, J., Pourrezai, K., Izzetoglu, K., Barroso, J.M. y Martin, M. and Dominguez- Morales, R. (2007). Does dorsolateral prefrontal cortex (DLPFC) activation return to baseline when sexual stimuli cease? The role of DLPFC in visual sexual stimulation. *Neurosci. Let*, 416, 55-60.

Lincoln, D.W., Hentzen, K., Hin, T., Van der Schoot, P., Clarke, G. and Summerlee, A.J.S. (1980). Sleep: A prerequisite for reflex milk ejection in the rat. *Exp. Brain Res.*, 38, 151-162.

Lorberbaum, J.P., Newman, J.D., Horwitz, A.R., Dubno, J.R., Bruce Lydiard, R., Hamner, M.B., Bohning, D.E. and George, M.S. (2002). A potential role for thalamocingulate circuitry in human maternal behavior. *Biol. Psychiatry,* 51, 431-445.

López da Silva, F.H. (1991). Neural mechanisms underlying brain waves from neural membranes to networks. *Electroenceph. Clin. Neurophysiol.,* 79, 81-93.

López, H.H. and Ettenberg, A. (2001). Dopamine antagonism attenuates the unconditioned incentive value of estrous female cues. *Pharmacol. Biochem. Behav.,* 68, 411-416.

Lubar, J.F., Herrman, T.J., Moore, D.R. and Shouse, M.N. (1973). Effect of septal and frontal ablations on species typical behavior in the rat. *J. Comp. Physiol. Psychol.,* 83, 260-270.

Madlafousek, J., and Hlinak, Z. (1977). Sexual behavior of the female laboratory rat: Inventory, patterning, and measurement. *Behaviour,* 63, 129–174.

Magnusson, J.E. and Fleming, A.S. (1995). Rat pups are reinforcing to the maternal rat: role of sensory cues. *Psychobiology*, 23, 69-75.

Maras, P.M. and Petrulis, A. (2006). Chemosensory and steroid-responsive regions of the medial amygdala regulate distinct aspects of opposite-sex odor preference in male Syrian hamsters. *Eur. J. Neurosci.,* 24, 3541-3552.

Matthews, T.J., Grigore,M., Tang, L., Doat, M., Kow, L.M. and Pfaff, D.W. (1997). Sexual reinforcement in the female rat. *J. Exp. Anal. Behav.*, 68, 399-410.

Mattson, B.J., Williams, S., Rosenblatt, J.S. and Morrell, J.I. (2001). Comparison of two positive reinforcing stimuli: pups and cocaine throughout the postpartum period. *Behav. Neurosci.*, 115, 683-694.

McClintock, M.K. and Adler, N.T. (1978). The role of the female during copulation in wild and domestic norway rats *(Rattus norvegicus). Behaviour*, 67, 67–96.

McIntosh R., Barfield R. and Thomas D. (1984). Electrophysiological and ultrasonic correlates of reproductive behavior in the male rat. *Behav. Neurosci.*, 98,1100-1103.

Mead, L.A. and Vanderwolf, C.H. (1992). Hippocampal electrical activity in the female rat: The estrous cycle, copulation, parturition, and pup retrieval. *Behav. Brain Res.*, 50, 105-113.

Meyerson, B.J. and Lindstrom, L.H. (1973). Sexual motivation in the female rat. A methodological study applied to the investigation of the effect of estradiol benzoate. *Acta Physiol. Scand Suppl.*, 389, 1-80.

Michal, E.K. (1973). Effects of limbic lesions on behavior sequences and courtship behavior of male rats (Rattus norvergicus). *Behavior*, 44:264-285.

Michael R., Holbrooke C., and Weller. C. (1977). Telemetry and continuous energy analysis of hypothalamic EEG changes in female cats during intromission by the male. *Psychoneuroendocrinology*, 2, 287-301.

Miceli, M.O., Fleming, A.S. and Malsbury, C.W. (1983). Disruption of maternal behavior in virgin and postparturient rats following sagittal plane knife cuts in the preoptic area-hypothalamus. *Behav. Brain Res.*, 9,337-360.

Mirenowicz, J. and Schultz, W. (1996). Preferential activation of midbrain dopamine neurons by appetitive rather than aversive stimuli. *Nature,* 379, 449-451.

Mishkin, M. (1964). Perseveration of central sets after frontal lesions in monkeys. In JM Warren., K. Akert, (eds). *The frontal granular cortex and behavior.* McGraw-hill, New York. pp. 219-241.

Mogenson, G.J., Jones, D.L. and Yim, C.Y., (1980). From motivation to action: Functional interface between the limbic system and the motor system. *Prog. Neurobiol.,* 14, 69-97.

Moncho-Bogani, J., Martinez-Garcia, F., Novejarque, A. and Lanuza, E. (2005). Attraction to sexual pheromones and associated odorants in female mice involves activation of the reward system and basolateral amygdala. *Eur. J. Neurosci.,* 21, 2186-2189.

Morgan, H.D., Watchus, J.A., Milgram, N.W. and Fleming, A.S. (1999). The long lasting effects of electrical stimulation of the medial preoptic area and medial amygdala on maternal behavior in female rats. *Behav. Brain Res.*, 99, 61-73.

Morgan, M.A. and Pfaff, D.W. (2002). Estrogen´s effects on activity, anxiety, and fear in two mouse strains. *Behav. Brain Res.,* 132, 85-93.

Mosovich A. and Tallaferro A. (1954). Studies on EEG and sex function orgasm. *Diseases of the Nervous System*, 15,218-220.

Newman, S.W. (1999). The medial extended amygdale in male. *Ann. NY Acad Sci.,* 877, 242-257.

Newman, S.W. (2002). Pheromonal signals access the medial extended amygdale: one node in a proposed social nehabior network. In DW Pfaff., AP Arnold., AM Etgen., SE Fahrbach., and Rubin, R.T. (eds.) *Hormones, brain and behavior.* Vol. 2, Academic Press, San Diego. pp. 17-32.

Nishino, H., Ono, T., Muramoto, K., Fukuda, M. and Sasaki, K. (1986). Neuronal activity in the ventral tegmental area (VTA) during motivated bar press feeding in the monkey. *Brain Res.*, 413, 302-313.

Numan, M. (1974). Medial preoptic area and maternal behavior in the female rat. *J. Comp. Physiol. Psychol.*, 87, 746-759.

Numan, M., Rosenblatt, J.S. and Komisaruk, B.R. (1977). Medial preoptic area and onset of maternal behavior in the rat. *J. Comp. Physiol. Psychol.*, 91, 146-164.

Numan, M. and Insel, T.R. (2003).The neurobiology of parental behavior. *Hormones Brain and behavior.* Springer-Verlag, New York.

Oldenburger, W.P., Everitte B.J., and DeJonge, F.H. (1992). Conditioned place preference induced by sexual interactionin female rats. *Horm. Behav.*, 26, 214–228.

Olds, J. (1973). Multiple unit recording from behaving rats. In RF Thompson., MM Patterson, (eds). *Bioelectric recording techniques.* Vol. 1A. Academic Press, New York. pp. 165-198.

Oley, N.N. and Slotnick, B.M. (1970). Nesting material as a reinforcement for operant behavior in the rat. *Psychonom. Sci.,* 21, 41-43.

Paredes, R.G., and Alonso, A. (1997). Sexual behavior regulated (paced) by the female induces conditioned place preference. *Behav. Neurosci.*, 111, 123–128.

Paredes, .R.G, and Baum, M.J. (1995). Altered sexual parter preference in male ferrets given excitotoxic lesions of the preoptic area antrerior hypothalamus. *J. Neurosci.,* 15, 6630.

Paredes, R.G., Highland, L. and Karam, P. (1993). Socio-Sexual behaviour in male rats after lesions of the medial preoptic area: evidence of reduced sexual motivation. *Brain Res.,* 618, 271-276.

Paredes, R.G., Tzschentke, T. and Nakach, N. (1998). Lesions of the medial preoptic area/anterior hypothalamus (MPOA/AH) modify partner preference in male rats. *Brain Res.,* 813, 1-8.

Passingham, R. (1993). *The frontal lobe*

Pfaff, D.W. and Sakuma, Y. (1979). Facilitation of the lordosis reflex of female rats from the ventromedial nucleus of the hypothalamus. *J. Physiol.*, 288, 189-202.

Pfaff, D.W. and Agmo, A. (2002). Reproductive motivation. In R Gakkistel., H Pashler, (eds). *Steven´s Handbook of Experimental Psychology, Learning, Motivation and Emotion.* Wiley, (chapter 17).

Picazo, O. and Fernandez-Guasti, A. (1993). Changes in experimental anxiety during pregnancy and lactation. *Physiol. Behav.*, 54, 295-299.

Pietras, J.R. and Moulton, G.D. (1974). Hormonal influences on odor detection in rats: Changes associated with the estrous cycle, pseudopregnancy, ovariectomy, and administration of testosterone propionate. *Physiol. Behav.*, 12, 475-491.

Pryce, C.R., Dobeli, M. and Martin, R.D. (1993). Effects of sex steroids on maternal motivation in the common marmoset (*Callithrix jacchus*): development and application of na operant system with maternal reinforcement. *J. Comp. Psychol.,* 107, 99-115.

Reilly, D.R., Delva, N.J., and Hudson, R.W. (2000). Protocols for the use of cyproterone, medroxyprogesterone, and lwuprolide in the treatment of paraphilia. *Can. J. Psychiat.,* 45, 559-563.

Robinson, J.T. (1954). The necessity of progesterone with estrogen for the induction of recurent estrus in the ovariectomixed ewe. *Endocrinology,* 55, 403-408.

Rolls, E.T. (2000). The orbitofrontal cortex and reward. *Cereb. Cortex.* 10, 284-294.

Rolls, E.T. (2004). The functions of the orbitofrontal cortex. *Brain and Cognition,* 55, 11-29.

Rolls, E.T., Treves, A. (1998). *Neural networks and brain functions.* Oxford, Oxford University Press.

Rose, J.E. and Woolsey, C.N. (1948). The orbitofrontal cortex and its connections with the mediodorsal nucleus in rabbit, sheep and cat. *Res. Publ. Assn. Nerv. Dis.,* 27, 210-232.

Rosen, R.C., Goldstein, L., Scoles, V. and Lazarus, C. (1986). Psychophysiologic correlates of nocturnal penile tumescence in normal males. *Psychosom Med.*, 48, 423-429.

Rosenblatt, J.S. and Aronson, L.R. (1958). The decline of sexual behavior in male cats after castration with special reference to the role of prior sexual experience. *Behavior*, 12, 285-338.

Rowe D.W., and Erskine M.S. (1993). c-fos Proto-oncogene activity induced by mating in the preoptic area, hypothalamus and amygdala in the female rat: role of afferent input via the pelvic nerve. *Brain Research,* 621. 25-34.

Shaw, J.C. (1984). Correlation and coherence analysis of the EEG. *A tutorial review Int. J. Psychophysiol*, 1,255-266.

Shwartz, A. and Whalen R. (1965). Amygdala activity during sexual behavior in the male cat. *Life Sciences*, 4, 1359-1365.

Sheffield, F.D., Wulff, J.J. and Backer, R. (1951) Reward value of copulation without sexual drive reduction. *J. Comp. Physiol. Psychol.*, 44, 3-8.

Sheffield, F.D., Roby, T.B. and Campbell, B.A. (1954). Drive reduction versus consummatory behavior as determinants of reinforcement. *J. Comp. Physiol. Psychol*, 47, 349-354.

Shimamura, A.P., Janowski, J.S. and Squire, L.R. (1990) Memory for the temporal order of events in patients with frontal lobe lesions and amnesic patients. *Neuropsychologia*, 28, 803-813.

Slotnick, B.M., (1967). Disturbances of maternal behavior in the rat following lesions of the cingulated cortex. *Behavior* 29, 204-236.

Soulairac, A. and Soulairac, M.L. (1972). Action des substances neurostimulantes sur le comportement sexual du rat mâle après lésions du cortex cérébral. *J. Physiol.*, 65, Suppl. 3, 504.

Stamm, J.S., (1955). The function of the medial cerebral cortex in maternal behavior in rats. *J. Comp. Physiol. Psychol.*, 48, 347-356.

Stern, J.M. (1989). Maternal behavior: sensory, hormonal and neural determinants. In FR Brush., S Levine, (eds). *Psychoendocrinology.* Academic Press, New York. pp. 103-226.

Stern, J. M., (1996). Somatosensation and maternal care in Norway rats. In JS Rosenblatt., CT Snowdon, (eds). Parental care, evolution, mechanisms and adaptative significance. *Advances in the study of behavior.* vol. 25, Academic Press, New York. pp. 243-294.

Stern, J.M. and MacKinnon, D.A. (1976). Postpartum, hormonal, and non-hormonal induction of maternal behavior in rats: effects on T-maze retrieval of pups. *Horm. Behav.*, 7: 305-316.

Stoléru Serge, M.D., Grégoire, M.C., Gerard, D., Decety, J., Lafarge, E., Cinotti, L., Lavenne, F., Le Bars, D., Vernet-Maury, E., Rada, H., Collet, C., Mazoyer, B., Forest, M.G., Magnin, F., Spira, A. and Comar, D. (1999). Neuroanatomical correlates of visually evoked sexual arousal in human males. *Arch. Sex Behav.*, 28(1), 1-21.

Swanson, L.W. and Mogenson, J. (1981). Neural mechanisms for the functional coupling of autonomic, endocrine and somatomotor responses in adaptative behavior. *Brain Res. Rev.*, 3: 1-34.

Takenouchi, K., Nishijo, H., Uwano, T., Tamura, R., Takigawa, M. and Ono, T. (1999). Emotional and behavioral correlates of the anterior cingulated cortex during associative learning in rats. *Neuroscience*, 93: 1271-1287.

Timberlake, W. and Silva, K.M. (1995). Appetitive behavior in ethology, psychology, and behavior systems. In NS Thompson, (ed). *Perspectives in ethology*, Vol. 11. Behavioral design Plenum Press, New York. pp 211-253.

Thompson, A.C. and Kristal, M.B. (1996). Opioid stimulation in the ventral tegmental area facilitates the onset of maternal behavior in rats. *Brain Res.*, 743, 184-201.

Tucker, D.M. and Dawson S.L. (1984). Asymmetric EEG changes as method actors generated emotions. *Biol. Psychol.*, 19, 63-75.

Tzschentke, T.M. (2000). The medial prefrontal cortex as a part of the brain reward system. *Amino Acids*, 19, 211-219.

Uylings, H.B.M. and Van Eden, C.G. (1990). Qualitative and quantitative comparison of the prefrontal cortex in rat and in primates, including humans. In HBM Uylings., CG Van

Eden., JPC de Bruin., MA Corner., and MPG Feenstra (eds). *The prefrontal cortex: Its structure, function and pathology*. Prog Brain Res Elsevier, Amsterdam. 85: 31-62.

Uylings, H.B.M., Groenewegen, H.J. and Kolb, B. (2003). Do rats have a prefrontal cortex? *Behav. Brain Res.,* 146, 3-17.

Wallen, K. (1990). Desire and ability: hormones and the regulation of female sexual behavior. *Neurosci. Biobehav. Rev.*, 14, 233-241.

Walsh, C., Fleming, A.S., Lee, A. and Magnusson, J (1996). The effects of olfactory and somatosensory desensitization on fos-like immunoreactivity in the brains of pup-exposed postpartum rats. *Behav. Neurosc.*, 110, 1-20.

Whalen, R.E. (1961). Effects of mounting without intromission and intromission without ejaculation on sexual behavior and maze learning. *J. Comp. Physiol. Psychol.*, 54, 409-415.

Whalen, R.E., Beach, F.A. and Kuehn, R.E. (1961). Effects of exogenous androgen on sexually responsive and unresponsive male rats. *Endocrinology,* 68, 373-380.

Willie, R. and Beler, K.M. (1989). Castration in Germany. *Annual Sex Research* 2, 103-133.

White, N.R., and Barfield, R.J. (1989). Playback of female rat ultrasonic vocalizations during sexual behavior. *Physiol. Behav.*, 45, 229–233.

Wood , R.I. (1998). Integration of chemosensory and hormonal input in the male Syrian hamster brain. *Ann. NY Acad. Sci.*, 855, 362-372.

Xiao, K., Kondo, Y. and Sakuma, Y. (2005). Differencial regulation of female rat olfactory preference and copulatory pacing by the lateral septum and medial preoptic area. *Neuroendocrinology,* 81, 56-62.

Yonemori, M., Nishijo, H., Uwano, T., Tamura, R., Furuta, I., Kawasaki, M., Takashima, Y. and Ono, T. (2000). Orbital cortex neuronal responses during an odor-based conditioned associative task in rats. *Neuroscience*, 95,691-703.

Zitrin A., Jaynes, J. and Beach, F.A. (1956) Neural mediation of mating in male cats. III. Contributions of occipital, parietal and temporal cortex. *J. Comp. Neurol.*, 105, 111-122.

In: Encyclopedia of Neuroscience Research
Editors: Eileen J. Sampson and Donald R. Glevins
ISBN 978-1-61324-861-4

Chapter IV

From Conflict to Problem Solution: The Role of the Medial Prefrontal Cortexin the Learning of Memory-Guidedand Context-Adequate Behavioral Strategies for Problem-Solving in Gerbils

Holger Stark
Leibniz Institute for Neurobiology,
Brenneckestr. 6, D-39118 Magdeburg, Germany

Abstract

We conducted an investigation of the way in which animals acquire problem-solving behavioral strategies. The focus was on the search for sequential feedback rules that guide an individual along the stepwise learning process from the initial conflict generated by a novel situation to the context-adequate, problem-solving and, therewith, successful behavioral strategy.

The medial prefrontal cortex plays a complex role in acquisition and storage of behavioral strategies. It is an important part of the internal reward system, inducing a motivational drive for the development of a memory-guided, goal-directed behavioral strategy. This view is also strongly supported by our work on the learning-related dynamics of prefrontal dopamine release.

Using Mongolian gerbils (Meriones unguiculatus) as an animal model, we found that the formation of an avoidance strategy in a shuttle-box during the stage of avoidance learning is accompanied by a strong increase of the activity in the prefrontal dopamine system. Cortical theta activation was found to be, overall, negatively correlated with the formation of avoidance strategy during learning. In learning stages preceding the stage of avoidance learning we found different decreases of theta activity. However, when the animals entered the stage of avoidance learning, i.e., the phase when prefrontal dopamine release increases, theta activity increased again as well. A detailed analysis of theta activation revealed unique events in single trials marking the transition between behaviorally-defined learning stages and reflecting the actual change of the task-context for the animal.

Based on the correlation between behavioral data and data of learning-relevant physiological parameters, i.e., prefrontal dopamine and theta activation, we discuss cognitive processes like feedback learning, error processing, planning, decision making, and short-term memory as components of working memory functions during stepwise learning. The detailed analysis of learning processes in simple avoidance conditioning paradigms is a prerequisite for a deeper understanding of information processing during the goal-directed formation of behavioral strategies.

Introduction

The tremendous capability of living creatures to individually acquire and store information ensures their survival by fast and adequate adaptation to their environment. Therefore, learning and memory have multiple facets and involve a complex and distributed network in the brain [Eichenbaum, 2008]. Our contribution in the field of cognition will be limited to observations on cortical and prefrontal dopamine as well as cortical theta activity during aversive learning and retrieval. Our hypotheses are derived from correlations between the behavior of learning animals and the aforementioned physiological learning-dependent observables. Nevertheless, we are dealing here with fundamental processes of individual acquisition and storage of information. The discussion will focus on two points: (1) how the learning process is initiated, and (2) the way in which the learner forms a goal for guiding the learning process.

This chapter is composed in the form of a lab report to illustrate that the sequence of experiments and discussions presents a learning process itself driven by steadily arising scientific questions.

Animal Model

The animal model we used allows a stringent formation of both simple and more complex behavioral strategies by the individuals. Parallel to the conditioning procedure, we recorded physiological, learning-relevant parameters like prefrontal dopamine release and theta activation to interpret our behavioral observations. We chose Mongolian gerbils because the natural abilities, i.e., attention, cognitive flexibility, and consciousness of that small desert rodents make them appropriate for fast conditioning in learning models like the shuttle-box., one-year-old male animals were trained to develop aversively-motivated avoidance and discrimination strategies over several days and were subjected to parallel long-term brain microdialysis or recording of electroencephalograms over the right prefrontal cortex as a learning-relevant brain area.

Both avoidance and discrimination learning in the shuttle-box represent operant conditioning paradigms including aversively-motivated classical conditioning [Riess, 1972; Wasserman and Miller, 1997]. Conditioned tone stimuli (CS) were paired with footshock punishment as an unconditioned stimulus (UCS). In the individual animal, the footshock initiates a conflict. It can either simply tolerate this punishment or find a strategy to escape from it. Thus, as a first important issue (1) the origin of the learning process is a conflict realized by the individual, which represents an actual detraction from familiar existence conditions in the present surroundings. The general goal of the learning individual is then to re-establish convenient environmental life conditions. This delivers, so to speak, the "potential energy" for the learning process in which the individual adapts to the changed environmental situation. In the present experiments the strength of the conflict is related to the individual's inability to tolerate the punishment. It thus depends on the strength of the footshock and on its stringency in the behavioral context, but also on a subjective evaluation of the conflict by the animal. Furthermore, the animal has no chance to flee from the uncomfortable situation of footshock punishment during shuttle-box learning. Therefore, the only way to minimize the conflict strength is to acquire an adequate behavioral strategy related to the shuttle-box context. The animal creates coordinated associations making use of the formerly indifferent, meaningless tone stimuli paired with the unconditioned footshock. During learning, the tones become conditioned stimuli carrying after learning the matured conditioned avoidance response, or containing the information for the immediately following decision making in the case of acquired discrimination strategy, respectively. The stepwise formation of essential associations requires cognitive performances such as comparing of facts, reasoning, comprehension, planning, error processing and decision making, and feedback processing.

At the beginning of learning the animal does not know the concrete behavioral strategy that solves the conflict. Therefore, the question remains in which way an individual can find out by itself how to direct its behavior towards the concrete goal, in order to cope with the conflicting situation. We hypothesize that during stepwise learning the gerbil evaluates trial outcome by trial outcome for conflict attenuation caused by its own successful actions in the shuttle-box. Any realized conflict reduction will be perceived as a useful step in the direction of conflict solving. Hence, the second important issue of this report (2) concerns the realization of useful actions with respect to conflict reduction by an internal emotional reward percept that provides the key information to reinforce and to rule the development of a behavioral strategy for the solution of the problem. Therefore, we propose that the internal reinforcement as the motivational side of learning essentially contributes to the goal-directed formation of a behavioral strategy.

Transient increase of activity in cortical and prefrontal dopamine systems indicate the formation of relevant associations underlying a successful behavioral strategy

Differential Activation of the Dopamine System in Auditory Cortex in Trained, Pseudotrained, and Auditory Control Animals during Avoidance Conditioning

The first idea about the learning stage-dependent involvement of cortical dopamine systems in the formation of behavioral strategies in gerbils we derived from investigations of the auditory cortex of gerbils [Stark et al., 1997]. This brain area has been shown to be a site of widespread neuronal learning processes especially in the context of auditory conditioning behavior with behavioral and motivational relevance of incoming information [Scheich and Zuschratter, 1995; Zuschratter et al., 1995; Ohl and Scheich, 1996].

We investigated whether the dopaminergic and serotonergic systems are involved in auditory cortex learning. Chronic microdialysis from the auditory cortex was obtained before, during and after daily footshock avoidance training. Levels of extracellular homovanillic acid (HVA) as metabolite of the dopaminergic transmission as well as 5-hydroxyindoleacetic acid (5-HIAA) as metabolite of the serotonergic transmission were analyzed as estimates of the responses of both systems in the auditory cortex during acoustic learning in the shuttle-box. The basal levels of dopamine and serotonin in the microdialysis samples of most cases were below the reliable detection limit [Stark et al., 1997].

In the group of trained animals subjected to paired tone-footshock combinations, the values of the dopamine metabolite HVA revealed a transient strong increase of the activity of cortical dopamine system shortly after the start of the first training session, i.e. in the stage of strong increase of conditioned avoidance responses (Figure 1).

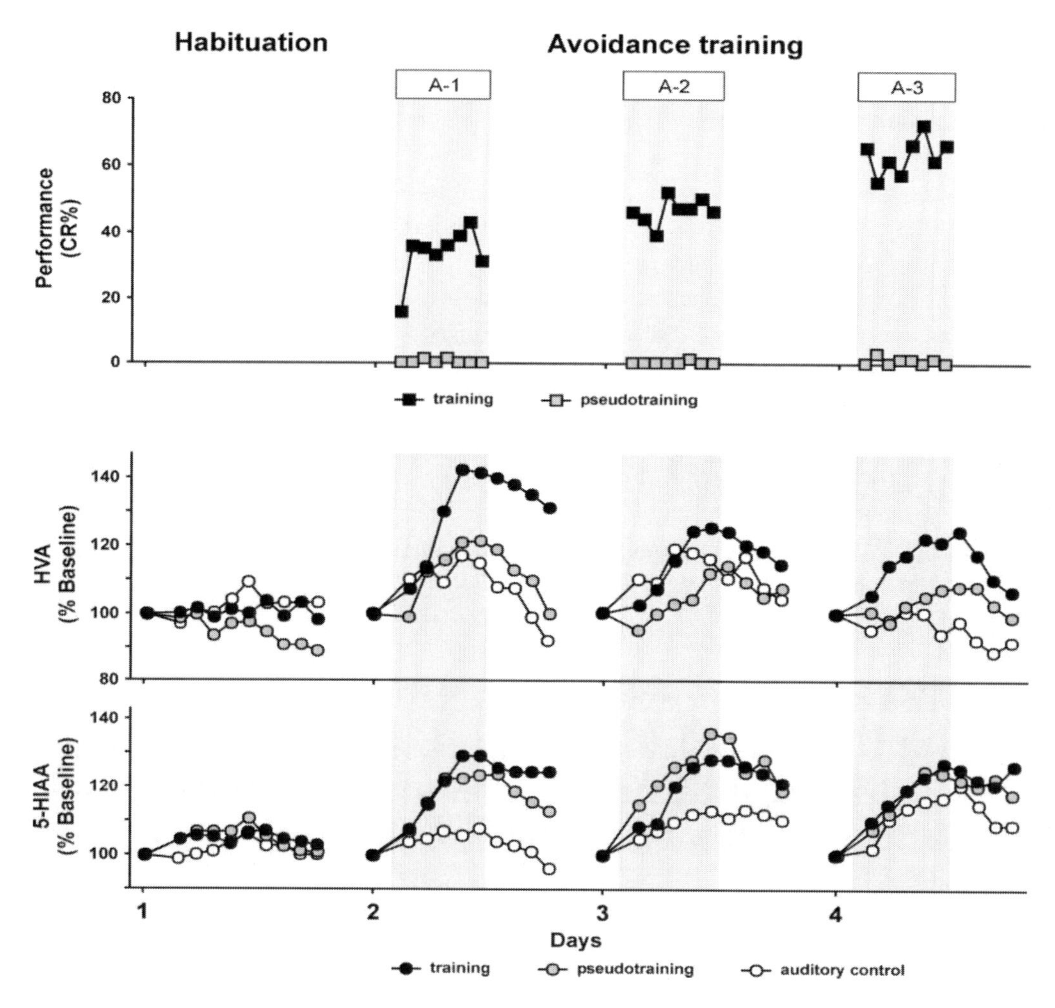

Figure 1. Temporal profiles of averaged learning curves (above) during the first day (unstimulated habituation), the first avoidance training session A-1 on the second day and the following two training sessions A-2 and A-3 on days 3 and 4 (retrieval), and of the relative content of HVA (middle) and 5-HIAA (bottom) in 15-min microdialysis samples from auditory cortex of trained, pseudotrained (no contingency between tones and footshocks), and auditory control animals (only tones without footshock).

The response of the dopaminergic system appeared to reflect the initial formation of behaviorally relevant associations exclusively during the first training day. In comparison, only a moderate increase in cortical dopamine was found in pseudotrained and in auditory control animals in the first training session. Whereas auditory controls received only the tones, footshocks were presented randomly within the next intertrial interval in the pseudotrained animal group. The duration of the footshock used thereby matched the footshock duration applied to the trained animals. In training sessions 2 and 3 all three animal groups showed only a moderate increase in the dopamine metabolite HVA.

The serotonergic response appeared to be correlated with the stress level of animals. The 5-HIAA values were increased in trained and pseudotrained animals in comparison to auditory controls without footshock punishment (Figure 1). The serotonin system seems to be rather involved in the stressful aspect of the situation by participating in the regulation of anxiety and defensive behavior [Deakin and Graeff, 1991; Graeff, 1993; Glass et al., 1993; Young, 1993]. From the HVA data we conclude that (1) the strong increase of cortical dopamine activity is correlated only with the formation of the relevant associations between tone stimulus and footshock connected with further associations like instrumental responses required for a context-adequate, conflict-solving behavioral strategy [Sawaguchi et al., 1990; Imperato et al., 1992; Goldman-Rakic, 1992; Young, 1993]. Supporting this assumption, it was previously shown in gerbil primary auditory cortex that pairing of a tone with an aversive stimulus led within minutes to specific changes of frequency receptive fields of about 60 % of excited neurons [Ohl and Scheich, 1996]. Suggestions that dopamine is acting as an "internal reinforcer" necessary only as long as the initiation of an appropriate behavioral strategy is in progress came previously from results in medial prefrontal cortex [Brozoski et al., 1979; Sawaguchi and Goldman-Rakic, 1991; Sokolowski et al., 1994], nucleus accumbens [Salamone, 1994], and parietal cortex [Schultz et al., 1993]. After the formation of a matured behavioral strategy, the activity of the dopamine system is only moderately increased in further avoidance training sessions. Therefore, the retrieval of the avoidance strategy in a shuttle-box (2) seems to be limited to the detection of the conditioned tone stimulus with a lower demand on the cognitive system. A strong increase of prefrontal dopamine was missing in pseudotrained animals with unpaired tone–footshock presentations. We found in gerbils subjected to an escape conditioning with a gap between a 5-s tone offset and the footshock onset of 20 s no increase of the prefrontal dopamine release (unpublished results). So we conclude that the animals had no possibility to associate the key stimuli due to a lacking temporal contiguity [Bischof et al., 2000]. (3) Non-specific stress like footshock punishment does not strongly increase the activity of dopamine system [Herman et al., 1982; Abercrombie et al., 1989; Hernandez and Hoebel, 1990; Imperato et al., 1992]. The moderate increases of cortical dopamine in pseudotrained animals receiving the same number of auditory and electric stimuli as trained animals, and in auditory control animals subjected exclusively to tone stimuli reflects presumably the initial novelty of the applied stimuli and, in the course of training sessions, a readiness for possible associative tasks. Thus, the dopamine system of the sensory auditory cortex is responsive to associative learning, i.e. (4) the function of this brain area seems not to be limited to perception of incoming auditory stimuli. Rather, the sensory field of auditory cortex and the medial prefrontal cortex with its central executive functions [Atkinson and Shiffrin, 1968, Baddeley, 1998] seem to be functionally linked by their dopamine systems during information processing.

Selective Activation of the Prefrontal Dopamine System in the Stage of Strong Increase of Performance during Avoidance Conditioning

From investigations in the auditory cortex the question arose whether a transient increase in dopamine system also occurred in medial prefrontal cortex during formation of key associations in avoidance conditioning. Microdialysis from right medial prefrontal cortex and recordings of learning behavior in the shuttle-box from gerbils subjected to avoidance training sessions were performed. Thereby, individual learning progress was monitored and related to dopamine levels [Stark et al., 1999]. It was of particular interest, whether prefrontal dopamine increase was correlated with the establishment or with the retrieval of the avoidance strategy.

Three behavioral parameters, the attention responses in terms of movement arrest of animals after CS onset as indication of auditory signal detection, the reaction time from tone onset to change of shuttle-box compartment, and the avoidance performance as percent conditioned responses to the conditioned tone stimulus, were used to divide post hoc individual performances into three behavioral stages: (1) low improvement of performance, (2) strong improvement of performance, and (3) constant high performance.

In the stage of low improvement of performance, there was a steep increase in the number of attention responses from initially low levels, a moderate decrease of reaction times, and a moderate increase of avoidance responses to less than 20% (Figure 2). During this stage, the initially indifferent tone became a meaningful tone signal, announcing the footshock, although the tone-dependent avoidance strategy was still undeveloped. There was no increase in prefrontal dopamine at this stage (Figure 2).

During strong improvement of performance the percentage of attention responses was already between 75 and 100%. The reaction times became shorter and jittered around the critical limit of 5 s, i.e. the jumps occurred frequently already during the presentation of the conditioned stimulus leading to footshock avoidance. Animals in this stage relevant for formation of the avoidance strategy showed now a stronger increase in performance (Figure 2). During association of the tone signal with the unconditioned footshock resulting in the avoidance response, prefrontal dopamine steadily and significantly increased until the middle of the training session and dropped before the session was finished (Figure 2).

In the last stage of constant high performance, animals always showed attention responses, reaction times before footshock onset, and a stable high level of avoidance responses (Figure 2). This behavioral stage characterized by the recall of the avoidance strategy showed basal dopamine levels again (Figure 2). The results confirm our hypothesis that the strong increase of dopamine due to the formation of relevant associations for strategy formation was also found in prefrontal cortex.

The temporal requirement in the acquisition of the avoidance strategy appeared to be highly individual. Aiming at a better time resolution for the correlation between individual behavior and dopamine content in dialysates, we exemplary collected dialysates from the right medial prefrontal cortex during the avoidance training session over 30 trials from trial to trial at 2-min intervals in a single experiment [Stark et al., 2000] (Figure 3).

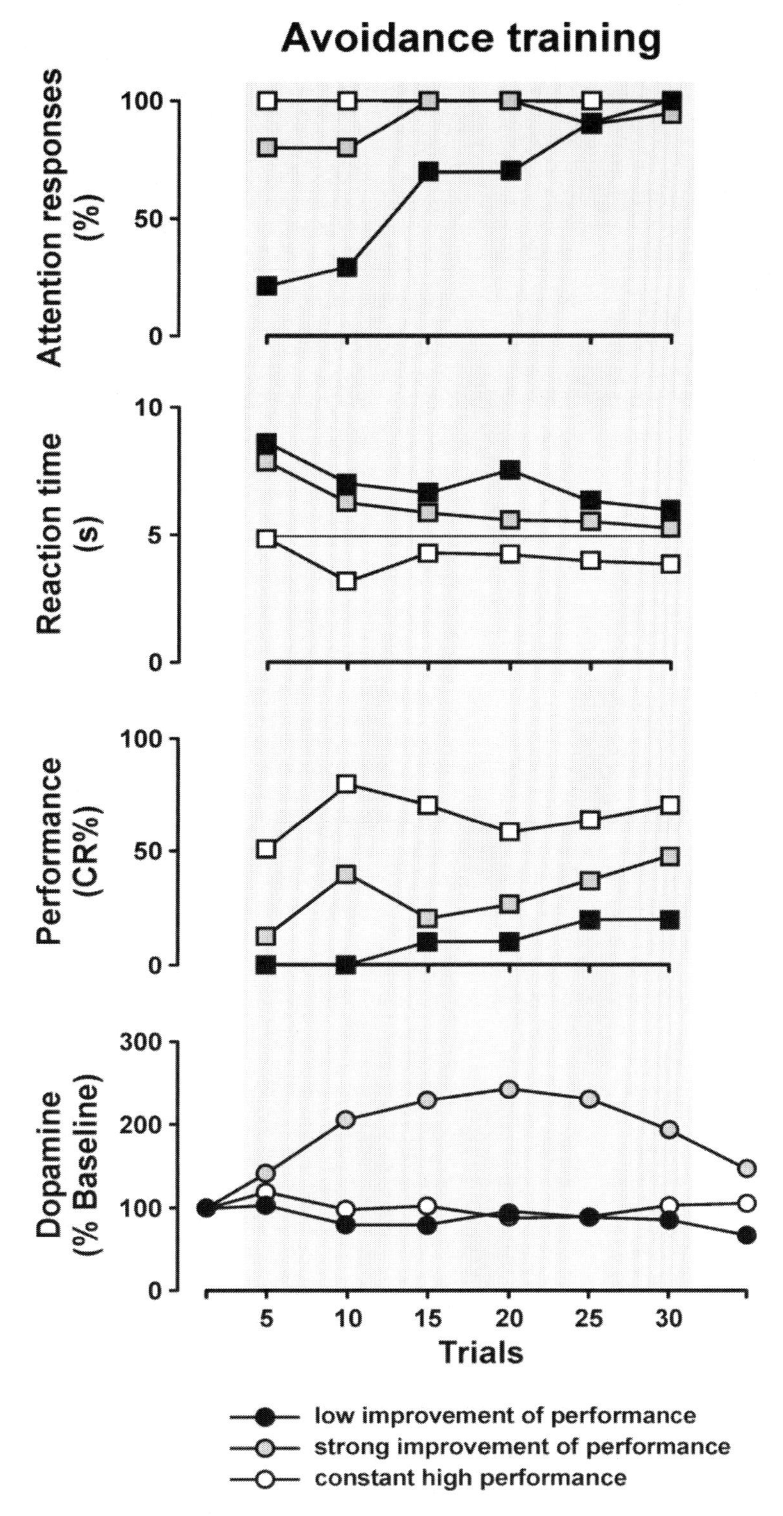

Figure 2. Profiles of the number of attention responses, the reaction times, the performances, and of the relative content of dopamine in dialysates from medial prefrontal cortex of gerbils subjected to avoidance training in the shuttle-box during the behavioral stages of low improvement of performance, of strong improvement of performance, as well as of constant high performance. During the stage of strong improvement of performance a transient strong increase of dopamine values was evident.

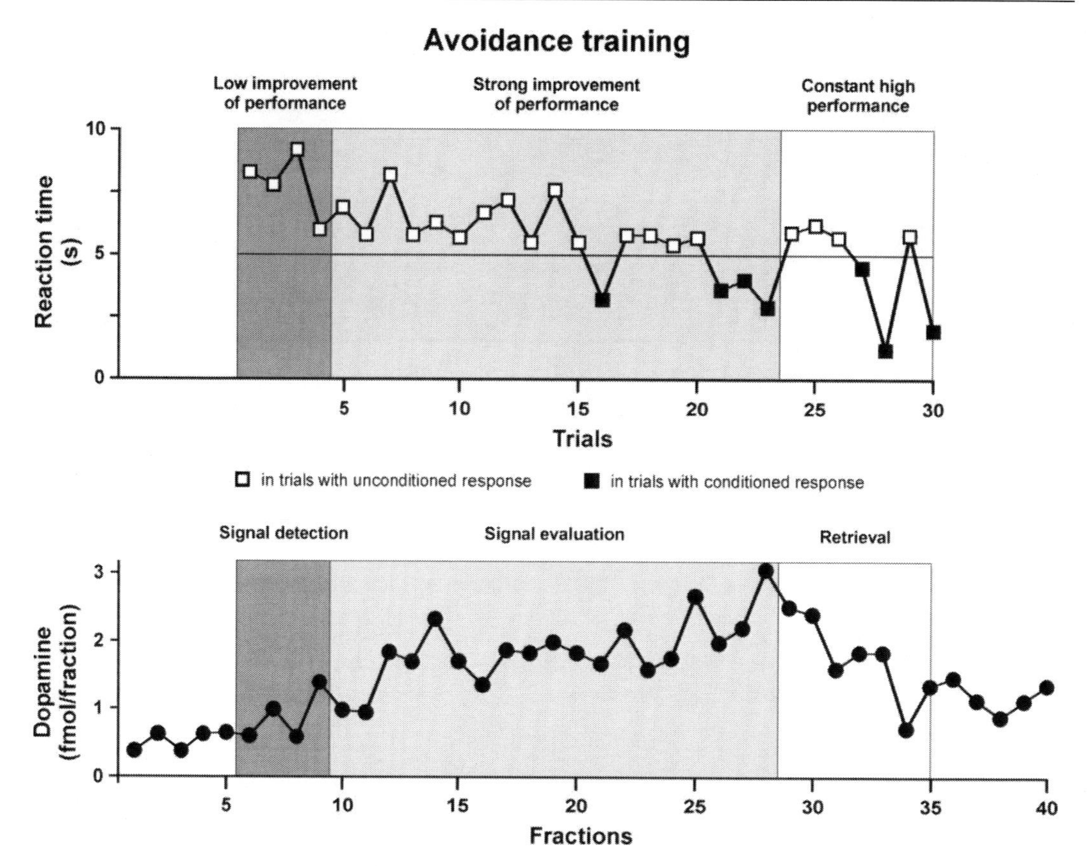

Figure 3. Individual temporal profiles of reaction times and dopamine values of microdialysis samples from prefrontal cortex in the course of avoidance conditioning in the shuttle-box. A reaction time longer than 4.9 s indicates an unconditioned escape response and shorter than 5.0 s a conditioned avoidance response, respectively. Correlation between reaction times and dopamine values from 2-min trial to the next 2-min trial converge into the division of avoidance strategy formation into three learning stages (1) "signal detection", (2) "signal evaluation", and (3) "retrieval of avoidance strategy".

For dopamine measurement in the 2-min dialysate fractions a highly sensitive HPLC analysis was developed. This closer correlation between individual behavioral measures and the learning relevant prefrontal dopamine release in single learning animals enabled us to interpret the cognitive background of the defined behavioral stages. From the first trial to about trial 4 of the avoidance training session we found reaction times far from the 5-s footshock onset time without avoidance responses as well as dopamine values close to the basal level. This first behavioral stage of "Low improvement of performance" [Stark et al., 1999] from trial 1 to 4 was probably characterized by promotion of the initially indifferent tone stimulus to a meaningful signal which announced the following unconditioned footshock. However, this essential key association as prerequisite for further learning still lacks information about the temporal contingency between tone onset and footshock onset, the clear direction for the escape response, namely the change into the contralateral shuttle-box compartment, as well as many instrumental requirements for avoidance behavior. This information, however, is crucial for the animal to build up a successful behavioral strategy to cope with the footshock. Accordingly, the behavioral stage of "Low improvement of performance" was defined as the cognitive stage of (1) "Signal detection". From trial 5 the reaction times became continuously shorter up to trial 23 including reaction times lower than 5 s. In contrast, the dopamine values steadily increased until trial 23. At the end of the second behavioral stage of "Strong improvement of performance" the frequency of responses with a reaction time lower than 5.0 s was already high and the increasing dopamine values reached their maximum. We assume that by about trial 24 the gerbil had acquired the essence of the behavioral strategy which includes the acquisition of further key associations.

The individual first learned to escape into the safe contralateral shuttle-box compartment after footshock onset. At the same time, the animal had started to test the temporal contingency between tone and footshock onset to recognize the possibility to avoid the footshock up to 4.9 s after tone onset. Thereby, the perception of a conditioned avoidance response after the execution of a trial with a reaction time lower than 5.0 s presumably was the most important association for the animal to acquire the avoidance strategy. At this point the individual was qualified to realize that the conflict could be solved by own goal-directed behavior. The behavior in the stage of "Strong improvement of performance" comprises the cognitive side of (2) "Signal evaluation", i.e. the tone signal was attributed with the new meaning signaling the possible escape from or avoidance of the footshock. Note that a clear separation of the stage of "escape learning" from the stage of "avoidance learning" was impossible at that time, due to the fact that a close correlation between behavioral measures and reliable prefrontal dopamine values from dialysates was very limited in this respect. For this distinction, another experimental approach was necessary (see below). From trial 24 in the stage of "Constant high performance" the gerbil started to retrieve the avoidance strategy, and the dopamine values started to attenuate. Presumably, there was no further need of forming additional key associations. Now the behavioral strategy was matured and stored in the memory. The cognitive reflection of this stage of "Constant high performance" may be described best as cognitive stage of (3) "Retrieval of avoidance strategy".

The Role of the Cortical Dopamine System in Cognition

Many aspects of adaptive behavior such as emotional states and motivational and cognitive functions interact with the dopamine system of medial prefrontal cortex. Prefrontal dopamine efflux was increased in response to arousal, stressful situations and appetitive stimuli [Abercrombie et al., 1989; Hernandez and Hoebel, 1990; Broersen et al., 1995; Feenstra and Botterblom, 1996; Spanagel and Weiss, 1999; Feenstra, 2000; Wightman and Robinson, 2002]. Electrophysiological studies of midbrain dopamine neurons terminating in the medial prefrontal cortex revealed that dopamine is integrated in functions of prediction and reward [Fibiger and Phillips, 1986; Koob, 1992; Schultz et al., 1993; Schultz et al., 1997; Schultz, 2002]. Cognitive aspects like attention, cognitive flexibility and consciousness as well as working memory functions are also mediated by dopamine in medial prefrontal cortex [Atkinson and Shiffrin, 1968; Goldman-Rakic, 1992; Goldman-Rakic, 1995; Watanabe et al., 1997; Baddeley, 1998]. Cognitive functions are disturbed by dopamine depletion or D1-receptor blockade [Brozoski et al., 1979; Sawaguchi et al., 1990; Sawaguchi and Goldman-Rakic, 1994; Sokolowski et al., 1994; Seamans et al., 1998]. Dopamine agonists can improve attention and impaired working memory [Arnsten et al., 1994; Granon et al., 2000].

The working memory based on short-term memory principles is vulnerable to dopamine loss [Goldman-Rakic, 1992]. The hypothesis of working memory implies that a recent cognitive information is temporarily kept in such a way that it can be related to the total experience stored in the long term memory as well as to the succeeding input of information

[Atkinson and Shiffrin, 1968; Baddeley, 1998]. Presumably, the significance of the working memory system extends far beyond the involvement in specialized working memory tasks like matching or non-matching to sample tasks. It is assumed that this system is used for the performance of many cognitively demanding processes involved in comprehension and problem solving [Eysenck et al., 1997].

Obviously, in the stage of "Signal evaluation" animals acquired the most important information for the formation of the shuttle-box avoidance strategy, i.e. they learned to avoid the punishment by the unconditioned stimulus changing the compartment within the period between onset of the conditioned stimulus and onset of the unconditioned footshock [Fuster, 1989]. We assume that hereby the previously accumulated experience with respect to the developing avoidance strategy is repeatedly compared, combined and completed by continuing feedback information from recent trials. After execution of the recent trial the animal extracts the actual information about an executed unconditioned escape response with footshock punishment or a conditioned avoidance response [Sternberg, 1969] leading to „memory-guided performance" [Goldman-Rakic, 1992] stepwise structured by single trials [Rescorla, 1988]. We assume further that during "Signal evaluation" short-term memory mechanisms could be utilized as proposed for working memory tasks [Atkinson and Shiffrin, 1968; Eysenck et al., 1997; Baddeley, 1998], i.e. to hold the feedback information of recent trials for adequate association with stored experience about shuttle-box learning. This feedback is necessary for comparing the outcome of conditioned compartment changes without footshock with compartment change triggered by footshock. It is further involved in making decisions like "jump before 4.9 s" or "wait for footshock", in planning, and in error processing. All these different sub-processes play an important role in experiments designed as delayed alternation tasks accompanied by strong prefrontal dopamine release. During such a delayed alternation task a prominent prefrontal dopamine increase in monkeys indicating the involvement of working memory was also observed [Watanabe et al., 1997]. The effective learning of an operant discrimination task was accompanied by a significantly increased prefrontal dopamine efflux in rats [Yamamuro et al., 1994; Izaki et al., 1998]. The prominent prefrontal dopamine increase in certain stages of avoidance conditioning in gerbils corresponded to prefrontal dopamine increase in monkeys during a delayed alternation task indicating the involvement of working memory principles [Watanabe et al., 1997]. During retrieval of the established shuttle-box avoidance strategy, the activity of prefrontal dopamine system was strongly attenuated. Therefore, we conclude that working memory principles are presumably not required anymore, i.e. in the stage of retrieval the tone signal implements the full behavioral strategy, and the tone onset triggers a procedure related to reference memory. However, the retrieval of a strategy implicating the "delayed alternation task" including decision making requires also working memory processes [Watanabe et al., 1997].

Thus, the working memory is considered as a tool for memory-guided information processing, i.e. with respect to avoidance learning the correct association between the incoming information of recent trials with information stored in long-term memory. Moreover, the stepwise formation of a behavioral strategy implicates that the animal rules its behavior to direct it towards the goal of learning.

Findings that the activity of prefrontal and further dopamine systems are increased by appetitive stimuli [Hernandez and Hoebel, 1990; Schultz, 1992; Feenstra and Botterblom, 1996] and that dopamine could play a role in internal reinforcement for optimization of learning processes [Schultz et al., 1997; Schultz, 2002; Fiorillo et al., 2003] may lead to the assumption that working memory principles in medial prefrontal cortex are presumably involved in the interaction with dopamine systems of limbic system signaling the individual a correct compartment change before footshock onset as a conflict-solving „internal reward“ and in this way as conditioned response [Koob, 1992; Young, 1993]. The question remains in which areas of the brain these signals are initiated that direct the development of a behavioral strategy towards a goal by reward related reinforcement of correct responses.

Activation of Prefrontal Dopamine System during Behavioral Strategy Changes from Avoidance to Discrimination

In summary, what we can learn from the described experiments is that (1) the prefrontal dopamine system is mainly activated presumably in a stage of the formation of key associations related to the recognition of the temporal contingency between the conditioned tone stimulus and the following unconditioned footshock as well as to the recognition of a conditioned avoidance response. Assumingly, working memory principles are markedly involved in this type of information processing indicated by transient nevertheless strong increase of activity of prefrontal dopamine system. (2) During retrieval of the matured avoidance strategy a remarkable activation of prefrontal dopamine system was lacking. Presumably, the retrieval of the avoidance strategy as a signal-coded procedure does not extensively require working memory principles. We hypothesized that the acquisition of a new behavioral strategy is generally accompanied by this extra prefrontal dopamine release.

A further experiment aimed at testing this hypothesis. For the acquisition of a pre-experience, gerbils were trained in the shuttle-box to generalize two conditioned tone stimuli a1 and a2 associated with the GO-condition. Relearning was induced with discrimination training (GO- vs. NO-GO-trials) [Wetzel et al., 1998] by changing the condition of tone a2 from GO to NO-GO. Using microdialysis, medial prefrontal dopamine efflux was compared between the retrieval of the accomplished avoidance response to the two tone stimuli and the relearning requiring discrimination behavior.

After avoidance pre-training the animals were familiar with the complete shuttle-box context including conditioned tone and unconditioned footshock stimuli. By this, arousal and stress as possible sources of dopamine increase [Abercrombie et al., 1989; Feenstra and Botterblom, 1996; Feenstra, 2000] were minimized in the relearning part of the experiment. Thus, any strong prefrontal dopamine increase should relate to the relearning, i.e. to the formation of associations for the new discrimination behavior. Compared to the avoidance learning the rather difficult relearning of discrimination behavior is assumed to be cognitively more demanding. Accumulation of partial experiences, comparing of events, comprehension of their relationships, reasoning, decision making, error processing, and planning could be again performed by working memory principles [Stark et al., 2004]. The reversal of GO-condition associated to tone stimulus a2 entailed important consequences for the animals. Gerbils were subjected to an initial misguidance and concurrently to a loss of successful avoidance of punishment. The question arose in which way an animal could develop by itself the goal-directed behavioral strategy for successful discrimination.

After extensive avoidance pre-training and insertion of microdialysis probe into medial prefrontal cortex the gerbils showed an avoidance performance of about 80 $CR^{GO}\%$ to both tone stimuli considered as sufficient pre-experience with the stimuli and the paradigm. Both tone signals a1 and a2 were generalized by the animals as verified by randomized application. No significant differences in dopamine profiles between post hoc divided strongly and weakly relearning animals were found. Prefrontal dopamine increased moderately during retrieval of avoidance response in sessions A1-5 and A2-5 [Stark et al., 2004] (Figure 4).

In discrimination training session D-1 all gerbils responded incorrectly to tone a2 by crossing the hurdle as expected from their avoidance pre-experience and were unexpectedly subjected to a footshock punishment. Therefore, any increase of $CR^{NO\text{-}GO}\%$ started from a low level of performance. But also the GO-performance $CR^{GO}\%$ of both strongly and weakly relearning animals was strongly reduced in the first discrimination training session D-1 compared with the last avoidance training session A2-5 indicating that the meaning of GO-response of an avoidance strategy is completely different from the meaning of GO-response of a discrimination strategy from the cognitive point of view. Presumably, the recent experiences from NO-GO-trial outcomes to tone a2 induced wrong associations in planning of actions concerning the next GO-trials to tone a1. Consequently, the gerbils did not change the shuttle-box compartment to tone a1 and were, therefore, punished again. The formerly high GO-performance of strongly and weakly relearning individuals during the last avoidance training sessions A1-5 and A2-5 collapsed although the GO-condition associated with tone a1 was formally retained.

However, strongly relearning individuals increased significantly their GO-performance during discrimination training sessions D-2 and D-3 resulting in an increased discrimination performance in contrast to weakly relearning animals (Figure 4 top). Obviously, weakly relearning animals could not differentiate between the conditions of the two tone signals, and changed the shuttle-box compartment in the GO-trials mainly on footshock onset. Presumably, weakly relearning animals regressed to an escape strategy from footshock, an initial behavior that had been overcome during avoidance training in the pre-experience period. Consequently, they missed important components of discrimination learning by passively staying in the start compartment of the shuttle-box. Therefore, we assume that they kept inflexibly the generalized GO-condition associated to tone a2 and were not able to substitute it for the NO-GO-condition.

Despite similar behavior of all animals in session D-1 the strongly relearning individuals were obviously able to start instantly a complex relearning process based on an effective recombination of the signal tones a1 and a2 as well as their associated GO- and NO-GO-conditions to establish the essential associations for goal-directed learning. Probably, the rapid abolishment of the generalized GO-condition associated to both tone stimuli was the prerequisite for the following differential re-evaluation with respect to the tones a1 and a2 which were now differentially associated with GO- or NO-GO-conditions.

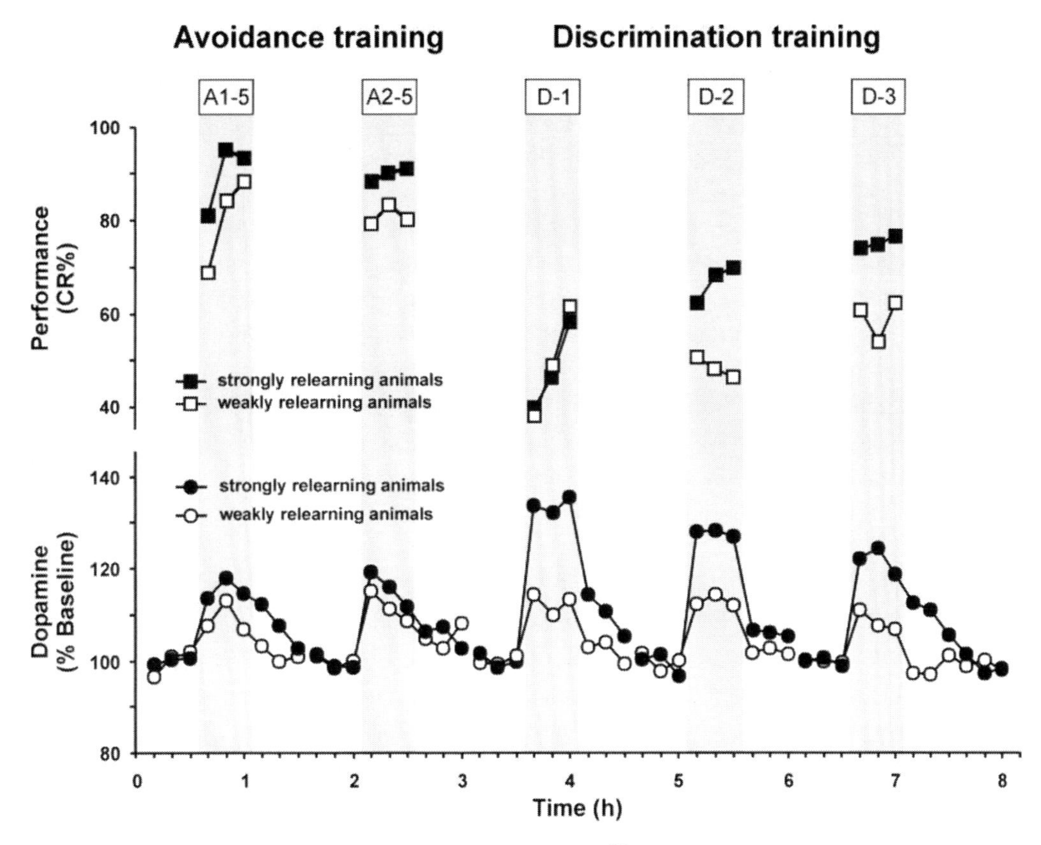

Figure 4. Temporal profiles of avoidance performances ($CR^{GO}\%$) and discrimination performances ($CR^{GO}\% + CR^{NO\text{-}GO}\%$) parallel to dopamine values in post hoc divided strongly relearning and weakly relearning animals. The last sessions for retrieval of avoidance strategy A1-5 and A2-5 and the discrimination training sessions D1 through D3 on the same day are covered. Top: Both strongly and weakly relearning animals retrieved the avoidance strategy in sessions A1-5 and A2-5. After change from GO- to NO-GO-meaning associated to tone a2 in discrimination training the performance in session D-1 dropped to about chance level. No differences between the two groups were found. The discrimination performance in strongly relearning individuals significantly increased in sessions D-2 and D-3 in comparison to weekly relearning individuals. Bottom: Relative prefrontal dopamine content in dialysate samples of strongly and weakly relearning animals across the avoidance training sessions A1-5 and A2-5 and the discrimination training sessions D-1, D-2, and D-3. In contrast to weakly relearning animals the strongly relearning individuals showed a significant extra dopamine increase during the sessions D-1, D-2, and D-3. Note that the extra dopamine increase was maximally in strongly relearning animals in session D-1 while there were no differences in discrimination behavior between strongly and weakly relearning animals.

Thus, only strongly relearning animals were able to realize with increasing distinctness the context-adequate consequences of tones a1 and a2 in GO- and NO-GO-trials during initial discrimination training.

The recombination of the two tone signals, the footshock, and the shuttle-box compartments as well as the temporal relationships between tone signals and footshock punishment [Barnet et al., 1991; Barnet and Miller, 1996; Meck, 1996] had to be established. The increasing frequency of correct GO- and NO-GO-responses resulted in decreasing punishments as consequence of their own actions in the shuttle-box. This process could mediate in strongly relearning individuals internally a reward percept which signals that the development of the discrimination strategy is well oriented to the goal of conflict solution.

This percept could guide the development of the behavioral strategy as internal reinforcement as already mentioned above in context of the establishment of the avoidance strategy.

The different cognitive performances of strongly and weakly relearning animals seemed to be reflected by strong or weak prefrontal dopamine release, respectively. After a moderate dopamine release in sessions A1-5 and A2-5 prefrontal dopamine was strongly elevated in the sessions D-1 through D-3 in strongly relearning animals in contrast to weakly relearning animals. Dopamine became attenuated during sessions D-2 and D-3 (Figure 4 bottom). These results confirmed the hypothesis that the acquisition of a new behavioral strategy causes an extra dopamine release in medial prefrontal cortex indicating the involvement of working memory [Goldman-Rakic, 1992; Goldman-Rakic, 1995; Baddeley, 1998; Durstewitz et al., 1999; Dreher and Burnod, 2002; Deco and Rolls, 2003]. The strong activation of prefrontal dopamine system in session D-1 indicated a type of information processing in strongly relearning animals which resulted in a subsequent rapid development of discrimination behavior in the following sessions D-2 and D-3.

Generally, a learning process stops when the appropriate behavior is acquired, i.e. when the trial outcomes are estimated by the individual as context-adequate in terms of conflict solution. The attenuation of the extra dopamine release in strongly relearning individuals in sessions D-2 and D-3 could indicate a reduction of the demand on the working memory system at an advanced stage of the development of the discrimination strategy. At this stage a partial retrieval of the conditioned response to conditioned tone stimuli is presumably permitted from the by then memorized discrimination strategy. However, the retrieval of the discrimination strategy requires always working memory because the animal has to decide in each trial whether the detected tone signal is associated with a GO- or a NO-GO-meaning, respectively, resulting in a certain strength of prefrontal dopamine release. In comparison, the retrieval of the avoidance response in sessions A1-5 and A2-5 resembled rather a sensory-guided task based only on the detection of the tone signal and was accompanied by an only moderate dopamine increase [Watanabe et al., 1997; Stark et al., 1999; Stark et al., 2000].

The relearning experiment confirmed (1) the coincidence of the formation of important associations and of transient extra prefrontal dopamine release at the beginning of relearning, i.e. the development of a new behavioral strategy indicating working memory functions. Obviously, the perception of an internal reward can guide the individual towards the goal of learning. The results obtained so far led us inescapably to the assumption that (2) the dopamine systems are learning stage-dependent tuned for fast and goal-directed adaptation of an individual to the requirements of environment.

Division of the Avoidance Strategy Formation into Discrete Learning Dependent Stages by Characteristic Changes of Cortical Theta Activity

From the aversive motivated learning experiments – avoidance as well as discrimination conditioning in the shuttle-box - we conclude that (1) the learning process is initiated by the unconditioned footshock punishment evoking a conflict in the individual. (2) The activity of cortical and prefrontal dopamine systems seems to increase during maturation of the avoidance or discrimination strategy, respectively, by associating the conditioned tone stimulus with the unconditioned footshock to build up a reliable behavioral strategy. Presumably in this stage, working memory with attributes like error processing, planning, decision making, and feedback processing is highly involved. The question arises whether there exist learning stage dependent signals governing the development of an adequate behavioral strategy for conflict solving as the goal of learning. Therefore, we used recording of electroencephalograms during avoidance conditioning with following theta analysis to warrant a sufficient temporal resolution for a close correlation between behavioral and learning relevant physiological data, a prerequisite for detection of such strategy-ruling signals.

Theta activity defined by the frequency range from 4 to 8 Hz in accordance with other studies in the gerbil [Bonner and Martinez, 1975], the rat [Krügel et al., 2003], or the mouse [Seidenbecher et al., 2003] appears to be a correlate for the intensity of information processing specifically during associative learning [Balleine and Curthoys, 1991; Asada et al., 1999; Ishii et al., 1999]. Accordingly, if the acquisition of the avoidance strategy consists of stages defined by behavioral and physiological observables, the stepwise learning could be reflected by fast changes of theta activation. The basic hypothesis of the experiment was that the strength of theta activity may characterize the course of intensity of information processing from the start of learning up to the retrieval of the matured behavioral strategy.

Generally, theta activity seems to be involved in brain states of arousal and consciousness specifically appearing during associative learning [Asada et al., 1999; Ishii et al., 1999]. Physiological studies in rat and human have also indicated that mental tasks involving working memory are accompanied by synchronized neuronal activity in the theta-frequency range [Givens, 1996; Sarnthein et al., 1998]. Oscillations of brain potentials between 4 and 8 Hz are generated in subjects during decision making [Jacobs et al., 2006]. Hippocampal theta activity involved in communication between prefrontal cortex and hippocampus plays a crucial role in encoding of information in working memory which is later retrieved for use in problem solving [Givens, 1996; Sarnthein et al., 1998; Jensen, 2005; Jones and Wilson, 2005]. The maintenance of hippocampal theta activity is essential in initial learning in rats [McNaughton et al., 2006]. Theta rhythms generated in hippocampus and cortex may nevertheless link hippocampal and cortical neuron ensembles that participate in representing different aspects of the training situation in discriminative avoidance conditioning [Talk et al., 2004]. Theta synchronization in humans coordinates the central executive circuits, including medial prefrontal cortex [Ishii et al., 1999; Mizuhara and Yamaguchi, 2007]. Together, such results suggest that theta activity as correlate for the intensity of information processing may also reflect learning during the formation of behavioral strategies that require a continuous updating of information.

We found an overall negative correlation of theta activity in the intertrial interval after compartment change with individual learning progress [Stark et al., 2007]. Beyond this general trend, a detailed comparison of actual theta activity with behavioral events in a trial by trial analysis could disclose markers for the formation of relevant associations and for the separation of stages of the learning process [Stark et al., 2008]. For a detailed analysis the theta activity in the intertrial epochs after compartment change until the start of a new trial was chosen. In these trial epochs an animal can update the previous cumulative behavioral feedback experience with the actual outcome of the preceding trial. The result of this processing may be used behaviorally for planning in the next trials. We assumed that during this epoch, the intensity of information processing in the course of learning is reflected most prominently.

As a fundamental result, the process of avoidance strategy formation was divided into four cognitive stages by unique marker trials (Figure 5). After an initial theta activity maximum the sharpest decline in theta power between any two trials (two big open circles on the left in Figure 5) was defined as the transition from stage 1 "signal detection" to stage 2 "escape learning" which started at the marker trial with lower theta activation. Subsequently, the theta power decreased gradually over a number of trials accompanied by a decrease of latency of the unconditioned responses after footshock. Across the trials with unconditioned responses during "escape learning" animals already showed reaction times shorter than 5 s, i.e. with reactions before the onset of the footshock, in some interspersed trials. While these singular events initially remained without consequences in the theta power there was always found one trial in each animal where such an avoidance of the footshock was followed by a sudden marked increase of theta power as referred to theta power in the preceding trials with reaction times shorter than 5 s. This trial was used as marker for the first conditioned response (big filled circle inside left in Figure 5). Subsequently, the number of conditioned responses to the tones increased and their response latencies steeply shortened even though there were still intermingled unconditioned responses. After the marker trial with a conspicuous increase of theta power the previous trend of decrease of theta power was reversed to an increase in trials with an unconditioned response. We consider this reversal at the marker trial as the boundary between stage 2 "escape learning" and stage 3 "avoidance learning". The reliability of this division was confirmed by behavioral data. The number of trials with a reaction time shorter than 5 s was lower than the number of trials with unconditioned responses before the marker trial and, vice versa, after the marker trial. During stage 3 "avoidance learning" mistakes in terms of unconditioned responses with footshocks were still made but with decreasing tendency concomitant with an increase of theta power after these mistakes. The stages "escape learning" and "avoidance learning" as result of theta activity analysis replaced the non-differentiated stage "signal evaluation" determined earlier using prefrontal dopamine values [Stark et al., 2000]. At some maximum of theta power (second maximum; big open circle in the center in Figure 5) this trend reversed in stage 4 "retrieval of avoidance response" to a slow decrease of theta activity after the remaining occasional mistakes. Assumingly, the conjunction of a mistake with a subsequent maximum of theta power represented the crucial event separating stage 3 from stage 4. The post hoc determination of stages in avoidance conditioning was performed by combining of theta power analysis with behavioral analysis. The four stages appear to be clearly distinguishable using the obtained behavioral events.

The close correlation between behavioral data and theta activations determined with high temporal precision during avoidance learning allows us to interpret our finding with respect to the already described cognitive learning stages. At the start of avoidance training in the stage of "signal detection" the tone stimuli are still meaningless for the animals, and they just try to flee from the sudden footshock punishment in an apparently uncoordinated and confused fashion. Analysis of video recordings revealed the strong excitement of the animals caused by the high degree of uncertainty. The theta power within these initial trials reaches the first and highest maximum indicating strong information processing by the animals. Assumingly, stress is involved in these first trials due to unpredictability of the context and drives the animal to search the contralateral shuttle-box compartment to escape the footshock punishment.

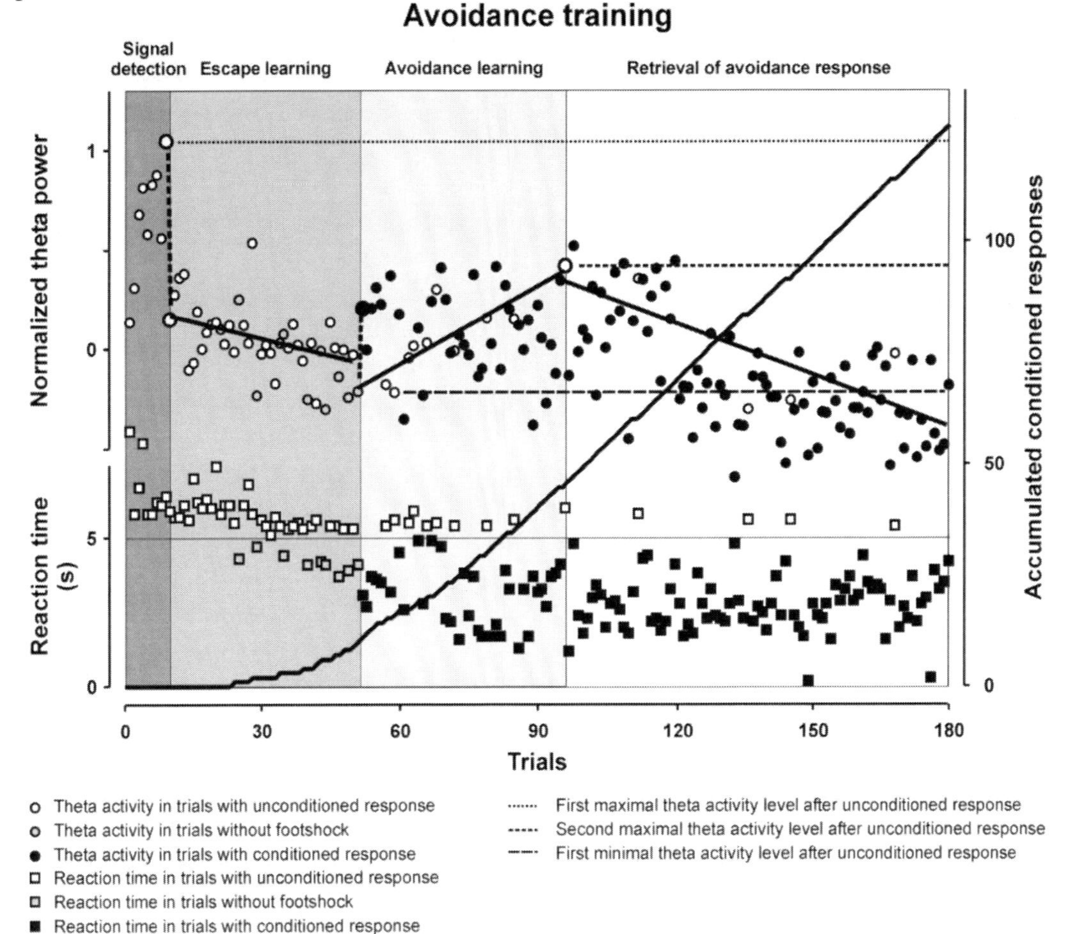

Figure 5. Stage-dependent theta activity in the trial periods after compartment change and corresponding reaction times after tone onset in a gerbil during avoidance conditioning in the shuttle-box. Upper part shows the normalized theta power in the period after compartment change in 180 successive trials with unconditioned responses (open circles) and conditioned responses (filled circles) plotted over the stages 1 "signal detection", 2 "escape learning", 3 "avoidance learning", and 4 "retrieval of avoidance response". After an initial theta power maximum (upper thin dotted line) the sharpest decline of theta power between trial 9 and trial 10 (big open circles connected by fat dotted line, falling) is defined as the event that separates the stages 1 and 2. The regression line of gradually falling theta power in stage 2 "escape learning" is computed from values in trials with unconditioned responses (fat line, falling). The beginning of stage 3 "avoidance learning", by our definition, is marked by the trial 52 with a reaction time lower than 5 s that entails a conspicuous increase of theta power; assumingly the first conditioned response (big filled circle, fat dotted, rising line indicating marked theta power increase). Note that previous trials with reaction time lower than 5 s do not show a substantial increase of theta power. In the neighborhood there is always a trial that shows the absolute theta power minimum of all unconditioned responses (lower thin dotted line connecting to the open circle). The regression line of gradually increasing theta power in stage 3 "avoidance learning" is computed from values in trials with unconditioned responses (fat solid line, rising). The line stops at the trial 96 with an unconditioned response that forms the second theta power maximum between stages 3 "avoidance learning" and 4 "retrieval of avoidance response" (big open circle; middle thin dotted line), defined as the end of stage 3. Lower part shows reaction times in trials with unconditioned responses

(open squares) and conditioned responses (filled squares). The horizontal 5-s line (footshock onset) separates unconditioned and conditioned responses.

Nevertheless, after several hurdle crossings animals can make the sudden experience, that the contralateral compartment is safe. As can be observed in the video recordings, this immediately leads to a coordinated escape behavior, i.e. after compartment change animals turn already during the intertrial interval towards the hurdle awaiting the next footshock punishment. The sharp decrease of theta power from one trial to the next may indicate that the demand of information processing is strongly reduced because the tone stimulus just became a conditioned signal announcing the footshock. By this, the initial stress as possible source of theta activation is largely reduced [Coover et al., 1973; Knardahl and Murison, 1989; Shors et al., 1997; van der Borght et al., 2005]. The association of the tone stimulus with the following footshock enhances strongly the predictability of the punishment. We assume that the sharp decline of theta power indicates the transition from completed stage 1 "signal detection" to the following stage 2 "escape learning". Paired tone and footshock stimuli in further trials optimize the instrumental response in animals, speeding up the reaction times. The perfection of the escape strategy is accompanied by a gradual drop in theta power in these trials with unconditioned responses. During "escape learning" animals occasionally change the compartment already before footshock onset without salient change of theta power compared with theta power in preceding trials with unconditioned responses. Presumably, the animals do not notice yet the contingency between the early reaction time and drop-out of footshock. Several such compartment changes before footshock seem to be required until the animals perceive the contingency between tone onset and the possibility of footshock avoidance. It is assumingly the trial in which the theta power markedly increases, in which the first conditioned response is performed by the animal. At the same time the theta power in trials with unconditioned responses adjacent to the first conditioned response reaches a first minimum. The first conditioned response to tone onset with elevated theta power marks probably the beginning of stage 3 "avoidance learning". After the first conditioned response the animals react more and more to tone onset and no longer to footshock onset, and thus receive the reward by "success" of footshock avoidance more and more frequent. From the video recordings it can be seen, that immediately after tone onset, animals are positioned in front of the hurdle. Often they move the head back and forth as external sign for internal indecision to jump or to stay. At the beginning of this stage, the poorly defined contingency between tone and footshock onset is subjected to an intensive error processing, decision making, and planning indicated by a renewed increase of theta activation immediately after remaining unconditioned responses up to a second maximum of theta activity, i.e. remaining mistakes become the main focus of attention and processing activity interpreted as learning from mistakes [Frank et al., 2004]. In this error processing the footshock punishment as a consequence of exceeding the 5-s latency after tone onset is compared to the consequence of the growing experience to change the shuttle-box compartment with respect to the time schedule followed by footshock avoidance. Feedback processing using short term memory could serve to compare the plan [Tanji and Hoshi, 2001; Tanji et al., 2007] - change the compartment or stay - with the outcome of action – footshock avoidance or punishment. In this respect the individual learns stepwise by unconditioned responses that the compartment change pursued to the reaction time proves the plan and that the conditioned response is the goal of learning. In this way the individual may perform the goal-directed and memory-guided learning process [Goldman-Rakic, 1992; Goldman-Rakic, 1995].

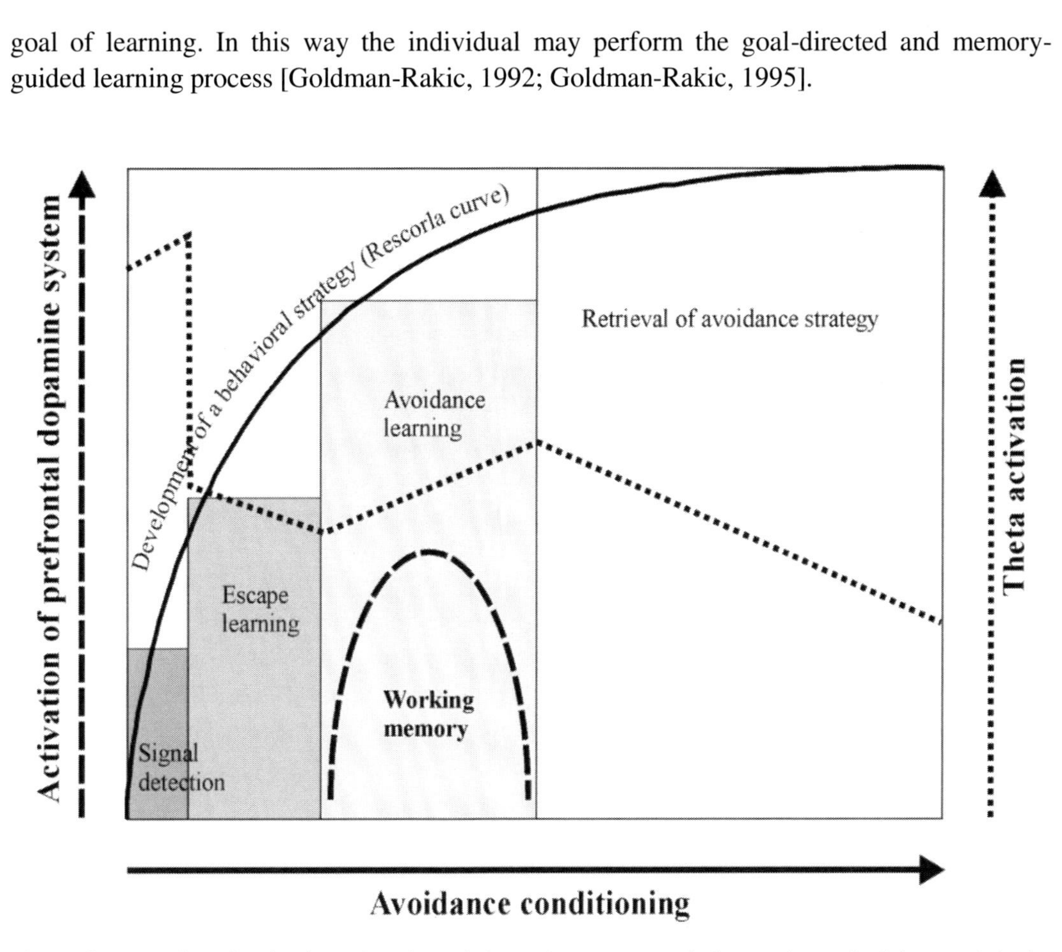

Figure 6. Dynamics of activation of prefrontal dopamine system and changes in cortical theta activity in the course of aversively motivated associative conditioning. The transient increase of prefrontal dopamine was found in the stage of avoidance learning during formation of relevant associations underlying a goal-directed, i.e. problem-solving behavioral strategy. During this stage presumably processes of internal reinforcement are involved to rule the behavioral strategy reducing stepwise the at the beginning of learning evoked conflict. Theta power presumably reflecting the current demand of information processing decreases differently in the stages “signal detection” and “escape learning” in trials with unconditioned responses. Theta activity increases up to a second theta activity maximum in the stage of “avoidance learning” presumably demanding working memory. In the following stage of “retrieval of avoidance strategy” theta activity steadily decreases. The correlation of dopamine and theta activity data with recorded behavioral events during associative learning allows the division of the development of the tested avoidance strategy into discrete learning dependent stages.

In that trial in which the animal experiences the congruity between the plan to reduce sufficiently the conflict and the goal of learning to avoid the footshock theta activity conspicuously increases at the first conditioned response. Possibly, we characterized the physiological substrate of suddenly increasing theta activation as indicator for internal reinforcement. However, in which way this crucial signal is produced is an open question requiring further investigations. Assumingly, dopamine systems of cortical and limbic brain areas are involved [Dayan and Balleine, 2002; Salamone and Correa, 2002; Bouret and Sara, 2004; Dalley et al., 2004; Nieuwenhuis et al., 2004; Kim et al., 2006; Izhikevich, 2007; Doya, 2008; Rushworth and Behrens, 2008]. The renewed increase of theta activity in trials with unconditioned responses during "avoidance learning" up to the second theta maximum presumably reflects an increasing demand on decision making, planning and feedback processing as attributes of working memory. The stage of "avoidance learning" is accompanied by further shortening of reaction times. We assume that the trial with unconditioned response showing the second maximum of theta activity marks the end of "avoidance learning", i.e. the temporal contingencies of the successful avoidance strategy have been acquired.

In the following stage 4 "retrieval of avoidance strategy" the theta power in further sporadic trials with unconditioned responses as a consequence of a "sloppy response" is steadily decreased obviously indicating a decreasing demand of information processing. Alternatively, sporadic trials with unconditioned responses which are normal even in highly trained animals may be interpreted as a check of still valid behavioral contingencies. The omission of footshocks would induce an extinction as a relearning process. The described markers used for division of the avoidance learning into different stages are found in all animals. However, the duration of the stages is individually different [Stark et al., 2008].

The hierarchy of stages and the processes within these stages during avoidance conditioning show the accumulation of partial experiences from trial to trial to a context-adequate memory-guided mature avoidance strategy [Rescorla, 1988; Goldman-Rakic, 1992]. However, very different aspects of the structure of the shuttle-box paradigm are learned before the avoidance response is fully consolidated. The demand of information processing as presumably reflected by theta activity shows several rising and falling phases with singularities in between. Our detailed trial by trial analysis suggests that experiences from two or few successive trials at a given stage of learning determine the further course of information processing under different aspects. This is the case for tone-shock association, for escape optimization, for the initial conditioned response, i.e. the first success, and for the subsequent clarification of remaining mistakes. As previously suggested [Sternberg, 1969; Goldman-Rakic, 1992; Stark et al., 2000; Luu et al., 2004; Krigolson and Holroyd, 2006] the demand of information processing is particularly strong during error analysis. But as judged from the theta activity being highest during initial tone-shock association in the stage of "signal detection", we assume that theta activation reflects the load of information processing in a rather comprehensive way including the state of arousal. In comparison the prefrontal dopamine system seems to be exclusively activated during learning the importance of associated tone signal and footshock stimulus responsible for the successful behavioral strategy.

Conclusion

Simultaneous recording of behavioral data and physiological data like dopamine release and theta activations from the beginning of a learning process up to the retrieval of memorized acquired facts and procedures allowed the investigation of the formation of behavioral strategies and revealed a cascade of discrete learning-dependent stages (Figure 6).

The activation of prefrontal dopamine system seems to reflect information processing related to the stage of formation of relevant associations for a successful behavioral strategy in which the animal realizes problem solving as the consequence of its own successful action.

Working memory principles seem to be involved with attributes like error processing, decision making, planning, and feedback processing. Associative learning always demands working memory. Also, during the retrieval of behavioral strategies requiring decision making, working memory is activated. However, the retrieval of a memorized procedure requires obviously no activation of working memory. Cortical theta activity presumably indicates more general demand of information processing, inclusively, the state of arousal to certain moments during stage-dependent learning. The dynamics of dopamine and theta activity data related to behavioral events converge in the assumption that the formation of a behavioral strategy may be divided into discrete learning-dependent stages.

The learning process is initiated by a conflict. Seemingly, in the stage of association of key stimuli activating working memory, a stepwise conflict reduction, which is critical for finding the sufficient and context-adequate problem solution as goal of learning, is performed. We assume that the reflection of successful punishment reduction by the animal acts as an "emotional ruler" to determine the direction of successful development of the behavioral strategy, i.e., animals learn from mistakes. Possibly, the trial-specific theta activation at the first conditioned response indicates a relevant reinforcing signal. Nevertheless, up to now many reports from literature about internal reinforcement may not explain comprehensively the source, the formation and the moment of triggering of the learning stage-dependent signal for internal reinforcement.

The increase in activity of the prefrontal dopamine system during the formation of relevant associations by working memory functions as prerequisite for the presumable perception of an internal reward to guide the individual to the goal of learning indicates that dopamine systems are involved in goal-directed learning-dependent reinforcement.

Acknowledgments

I thank Thomas Rothe and Matthias Deliano for critical reading of the manuscript.

References

Abercrombie, E. D., Keefe, K. A., DiFrischia, D. S., and Zigmond, M. J. (1989). Differential effect of stress on in vivo dopamine release in striatum, nucleus accumbens, and medial frontal cortex. *J. Neurochem.*, 52, 1655-1658.

Arnsten, A. F. T., Cai, J. X., Murphy, B. L., and Goldman-Rakic, P. S. (1994). Dopamine D1 receptor mechanisms in the cognitive performance of young adult and aged monkeys. *Psychopharmacology* (Berl), 116, 143-151.

Asada, H., Fukuda, Y., Tsunoda, S., Yamaguchi, M., and Tonoike, M. (1999). Frontal midline theta rhythms reflect alternative activation of prefrontal cortex and anterior cingulate cortex in humans. *Neurosci. Lett.*, 274, 29-32.

Atkinson, R. C. and Shiffrin, R. M. (1968). Human memory: a proposed system and its control processes. In K. W. Spence and J. T. Spence (Eds.), *The Psychology of Learning and Motivation,* Vol.2 (pp. 89-195). London: Academic Press.

Baddeley, A. (1998). Recent developments in working memory. *Curr. Opin. Neurobiol.*, 8, 234-238.

Balleine, B. W. and Curthoys, I. S. (1991). Differential effects of escapable and inescapable footshock on hippocampal theta activity. *Behav. Neurosci.*, 105, 202-209.

Barnet, R. C., Arnold, H. M., and Miller, R. R. (1991). Simultaneous conditioning demonstrated in second order conditioning: evidence for similar associative structure in forward and simultaneous conditioning. *Learn. Motiv.*, 22, 253-268.

Barnet, R. C. and Miller, R. R. (1996). Temporal encoding as a determinant of inhibitory control. *Learn. Motiv.*, 27, 73-91.

Bischof, A., Stark, H., Wagner, T., and Scheich, H. (2000). The inhibitory influence of an acquired escape strategy on subsequent avoidance learning in gerbils. *Neurosci. Lett.*, 281, 175-178.

Bonner, S. J. and Martinez, J. L. (1975). The early development of the EEG in the gerbil. *Behav. Biol.,* 14, 367-372.

Bouret, S. and Sara, S. J. (2004). Reward expectation, orientation of attention and locus coeruleus-medial frontal cortex interplay during learning. *Eur. J. Neurosci.,* 20, 791-802.

Broersen, L. M., Heinsbroek, R. P. W., de Bruin, J. P. C., Uylings, H. B. M., and Olivier, B. (1995). The role of the medial prefrontal cortex of rats in short-term memory functioning: further support for involvement of cholinergic, rather than dopaminergic mechanisms. *Brain Res.,* 674, 221-229.

Brozoski, T. J., Brown, R. M., Rosvold, H. E., and Goldman, P. S. (1979). Cognitive deficit caused by regional depletion of dopamine in prefrontal cortex of rhesus monkey. *Science,* 205, 929-932.

Coover, G. D., Ursin, H., and Levine, S. (1973). Plasma-corticosterone levels during active-avoidance learning in rats. *J. Comp. Physiol. Psychol.*, 82, 170-174.

Dalley, J. W., Cardinal, R. N., and Robbins, T. W. (2004). Prefrontal executive and cognitive functions in rodents: neural and neurochemical substrates. *Neurosci. Biobehav. Rev.*, 28, 771-784.

Dayan, P. and Balleine, B. W. (2002). Reward, motivation, and reinforcement learning. *Neuron*, 36, 285-298.

Deakin, J. F. W. and Graeff, F. G. (1991). 5-HT and mechanisms of defence. *J. Psychopharmacol.*, 5, 305-315.

Deco, G. and Rolls, E. T. (2003). Attention and working memory: a dynamical model of neuronal activity in the prefrontal cortex. *Eur. J. Neurosci.*, 18, 2374-2390.

Doya, K. (2008). Modulators of decision making. *Nat. Neurosci.*, 11, 410-416.

Dreher, J. C. and Burnod, Y. (2002). An integrative theory of the phasic and tonic modes of dopamine modulation in the prefrontal cortex. *Neural. Netw.*, 15, 583-602.

Durstewitz, D., Kelc, M., and Güntürkün, O. (1999). A neurocomputational theory of the dopaminergic modulation of working memory functions. *J. Neurosci.*, 19, 2807-2822.

Eichenbaum, H. (2008). *Learning and memory*. New York, NY: Norton and Company.

Eysenck, M. W., Ellis, A., Hunt, E., and Johnson-Laird, P. (1994). *The Blackwell Dictionary of Cognitive Psychology* (pp. 333-335). Oxford: Blackwell.

Feenstra, M. G. P. and Botterblom, M. H. A. (1996). Rapid sampling of extracellular dopamine in the rat prefrontal cortex during food consumption, handling and exposure to novelty. *Brain Res.*, 742, 17-24.

Feenstra, M. G. P. (2000). Dopamine and noradrenaline release in the prefrontal cortex in relation to unconditioned and conditioned stress and reward. *Prog. Brain Res.,* 126, 133-163.

Fibiger, H. C. and Phillips, A. G. (1986). Reward, motivation, cognition: psychobiology of mesotelencephalic dopamine systems. In F. E. Bloom (Ed.), *Handbook of Physiology. The Nervous System.* Intrinsic Regulatory Systems of the Brain (pp. 647-675). Bethesda, MD: Am. Physiol. Soc.

Fiorillo, C. D., Tobler, P. N., and Schultz, W. (2003). Discrete coding of reward probability and uncertainty by dopamine neurons. *Science*, 299, 1898-1902.

Frank, M. J., Seeberger, L. C., and O'Reilly, R. C. (2004). By carrot or by stick: cognitive reinforcement learning in parkinsonism. *Science*, 306, 1940-1943.

Fuster, J. M. (1989). *The Prefrontal Cortex: Anatomy, Physiology, and Neuropsychology of the Frontal Lobe* (2 ed., pp. 179-187). New York, NY: Raven Press.

Givens, B. (1996). Stimulus-evoked resetting of the dentate theta rhythm: relation to working memory. *Neuroreport,* 8, 159-163.

Glass, J. D., Hauser, U. E., Randolph, W. W., Rea, M. A., and De Vries, M. J. (1993). In vivo microdialysis of 5-hydroxyindoleacetic acid and glutamic acid in the hamster suprachiasmatic nuclei. *Amer. Zool.*, 33, 212-218.

Goldman-Rakic, P. S. (1992). Dopamine-mediated mechanisms of the prefrontal cortex. *Semin. Neurosci.,* 4, 149-159.

Goldman-Rakic, P. S. (1995). Cellular basis of working memory. *Neuron*, 14, 477-485.

Graeff, F. G. (1993). Role of 5-HT in defensive behavior and anxiety. *Rev. Neurosci.*, 4, 181-211.

Granon, S., Passetti, F., Thomas, K. L., Dalley, J. W., Everitt, B. J., and Robbins, T. W. (2000). Enhanced and impaired attentional performance after infusion of D1 dopaminergic receptor agents into rat prefrontal cortex. *J. Neurosci.*, 20, 1208-1215.

Herman, J. P., Guillonneau, D., Dantzer, R., Scatton, B., Semerdjian-Rouquier, L., and Le Moal, M. (1982). Differential effects of inescapable footshocks and of stimuli previously paired with inescapable footshocks on dopamine turnover in cortical and limbic areas of the rat. *Life Sci.*, 30, 2207-2214.

Hernandez, L. and Hoebel, B. G. (1990). Feeding can enhance dopamine turnover in the prefrontal cortex. *Brain Res. Bull.*, 25, 975-979.

Imperato, A., Puglisi-Allegra, S., Scrocco, M. G., Casolini, P., Bacchi, S., and Angelucci, L. (1992). Cortical and limbic dopamine and acetylcholine release as neurochemical correlates of emotional arousal in both aversive and non-aversive environmental changes. *Neurochem. Int.,* 20 Suppl, 265S-270S.

Ishii, R., Shinosaki, K., Ukai, S., Inouye, T., Ishihara, T., Yoshimine, T., Hirabuki, N., Asada, H., Kihara, T., Robinson, S. E., and Takeda, M. (1999). Medial prefrontal cortex generates frontal midline theta rhythm. *Neuroreport,* 10, 675-679.

Izaki, Y., Hori, K., and Nomura, M. (1998). Dopamine and acetylcholine elevation on lever-press acquisition in rat prefrontal cortex. *Neurosci. Lett.*, 258, 33-36.

Izhikevich, E. M. (2007). Solving the distal reward problem through linkage of STDP and dopamine signaling. *Cereb. Cortex*, 17, 2443-2452.

Jacobs, J., Hwang, G., Curran, T., and Kahana, M. J. (2006). EEG oscillations and recognition memory: Theta correlates of memory retrieval and decision making. *Neuroimage,* 32, 978-987.

Jensen, O. (2005). Reading the hippocampal code by theta phase-locking. *Trends Cogn. Sci*, 9, 551-553.

Jones, M. W. and Wilson, M. A. (2005). Theta rhythms coordinate hippocampal-prefrontal interactions in a spatial memory task. *PLoS Biol.,* 3, e402.

Kim, H., Shimojo, S., and O'Doherty, J. P. (2006). Is Avoiding an Aversive Outcome Rewarding? Neural Substrates of Avoidance Learning in the Human Brain. *PLoS Biol.,* 4, e233.

Knardahl, S. and Murison, R. (1989). Plasma corticosterone and renin activity during two-way active avoidance learning in spontaneously hypertensive and Wistar-Kyoto rats. *Behav. Neural. Biol.,* 51, 389-400.

Koob, G. F. (1992). Dopamine, addiction and reward. *Semin. Neurosci.,* 4, 139-158.

Krigolson, O. E. and Holroyd, C. B. (2006). Evidence for hierarchical error processing in the human brain. *Neuroscience,* 137, 13-17.

Krügel, U., Kittner, H., Franke, H., and Illes, P. (2003). Purinergic modulation of neuronal activity in the mesolimbic dopaminergic system in vivo. *Synapse*, 47, 134-142.

Luu, P., Tucker, D. M., and Makeig, S. (2004). Frontal midline theta and the error-related negativity: neurophysiological mechanisms of action regulation. *Clin. Neurophysiol.*, 115, 1821-1835.

McNaughton, N., Ruan, M., and Woodnorth, M. A. (2006). Restoring theta-like rhythmicity in rats restores initial learning in the Morris water maze. *Hippocampus*, 16, 1102-1110.

Meck, W. H. (1996). Neuropharmacology of timing and time perception. *Cogn. Brain Res.*, 3, 227-242.

Mizuhara, H. and Yamaguchi, Y. (2007). Human cortical circuits for central executive function emerge by theta phase synchronization. *Neuroimage*, 36, 232-244.

Nieuwenhuis, S., Holroyd, C. B., Mol, N., and Coles, M. G. H. (2004). Reinforcement-related brain potentials from medial frontal cortex: origins and functional significance. *Neurosci. Biobehav. Rev.*, 28, 441-448.

Ohl, F. W. and Scheich, H. (1996). Differential frequency conditioning enhances spectral contrast sensitivity of units in auditory cortex (field AI) of the alert Mongolian gerbil. *Eur. J. Neurosci.*, 8, 1001-1017.

Rescorla, R. A. (1988). Pavlovian conditioning. It's not what you think it is. *Am. Psychol.*, 43, 151-160.

Riess, D. (1972). Vicarious conditioned acceleration: successful observational learning of an aversive Pavlovian stimulus contingency. *J. Exp. Anal. Behav.,* 18, 181-186.

Rushworth, M. F. and Behrens, T. E. (2008). Choice, uncertainty and value in prefrontal and cingulate cortex. *Nat. Neurosci.*, 11, 389-397.

Salamone, J. D. (1994). The involvement of nucleus accumbens dopamine in appetitive and aversive motivation. *Behav. Brain Res.*, 61, 117-133.

Salamone, J. D. and Correa, M. (2002). Motivational views of reinforcement: implications for understanding the behavioral functions of nucleus accumbens dopamine. *Behav. Brain Res.,* 137, 3-25.

Sarnthein, J., Petsche, H., Rappelsberger, P., Shaw, G. L., and von Stein, A. (1998). Synchronization between prefrontal and posterior association cortex during human working memory. *Proc. Natl. Acad. Sci. USA*, 95, 7092-7096.

Sawaguchi, T., Matsumura, M., and Kubota, K. (1990). Effects of dopamine antagonists on neuronal activity related to a delayed response task in monkey prefrontal cortex. *J. Neurophysiol.,* 63, 1401-1412.

Sawaguchi, T. and Goldman-Rakic, P. S. (1991). D1 dopamine receptors in prefrontal cortex: involvement in working memory. *Science*, 251, 947-950.

Sawaguchi, T. and Goldman-Rakic, P. S. (1994). The role of D1-dopamine receptor in working memory: local injections of dopamine antagonists into the prefrontal cortex of rhesus monkeys performing an oculomotor delayed-response task. *J. Neurophysiol.*, 71, 515-528.

Scheich, H. and Zuschratter, W. (1995). Mapping stimulus features and meaning in gerbil auditory cortex with 2-deoxyglucose, and c-Fos antibodies. *Behav. Brain Res.*, 66, 195-205.

Schultz, W. (1992). Activity of dopamine neurons in the behaving primate. *Semin. Neurosci.*, 4, 129-139.

Schultz, W., Apicella, P., and Ljungberg, T. (1993). Responses of monkey dopamine neurons to reward and conditioned stimuli during successive steps of learning a delayed response task. *J. Neurosci.,* 13, 900-913.

Schultz, W., Dayan, P., and Montague, P. R. (1997). A neural substrate of prediction and reward. *Science*, 275, 1593-1599.

Schultz, W. (2002). Getting formal with dopamine and reward. *Neuron*, 36, 241-263.

Seamans, J. K., Floresco, S. B., and Phillips, A. G. (1998). D1 receptor modulation of hippocampal-prefrontal cortical circuits integrating spatial memory with executive functions in the rat. *J. Neurosci.,* 18, 1613-1621.

Seidenbecher, T., Laxmi, T. R., Stork, O., and Pape, H. C. (2003). Amygdalar and hippocampal theta rhythm synchronization during fear memory retrieval. *Science*, 301, 846-850.

Shors, T. J., Gallegos, R. A., and Breindl, A. (1997). Transient and persistent consequences of acute stress on long-term potentiation (LTP), synaptic efficacy, theta rhythms and bursts in area CA1 of the hippocampus. *Synapse*, 26, 209-217.

Sokolowski, J. D., McCullough, L. D., and Salamone, J. D. (1994). Effects of dopamine depletions in the medial prefrontal cortex on active avoidance and escape in the rat. *Brain Res.,* 651, 293-299.

Spanagel, R. and Weiss, F. (1999). The dopamine hypothesis of reward: past and current status. *Trends Neurosci.,* 22, 521-527.

Stark, H. and Scheich, H. (1997). Dopaminergic and serotonergic neurotransmission systems are differentially involved in auditory cortex learning: a long-term microdialysis study of metabolites. *J. Neurochem.*, 68, 691-697.

Stark, H., Bischof, A., and Scheich, H. (1999). Increase of extracellular dopamine in prefrontal cortex of gerbils during acquisition of the avoidance strategy in the shuttle-box. *Neurosci. Lett.,* 264, 77-80.

Stark, H., Bischof, A., Wagner, T., and Scheich, H. (2000). Stages of avoidance strategy formation in gerbils are correlated with dopaminergic transmission activity. *Eur. J. Pharmacol.,* 405, 263-275.

Stark, H., Rothe, T., Wagner, T., and Scheich, H. (2004). Learning a new behavioral strategy in the shuttle-box increases prefrontal dopamine. *Neuroscience*, 126, 21-29.

Stark, H., Rothe, T., Deliano, M., and Scheich, H. (2007). Theta activity attenuation correlates with avoidance learning progress in gerbils. *Neuroreport*, 18, 549-552.

Stark, H., Rothe, T., Deliano, M., and Scheich, H. (2008). Dynamics of cortical theta activity correlates with stages of auditory avoidance strategy formation in a shuttle-box. *Neuroscience,* 151, 467-475.

Sternberg, S. (1969). The discovery of processing stages: extensions of Donder's method. *Acta Psychol.*, 30, 276-315.

Talk, A., Kang, E., and Gabriel, M. (2004). Independent generation of theta rhythm in the hippocampus and posterior cingulate cortex. *Brain Res.,* 1015, 15-24.

Tanji, J. and Hoshi, E. (2001). Behavioral planning in the prefrontal cortex. *Curr. Opin. Neurobiol., 11,* 164-170.

Tanji, J., Shima, K., and Mushiake, H. (2007). Concept-based behavioral planning and the lateral prefrontal cortex. *Trends Cogn. Sci., 11,* 528-534.

Van der Borght, K., Meerlo, P., Luiten, P. G. M., Eggen, B. J. L., and Van der Zee, E. A. (2005). Effects of active shock avoidance learning on hippocampal neurogenesis and plasma levels of corticosterone. *Behav. Brain Res.*, 157, 23-30.

Wasserman, E. A. and Miller, R. R. (1997). What's elementary about associative learning? *Annu. Rev. Psychol.,* 48, 573-607.

Watanabe, M., Kodama, T., and Hikosaka, K. (1997). Increase of extracellular dopamine in primate prefrontal cortex during a working memory task. *J. Neurophysiol.*, 78, 2795-2798.

Wetzel, W., Wagner, T., Ohl, F. W.,Scheich, H. (1998). Categorical discrimination direction in frequency-modulated tones by Mongolian gerbils. *Behav. Brain Res.*, 91, 29-39.

Wightman, R. M. and Robinson, D. L. (2002). Transient changes in mesolimbic dopamine and their association with 'reward'. *J. Neurochem.*, 82, 721-735.

Yamamuro, Y., Hori, K., Iwano, H., and Nomura, M. (1994). The relationship between learning performance and dopamine in the prefrontal cortex of the rat. *Neurosci. Lett.*, 177, 83-86.

Young, A. M. J. (1993). Intracerebral microdialysis in the study of physiology and behaviour. *Rev. Neurosci.,* 4, 373-395.

Zuschratter, W., Gass, P., Herdegen, T., and Scheich, H. (1995). Comparison of frequency-specific c-Fos expression and fluoro-2-deoxyglucose uptake in auditory cortex of gerbils (Meriones unguiculatus). *Eur. J. Neurosci.*, 7, 1614-1626.

In: Encyclopedia of Neuroscience Research
Editors: Eileen J. Sampson and Donald R. Glevins ISBN 978-1-61324-861-4

Chapter V

The Orbitofrontal Cortex and Emotional Decision-Making: The Neglected Role of Anxiety

Sabine Windmann* * *and Martina Kirsch
Goethe University, Department of General Psychology II
Institute of Psychology and Sports Sciences
Mertonstr.17, 60054 Frankfurt/Main, Germany

Abstract

The orbital part of the prefrontal cortex is known to be crucially involved in emotional decision-making. Its function seems to be, first, to associate affective value with behavioral bias (i.e., approach or withdrawal) in complex choice situations, and secondly, to flexibly control and transiently inhibit affective and behavioral impulses. Dysfunction of this region due to damage or immaturity of OFC cells leads to impulsiveness, behavioral perseveration, and reduced sensitivity for risk, as observed in children as well as various patients groups. We review the evidence with a focus on the much less investigated question of whether and how *hyper*activation (or hyperfunctioning of modules and processes) involving OFC leads to the opposite behavioral tendencies, namely risk-aversion, enhanced inhibition, undecidedness and higher risk sensitivity including heightened anxiety. We identify two perspectives with opposing views onto this issue, and report some unpublished studies of populations with children and patients with anxiety disorders using the Iowa gambling task, a task that is presumed to be sensitive to OFC function. We find that, if anything, anxiety has the same beneficial effects onto performance as cortical maturation. Although effects of anxiety are much weaker and less conclusive than those of age during childhood, it seems that there are situations in which enhanced anxiety can support advantageous decision-making. Our results highlight the important role of orbitofrontal cortex in integrating approach- and withdrawal-tendencies mediated by mesolimbic and mesostriatal structures on the one hand and higher cognitive centres on the other, with the latter ones recruiting more dorsal

* Phone: +44 69/798-22905; Fax: +44 69/798-23457; Email: Sabine.Windmann@psych.uni-frankfurt.de

parts of the prefrontal cortex and allowing for conscious thinking, reasoning, and problem-solving. Behavioral interventions as in education, cognitive training, and psychotherapy are most effective when they take into account the motivational and evaluative function of this region.

The ventral surface of the frontal lobe is known as the orbitofrontal cortex (OFC) as it is located above the eyeballs and their muscles. It comprises Brodman areas14 medially and 13/25 caudally as well as areas 11 and 12 around the ventral convexity. As part of the prefrontal cortex, these regions are among the last to mature during ontogenesis, to be completed usually only during early adulthood (e.g., Diamond, 2002; Fuster, 2008; Schore, 1999).

Functions of the OFC have been linked to higher emotion control in the broadest sense. Animal and neuroimaging work has shown that on the input side, the OFC responds to primary as well as secondary rewards and punishments of various sensory modalities and mediates anticipation of such events (Breiter, Aharon, Kahneman, Dale, and Shizgal, 2001; Elliott, Dolan, and Frith, 2000; Gottfried, O'Doherty, and Dolan, 2003; Kirsch, Schienle, Stark, Sammer, Blecker, Walter, Ott, Burkart, and Vaitl, 2003; Knutson, Fong, Adams, Varner, and Hommer, 2001; O'Doherty, 2004; Rogers, Owen, Middleton, Williams, Pickard, Sahakian, and Robbins,1999; Rolls, 1996, 2002; Rolls, Kringelbach, and de Araujo, 2003; Rolls, O'Doherty, Kringelbach, Francis, Bowtell, and McGlone, 2003; Small, Zatorre, Dagher, Evans, and Jones-Gotman, 2001). On the output side, the OFC is involved in controlling the flexible adjustment of behavior according to expected and changing outcomes (Baxter, Parker, Lindner, Izquierdo, and Murray, 2000; Bechara, 2004; Damasio, 1996; Fellows, 2007; O'Doherty, 2007; O'Doherty, Critchley, Deichmann, and Dolan, 2003; Pickens, Saddoris, Gallagher, and Holland, 2005). Thus, the region is thought to be involved in the dynamical linking of higher affective evaluation with behavior control.

Response control in accordance with higher affective evaluation often involves inhibition of primary, conditioned, or otherwise automatic affective responses, particularly fear. It is mostly relevant in situations that were previously dangerous or risky, or otherwise resulted in negative outcome, but have changed and are now useful, attractive response options. For instance, a food source that had been contaminated before may have been cleared and now provide good quality food. An elevator that got stuck a couple of days ago can have been repaired and might be functioning properly again, to be preferred to the stairs. A person who was upset and angry when we met them the first time may be more relaxed when we meet them again and actually turn out to become a very good friend. In all these situations, what needs to be inhibited is feelings of fear, risk, or anxiety (and the accompanying withdrawal behaviors) that are no longer adaptive motives as the circumstances have changed.

The orbitofrontal cortex with its bilateral connections to the amygdala on the one hand and goal-related thinking centres like dorsolateral prefrontal cortex on the other is best suited to mediate these functions. In fact, animal electrophysiology as well as human imaging studies have often demonstrated the crucial role of OFC subregions in extinguishing conditioned fear, recovering extinguished fear, regulating anticipatory emotions in general, and inhibiting previously established responses (Milad and Quirk, 2002; Schoenbaum, Setlow, Saddoris, and Gallagher, 2003). Accordingly, damage to the OFC leads to ignorance towards danger and threat, and to inflexible behavior that is not dynamically adjusted to changing outcome contingencies in the environment. Neuropsychology has demonstrated many times how patients with damage to regions of the OFC fail to control affect, to inhibit impulsive behavior, and to choose the behavioral option that is in the best of their long-term interests in complex decision-making tasks as well as everyday life situations. The famous case of Phineas Gage, alongside more systematic patient studies, illustrates this point (Bechara, Damasio, Tranel, and Damasio, 1997; Bechara, Tranel, and Damasio, 2000; Damasio, 1994).

The same tendency can be found in children and adolescents whose OFC neurons and connections are still underdeveloped; these show increased impulsiveness (Chambers and Potenza, 2003), reduced inhibition (Bunge, Dudukovic, Thomason, Vaidya, and Gabrieli, 2002), and a choice pattern in complex decision-making tasks like the Iowa gambling task (see below) that is similar to that of patients with OFC lesions (Crone, Bunge, Latenstein, and van der Molen, 2005; Hooper, Luciana, Conklin, and Yarger, 2004; Letho and Elorinne, 2003). Like patients with OFC dysfunction, the children tend to choose risky behavioral options without much worry or care, alongside reduced anticipatory physiological reactions (like skin conductance response) that normally accompany such decisions. The behaviour resembles that found in individuals with antisocial tendencies such as substance abuse, pathological gambling, and psychopathy, conditions which therefore have also been linked to orbitofrontal dysfunction (Forbush, Shaw, Graeber, Hovick, Meyer, Moser, Bayless, Watson, and Black, 2008; van Honk, Hermans, Putman, Montagne, and Schutter, 2002). More generally, failure to experience a healthy amount of fear and to appropriately process risk and punishment signals are considered part of psychopathy (Blair, 2008; Sommer, Hajak, Dohnel, Schwerdtner, Meinhardt, and Muller, 2006; van Honck and Schutter, 2006).

How then does the intact orbitofrontal cortex "know" what amount of fear is healthy and appropriate? Normally, this information should be deduced from behavioral outcomes and is mediated by interaction of various brain regions, including amygdale and striatum. Successful behavior, that is, decisions whose rewarding outcomes exceed expectations, should reinforce existing stimulus-response associations, while lack of success or unexpected misfortune (punishment) should weaken these connections and bias withdrawal behavior. This is the central assumption of temporal difference error reinforcement learning theories (McClure, Daw, and Montague, 2003; Schultz, Dayan, and Montague, 1997). However, in a damaged, immature (as in children), or otherwise malfunctioning OFC that is less flexible and plastic than normal, the system may get stuck in a state where behavioral options with the strongest associations are repeatedly activated, regardless of outcome, while alternative behaviors are never tried, and their outcomes never experienced. To some degree, this conflict between exploitation (continuing to choose one particular behavioral option) and exploration (choosing alternative behaviors to see whether these lead to better outcomes) is present in all individuals. However, in cases with diminished functionality of OFC circuits, the right balance between these two tendencies could be shifted towards exploitation, regardless of long-term costs. As a matter of fact, patients with damage to the OFC show perseveration, a reduced capability to switch from a previously reinforced behavior to novel behaviors (Fellows and Farah, 2003, 2005).

From a purely psychofunctional point of view, if malfunctioning of OFC regions leads to perseveration, impulsivity, and reduced sensitivity towards fear and risk, then "hyper-functioning" of those same modules should lead to enhanced inconsistency, higher avoidance of risk, heightened concern about long-term outcomes, and a generally heightened level of fear or worries. This notion is reminiscent of what many decades ago the German psychiatrist

Lungwitz (1955, Becker, 1993) described as a manifestation of 'hypertrophy of anxiety feeling cells'. In his conception, termed the "Theory of Psychobiology", there are five types of feeling cells, all of which are seen in connection with their afferent and efferent connections including peripheral autonomic nerves and effector systems (as part of what is called a "reflex system"). The type of feeling cells that are concerned with anxiety mediate the feeling of "being cornered, beset, of compulsion, of inhibition, of astonishment, of defiance, of withdrawal, shame, shyness, carefulness, care, etc." (Becker, 1993, p. 41). 'Hypertrophy' of those cells is described as the psychobiological basis of anxiety neuroses. The condition lets the patients' cognitive styles and attitudes towards the world regress towards a more self-absorbed and less sophisticated way of thinking that Lungwitz described as 'infantilistic' and based on immature (or underdeveloped) reflex systems. This theory would predict that due to their "immature" feeling cells (perhaps situated in OFC), patients with anxiety neurosis think and behave similarly to children (whose OFC neurons are not mature yet). Clearly, Lungwitz did not have any knowledge of modern neuroscience and of the precise role and function of neurons (he seems to identify these with a muscle rather than with an information processing device), but at a very functional-descriptive level, his conception fits with existing evidence of the role of dysfunctional OFC sensitivity for risk and inhibition of impulsiveness in both children and patients with anxiety disorders (c.f., Kalin, Shelton, and Davidson, 2007).

What happens according to the Theory of Psychobiology when children are neurotic themselves? Lungwitz proposed that neuroses of childhood do not differ in principle from neuroses of adults: Neurotic children with high anxiety show a yet more "infantilistic" and evolutionarily "primitive" thinking style than do their healthy peers because of underdeveloped anxiety reflex systems (see Becker, 1993, page 107). Thus, not only do the effects of neurotic anxiety and young age lie on the same dimension (as they both rely on immature "reflex systems"), they are also presumed to have additive effects. When applied to a decision making task that tests OFC function, the theory would predict that highly anxious children perform worst, followed by healthy children and adult patients with anxiety disorders, who both perform worse than do healthy adults.

The theory also knows of intermediate forms of anxiety that lie in between healthy and "neurotic" states, suggesting that Lungwitz held a dimensional perspective onto mental illnesses rather than a categorical one with discrete boundaries between healthy and ill, in line with modern psychological theories of anxiety (e.g., Endler and Kocovski, 2002; Hansell and Damour, 2008). These intermediate conditions are referred to as nervousness: "'Nervousness' is commonly used as a description of mild forms of neurosis; namely, degrees of excitability of the reflex system approaching the normal, but already of pathological grade. Less frequently, it is used to describe still normal degrees of excitability of the reflex systems that approach the abnormal, especially in the anxiety and pain reflex systems" (Becker, 1993, p. 9). Consequently we would expect to find not only individuals with pathological anxiety, but also with heightened levels of normal anxiety who are on the borderline between healthy and ill to behave more like children, albeit not as much as patients with anxiety disorders.

Modern theories of anxiety disorders describe the patients as overly sensitive to risk and threat (Clark, 1986; Ehlers, Margraf, Davies, and Roth, 1988; McNally, 1994, 1995, 1997; Williams, Watts, MacLeod, and Mathews, 1997). The idea is that the patients hold an abnormal attentional bias, higher sensitivity, and higher anticipatory fear towards threat signals that can be either internal (body) or external (world). It has been speculated many times that imbalanced interaction between limbic centres, including the amygdala, and prefrontal regions, particularly orbitofrontal cortex, could be involved in the underlying neural pathology (Davidson, 2002; Gorman, Kent, Sullivan, and Coplan, 2000; LeDoux and Müller, 1997; Milad and Rauch, 2007; Windmann, 1998; Windmann, Sakhavat, and Kutas, 2002). These descriptions have been made independently of Damasios (1994, 1996) and Becharas (2004) work on the role of somatic markers in risk anticipation.

Recent neuroimaging studies have begun to confirm the speculations on the role of OFC in anxiety disorders (Domschke, Ohrmann, Braun, Suslow, Bauer, Hohoff, Kersting, Engelien, Arolt, Heindel, Deckert, and Kugel, 2008; Monk, Nelson, McClure, Mogg, Bradley, Leibenluft, Blair, Chen, Charney, Ernst, and Pine, 2006; Monk, Telzer, Mogg, Bradley, Mai, Louro, Chen, McClure-Tone, Ernst, and Pine, 2008). The patients tend towards higher activation or higher involvement of OFC regions, contrary to patients with OFC lesions or psycho-/sociopathy. Healthy individuals under induction of "worry" show the same tendency (Hoehn-Saric, Lee, McLeod, and Wong, 2005) as do individuals who are high at risk for anxiety and depression when reading negative emotional words (Wolfensberger, Veltman, Hoogendijk, De Ruiter, Boomsma, and de Geus, in press). This further adds to the notion that there is a continuum between normal (state) and pathological (trait) forms of anxiety.

If patients with damage to OFC and patients are more prone to risky and impulsive behavior that is not in the best of their long-term interest, and patients with anxiety disorder do the opposite due to increased activation/recruitment of OFC circuits, then this difference between the two groups should manifest in decision-making tasks designed to test OFC function. One of those tasks is the Iowa gambling tasks established by Bechara, Damasio, Damasio, and Anderson (1994). This refers to a card playing game where the subject is asked to collect points by drawing cards from four decks, two of which return 100 points whereas the other two return only 50 points with each card, which the subjects quickly learns. However, what takes longer to appreciate is that some of the cards with the higher returns are associated with severe losses, up to 1250 points, or very frequent losses, such that the expectancy value is negative (-25 points per card). It is therefore not advantageous in the long run to draw cards from those decks (the "bad" decks). The cards from the other two decks are associated with occasional losses too, but frequency and/or amounts of the losses are fewer so that the expectancy value is still positive (+25 points per card, "good" decks). Players who want to maximize their profit on the task should therefore draw from the cards that at first appear to be associated with only small wins, and avoid the risky bad decks even though their immediate return value appears high.

Healthy individuals do indeed learn these associations and gradually increase their payoff across blocks of trials (Bechara et al., 1997). However, patients with damage to OFC are unable to do that; they keep drawing from the bad decks even though many of themunderstand at some point that these are risky choices. They also fail to develop skin conductance responses when drawing from the bad decks indicating that their emotional system does not signal to them the presence of imminent risk. Children, due to immature prefrontal circuits, show similarly impulsive choice behavior on this task (Crone and van der Molen, 2004, 2007).

How do patients with anxiety disorders behave in the Iowa gambling task? Lungwitz (1955) would have predicted that their performance is reduced, like the ones of the children, due to their "infantilistic" thinking styles. On the other hand, modern cognitive theories on anxiety disorders as well as neuroscience evidence pointing to a tendency towards "hyperactivation" of OFC regions in these individuals would probably predict the opposite, namely, that they learn the task better than do healthy adults due to their enhanced sensitivity to risk (see also Must, Szabó, Bódi, Szász, Janka, and Kéri, 2006; Smoski, Lynch, Rosenthal, Cheavens, Chapman, and Krishnan, 2008, for a similar logic in depressed individuals).

In a series of experiments, we failed to find any consistent differences between patients with highly somatic anxiety disorders (i.e., panic disorder) and healthy control subjects in the original and an inverted version (see Bechara et al., 2000) of the Iowa gambling task (Guse, 2006; Wischniewski, 2007), although we did confirm in a study with healthy subjects that both these tasks activate medial and lateral orbitofrontal cortex (Windmann, Kirsch, Mier, Stark, Walter, Güntürkün, and Vaitl, 2006). This failure to find any differences between the groups could at least in part be due to the chosen payoff values used in the card game which followed very literally the procedures described in Bechara et al. (2000, Figure 1). It was only later during personal communication with Antoine Bechara (the latest on 15/10/2007) that we learnt that he had used different payoff amounts in his studies that were dynamically adapted to subject's decisions, as in the commercial version of the Iowa gambling task. We are now working with these probabilistic payoffs in present and future studies.

Another possibility is that panic disorder and generalized anxiety disorder are not based on the same mechanisms, and that only generalized anxiety disorder is related to the hypothesized effects of OFC 'hyperactivation'. We are currently exploring this possibility in ongoing studies with a computerized version of the Iowa gambling task. However, one thing we did find is that children with high anxiety showed higher performance in the Iowa gambling task than their healthy peers (Kirsch, Kirsch, Mier, and Windmann, 2007). Although the effect is relatively weak, it seems to conform to some other reports in the literature (Garon, Moore, and Waschbusch, 2006; Schmitt, Brinkley, and Newman, 1999). Surprisingly though, another study reported in the literature found the contrary effects of anxiety on performance in the Iowa gambling task in healthy adults (Miu, Heilman, and Houser, 2008). The crucial factor explaining the inconsistency could be the question of whether enhanced anxiety and risk sensitivity can be directed to or linked with the task at hand, or, alternatively, comes from other sources and distracts participants during task performance, as some authors have speculated (Lösel and Schmucker, 2004; Preston, Buchanan, Stansfield, and Bechara, 2007). Clearly, more research is needed on that side of the spectrum of abnormal OFC function.

Depending on the outcome of this research it would be recommendable to reconsider in education, psychotherapy, and other interventions that in some populations, OFC dysfunction may be causal to the underlying problem. If this is the case, then the intervention program would be most successful if it contained measures to help, first, accurate and clear affective evaluation of situations and choice options, and second, self-controlled adjustments in behavior. For instance, for children and individuals with deficits in OFC function like ADHD, impulsive personality disorder, and addiction, it could be helpful to make incentives and long-term benefits of the desired behavior more transparent to increase consistency in decision-making. An example would be attaching a picture of a wanted camera on the wallet or the piggybank to foster the long-term motivation to save the required money, or engaging in imagination of the beneficial effects of a proper meal at home before opening the bag of potato chips in the car. An example of an intervention measure helping to self-control behavior would be the employment of a reinforcement scheme according to which reinforcers accumulate in a nonlinear fashion over time: The longer the individual inhibits impulsive responses, the more reinforcement (e.g., money) they will gain until they have reached their long-term goal.

For patients with negative affect like anxiety and depression, where OFC mediated functions may be overly sensitive, the problem seems to be not so much the inhibition of behavior but, on the contrary, motivating and initiating behavior. These patients are known to often overgeneralize their negative evaluation of stimuli and response options and as a result become quite passive and socially isolated. The reason is that they do not find the available behavioral options attractive and/or safe enough to promote activity. The patients should therefore be taught how to better differentiate between actually harmful and actually harmless stimuli/situations (Windmann, 1998; Windmann et al., 2002), as proposed by cognitive therapy (Beck, 1970, 1976; Beck, Emery, and Greenberg, 2005). Learning how to relax probably needs to be part of this program, if it is true that the patients' abnormally high risk perception is further supported by increased sensitivity to bodily fear signals (Clark, 1986; Ehlers et al., 1988). In addition, these patients could be motivated to act on the basis of revised affective evaluations by helping them focus on (and keep record of) expected and experienced positive outcomes, such as, keeping a diary of the events and activities that they did not want to take part in at first, but in the end were glad having joined. Some of these measures are part of the classical behavioral intervention program designed by Lewinsohn, Clarke, Hops, and Andrews (1990). A combination of both cognitive and behavioral elements would probably be most successful paired with psychoactive medication in select cases.

Acknowledgment

Work described in this article was supported by a research grant from the Lungwitz foundation, Berlin (Germany).

References

Baxter, M. G., Parker, A., Lindner, C. C., Izquierdo, A. D., and Murray, E. A. (2000). Control of response selection by reinforcer value requires interaction of amygdala and orbital prefrontal cortex. *J. Neurosci., 20*(11), 4311-4319.

Bechara, A. (2004). The role of emotion in decision-making: evidence from neurological patients with orbitofrontal damage. *Brain Cogn., 55*(1), 30-40.

Bechara, A., Damasio, A. R., Damasio, H., and Anderson, S. W. (1994). Insensitivity to future consequences following damage to human prefrontal cortex. *Cognition, 50*(1-3), 7-15.

Bechara, A., Damasio, H., Tranel, D., and Damasio, A. R. (1997). Deciding advantageously before knowing the advantageous strategy. *Science, 275*(5304), 1293-1295.

Bechara, A., Tranel, D., and Damasio, H. (2000). Characterization of the decision-making deficit of patients with ventromedial prefrontal cortex lesions. *Brain, 123 (Pt 11)*, 2189-2202.

Beck, A. T. (1970). Cognitive therapy: Nature and relation to behavior therapy. *Behav. Ther.* (1), 184-200.

Beck, A. T. (1976). *Cognitive Therapy and the Emotional Disorders*. New York: International Universities Press.

Beck, A. T., Emery, G., and Greenberg, R. L. (2005). *Anxiety Disorders and Phobias: A Cognitive Perspective*. New York: Basic Books.

Becker, R. (Ed.). (1993). *Hans Lungwitz: Psychobiology and Cognitive Therapy of the Neuroses*. Basel: Birkhäuser Verlag.

Blair, R. J. (2008). Review. The amygdala and ventromedial prefrontal cortex: functional contributions and dysfunction in psychopathy. *Philos. Trans R Soc. Lond B Biol. Sci.*

Breiter, H. C., Aharon, I., Kahneman, D., Dale, A., and Shizgal, P. (2001). Functional imaging of neural responses to expectancy and experience of monetary gains and losses. *Neuron, 30*(2), 619-639.

Bunge, S. A., Dudukovic, N. M., Thomason, M. E., Vaidya, C. J., and Gabrieli, J. D. (2002). Immature frontal lobe contributions to cognitive control in children: evidence from fMRI. *Neuron, 33*(2), 301-311.

Chambers, R. A., and Potenza, M. N. (2003). Neurodevelopment, impulsivity, and adolescent gambling. *J. Gambl. Stud, 19*(1), 53-84.

Clark, D. M. (1986). A cognitive approach to panic. *Behav. Res. Ther., 24*(4), 461-470.

Crone, E. A., Bunge, S. A., Latenstein, H., and van der Molen, M. W. (2005). Characterization of children's decision making: sensitivity to punishment frequency, not task complexity. *Child Neuropsychol, 11*(3), 245-263.

Crone, E. A., and van der Molen, M. W. (2004). Developmental changes in real life decision making: performance on a gambling task previously shown to depend on the ventromedial prefrontal cortex. *Dev. Neuropsychol., 25*(3), 251-279.

Crone, E. A., and van der Molen, M. W. (2007). Development of decision making in school-aged children and adolescents: evidence from heart rate and skin conductance analysis. *Child Dev., 78*(4), 1288-1301.

Damasio, A. R. (1994). *Descartes' Error: Emotion, Reason, and the Human Brain*. New York: Putnam Pub Group.

Damasio, A. R. (1996). The somatic marker hypothesis and the possible functions of the prefrontal cortex. *Philos Trans R Soc Lond B Biol Sci, 351*(1346), 1413-1420.

Davidson, R. J. (2002). Anxiety and affective style: role of prefrontal cortex and amygdala. *Biol. Psychiatry, 51*(1), 68-80.

Diamond, A. (2002). Normal Development of Prefrontal Cortex, from Birth to Young Adulthood: Cognitive Functions, Anatomy and Biochemistry. In D. T. Stuss and R. T. Knight (Eds.), *Principles of Frontal Lobe Function*. Oxford: Oxford University Press.

Domschke, K., Ohrmann, P., Braun, M., Suslow, T., Bauer, J., Hohoff, C., Kersting, A., Engelien, A., Arolt, V., Heindel, W., Deckert, J., and Kugel, H. (2008). Influence of the catechol-O-methyltransferase val158met genotype on amygdala and prefrontal cortex emotional processing in panic disorder. *Psychiatry Res., 163*(1), 13-20.

Ehlers, A., Margraf, J., Davies, S., and Roth, W. T. (1988). Selective Processing of Threat Cues in Subjects with Panic Attacks. *Cognition and Emotion, 2*, 201 – 219.

Elliott, R., Dolan, R. J., and Frith, C. D. (2000). Dissociable functions in the medial and lateral orbitofrontal cortex: evidence from human neuroimaging studies. *Cereb. Cortex, 10*(3), 308-317.

Endler, N. S., and Kocovski, N. L. (2002). Personality disorders at the crossroads. *J. Personal. Disord., 16*(6), 487-502.

Fellows, L. K. (2007). The role of orbitofrontal cortex in decision making: a component process account. *Ann. N Y Acad. Sci., 1121*, 421-430.

Fellows, L. K., and Farah, M. J. (2003). Ventromedial frontal cortex mediates affective shifting in humans: evidence from a reversal learning paradigm. *Brain, 126*(Pt 8), 1830-1837.

Fellows, L. K., and Farah, M. J. (2005). Different underlying impairments in decision-making following ventromedial and dorsolateral frontal lobe damage in humans. *Cereb. Cortex, 15*(1), 58-63.

Fuster, J. (2008). *The Prefrontal Cortex*. London: Academic Press.

Garon, N., Moore, C., and Waschbusch, D. A. (2006). Decision making in children with ADHD only, ADHD-anxious/depressed, and control children using a child version of the Iowa Gambling Task. *J Atten Disord, 9*(4), 607-619.

Gorman, J. M., Kent, J. M., Sullivan, G. M., and Coplan, J. D. (2000). Neuroanatomical hypothesis of panic disorder, revised. *Am. J. Psychiatry, 157*(4), 493-505.

Gottfried, J. A., O'Doherty, J., and Dolan, R. J. (2003). Encoding predictive reward value in human amygdala and orbitofrontal cortex. *Science, 301*(5636), 1104-1107.

Guse, B. (2006). *Entscheidungsverhalten bei Personen mit klinisch relevanter Angst.* Unpublished Internal Report [Diploma Thesis], University of Bochum, Germany.

Forbush, K.T., Shaw, M., Graeber, M.A., Hovick, L., Meyer, V.J., Moser, D.J., Bayless, J., Watson, D., and Black, D.W. (2008). Neuropsychological characteristics and personality traits in pathological gambling. *CNS Spectr, 13(4),* 306-315.

Hansell, J. H., and Damour, L. (2008). *Abnormal Psychology*. New York: Wiley.

Hoehn-Saric, R., Lee, J. S., McLeod, D. R., and Wong, D. F. (2005). Effect of worry on regional cerebral blood flow in nonanxious subjects. *Psychiatry Res., 140*(3), 259-269.

Hooper, C.J., Luciana, M., Conklin, H.M., Yarger, R.S. (2004). Adolescents' performance on the Iowa Gambling Task: implications for the development of decision making and ventromedial prefrontal cortex. *Dev. Psychol., 40 (6),* 1148-1158.

Kalin, N. H., Shelton, S. E., and Davidson, R. J. (2007). Role of the primate orbitofrontal cortex in mediating anxious temperament. *Biol. Psychiatry, 62*(10), 1134-1139.

Kirsch, M., Kirsch, P., Mier, D., and Windmann, S. (2007). Entscheidungsverhalten von Kindern in der IOWA-Gambling-Task: Der Einfluss von Alter und Ängstlichkeit. *Zeitschrift für Neuropsychologie, 18*, 126.

Kirsch, P., Schienle, A., Stark, R., Sammer, G., Blecker, C., Walter, B., Ott, U., Burkart, J., and Vaitl, D. (2003). Anticipation of reward in a nonaversive differential conditioning paradigm and the brain reward system: an event-related fMRI study. *Neuroimage, 20*(2), 1086-1095.

Knutson, B., Fong, G. W., Adams, C. M., Varner, J. L., and Hommer, D. (2001). Dissociation of reward anticipation and outcome with event-related fMRI. *Neuroreport, 12*(17), 3683-3687.

Ledoux, J. E., and Muller, J. (1997). Emotional memory and psychopathology. *Philos. Trans R Soc Lond B Biol. Sci., 352*(1362), 1719-1726.

Lehto, J.E. and Elorinne, E. (2003). Gambling as an executive function task. *Appl. Neuropsychol.* 10(4), 234-238.

Lewinsohn, P. M., Clarke, G. N., Hops, H., and Andrews, J. (1990). Cognitive-behavioral treatment for depressed adolescents. *Behav. Ther., 21*, 385-401.

Losel, F., and Schmucker, M. (2004). Psychopathy, risk taking, and attention: a differentiated test of the somatic marker hypothesis. *J. Abnorm. Psychol., 113*(4), 522-529.

Lungwitz, H. (1955). *Die Neurosenlehre – Die Erkenntnistherapie [2 Bände]*. Berlin: de Gruyter.

McClure, S. M., Daw, N. D., and Montague, P. R. (2003). A computational substrate for incentive salience. *Trends Neurosci., 26*(8), 423-428.

McNally, R. J. (1994). *Panic Disorder: A Critical Analysis*. New York: Guildford Press.

McNally, R. J. (1995). Automaticity and the anxiety disorders. *Behav. Res. Ther., 33*(7), 747-754.

McNally, R. J. (1997). Memory and anxiety disorders. *Philos. Trans R Soc. Lond B Biol. Sci., 352*(1362), 1755-1759.

Milad, M. R., and Quirk, G. J. (2002). Neurons in medial prefrontal cortex signal memory for fear extinction. *Nature, 420*(6911), 70-74.

Milad, M. R., and Rauch, S. L. (2007). The role of the orbitofrontal cortex in anxiety disorders. *Ann. N Y Acad. Sci., 1121*, 546-561.

Miu, A. C., Heilman, R. M., and Houser, D. (2008). Anxiety impairs decision-making: psychophysiological evidence from an Iowa Gambling Task. *Biol. Psychol., 77*(3), 353-358.

Monk, C. S., Nelson, E. E., McClure, E. B., Mogg, K., Bradley, B. P., Leibenluft, E., Blair, R. J., Chen, G., Charney, D. S., Ernst, M., and Pine, D. S. (2006). Ventrolateral prefrontal cortex activation and attentional bias in response to angry faces in adolescents with generalized anxiety disorder. *Am. J. Psychiatry, 163*(6), 1091-1097.

Monk, C. S., Telzer, E. H., Mogg, K., Bradley, B. P., Mai, X., Louro, H. M., Chen, G., McClure-Tone, E. B., Ernst, M., and Pine, D. S.(2008). Amygdala and ventrolateral prefrontal cortex activation to masked angry faces in children and adolescents with generalized anxiety disorder. *Arch. Gen. Psychiatry, 65*(5), 568-576.

Must, A., Szabo, Z., Bodi, N., Szasz, A., Janka, Z., and Keri, S. (2006). Sensitivity to reward and punishment and the prefrontal cortex in major depression. *J. Affect Disord., 90*(2-3), 209-215.

O'Doherty, J., Critchley, H., Deichmann, R., and Dolan, R. J. (2003). Dissociating valence of outcome from behavioral control in human orbital and ventral prefrontal cortices. *J. Neurosci., 23*(21), 7931-7939.

O'Doherty, J. P. (2004). Reward representations and reward-related learning in the human brain: insights from neuroimaging. *Curr. Opin. Neurobiol., 14*(6), 769-776.

O'Doherty, J. P. (2007). Lights, camembert, action! The role of human orbitofrontal cortex in encoding stimuli, rewards, and choices. *Ann. N Y Acad. Sci., 1121*, 254-272.

O'Doherty, J. P., Dayan, P., Friston, K., Critchley, H., and Dolan, R. J. (2003). Temporal difference models and reward-related learning in the human brain. *Neuron, 38*(2), 329-337.

Pickens, C. L., Saddoris, M. P., Gallagher, M., and Holland, P. C. (2005). Orbitofrontal lesions impair use of cue-outcome associations in a devaluation task. *Behav. Neurosci., 119*(1), 317-322.

Preston, S. D., Buchanan, T. W., Stansfield, R. B., and Bechara, A. (2007). Effects of anticipatory stress on decision making in a gambling task. *Behav. Neurosci., 121*(2), 257-263.

Quirk, G. J., Russo, G. K., Barron, J. L., and Lebron, K. (2000). The role of ventromedial prefrontal cortex in the recovery of extinguished fear. *J. Neurosci., 20*(16), 6225-6231.

Rogers, R. D., Owen, A. M., Middleton, H. C., Williams, E. J., Pickard, J. D., Sahakian, B. J., and Robbins, T. W. (1999). Choosing between small, likely rewards and large, unlikely rewards activates inferior and orbital prefrontal cortex. *J. Neurosci., 19*(20), 9029-9038.

Rolls, E. T. (1996). The orbitofrontal cortex. *Philos. Trans R Soc. Lond B Biol. Sci., 351*(1346), 1433-1443; discussion 1443-1434.

Rolls, E. T. (2002). The functions of the orbitofrontal cortex. In D. T. Stuss, and Knight, R.T. (Ed.), *Principles of frontal lobe function*. Oxford, UK: Oxford University Press.

Rolls, E. T., Kringelbach, M. L., and de Araujo, I. E. (2003). Different representations of pleasant and unpleasant odours in the human brain. *Eur. J. Neurosci., 18*(3), 695-703.

Rolls, E. T., O'Doherty, J., Kringelbach, M. L., Francis, S., Bowtell, R., and McGlone, F. (2003). Representations of pleasant and painful touch in the human orbitofrontal and cingulate cortices. *Cereb. Cortex, 13*(3), 308-317.

Schmitt, W. A., Brinkley, C. A., and Newman, J. P. (1999). Testing Damasio's somatic marker hypothesis with psychopathic individuals: risk takers or risk averse? *J. Abnorm. Psychol., 108*(3), 538-543.

Schoenbaum, G., Setlow, B., Saddoris, M. P., and Gallagher, M. (2003). Encoding predicted outcome and acquired value in orbitofrontal cortex during cue sampling depends upon input from basolateral amygdala. *Neuron, 39*(5), 855-867.

Schore, A. N. (1999). *Affect Regulation and the Origin of the Self: The Neurobiology of Emotional Development*. Philadelphia: Lawrence Erlbaum Associates Inc.

Schultz, W., Dayan, P., and Montague, P. R. (1997). A neural substrate of prediction and reward. *Science, 275*(5306), 1593-1599.

Small, D. M., Zatorre, R. J., Dagher, A., Evans, A. C., and Jones-Gotman, M. (2001). Changes in brain activity related to eating chocolate: from pleasure to aversion. *Brain, 124*(Pt 9), 1720-1733.

Smoski, M. J., Lynch, T. R., Rosenthal, M. Z., Cheavens, J. S., Chapman, A. L., and Krishnan, R. R. (2008). Decision-making and risk aversion among depressive adults. *J. Behav. Ther. Exp. Psychiatry*.

Sommer, M., Hajak, G., Dohnel, K., Schwerdtner, J., Meinhardt, J., and Muller, J. L. (2006). Integration of emotion and cognition in patients with psychopathy. *Prog. Brain Res., 156*, 457-466.

van Honk, J., Hermans, E.J., Putman, P., Montagne, B., and Schutter, D.J. (2002). Defective somatic markers in sub-clinical psychopathy. *Neuroreport, 13(8),* 1025-1027.

van Honk, J., and Schutter, D. J. (2006). Unmasking feigned sanity: a neurobiological model of emotion processing in primary psychopathy. *Cognit Neuropsychiatry, 11*(3), 285-306.

Williams, J. M. G., F. N. Watts, C. MacLeod and A. Mathews. (1997). *Cognitive Psychology and Emotional Disorders*. Chichester: Wiley.

Windmann, S. (1998). Panic disorder from a monistic perspective: integrating neurobiological and psychological approaches. *J. Anxiety Disord, 12*(5), 485-507.

Windmann, S., Kirsch, P., Mier, D., Stark, R., Walter, B., Güntürkün, O., and Vaitl, D. (2006). On framing effects in decision making: linking lateral versus medial orbitofrontal cortex activation to choice outcome processing. *J. Cogn. Neurosci., 18*(7), 1198-1211.

Windmann, S., Sakhavat, Z., and Kutas, M. (2002). Electrophysiological evidence reveals affective evaluation deficits early in stimulus processing in patients with panic disorder. *J. Abnorm. Psychol., 111*(2), 357-369.

Wischniewski, J. (2006). *Treffen Panikpatienten Entscheidungen anders? Eine Untersuchung mit der Iowa Gambling Task.* Unpublished Internal Report [Diploma Thesis], University of Bochum, Germany.

Wolfensberger, S. P., Veltman, D. J., Hoogendijk, W. J., De Ruiter, M. B., Boomsma, D. I., and de Geus, E. J. (in press). The neural correlates of verbal encoding and retrieval in monozygotic twins at low or high risk for depression and anxiety. *Biol. Psychol.*

In: Encyclopedia of Neuroscience Research
Editors: Eileen J. Sampson and Donald R. Glevins
ISBN 978-1-61324-861-4

Chapter VI

PEEF: Premotor Ear-Eye Field. A New Vista of Area 8B

Bon Leopoldo, Marco Lanzilotto and Cristina Lucchetti
Department of Biomedical Sciences,
University of Modena and Reggio Emilia,
Via Campi 287, 41100 Modena, Italy

Abstract

In macaque monkeys, area 8B may be considered cytoarchitectonically as a transitional area between the granular area 9, rostrally, and the rostral part of the dorsal agranular area 6 (FC or F7), caudally. This area is anatomically connected with auditory cortical areas, cerebellum and superior colliculus. Microstimulation of area 8B evokes ear movements or eye movements in some sites; in other sites, it evokes both ear and eye movements by varying the intensity of electric stimulation. Unit activity recording shows that neurons in area 8B have a role in encoding different auditory environmental stimuli and movements of the ear and eye. In addition, fixation of a visual stimulus, when attention is engaged, modulates the discharge of auditory environmental neurons, and sensory-motor neurons.

Current functional and anatomical evidences strongly support the proposal that area 8B is a specific Premotor Ear-Eye Field (PEEF) involved in the recognition of auditory stimuli and in orienting processes. Moreover, activation by specific environmental auditory stimuli, suggests that PEEF is an important node in the hierarchical organisation of the auditory network. The inhibitory effects on neural discharge of auditory and auditory-motor cells might be a consequence of the engagement of attention during visual fixation. This phenomenon may be the expression of a combination of a covert orienting of attention relative to eye effectors, and an overt orienting of attention relative to ear effectors. It seems that attention may affect more than one channel at the same time during a conditioned task or during natural behaviour. Then the PEEF, based on its connections and functional characteristics may be involved in three circuits: 1) an auditory circuit, 2) an oculomotor circuit, 3) and a fronto-cerebellar circuit.

I. Introduction

The prefrontal cortex is the rostral part of the frontal cortex. It is subdivided in many cytoarchitectonic areas with different functions. The prefrontal cortex is more developed in non-human and human primates than in other animals. The anatomical interconnections and physiological data suggest that the prefrontal cortex is involved in high-order brain functions and in the organization of motor behaviour (Tanji and Hoshi 2008).

In this review we will talk about area 8B, which has remained relatively an unexplored part of the prefrontal cortex. Area 8B, together with area 6 and 8A, based on cytoarchitecture, anatomical connections and electrophysiological data, may be considered as part of the premotor cortex (Pandya and Yeterian 1985, Mann et al. 1988, Passingham 1993, Bon and Lucchetti 1994, Lucchetti et al. 1998, Moschovakis et al. 2004) and area 8B may be considered a “premotor ear-eye field” (PEEF) (Lucchetti et al. 2008).

II. Anatomical Background

Cytoarchitecture

In macaque monkeys area 8B may be considered cytoarchitectonically as a transitional area between the granular area 9, rostrally, and the rostral part of the dorsal agranular area 6 (FC or F7), caudally (Von Bonin and Bailey 1947). The first scientist to identify this area was Walker (1940), and successively Von Bonin and Bailey (1947) and Matelli et al. (1991) confirmed this statement. Some neuroanatomists considered area 8B as the dorso-mesial caudal part of area 9 (Barbas and Pandya 1989, Pandya and Yeterian 1990, Barbas et al. 1999). Petrides and Pandya (1999) re-examined, in a comparative cytoarchitectonic analysis the dorsolateral prefrontal cortex of the human and the macaque brain. They identified an area interposed between area 9 and the dorsal part of area 6 and following Walker (1940) they defined this area as 8B. In the original paper they described this area as follows:

> “Its posteroventral part is located in the rostralmost section of the superior limb of the arcuate sulcus. Dorsally, it extends up to the midline, and it continues medially as far as the border of the cingulate sulcus. In this area, layer II is well defined and blends gently with layer III. Layer III is sparse and contains small- to medium-sized pyramidal neurons; only a few neurons in this layer are as large as those encountered in 8A. Layer IV is poorly developed and layer V almost blends with layer VI. A few darkly stained cells can be distinguished in layer Va.”

Neuroanatomy

Anatomical investigations showed the complex connections of area 8B with cortical and subcortical structures. The study by Barbas and Pandya (1989) showed that dorsal prefrontal areas (9, 8, 46) are strongly interconnected and area 9 sends a robust projection toward dorsal area 6. Recent anatomical investigations showed that area 8B sends robust projections to SEF (Luppino et al. 2003, Wang et al. 2005).

Barbas and Mesulam (1981) showed prominent projections from auditory association cortices: in particular from the superior temporal gyrus and part of the dorsal bank of the superior temporal sulcus, to the rostral part of area 8A and contiguous area 8B. Subsequent investigations pointed out the reciprocal connections between auditory association area AA1 and area 8A, and between area AA2 and dorsal prefrontal cortex areas 9 and 10 (Pandya and Yeterian 1990, Barbas et al. 1999). Quantitative investigations have addressed the anatomical interactions between prefrontal areas 9 and 10 and inhibitory neurons in superior temporal cortices involved in auditory perception (Barbas et al. 2005).

It has been suggested that the prefrontal cortex is devoted to the integration of "what" and "where" in non human-primates for the visual cortical system: neurons that are object-tuned, location-tuned or are have both what and where tuning were found in the dorsolateral and ventrolateral prefrontal cortex (Rao et al. 1997). Using anatomical and electrophysiological recordings methods, Romanski et al. (1999) also identified two different auditory streams in non-human primates, which originate from the caudal and rostral auditory belt cortex and project to different areas of the frontal lobe. The rostral belt is reciprocally connected to the frontal pole (area 10), rostral principal sulcus (area 46) and ventral prefrontal region (areas 12 and 45), whereas the caudal belt is mainly connected with caudal principal sulcus (area 46), frontal eye fields (area 8A) and with area 8B. They suggested that two streams originate from caudal and rostral auditory cortices and target areas associated with spatial "where" and non spatial "what" processes in the prefrontal cortex. Moreover, investigations using fMRI and ERPs (event related potentials) methods have shown both streams for sound identity and sound spatial location in humans, suggesting that ventral and dorsal prefrontal areas are involved in representing the "what" and "where" respectively (Alain et al. 2001)

In addition, the prefrontal cortex is highly connected with the parietal lobe. In particular the medial surface of parietal lobe area PGm sends projections to the rostral part of dorsal area 6, area 8, dorsal areas 46 and 9, while area 9 sends back projections to area PGm (Pandya and Yeterian 1990). Electrical stimulation of area PGm evokes eye and ear movements in non human-primates (Thier and Andersen 1998). Kelly and Strick (2003) showed anatomical cerebro-cerebellum connections through the basilar pons and thalamus. The strongest projections to the basilar pons derive from the dorsolateral prefrontal convexity, including rostral area 8A, area 9 and dorsal area 46. Area 8B sends projections to the paramedian nucleus, peripeduncular nucleus and nucleus reticularis tegmentum ponti (NRTP) (Schmahmann and Pandya 1997). Middleton and Strick (2001) showed that area 9 and area 46 receive a consistent input from the cerebellum through thalamic regions, and proposed distinct cerebellum output channels to the prefrontal and motor cortices.

Auditory neurons were described in different part of cerebellum: in the middle part of the vermis (Snider and Stowell 1944; Fadiga and Pupilli 1964), in the nuclei interpositi and dentate nucleus (Mortimer 1975). The dentate nucleus, as shown by Lynch et al. (1994), projects to area 8B. Moreover, areas 8A, 8B and 9 project to the deep layers of the superior colliculus and pretectum (Fries 1984, Tanila et al. 1993). The deep layers of the superior colliculus project to the reticular formation and to the cervical spinal cord where the neck is represented (Raybourn and Keller 1977, Kuypers 1981). Electrophysiological evidence shows a clear relationship between superior colliculus stimulation and head movements in the monkey (Corneil et al. 2002). The projections from the cerebellum (vermis), the superior and the inferior colliculus and auditory cortex onto the pontine nulei, suggest a role of the dorsolateral pontine nulei in the integration of visual and auditory information. It appears that this complex system of multiple loops of cerebellar connections with the presence of auditory neurons represent a substantial sensory input for the cerebellar motor control.

The dorsomedial frontal cortex (areas 9,8B, 6 and 4) receives dopaminergic projections from cells of type A9 located dorsally to the substantia nigra pars compacta and to the retro rubral area, from cells of type A10 in the parabrachial pigmented nucleus and linear nuclei. This suggests that area 8B is part of the mesofrontal dopamine system implicated in psychiatric disorders (schizophrenia, psychotic status) (Williams and Goldman-Rakic 1998). Finally, the injection of rabies virus into lateral rectus muscle of the macaque monkey showed the presence of labelled neurons in area 8B and 9, as well as in area 8A (FEF), area F7 (SEF) and F6 (Moschovakis et al 2004) (Fig. 1).

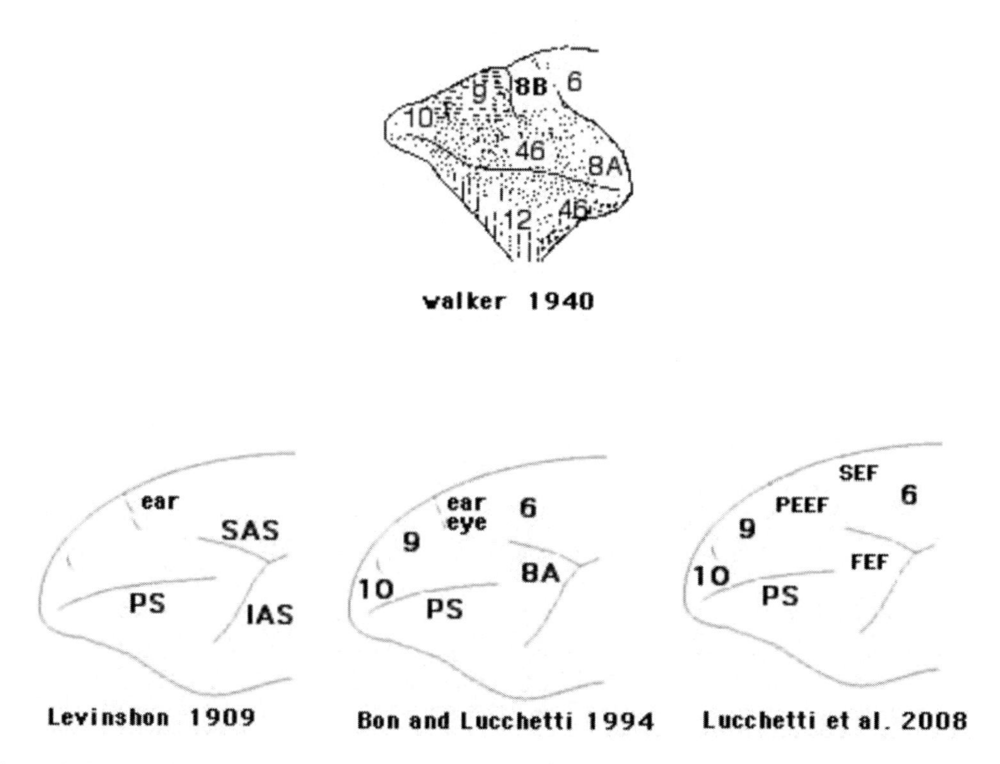

Figure 1. Cytoarchitectonic map of Walker (1940). Stimulation studies of Levinshon (1909). Representation of the stimulation and recording study of Bon and Lucchetti 1994. The actual vision of frontal fields by Lucchetti et al. (2008).

III. Lesions Studies

Many studies of clinical human disorders following natural lesions, and experimental prefrontal lesions in monkeys showed the involvement of prefrontal cortex in personality, motor behaviour, working memory and in temporal encoding to execute a motor act (Harlow 1848, Fuster 1997).

Specific lesion studies showed that animals with prefrontal cortex damage exhibited a deficit in auditory discrimination tasks (Rosenkilde 1979) and that this deficit was manifested when the lesion included areas 8A and 8B (Wegener 1973). Localised lesions in area 8A

(FEF) gave rise to a transitory deficit in attention, eye movements and eye-head orienting toward a visual stimulus (Rizzolatti et al. 1983, van der Steen et al. 1986, Schiller et al 1987).

One case with lesion in the monkey, extending from area 8B throughout dorsal area 6, showed initially tilting neck movements. In the days following the lesion, the animal performed saccades and fixation tasks correctly in all directions, while contralateral arm reaching movements were severely impaired. Pleasant stimuli, presented in the ipsilateral or contralateral hemifield, readily drew the attention of the animal. If the same stimuli were presented simultaneously in both hemifields, the monkey oriented itself only toward the ipsilateral one. Aversive stimuli evoked an aggressive reaction only when the stimulus was localized in the ipsilateral hemifield. The animal clearly neglected the aversive stimulus presented in the contralateral hemifield. The animal recovered completely in 30 days. The postmortem examination revealed a lesion in the dorsomedial frontal cortex. The combined attentional and motor deficits suggest that area 6 and 8B, called dorsomedial frontal cortex (DMFC) may be involved in the preparation and execution of movements triggered by the affective meaning of the stimulus (Lucchetti et al. 1998).

On the other hand, combined lesions of dorsal area 46 and 9 produced different deficits in memory tasks. The dorsal prefrontal region receives visuospatial information from areas 8A, 6 and 46. The ventrolateral prefrontal cortex receives non-spatial visual information. Consequently, a lesion of the dorsal part of prefrontal cortex affects visuospatial input and does not affect nonspatial visual object information. Lesions limited to area 9 produced a mild impairment in memory task performance (Petrides 2000).

IV. Neurophysiology

Electrophysiological investigation of area 8B started with the experiments of Ferrier in 1875. In those experiments the author electrically stimulated a large frontal region: areas 9, 8B, 8A, 46 and dorsal area 6 and evoked eye movements. Later Levinsohn (1909) used electric stimulation to evoke eye movements in dorsal area 6 and ear movements in area 8B.

Many years later an explosion of investigations pointed out relevant neuronal characteristics of area 8A (FEF), and the rostral part of dorsal area 6 (SEF). For a detailed vision about FEF and SEF see Tehovnik et al. (2000), and about the cortical and subcortical networks see Lynch and Tian (2006).

Area 8B had been neglected until 1994 when experiments of microstimulation and unit activity recording in macaque monkey pointed out the role of this area in ear and eye motor control in spontaneous behaviours (Bon and Lucchetti 1994). Microstimulation in some sites evoked only ear movements or only eye movements; in other locations it evoked both ear and eye movements by varying the intensity of electrical stimulation. The electrically evoked ear movements were forward, or backward or oblique (upward-forward; upward-backward). The evoked eye movements were mostly fixed vector saccades, contralateral and with an upward orientation of about 45 degrees. If we consider only the sites where the threshold was equal to or lower than 50 μA, mostly of the evoked ear movements were position dependent (Fig. 2).

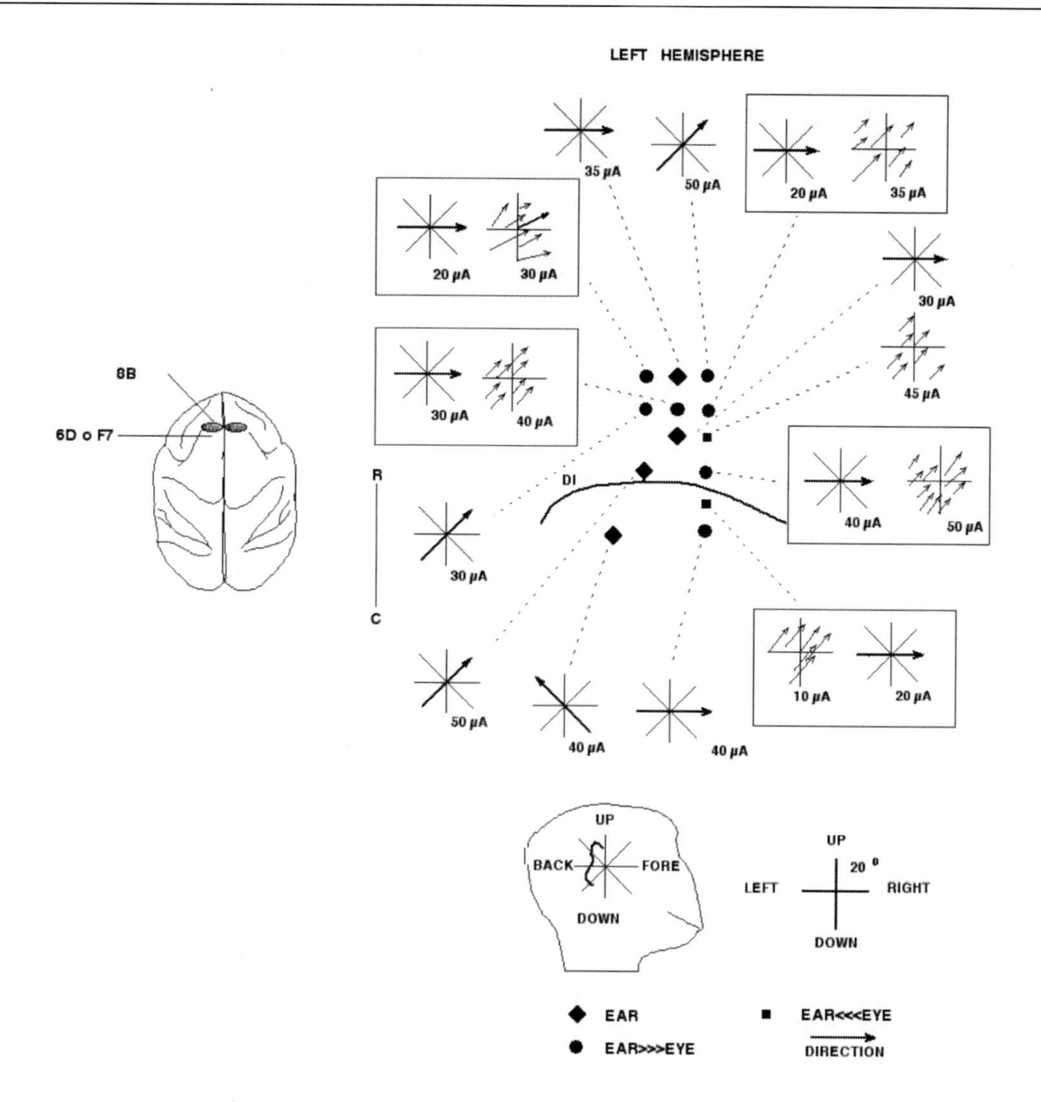

Figure 2. Microstimulation map of area 8B. Representation of ear, eye, and both ear and eye movements evoked with a current threshold equal or lower than 50 μA. DI: dimple. (R rostral, C caudal). Arrows represent the mean direction of ear and eye movements in each site. (Exp Brain Res, 102, 1994, 259-71, "Ear and eye representation in the frontal cortex, area 8b, of the macaque monkey: an elettrophysiological study", Bon L and Lucchetti C, Figure 2-Figure 5, With kind permission of Springer Science+ Business Media).

In addition, neurons discharged before ear movements, and before ear and eye movements, or only before eye movements. The majority of ear and ear-eye cells had a preferred direction for ear movements. Eighty-five percent of cells did not have a preferred direction for visual guided saccades and were active when the monkey made saccades toward the unlit targets (checking saccades). This type of saccades were found and defined previously in SEF (Bon and Lucchetti, 1992).

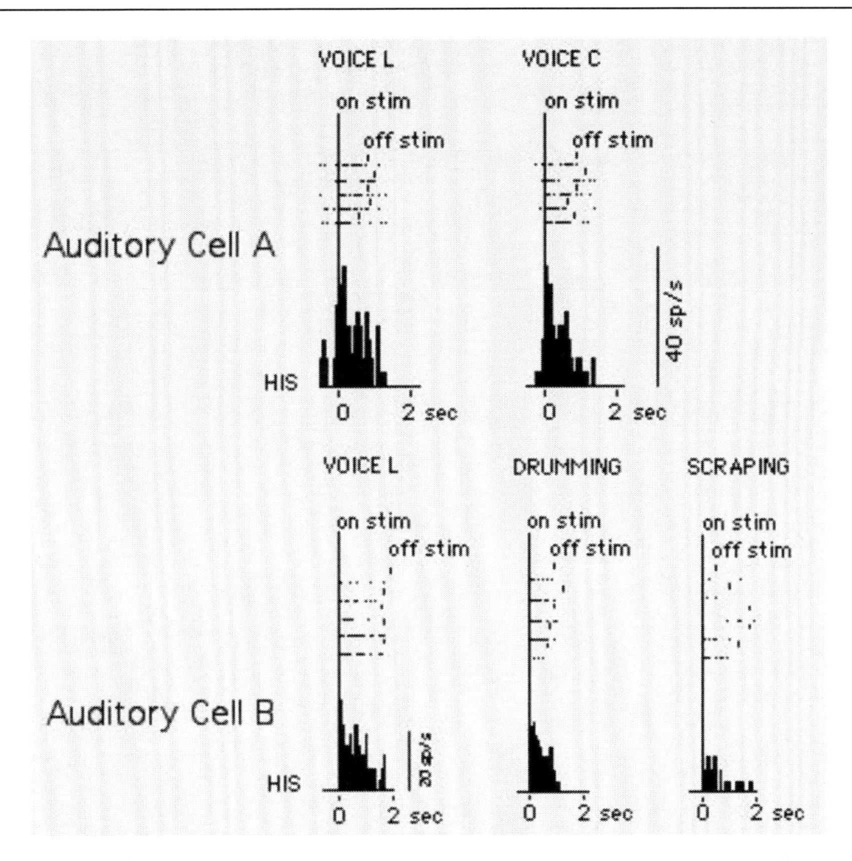

Figure 3. Example of auditory cell. A: auditory cell active for experimenters' voices. B: Auditory cell active for experimenter's voice and environmental stimuli. Raster: each dot is one spike. HIS: unit activity histogram. Bin width: 50 ms. (Exp Brain Res, 168, 2006, 441-9, "Auditory environmental cells and visual fixation effect in area 8B of macaque monkey", Bon L and Lucchetti C, Figure 2, With kind permission of Springer Science+ Business Media).

Further investigations showed new and interesting aspects for auditory, motor and attentional characteristics of neurons of area 8B (Bon and Lucchetti 2006; Lucchetti et al. 2008). Auditory neurons, were active only for environmental auditory stimuli: experimenter's voices, opening and closing of the lab doors, scraping of feet. However, the best stimuli were represented by the experimenters' voices (Fig.3).

In addition, some auditory cells showed activity directionally tuned with the stimulus. The stimulus in these cases was represented by the experimenters' footsteps. In other words, the cell activity increased or decreased when the experimenter moved toward the Faraday cage and viceversa when he/she moved away. Cells that showed activity synchronized with the onset of auditory stimuli and with the onset of an orienting saccade and/or an orienting ear movement were defined as "auditory-motor". For the sensory aspects of neural activity, the unit's discharge rates could be modulated in relation to one or more types of environmental stimuli. With some stimuli, this activity could be very weak and show rapid adaptation.

One part of auditory-motor cells presented a discharge synchronized with the onset of orienting movements of both the eye and ear (Fig. 4). Few of these neurons showed activity that correlated only with the onset of orienting eye movements, other neurons had activity that correlated only with the onset of orienting ear movements (Fig. 5).

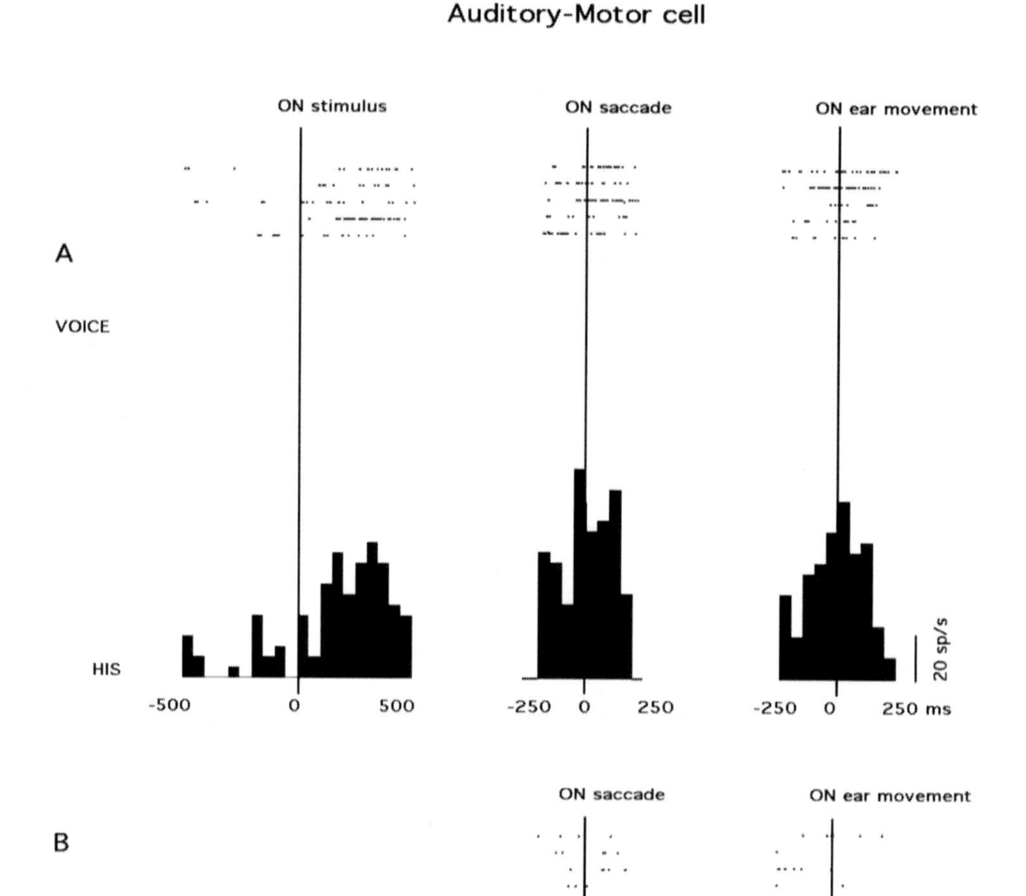

Figure 4. Example of auditory motor cell. A: This cell is active for auditory environmental stimuli, for orienting saccades, and orienting ear movements. B: In spontaneous condition the discharge is not synchronized with saccades or with ear movements. Raster: each dot is one spike. HIS: unit activity histogram. Bin width: 50 ms. (Exp Brain Res, 186, 2008, 131-41, "Auditory-motor and cognitive aspects in area 8B of macaque monkey's frontal cortex: a premotor ear–eye Weld (PEEF)", Lucchetti C et al. , Figure 3, With kind permission of Springer Science+ Business Media).

Those cells that did not show activity synchronized with the onset of auditory stimuli, but only with orienting ear movements, were defined as motor. The motor cells showed phasic, tonic, and phasic-tonic activity. An example of phasic-tonic activity is presented in Figure 6, where we can see a phasic discharge related to orienting ear movements and a tonic discharge in relation to ear position. These cells sometimes showed directional selectivity, forward and backward, and only a few cells discharged for oblique, upward and downward movements of the ear.

In addition different types of cells were found: "complex behaviour" cells showed inhibition of discharge by an auditory stimulus only when the ear was in a posterior position.

"Novel visual stimuli" cells discharged at the presentation of different objects (mirror, cans, tools). Finally, cells showed neural activity for the experimenter's body or for a white coat presented on a stick. Similar cells were described by Tanila et al. (1992) in the prefrontal cortex.

Auditory-Motor cell

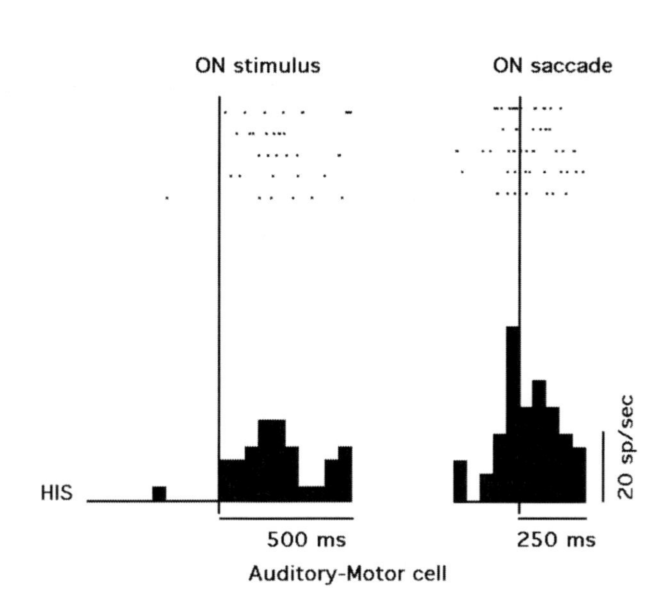

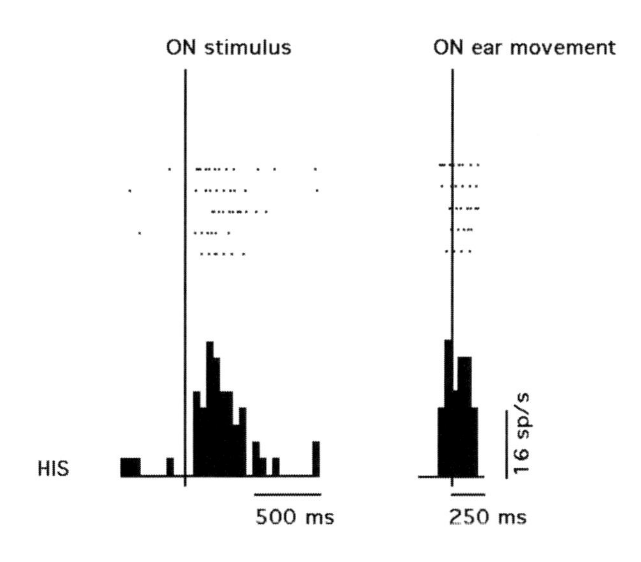

Figure 5. Auditory motor cells. Top: auditory-eye cell. Bottom: auditory-ear cell. Raster: each dot is one spike. HIS: unit activity histogram. Bin width: 50 ms. (Exp Brain Res, 186, 2008, 131-41, "Auditory-motor and cognitive aspects in area 8B of macaque monkey's frontal cortex: a premotor ear–eye Weld (PEEF)", Lucchetti C et al. , Figure 4-Figure 5, With kind permission of Springer Science+ Business Media).

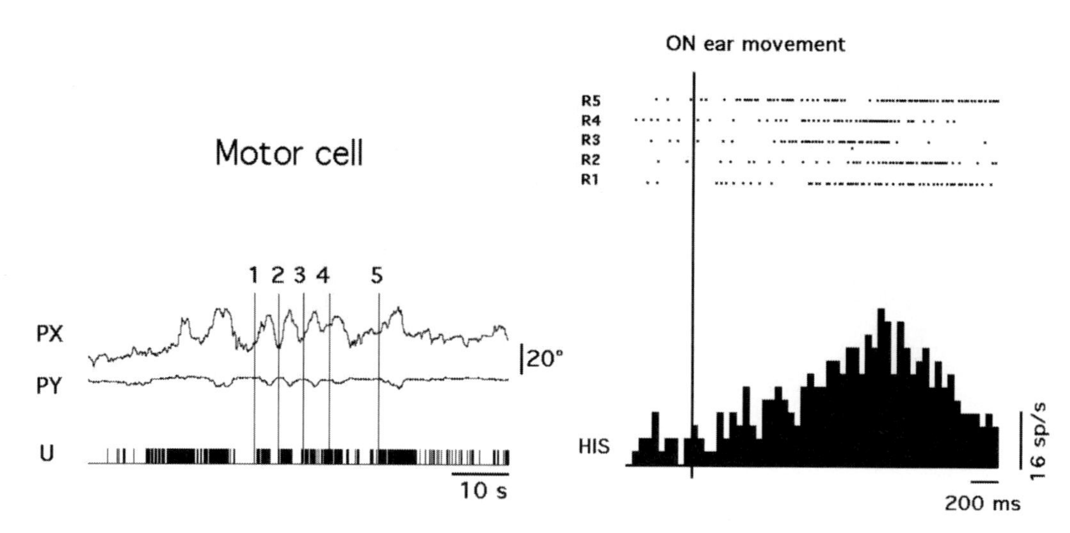

Figure 6. Example of motor cell. In this example, we can see a relationship to ear movement and a different tonic discharge in relation to ear position. PX: horizontal ear component, PY: vertical ear component. U: unit activity, each bar is one spike. Raster: each dot is one spike. HIS: unit activity histogram. Bin width: 50 ms. (Exp Brain Res, 186, 2008, 131-41, "Auditory-motor and cognitive aspects in area 8B of macaque monkey's frontal cortex: a premotor ear–eye Weld (PEEF)", Lucchetti C et al. , Figure 6, With kind permission of Springer Science+ Business Media).

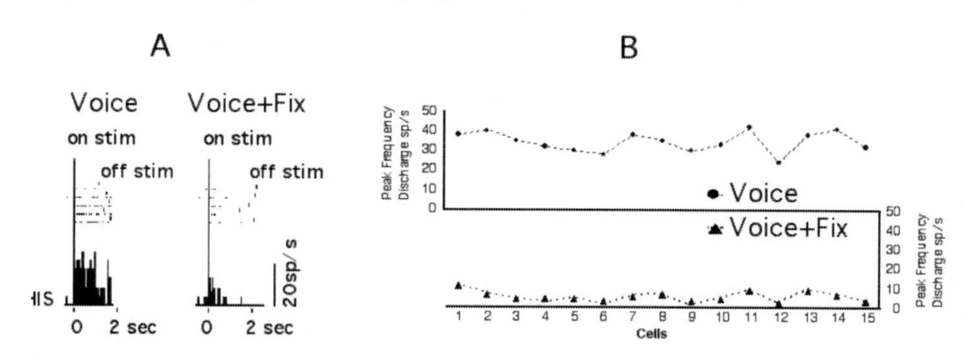

Figure 7. Inhibitory effect of visual fixation on auditory cells. Left: example of auditory cell (VOICE) and an inhibitory effect during visual fixation (FIX-VOICE). Each dot in the raster represents one spike. HIS: unit activity histogram. Bin width: 50 ms. Right: population analysis of 15 auditory cells influenced by fixation, Bi-Polar Chart Plot (Exp Brain Res, 168, 2006, 441-9, "Auditory environmental cells and visual fixation effect in area 8B of macaque monkey", Bon L and Lucchetti C, Figure 5-Figure 6, With kind permission of Springer Science+ Business Media).

Moreover, the auditory cells and the auditory-motor cells were tested when the animal performed a visual fixation in primary position in which the attention was engaged. Fifty percent of auditory cells showed a significantly lower activity when auditory stimuli were generated during the execution of visual fixation task, instead of when the same stimuli were generated in spontaneous conditions (Fig. 7). The same effect was present also in about fifty percent of the auditory-motor cells, related only to the orienting ear movement. These neurons showed a significantly lower activity level, in both their auditory and motor components, when auditory stimuli were presented during the execution of visual fixation task, as opposed to when the same stimuli were presented out of the task (Fig. 8).

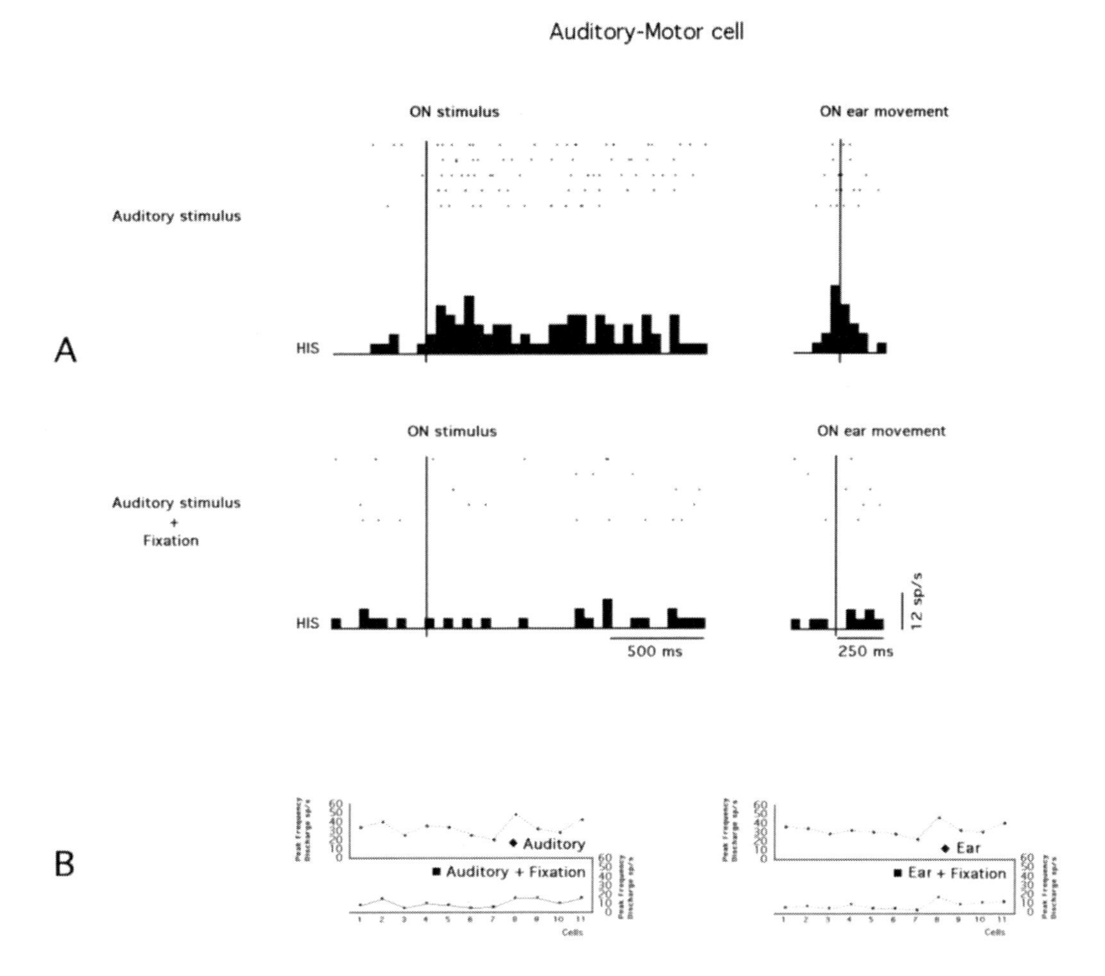

Figure 8. Inhibitory effect of visual fixation on auditory-motor cells. A: Example of auditory–motor cell in spontaneous condition and during visual fixation. In spontaneous condition, the activity is brisk for both auditory stimulus and ear movement, while during a visual fixation task the discharge is weak for both auditory stimulus and ear movement. Each dot in the raster represents one spike. HIS: unit activity histogram. Bin width: 50 ms. B: population analysis of 11 auditory-motor cells influenced by visual fixation. Left: auditory component, right: motor component (Exp Brain Res, 186, 2008, 131-41, "Auditory-motor and cognitive aspects in area 8B of macaque monkey's frontal cortex: a premotor ear–eye Weld (PEEF)", Lucchetti C et al. , Figure 8-Figure 9, With kind permission of Springer Science+ Business Media).

Conclusion

Considerations

If we observe the natural behaviour of an animal pointing to an another animal that may be a prey (cat and rat) and at the same time an auditory stimulus is produced, we may note two different reactions: the animal maintains its gaze towards the prey and orients its ears towards the auditory stimulus, or the animal orients both gaze and ears towards the auditory stimulus. A similar behaviour may be seen in humans, for example when we are reading a book and an external sound is produced, for example a message on a cellular phone, we may continue to read the book or we may orient the head toward the phone.

The electrophysiological investigations show that area 8B is involved in ear-eye motor control and in the auditory perception of environmental auditory stimuli. Other studies, carried out in the lateral belt area and in the core of the auditory cortex, connected to area 8B and other prefrontal cortices, showed neurons with discharge related to complex stimuli, vocalizations, and to faces and conspecific vocalizations (Rauschecker et al. 1995, Recanzone 2000, Ghanzanfar et al. 2005).

Three issues are relevant at this moment about area 8B (Bon and Lucchetti 1994, 2006, Lucchetti et al. 2008): a) the presence of auditory environmental neurons, b) the presence of auditory-motor neurons; c) the involvement of attentional processes.

A) Auditory Environmental Neurons

That environmental or naturalistic stimuli are more efficient has already been suggested by experimental evidence in the midbrain and inferior colliculus, and it has been proposed that this aspect is related to the intrinsic properties of the signal (Attias and Schreiner 1997, Escabì et al. 2003). More recent studies indicate that neurons of the lateral belt area, discharge better for complex stimuli and vocalisation, which may suggest, as argued by the authors, that this field is processing “what” information (Rauschecker et al. 1995, Recanzone 2000). Fuster et al. (2000) showed that neurons in area 8B, but also in area 9 and 6 were active when the animal performed correctly an association task between tone and colour.

The activity related to environmental auditory stimuli, found in area 8B, suggests a natural behavioural meaning in a cross-modal association between stimulus-subject (voice-experimenter) and between stimulus-motor act (footstep-walking). Even if largely speculative, this natural cross-modal association may represent an abstraction of auditory stimuli or alternatively may be an expression of auditory recognition based on emotion.

B) Auditory-Motor Neurons

Auditory-motor neurons, sensory and motor in nature, were tested without any behavioural constraints imposed by operant conditioning. These neurons were specific for environmental auditory stimuli and also showed properties of motor integration with two effectors: ear and/or eye. The animals oriented their gaze, in natural behaviour, turning their ears towards a stimulus presented in front of them or behind them. Accordingly, the neural activity could reflect a general aspect of orienting that involves the spatial localization of an acoustic stimulus.

The above evidences suggest that area 8B may be involved not only in the recognition of auditory stimuli but also in orienting processes (Romanski et al. 1999, Rao et al. 1997, Alain et al. 2001). In addition, activation for specific environmental auditory stimuli and the presence of auditory-motor and complex cells suggest that area 8B has an important place in the hierarchical organisation of the auditory system (Kaas and Hackett 2000, Wessinger et al. 2001). According to the model proposed by Kaas and Hackett (2000), area 8B may be considered together with areas 8A, 46, 12 and 10 at the fourth level of this hierarchical organisation of auditory system in primates.

C) Attentional Processes

Some considerations are necessary to explain the inhibitory effects that visual fixation tasks have on auditory and auditory-motor neurons. Similar results were found in the superior colliculus of the cat, when two visual stimuli were presented at the same time (Rizzolatti et al. 1974) and when visual fixation inhibited the auditory response (Populin and Yin 2002). Moreover, a similar effect has been found in the superior colliculus and in the posterior parietal cortex of the macaque monkey for visual and auditory peripheral stimuli (Bell et al. 2003, Gifford and Cohen 2004, Cohen et al. 2005). In the experiments of Rizzolatti et al. (1974), carried out with curare, both stimuli were visual. The authors explained the inhibitory effect as a consequence of orienting attention. Our experimental approach was similar to that of Bell et al. (2003), since one stimulus was visual and the other one is auditory. The inhibitory effect on neural discharge of both auditory and auditory-motor cells might be a consequence of the engagement of attention during visual fixation (Bon and Lucchetti 1997). This issue may be the expression of a combination of covert orienting of attention relative to eye position (i.e., nonfoveal locations), similar to the classic finding of Cherry (1953), and an overt orienting of attention relative to orientation of the ear. Accordingly, it seems that attention may affect more than one channel at the same time during a conditioned task or during natural behaviour (Turatto et al. 2004). Moreover, the inhibitory effect of auditory and auditory-motor neuron activity suggests that attention may filter or attenuate the auditory signals while a subject is orienting attention toward a visual stimulus (Treisman 1960). If we accept the attention as a link between sensory input and motor output, the involvement of attention in the gaze and ear motor control is also in agreement with the premotor theory of attention (Rizzolatti and Craighero 1998). All together, based on its functional characteristics and anatomical connections, area 8B may be part of the inhibitory network for gaze shift and ear orienting (Schiller and Tehovnik 2005).

Why "PEEF"?

There is a debate about whether area 8 (area 8A plus 8B) should be treated as a premotor or a prefrontal area. In macaque monkeys area 8B may be considered cytoachitectonically a transitional area between the granular area 9, rostrally, and the rostral part of the dorsal agranular area 6 caudally (Walker, 1940, Von Bonin a Bailey 1947, Matelli et al. 1991). Pandya and Yeterian (1985) used the term "premotor" to include both areas 8 and 6. Subsequently, other authors also treated area 6 and 8 as premotor cortex (Mann et al., 1988, Passingham, 1993, Bon and Lucchetti, 1994, Lucchetti et al. 1998, Lucchetti et al. 2008). In addition, the cytoarchitecture of area 8B differs from the granular area 8A (FEF), located latero-caudally to it in the prearcuate region (Barbas and Mesulam, 1981, Petrides and Pandya, 1994, Petrides and Pandya 1999). Transneuronal transfer of rabies virus and a functional imaging study clearly show that area 8B is involved with SEF in eye motor control (Moschovakis et al. 2004).

Area 8B is reciprocally connected with the auditory cortex (Barbas and Mesulam 1981), receives projections from the dentate nucleus of cerebellum via the thalamus, much like FEF (Lynch et al 1994), and projects to the superior colliculus, also like FEF (Fries 1984). Recent anatomical studies show that area 8B sends robust projections to (SEF) (Luppino et al. 2003, Wang et al. 2005).

The neural activity of area 8B is prevalently related to ear or to ear-eye movement and to auditory environmental stimuli (Bon and Lucchetti 1994, 2006). In contrast, neural activity in

SEF is related to eye movement, eye fixation, ear and neck movement (Schlag and Schlag-Rey 1987, Bon and Lucchetti 1990, 1991,1992, Tehovnik and Lee 1993, Tehovnik and Sommer 1997, Chen and Walton 2005), auditory stimuli (Schall 1991) and attentional processes (Bon and Lucchetti 1997). Neural activity in the FEF is related to eye movement, neck movement, auditory and visual stimuli and attentional processes (Schall 1997).

Microstimulation of area 8B predominantly evoked small fixed-vector saccades and ear movement, as well as ear-eye movement (Levinshon 1909, Bon and Lucchetti 1994). Microstimulation of SEF predominantly evoked goal directed saccades and, in a few penetrations, also ear movements (Schlag and Schlag-Rey 1987, Bon and Lucchetti 1992, Tehovnik et al. 2000). In contrast, microstimulation of FEF evoked small fixed-vector saccades in the inferior part, and large fixed-vector saccades in the superior part (Bruce et al. 1985).

Taking these results together, the principal difference between area 8B and SEF/FEF is that the neurons in area 8B are rarely related to saccades and visual fixation, and they are more related to ear movement. Moreover, microstimulation of area 8B predominantly evokes activation of ear muscles or a coactivation of ear and eye muscles.

In the first report by Bon and Lucchetti (1994) about area 8B, they said that this area "may be considered a rostral extension of supplementary eye field (SEF) or a different region". Considering, the functional differences between area 8B and both the SEF and FEF (area 8A), as well as the differences in cytoarchitecture and anatomical projections, we can now consider area 8B as a separate field, a Premotor Ear-Eye Field (PEEF).

Acknowledgments We wish to thank Dr, Helen Barbas for helpful suggestions in revising this review.

References

Alain, C; Arnott, SR; Hevenor, S; Graham, S; Grady, CL. "What" and "Where" in the human auditory system. *Proc. Natl. Acad. Sci. USA,* 2001 98, 12301-12306.

Attias, H; Schreiner, C. Temporal low-order statistics of natural sounds. In: Mozer MC, Jordan MI, Petsche T, editor. *Advances in neural information processes systems.* Cambridge MA: MIT Press; 1997; 27-33.

Barbas, H; Mesulam, MM. Organization of afferent input to subdivisions of area 8 in the rhesus monkey. *J. Comp. Neurol.*, 1981 200, 407-431.

Barbas, H; Ghashghaei, H; Dombrowski, SM; Rempel-Clower, NL. Medial prefrontal cortices are unified by common connections with superior temporal cortices and distinguished by input from memory-related areas in the rhesus monkey. *J. Comp. Neurol.,* 1999 410, 343-67.

Barbas, H; Medalla, M; Alade, O; Suski, J; Zikopoulos, B; Lera, P. Relationship of prefrontal connections to inhibitory systems in superior temporal areas in the rhesus monkey. *Cerebral Cortex,* 2005 15, 1356-1370.

Barbas, H; Pandya, DN. Architecture and intrinsic connections of the prefrontal cortex in the rhesus monkey. *J. Comp. Neurol,* 1989 286, 353-375.

Bell, AH; Corneil, BD; Munoz, DP; Meredith, MA. Engagement of visual fixation suppresses sensory responsiveness and multisensory integration in the primate superior colliculus. *Eur J. Neurosci.*, 2003 18, 2867-2873.

Bon, L; Lucchetti, C. Neurons signalling the maintenance of attentive fixation in frontal area 6aß of macaque monkey. *Exp. Brain Res.,* 1990 82, 231-233.

Bon, L; Lucchetti, C. Behavioral and motor mechanisms of dorsomedial frontal cortex in Macaca monkey. *Intern J. Neuroscience*, 1991 60, 187-193.

Bon, L; Lucchetti, C. The dorsomedial frontal cortex of the macaca monkey: fixation and saccade-related activity. *Exp. Brain Res.*, 1992 89, 571-580.

Bon, L; Lucchetti, C. Ear and eye representation in the frontal cortex, area 8b, of the macaque monkey: an electrophysiological study. *Exp. Brain Res.*, 1994 102, 259-271.

Bon, L; Lucchetti, C. Attentional-related neurons in the supplementary eye field of the macaque monkey. *Exp. Brain Res.*, 1997 113, 180–185.

Bon, L; Lucchetti, C. Auditory environmental cells and visual fixation effect in area 8B of macaque monkey. *Exp. Brain Res.,* 2006 168, 441-449.

Bruce, CJ; Goldberg, ME; Bushnell, MC; Stanton, GB. Primate frontal eye fields: II. Physiological and anatomic correlates of electrically evoked eye movements. *J. Neurophysiol,* 1985 54, 714-734.

Chen, LL; Walton, MMG. Head movement evoked by electric stimulation in the supplementary eye field of the rhesus monkey. *J. Neurophysiol.,* 2005 94, 4502-4519.

Cherry, EC. Some experiments on the recognition of speech with one and with two ears. *J. Acustical Soc of America,* 1953 25, 975-979.

Cohen, YE; Russ, BE; Gifford III, GW. Auditory processing in the posterior parietal cortex. *Behavioural and Cognitive Neuroscience Reviews*, 2005 4, 218-231.

Corneil, BD; Olivier, E; Munoz, DP. Neck muscle responses to stimulation of monkey superior colliculus. I. Topography and manipulation of stimulation parameters. *J. Neurophysiol.*, 2002 88, 1980-1999.

Escabì, MA; Miller, LM; Read, HL; Schreiner, CE. Naturalistic auditory contrast improves spectrotemporal coding in the cat inferior colliculus. *J. Neurosci.,* 2003 23, 11489-11504.

Fadiga, E; Pupilli, GC. Teleceptive components of the cerebellar functions. *Physiol Rev,* 1964 44, 432-486.

Ferrier, D. Experiments on the brains of monkeys. *Philos. Trans R Soc Lond B Biol. Sci.,* 1875 165, 433-488.

Fries, W. Cortical projections to the superior colliculus in the macaque monkey: a retrograde study using horseradish peroxidase. *J. Comp. Neurol.,* 1984 230, 55-76.

Fuster, JM. *The prefrontal cortex.* 3rd Edition. Philadelphia-New York: Lippincot-Raven publishers; 1997.

Fuster, JM; Bodner, M; Kroger, JK. Cross-modal and cross-temporal association in neurons of frontal cortex. *Nature,* 2000 405, 347-351.

Ghazanfar, AA; Maier, JX; Hoffman, KL; Logothetis, NK. Multisensory integration of dynamic faces and voices in rhesus monkey auditory cortex. *J. of Neuroscience,* 2005 18, 5004-5012.

Gifford III, GW; Cohen, YE. The effect of a central fixation light on auditory spatial responses in area LIP. *J. of Neurophysiol.,* 2004 91, 2929-2933.

Harlow, JM. Passage of an iron rod through the head. *Boston med surg J.,* 1848 39, 389-393.

Kaas, JH; Hackett, TA. Subdivision of auditory cortex and processing streams in primates. *Proc. Natl. Acad. Sci. USA*, 2000 97, 11793-11799.

Kelly, RM; Strick, PL. Cerebellar loops with motor cortex and prefrontal cortex of a nonhuman primate. *J. Neurosci.,* 2003 23, 8432-8444.

Kuypers, HGJM. Anatomy of the descending pathways. In: Brooks V, editor. *Handbook of Physiology, Vol 2.* Bethesda, MA: Am Physiol Soc; 1981; 63-86.

Levinsohn, G. Uber die beziehungen der grosshirnrinde beim affen zu den bewegungen des auges. *Graefes Archive for Clinical and Experimental Ophthalmology,* 1909 71, 313-378.

Lynch, JC; Hoover, JE; Strick, PL. Input to the primate frontal eye field from the substantia nigra, superior colliculus, and dentate nucleus demonstrated by transneuronal transport. *Exp. Brain Res.*, 1994 100, 181-186.

Lynch, JC; Tian, JR. Cortico-cortical networks and cortico-subcortical loops for the higher control of eye movements. *Progress in Brain research,* 2006 151, 461-501.

Lucchetti, C; Lanzilotto, M; Bon, L. Auditory-motor and cognitive aspects in area 8B of macaque monkey's frontal cortex: a premotor ear-eye field (PEEF). *Exp. Brain Res.,* 2008 186, 131-141.

Lucchetti, C; Lui, F; Bon, L. Neglect syndrome for aversive stimuli in a macaque monkey with dorsomedial frontal cortex lesion. *Neuropsychologia*, 1998 36, 251-257.

Luppino, G; Rozzi, S; Calzavara, R; Matelli, M. Prefrontal and agranular cingulate projections to the dorsal premotor areas F2 and F7 in the macaque monkey. *Eur. J. Neurosci.,* 2003 17, 559-578.

Mann, SE; Thau, R; Schiller, PH. Conditional task-related responses in monkey dorsomedial frontal cortex. *Exp. Brain Res.,* 1988 69, 460-468.

Matelli, M; Luppino, G; Rizzolatti, G. Architecture of superior and mesial area 6 and adiacent cingulate cortex in the macaque monkey. *J. Comp. Neurol.,* 1991 311, 445-462.

Middleton, FA; Strick, PL. Cerebellar projections to the prefrontal cortex of the primate. *J. Neuroscience,* 2001 21, 700-712.

Mortimer, JA. Cerebellar responses to teleceptive stimuli in alert monkeys. *Brain Research,* 1975 83, 369-390.

Moschovakis, AK; Gregoriou, GG; Ugolini, G; Doldan, M; Graf, W; Guldin, W; Hadjidimitrakis, K; Savaki, HE. Oculomotor areas of the primate frontal lobes: a transneuronal transfer of rabies virus and [^{14} C]-2-deoxyglucose functional imaging study. *J. Neurosci.*, 2004 24, 5726-5740.

Pandya, DN; Yeterian, EH. Architecture and connections of cortical association areas. In: Peters A, Jones EG, editor. *Association and auditory cortices*. New York: Plenum; 1985; 3-61.

Pandya, DN; Yeterian, EH. Prefrontal cortex in relation to other cortical areas in rhesus monkey: architecture and connections. *Progr. Brain Res.,* 1990 85, 63-94.

Passingham, R. *The frontal lobes and voluntary action*. Oxford: Oxford University Press; 1993.

Petrides, M; Pandya, DN. Comparative architectonic analysis of the human and the macaque frontal cortex. In: Boller F, Grafman J, editors. *Handbook of Neuropsychology, vol 9.*, Amsterdam: Elsevier; 1994; 17-57.

Petrides, M; Pandya, DN. Dorsolateral prefrontal cortex: comparative cytoarchitectonic analysis in the human and the macaque brain and corticocortical connections patterns. *European Journal of Neurosciences,* 1999 11, 1011-1036.

Petrides, M. The role of the mid-dorsolateral prefrontal cortex in working memory. *Exp. Brain Res.,* 2000 133, 44-54.

Populin, LC; Yin, TCT. Bimodal interactions in the superior colliculus of the behaving cat. *J Neuroscience,* 2002 22, 2826-2834.

Rao, SC; Rainer, G; Miller, EK. Integration of what and where in the primate prefrontal cortex. *Science,* 1997 276, 821-824.

Rauschecker, JP; Tian, B; Hauser, M. Processing of complex sounds in the macaque non primary auditory cortex. *Science,* 1995 268, 111-114.

Raybourn, MS; Keller, EL. Colliculoreticular organization in primate oculomotor system. *J. Neurophysiol.,* 1977 40, 861-878.

Recanzone, GH. Spatial processing in the auditory cortex of the macaque monkey. *Proc. Natl. Acad Sci USA,* 2000 97, 11829-11835.

Rizzolatti, G; Camarda, R; Grupp, LA; Pisa, M. Inhibitory effect of remote visual stimuli on visual responses of cat superior colliculus: spatial and temporal factors. *J. Neurophysiol.,* 1974 37, 1262-1275.

Rizzolatti, G; Matelli, M; Pavesi, G. Deficits in attention and movement following the removal of postarcuate (area 6) and prearcuate (area 8) cortex in macaque monkeys. *Brain,* 1983 106, 655-673.

Rizzolatti, G; Craighero, L. Spatial attention: mechanisms and theories. In: Sabourin M, Craik F, Robert M, editors. *Advances in psychological science vol. 2: biological and cognitive aspects.* Psychological Press; 1998; 171-198.

Romanski, LM; Bates, JF; Goldman-Rakic, PS. Auditory belt and parabelt projections to the prefrontal cortex in the rhesus monkey. *J. Comp. Neurol.,* 1999 403, 141-157.

Rosenkilde, CE. Functional eterogeneity of the prefrontal cortex in the monkey: a review. *Behav. Neural. Biol.,* 1979 25, 301-345.

Schall, JD. neural activity related to visual saccadic eye movements in the supplementary motor area of rhesus monkey. *J. Neurophysiology,* 1991 66, 530-558.

Schall, JD. Visuomotor areas of the frontal lobe in Cerebral Cortex Vol.12. In: Rockland KS, Kaas JH, PetersA, editors. *Extrastriate cortex in primates.* New York and London: Plenum Press; 1997; 527-616.

Schiller, PH; Sandell, JH; Maunsell, JHK. The effect of frontal eye field and superior colliculus lesions on saccadic latencies in the rhesus monkey. *J Neurophysiol,* 1987 57, 1033-1049.

Schiller, PH; Tehovnik, EJ. Neural mechanisms underlying target selection with saccadic eye movements. *Progress in Brain Research,* 2005 149, 157-171.

Schlag, J; Schlag-Rey, M. Evidence for a supplementary eye field. *J. Neurophysiol.,* 1987 57, 179-200.

Schmahmann, JD; Pandya, DN. Anatomic organization of the basilar pontine projections from prefrontal cortices in rhesus monkey. *J. Neurosci.,* 1997 17, 438-458.

Snider, RS; Stowell, A. The receiving areas of the tactile, auditory and visual systems in the cerebellum. *J. Neurophysiol.,* 1944 7, 331-357.

van der Steen, J; Russell, IS; James, GO. Effects of unilateral frontal eye-field lesions on eye-head coordination in monkey. *J. Neurophysiol.,* 1986 55, 696-713.

Tanila, H; Carlson, S; Linnankoski, I; Lindros, F; Kahila, H. Functional properties of dorsolateral prefrontal cortical neurons in awake monkey. *Behav. Brain Res.,* 1992 47, 169-180.

Tanila, H; Carlson, S; Linnankoski, I; Kahila, H. Regional distribution of functions in dorsolateral prefrontal cortex of the monkey. *Behav. Brain Res.,* 1993 53, 63-71.

Tanji, J; Hoshi, E. Role of the lateral prefrontal cortex in executive behavioral control. *Physiol. Rev.*, 2008 88, 37-57.

Tehovnik, EJ; Lee, KM. The dorsomedial frontal cortex of the rhesus monkey. Topographic representation of saccades evoked by electric stimulation. *Exp. Brain Res.,* 1993 96, 430-442.

Tehovnik, EJ; Sommer, MS. Electrically evoked saccades from the dorsomedial frontal cortex and frontal eye field: a parametric evaluation reveals differences between areas. *Exp. Brain Res.,* 1997 117, 369-378.

Tehovnik, EJ; Sommer, MA; Chou, IH; Slocum, WM; Schiller, PH. Eye fields in the frontal lobes of primates. *Brain Res. Rev.,* 2000 32, 413-448.

Thier, P; Andersen, RA. Eletrical microstimulation distinguishes distinct saccade-related areas in the posterior parietal cortex. *J. Neurophysiol.,* 1998 80, 1713-1735.

Treisman, AM. Contextual cues in selective listening. *Quarterly Journal of Experimental Psycholology: Human Perception and Performance,* 1960 12, 242-248.

Turatto, M; Galfano, G; Bridgeman, B; Umiltà, C. Space-independent modality-driven attentional capture in auditory, tactile and visual systems. *Exp Brain Res,* 2004 155, 301-310.

Von Bonin, G; Bailey, P. *The neocortex of macaca mulatta.* Urbana: University of Illinois: Press; 1947.

Walker, AE. A cytoarchitectural study of frontal area of the Macaque monkey. *J. Comp Neurol.*, 1940 73, 59-86.

Wang, YW; Isoda, M; Matsuzaka, Y; Shima, K; Tanji, J. Prefrontal cortical cells projecting to the supplementary eye field and presupplementary motor area in the monkey. *Neuroscience Research,* 2005 53, 1-7.

Wegener, JG. The sound localizing behavior of normal and brain damaged monkeys. *J. Auditory Res.*, 1973 13, 191-219.

Wessinger, CM; Van Meter, J; Tian, B; Van Lare, J; Pekar, J; Rauschecker, JP. Hierarchical organization of the human auditory cortex revealed by functional magnetic resonance imaging. *J. Cognitive Neurosci.,* 2001 13, 1-7.

Williams, SM; Goldman–Rakic, P. Widespread origin of the primate mesofrontal dopamine system. *Cerebral Cortex*, 1998 8, 321-345.

In: Encyclopedia of Neuroscience Research
Editors: Eileen J. Sampson and Donald R. Glevins
ISBN 978-1-61324-861-4

Chapter VII

Prefrontal Cortex: Its Roles in Cognitive Impairment in Parkinson's Disease Revealed by PET

Qing Wang[a,b]*, Kelly A. Newell[a], Peter T. H. Wong[d] and Ying Lu[c]

[a]Neurobiology Research Centre, School of Health Sciences, University of Wollongong, NSW 2522, Australia

[b]Department of Neurology, the Third Hospital of Sun Yat-Sen University, 600 Tianhe Road, Guangzhou, Guangdong 510630, P.R.China

[c]Department of Radiology and Biostatistics, University of California, San Francisco, CA 94143, USA

[d]Departments of Pharmacology, Yong Loo Lin School of Medicine, National University of Singapore 117597, Singapore

Parkinson's disease (PD) is a neurodegenerative disorder, classically characterized by the chronic loss of dopaminergic neurons primarily in substantia nigra pars compacta, which leads to a reduction of dopamine levels in the striatum. The motor symptoms do not become obvious until at least an 80% reduction in striatal dopamine levels. Recent works have demonstrated that cognitive impairment occurs even in the early course of PD which is correlated with dopaminergic dysfunctions in the prefrontal cortex (Brück, 2005b; Chudasama and Robbins, 2006) and not necessarily related to motor disorders. This commentary will discuss the current knowledge of the prefrontal cortex and its association with cognitive deficits in PD.

Increasing evidence show that profound dopamine depletion not only occurs in the striatum but also in the prefrontal cortex (Scatton, 1982), and this may be associated with the cognitive deficits in PD (Roberts, 1994; Goldman-Rakic, 1998; Kulisevsky, 2000). L-dopa is widely used for the treatment of PD and shows beneficial effects on the cognitive dysfunctions among PD patients. One study by Shohamy, based on electrophysiological and

* Corresponding author: Dr. Qing Wang, Neurobiology Research Centre, School of Health Sciences, University of Wollongong, NSW 2522, Australia. Phone: +61 2 4221 3199; Fax: +61 2 4221 4096; Email: denniswq@yahoo.com or dwang@uow.edu.au

computational examination, demonstrated that the loss of dopamine in PD led to specific learning impairment, and that enhancing dopamine levels with L-dopa treatment alleviated the impairment of feedback-based sequence learning (Shohamy, 2005). Mollion also showed that in two conditional associative learning tasks, the increase in response time and the decrease in errors found in L-dopa medicated PD patients compared to non-L-dopa medicated patients reflected enhancement of working memory, suggesting the L-dopa treatment improve the cognitive impairment in PD patients (Mollion, 2003). More evidence indicated that mesocortical regions, especially prefrontal cortex, may be the key locus for cognitive functional improvement following dopaminergic enhancement (Arnsten, 1998; Mattay, 2000; Mehta, 2000). Interestingly, L-dopa was also found to exert detrimental effects on cognition related functions like extinction learning (Czernecki, 2002) and probabilistic reversal learning (Cools, 2001). For these contrasting effects, Cools proposed that increases of dopamine transmission in prefrontal cortex and its reductions in striatum contribute to the cognitive stability; however, cognitive flexibility benefited from potentiated phasic striatal dopamine transmission and its reduction in the prefrontal cortex (Cools, 2006).

Literature shows that disruption of the prefrontal cortex in PD leads to considerable cognitive dysfunction, which is likely related to metabolic and cerebral blood flow changes in prefrontal cortex and associated with the striatal dysfunction in PD patients. Early in 1985, Globus examined regionally reduced cerebral blood flow using a 133Xenon inhalation technique, and evaluated cognitive dysfunction in PD patients (Globus, 1985). However, with technological advancement, positron emission tomography (PET) has become a very useful tool in investigating, among other things, the dopaminergic system. PET is an imaging method used to track the regional distribution and kinetics of chemical compounds in the living body by labelling them with short-lived positron-emitting isotopes. It was thc first and most precise technology to directly measure components of the dopamine system like dopamine uptake and release, dopamine transport, dopamine receptors, and cerebral blood flow in the living human brain (Volkow, 1996).

In the MPTP-induced Parkinson's disease monkey model, a significant decrease of blood flow and dopamine transporters was observed by PET in mesolimbicortical regions like putamen, caudate, and primary motor cortex (Brownell, 2003). Playford firstly used PET to examine regional cerebral blood flow in PD patients and found the obvious reduction in the prefrontal cortical cerebral blood flow when compared to control subjects (Playford, 1992). An obvious decrease of cerebral blood flow in bilateral prefrontal cortex of PD patients during central sensory processing was determined through PET (Boecker, 1999); whilst the dopamine agonist apomorphine profoundly induced decreases of regional cerebral blood flow in the primary motor cortex, the medial and dorsolateral prefrontal cortex as evidenced by PET (Hosey, 2005). Similarly, recent work using PET showed a significant correlation between L-dopa-induced blood flow decreases in the dorsolateral prefrontal cortex and cognitive functional changes in performance on the planning task and spatial working memory in PD patients (Cools, 2002). The work suggested that high-level cognitive impairment in PD patients could be ameliorated by L-dopa through inducing relative blood flow changes in the dorsolateral prefrontal cortex (Cools, 2002).

Consistent with Cools work (Cools, 2002)., with the help of fMRI, Mattay and colleagues (Mattay, 2002) demonstrated that prefrontal cortex, subserving working memory, displayed higher activation during the hypodopaminergic state, and increased activation in prefrontal cortex during the working memory task in the hypodopaminergic state was positively associated with errors in task performance. These results suggest that prefrontal cortex exert important roles in modulating cognitive function like working memory via mesocortical dopaminergic system, and hypodopaminergic state is likely related to the decreased efficiency of prefrontal cortical information processing.

In a study by Owen (Owen, 1998), a Tower of London planning task was performed to present abnormal spatial working memory in PD patients, which was associated with a regional decrease of cerebral blood flow as a metabolic marker in the basal ganglia. This study is similar to Dagher's work, showing abnormal blood flow in the basal ganglia of PD patients with `frontal' cognitive deficits (Dagher, 2001). Besides the L-dopa mediated cerebral blood flow changes in PD patients, PET study also showed an obvious atrophy associated with cognitive impairment which was observed in the right and the left prefrontal cortex in early stage PD patients (Brück, 2004).

With the help of PET, a profound reduction of Fluorodopa uptake in frontal cortex of PD patients was observed, and this reduction was positively correlated with cognitive deficits including verbal fluency, working memory, and attentional functioning reflecting frontal lobe function (Rinne, 2000). In contrast, using 6-[^{18}F]fluoro-l-dopa (Fdopa) as the tracer in PET, Brück observed an *upregulation* of cortical Fdopa uptake in early non-medicated PD patients, and indicated that the increase in dorsolateral prefrontal cortical Fdopa uptake was associated with cognitive deficits in early PD (Brück, 2005a). This increase of Fdopa uptake in prefrontal cortex might result from the compensatory process in the cortical-subcortical dopamine loops among PD patients (Cropley, 2006).

In addition to the dopamine transmission, various receptors in the cortical regions also contribute to cognitive deficits in PD. It is interesting to find that in animal studies dopamine D1 and D2 receptors, muscarinic receptors and serotonin receptors in prefrontal cortex were involved in regulating cognitive activity in various PD animal models (Tzavara, 2004; Wang, 2005; Cao, 2007). Clinical studies showed that dopamine D2/3 receptors in the dorsolateral prefrontal cortex of advanced PD patients were significantly decreased, and slightly declined in early stage of PD (Kaasinen, 2000). Meoni found that NMDA NR1 expression was decreased in the prefrontal cortex of PD patients, and indicated that this downregulation may result from the degeneration of the dopaminergic input to the prefrontal cortex in PD (Meoni, 1999). Also, it was shown that nicotinic ACh receptors were reduced in cortical regions of PD patients (Perry, 1993).

Prefrontal cortex is increasingly recognized to be the important brain region in mediating cognitive dysfunctions among different stages in the progression of PD. However, more work needs to explore how cognitive dysfunctions are affected through prefrontal cortex modulating interventions in PD. Future work should address the functional correlations of mesocortical and nigrostriatal regions in mediating cognitive impairment in PD. Moreover, experimental and preclinical data should be converted to clinical practice and provide novel strategies in slowing down the progression of cognitive deficits in PD patients.

References

Arnsten AFT (1998) Catecholamine modulation of prefrontal cortical cognitive function. *Trends Cogn. Sci.* 2:436-447.

Boecker H, Ceballos-Baumann, A., Bartenstein, P., Weindl, A., Siebner, H.R., Fassbender, T., Munz, F., Schwaiger, M., Conrad, B. (1999) Sensory processing in Parkinson's and Huntington's disease: investigations with 3D H(2)(15)O-PET. *Brain* 122 (Pt 9):1651-1665.

Brownell AL, Canales, K., Chen, Y.I., Jenkins, B.G., Owen, C., Livni, E., Yu, M., Cicchetti, F., Sanchez-Pernaute, R., Isacson, O. (2003) Mapping of brain function after MPTP-induced neurotoxicity in a primate Parkinson's disease model. *Neuroimage* 20(2):1064-1075.

Brück A, Aalto, S., Nurmi, E., Bergman, J., Rinne, J.O. (2005a) Cortical 6-[18F]fluoro-L-dopa uptake and frontal cognitive functions in early Parkinson's disease. *Neurobiol. Aging* 26(6):891-898.

Brück A, Aalto, S., Nurmi, E., Bergman, J., Rinne, J.O. (2005b) Cortical 6-[18F]fluoro-L-dopa uptake and frontal cognitive functions in early Parkinson's disease. *Neurobiol. Aging* 26(6):891-898.

Brück A, Kurki, T., Kaasinen, V., Vahlberg, T., Rinne, J.O. (2004) Hippocampal and prefrontal atrophy in patients with early non-demented Parkinson's disease is related to cognitive impairment. *J. Neurol. Neurosurg. Psychiatry* 75(10):1467-1469.

Cao J, Liu, J., Zhang, Q.J., Wang, T., Wang, S., Han, L.N., Li, Q. (2007) The selective 5-HT1A receptor antagonist WAY-100635 inhibits neuronal activity of the ventromedial prefrontal cortex in a rodent model of Parkinson's disease. *Neurosci. Bull.* 23(6):315-322.

Chudasama Y, Robbins TW (2006) Functions of frontostriatal systems in cognition: comparative neuropsychopharmacological studies in rats, monkeys and humans. *Biol. Psychol.* 73(1):19-38.

Cools R (2006) Dopaminergic modulation of cognitive function-implications for L-DOPA treatment in Parkinson's disease. *Neurosci. Biobehav. Rev.* 30(1):1-23 Review.

Cools R, Barker, R.A., Sahakian, B.J., Robbins, T.W. (2001) Enhanced or impaired cognitive function in Parkinson's disease as a function of dopaminergic medication and task demands. *Cereb. Cortex* 11(12):1136-1143.

Cools R, Stefanova, E., Barker, R.A., Robbins, T.W., Owen, A.M. (2002) Dopaminergic modulation of high-level cognition in Parkinson's disease: the role of the prefrontal cortex revealed by PET. *Brain* 125(Pt 3):584-594.

Cropley VL, Fujita, M., Innis, R.B., Nathan, P.J. (2006) Molecular imaging of the dopaminergic system and its association with human cognitive function. *Biol. Psychiatry* 59(10):898-907.

Czernecki V, Pillon, B., Houeto, J.L., Pochon, J.B., Levy, R., Dubois, B. (2002) Motivation, reward, and Parkinson's disease: influence of dopatherapy. *Neuropsychologia* 40(13):2257-2267.

Dagher A, Owen, A.M., Boecker, H., Brooks, D.J. (2001) The role of the striatum and hippocampus in planning: a PET activation study in Parkinson's disease. *Brain* 124(Pt 5):1020-1032.

Globus M, Mildworf, B., Melamed, E. (1985) Cerebral blood flow and cognitive impairment in Parkinson's disease. *Neurology* 35(8):1135-1139.

Goldman-Rakic PS (1998) The cortical dopamine system: role in memory and cognition. *Adv. Pharmacol.* 42:707-711.

Hosey LA, Thompson, J.L., Metman, L.V., van den Munckhof, P., Braun, A.R. (2005) Temporal dynamics of cortical and subcortical responses to apomorphine in Parkinson disease: an H2(15)O PET study. *Clin. Neuropharmacol.* 28(1):18-27.

Kaasinen V, Någren, K., Hietala, J., Oikonen, V., Vilkman, H., Farde, L., Halldin, C., Rinne, J.O. (2000) Extrastriatal dopamine D2 and D3 receptors in early and advanced Parkinson's disease. *Neurology* 54(7):1482-1487.

Kulisevsky J (2000) Role of dopamine in learning and memory: implications for the treatment of cognitive dysfunction in patients with Parkinson's disease. *Drugs Aging* 16(5):365-379.

Mattay VS, Callicott, J.H., Bertolino, A., Heaton, I., Frank, J.A., Coppola, R., Berman, K.F., Goldberg, T.E., Weinberger, D.R. (2000) Effects of dextroamphetamine on cognitive performance and cortical activation. *Neuroimage* 12(3):268-275.

Mehta M, Owen, A.M., Sahakian, B.J., Mavaddat, N., Pickard, J.D., Robbins, T.W. (2000) Ritalin and working memory modulation in humans: ritalin enhances working memory by modulating discrete frontal and parietal lobe regions in the human brain. *J. Neurosci.* 20:RC1-6.

Meoni P, Bunnemann, B.H., Kingsbury, A.E., Trist, D.G., Bowery, N.G. (1999) NMDA NR1 subunit mRNA and glutamate NMDA-sensitive binding are differentially affected in the striatum and pre-frontal cortex of Parkinson's disease patients. *Neuropharmacology* 38(5):625-633.

Mollion H, Ventre-Dominey, J., Dominey, P.F., Broussolle, E. (2003) Dissociable effects of dopaminergic therapy on spatial versus non-spatial working memory in Parkinson's disease. *Neuropsychologia* 41(11):1442-1451.

Owen AM, Doyon, J., Dagher, A., Sadikot, A., Evans, A.C. (1998) Abnormal basal ganglia outflow in Parkinson's disease identified with PET. Implications for higher cortical functions. *Brain* 121 (Pt 5):949-965.

Perry EK, Irving, D., Kerwin, J.M., McKeith, I.G., Thompson, P., Collerton, D., Fairbairn, A.F., Ince, P.G., Morris, C.M., Cheng, A.V., et al (1993) Cholinergic transmitter and neurotrophic activities in Lewy body dementia: similarity to Parkinson's and distinction from Alzheimer disease. *Alzheimer Dis. Assoc. Disord.* 7:69-79.

Playford ED, Jenkins, I.H., Passingham, R.E., Nutt, J., Frackowiak, R.S., Brooks, D.J. (1992) Impaired mesial frontal and putamen activation in Parkinson's disease: a positron emission tomography study. *Ann. Neurol.* 32(2):151-161.

Rinne JO, Portin, R., Ruottinen, H., Nurmi, E., Bergman, J., Haaparanta, M., Solin, O. (2000) Cognitive impairment and the brain dopaminergic system in Parkinson disease: [18F]fluorodopa positron emission tomographic study. *Arch. Neurol.* 57(4):470-475.

Roberts AC, De Salvia, M.A., Wilkinson, L.S., Collins, P., Muir, J.L., Everitt, B.J., Robbins, T.W. (1994) 6-Hydroxydopamine lesions of the prefrontal cortex in monkeys enhance performance on an analog of the Wisconsin Card Sort Test: possible interactions with subcortical dopamine. *J. Neurosci.* 14(5 Pt 1):2531-2544.

Scatton B, Rouquier, L., Javoy-Agid, F., Agid, Y. (1982) Dopamine deficiency in the cerebral cortex in Parkinson disease. *Neurology* 32(9):1039-1040.

Shohamy D, Myers, C.E., Grossman, S., Sage, J., Gluck, M.A. (2005) The role of dopamine in cognitive sequence learning: evidence from Parkinson's disease. *Behav. Brain Res.* 156(2):191-199.

Tzavara ET, Bymaster, F.P., Davis, R.J., Wade, M.R., Perry, K.W., Wess, J., McKinzie, D.L., Felder, C., Nomikos, G.G. (2004) M4 muscarinic receptors regulate the dynamics of cholinergic and dopaminergic neurotransmission: relevance to the pathophysiology and treatment of related CNS pathologies. *FASEB J.* 18(12):1410-1412.

Volkow ND, Fowler, J.S., Gatley, S.J., Logan, J., Wang, G.J., Ding, Y.S., Dewey, S. (1996) PET evaluation of the dopamine system of the human brain. *J. Nucl. Med.* 37(7):1242-1256.

Wang Q, Wang, P.H., McLachlan, C., Wong, P.T. (2005) Simvastatin reverses the downregulation of dopamine D1 and D2 receptor expression in the prefrontal cortex of 6-hydroxydopamine-induced Parkinsonian rats. *Brain Res.* 1045(1-2):229-233.

Peer-reviewer for this commentary

Dr. Penghua Wang
Department of Internal Medicine
Yale University School of Medicine
Post Box 208022, 300 Cedar Street
New Haven, CT 06510 USA
Tel: 1-203-737-5635, 203-785-4140
FAX: 1-203-785-3864
Email: penghua.wang@yale.edu

In: Encyclopedia of Neuroscience Research
Editors: Eileen J. Sampson and Donald R. Glevins
ISBN 978-1-61324-861-4

Chapter VIII

Noradrenergic Actions in Prefrontal Cortex: Relevance to AD/HD

Amy F. T. Arnsten *
Yale Medical School, New Haven, Connecticut, USA

Abstract

This chapter reviews the important role of prefrontal cortex (PFC) in the regulation of behavior and attention, and evidence that these processes are weakened in Attention-Deficit/Hyperactivity Disorder (AD/HD). Norepinephrine (NE) has been found to have a critical beneficial effect on PFC function through actions at post-synaptic, α2A-adrenoceptors, while high levels of NE release during stress impair PFC regulation of behavior through actions at α1 adrenoceptors. The PFC regulates its own NE input through projections to NE-containing locus coeruleus (LC) neurons. The LC fires according to arousal state and attentional demands. Most medications effective for the treatment of AD/HD powerfully influence NE transmission, either by blocking reuptake or mimicking its actions at α2A-adrenoceptors. The importance of NE to AD/HD is increasingly appreciated.

Lay Abstract

The symptoms of Attention-Deficit/Hyperactivity Disorder (AD/HD) - impaired regulation of attention and behavior- likely involve weaknesses in higher brain functions. The prefrontal cortex (PFC) is a higher brain region necessary for the appropriate control of behavior, attention and emotion. Imaging studies indicate this brain region is afflicted in AD/HD. The PFC is highly dependent on its neurochemical state, and requires optimal levels of norepinephrine for proper function. Norepinephrine is released in brain according to arousal state and attentional demands. Many AD/HD medications such as methylphenidate (Ritalin), amphetamine (Adderall), atomoxetine (Strattera) and guan-

* Correspondence to: Dr. Amy F.T. Arnsten. Department of Neurobiology, Yale Medical School, 333 Cedar St., New Haven, CT 06510; Email: amy.arnsten@yale.edu; Phone: 203-785-4431; Fax: 203-785-5263

facine, alter norepinephrine actions. It is likely that these medications act, at least in part, by providing a more optimal chemical environment for PFC regulation of behavior and attention.

Keywords: norepinephrine, attention, executive function, hyperactivity, adrenoceptor, guanfacine.

Introduction

Basic science has made great strides in understanding the neurochemical influences needed for optimal prefrontal cortical (PFC) regulation of behavior and attention. Although much research has focused on the important role of dopamine (DA), there is increasing appreciation that norepinephrine (NE) has a vital influence as well. NE plays an essential role in arousal and attention, and may orchestrate the interplay of posterior sensory cortices, subcortical structures mediating primitive responses, and PFC regulation of attention and behavior. The current chapter reviews this literature, and outlines its direct relevance to our understanding and treatment of Attention Deficit Hyperactivity Disorder (AD/HD).

Prefrontal Cortical Cognitive Dysfunction in AD/HD

Neuropsychological and imaging studies have established that PFC executive abilities are weakened in patients with AD/HD (reviewed in Arnsten et al., 1996; Barkley, 1997). The PFC guides behavior and attention using working memory, applying represented information to inhibit inappropriate actions or thoughts. These processes are the basis of the so-called executive functions, including regulation of attention, planning, impulse control, mental flexibility, and the initiation and monitoring of action. Lesions to the PFC produce symptoms such as forgetfulness, distractibility, impulsivity, perseveration, and disorganization.

The PFC is critical for the inhibition of both covert responses, i.e. attention regulation, and overt responses, e.g. impulse control. The role of PFC in attention regulation has been appreciated for many years. While the posterior cortices provide processing resources, allowing us to pay attention as it were, the PFC regulates what we attend to based on represented goals. The PFC inhibits responses to distracting stimuli and suppresses irrelevant thoughts such as proactive interference. Thus, patients with PFC lesions are easily distracted (Woods and Knight, 1986; Godefroy and Rousseaux, 1996), are impaired at gating sensory stimuli (Knight et al., 1989; Yamaguchi and Knight, 1990), have poor concentration and organization, and are more vulnerable to disruption from proactive interference (Thompson-Schill et al., 2002). PFC lesions impair the ability to sustain attention, particularly over long delays (Wilkins et al., 1987). Lesions of the dorsolateral PFC impair the ability to shift attentional set (Manes et al., 2002). PFC lesions also impair divided attention, and these attentional deficits have been associated with lesions in the left, superior PFC (Godefroy and Rousseaux, 1996). Similar results have been seen in animals, where PFC lesions impair attentional regulation in monkeys e.g. (Malmo, 1942; Bartus and Levere, 1977; Dias et al., 1996), and rats (Muir et al., 1996).

The PFC also allows us to regulate overt behaviors including locomotor activity. The right inferior PFC is particularly important for behavioral inhibition (reviewed in (Aron et al., 2004)). Both imaging (Konishi et al., 1999; Rubia et al., 2003) and lesion studies indicate that the right PFC in humans is critical for inhibitory abilities, e.g. performance of the Stop or Go-No Go tasks. The importance of the PFC to inhibitory control has also been shown in monkeys with lesion (Petrides, 1986), electrophysiological (Watanabe, 1986) and imaging studies (Morita et al., 2004). There is also a classic but often forgotten literature demonstrating that PFC lesions cause locomotor hyperactivity in monkeys (Kennard et al., 1941; French, 1959; Gross, 1963; Gross and Weiskrantz, 1964). Monkeys were often observed to circle in their cages following PFC lesions. Thus, some of the locomotor hyperactivity observed in AD/HD may arise from PFC dysfunction.

The Cellular Basis of PFC Cognitive Function

Single unit recording studies in monkeys have shown that PFC neurons are able to hold modality-specific information "on-line" over a delay and use this represented information to guide behavior in the absence of environmental cues (Goldman-Rakic, 1995). PFC neurons can also fire in relationship to an abstract rule that is used to govern action (Wallis et al., 2001). A unique feature of PFC neurons is their ability to maintain information in the presence of interference from distracting stimuli (Miller et al., 1993). Delay-related firing also can serve as the basis for behavioral inhibition, as examined in an anti-saccade task in which monkeys must look away from a remembered visual stimulus (Funahashi et al., 1993). Thus, delay-related activity is observed both when an animal must make a memory-guided action, and when an animal must withhold a prepotent response based on representational knowledge. This cellular marker for working memory has been very useful when examining the neuromodulatory influences of catecholamines on PFC cognitive function (see below).

PFC Dysfunction in AD/HD

Research over the last 20 or more years has established that PFC regulation of behavior is weaker in patients with AD/HD. AD/HD patients are impaired on tasks of behavioral inhibition, reward reversal and working memory that require PFC function e.g. (Itami and Uno, 2002; Bedard et al., 2003; McLean et al., 2004). In contrast, they are not impaired on tasks of covert attentional orienting that depend on posterior parietal association cortex (Swanson et al., 1991).

Conversely, stimulant medication has been found to improve spatial working memory, response inhibition, set-shifting and other PFC cognitive functions in both "normal" college students (Mehta et al., 2000) and in children and adults with AD/HD (Aron et al., 2003; Bedard et al., 2003; Mehta et al., 2004). Interestingly, in adults with AD/HD, childhood ratings of AD/HD (both self-reported and informant ratings) correlated with response to methylphenidate on the spatial working memory task (Turner et al., 2004). Thus, studies of spatial working memory performance are likely very relevant to the therapeutic effects of AD/HD medications.

Numerous structural imaging studies have shown reduced size of the PFC in AD/HD patients, particularly in the right hemisphere (Castellanos et al., 1996; Casey et al., 1997; Filipek et al., 1997; Giedd et al., 2001; Kates et al., 2002; Hill et al., 2003; Sowell et al., 2003). Imaging studies have also shown evidence of reduced blood flow or metabolism in PFC that correspond with poor PFC cognitive function (Rubia et al., 1999; Yeo et al., 2000). These functional deficits are normalized by stimulant medication (Vaidya et al., 1998). Imaging studies have shown more efficient dorsolateral PFC activity (BOLD) following stimulant treatments that improve spatial working memory, consistent with improved PFC cognitive function (Mehta et al., 2000). Thus, there is an excellent correspondence between AD/HD symptoms, neuropsychological deficits, PFC deficiency with imaging, and normalization with stimulant medication.

There is also suggestive evidence of reduced catecholamine inputs to the PFC in adults with AD/HD based on fluoro-dopa PET imaging (Ernst et al., 1998). These data are especially compelling, given that most medications used to treat AD/HD enhance DA and NE transmission. Although most genetic studies have focused on linkage of AD/HD with genes related to DA (the DA transporter, DA D1 and D4 receptors, and calcyon, a molecule that links D1 receptors to IP3/protein kinase C signaling), NE-related genes have also been associated with AD/HD. Most prominently, the gene for the synthetic enzyme for NE, dopamine β hydroxylase, has been associated with AD/HD (Daly et al., 1999; Roman et al., 2002), and with less consistency, the α2A adrenoceptor (Comings et al., 1999; Xu et al., 2001). It should also be remembered that the D4 receptor has very high affinity for NE (Van Tol et al., 1991) and thus should really be considered a catecholamine receptor. As described below, catecholamines have a critical influence on PFC cognitive function. Thus, it is logical to argue that many AD/HD symptoms may arise from genetic alterations in catecholamine transmission which lead to weakened PFC regulation of attention and behavior. These symptoms would be ameliorated by catecholamine-enhancing medications. An understanding of catecholamine influences on PFC function, including an appreciation for the conditions when NE would normally be released in cortex, can help provide a rational view of AD/HD symptoms and treatment.

NE's Important Roles in Arousal and Attention

The NE-synthesizing cells of the locus coeruleus (LC) project throughout the forebrain, and provide NE to the entire cortical mantle. These neurons are a critical portion of the arousal system, and have been shown to fire according to the sleep/waking state of the animal. LC neurons fire during alert waking,, slow their firing when the animal becomes drowsy and falls into slow wave sleep, and shut off completely during rapid eye movement (REM) sleep (Foote et al., 1980; Aston-Jones et al., 1999).

NE has direct alerting actions on cortical neurons (McCormick et al., 1991), and also has powerful indirect actions, exciting other ascending arousal systems in basal forebrain (Berridge and España, 2000; Berridge and Waterhouse, 2003) and thalamus (McCormick et al., 1991). Both the direct and indirect stimulation of cortical arousal occurs through α1 and β adrenergic receptor stimulation (ibid). In contrast, low levels of NE released in thalamus under drowsy conditions engage α2 receptors, which have higher affinity for NE, and reduce thalamo-cortical stimulation (Buzsaki et al., 1991). The thalamus has a very dense NE innervation, and this input contributes to information transfer through this structure. Thus, under conditions of sufficient NE release to engage α1 and β adrenergic receptors, information is faithfully transmitted from thalamus to cortex, while under conditions of low NE release when α2 receptors are engaged, thalamic neurons are put into a burst mode which prevents information transfer (McCormick et al., 1991). Thus, the varying affinities of NE for α2 vs. α1 or β adrenergic receptors acts as a switch to alter neuronal, and thus behavioral state.

Within the waking state, NE has a key role in modulating attention (reviewed in (Aston-Jones et al., 1999) and contributes to decision-making about relevant vs. irrelevant stimuli (Usher et al., 1999; Rajkowski et al., 2004). Aston-Jones and colleagues have shown that under optimal waking conditions (i.e. alert, unstressed), the LC exhibits modest tonic firing and pronounced phasic firing to relevant stimuli (Rajkowski et al., 1998; Rajkowski et al., 2004). Importantly to AD/HD, the LC also suppresses firing to irrelevant stimuli during these ideal conditions (ibid). In contrast, when animals are stressed or agitated, high levels of tonic activity are recorded from LC, and phasic firing is poorly regulated, including inappropriate responses to distracting stimuli (ibid). Given the extensive projections of NE axons, these response patterns would ramify throughout the neuroaxis, and have widespread effects on cortical processing as described below.

Noradrenergic Modulation of PFC Cognitive Function

NE has powerful, bimodal effects on the higher cognitive functions of the PFC. It has been appreciated for many years that DA has a critical influence on PFC cognitive functions (Brozoski et al., 1979), especially through its actions at the D1 receptor family (Sawaguchi and Goldman-Rakic, 1994). D1 receptor stimulation produces an inverted U dose response, whereby either too much or too little receptor stimulation impairs working memory in mice (Lidow et al., 2002), rats (Zahrt et al., 1997), monkeys (Arnsten and Goldman-Rakic, 1990, 1998), and humans (Kimberg et al., 1997; Egan et al., 2001). However, research indicates that NE has an equally important influence on PFC cognitive function, and this body of work has increasing relevance to AD/HD.

As with DA, NE has dual effects on PFC function depending on the amount of NE release. Moderate levels of NE have important, beneficial effects on PFC function, whereas high concentrations of NE released during stress contribute to impaired PFC function. However, in contrast to DA, the beneficial vs. detrimental actions can be dissociated by actions at different adrenoceptors: moderate levels of NE improve PFC function via α2A adrenoceptors, whereas high levels of NE released during stress impair PFC function by engaging lower affinity α1-adrenoceptors and possibly β1 adrenoceptors (reviewed in Arnsten 2000). This dichotomy is reminiscent of NE dual actions in thalamus. However, the NE innervation of PFC is far more delicate than that in thalamus, and it is likely that higher rates of LC firing (i.e. rates that occur during non-stressed waking) are needed in PFC to release sufficient NE to engage α2A adrenoceptors. Still higher tonic LC firing rates, e.g., during stress, appear to be needed for NE to engage the lower affinity α1 adrenoceptors. Thus, α1 adrenoceptor antagonists have no effect on working memory when infused into the PFC during non-stressful conditions (Li and Mei, 1994; Birnbaum et al., 1999), but protect PFC cognitive function under conditions of uncontrollable stress (Birnbaum et al., 1999). Conversely, infusions of α1 adrenoceptor agonists such as phenylephrine mimic the stress response and impair working memory (Arnsten et al., 1999; Mao et al., 1999). These detrimental actions have been observed at the cellular level as well, where iontophoresis of phenylephrine onto PFC neurons decreased the memory-related cell-firing necessary for working memory function (Birnbaum et al., 2004). These mechanisms are especially relevant to Post-Traumatic Stress Disorder (PTSD), which in children can produce symptoms of distractibility and poor impulse control similar to AD/HD. In this regard, it is noteworthy that the α1 adrenoceptor antagonist, prazosin, is now successfully used to treat PTSD in patients with PTSD (Raskind et al., 2003).

Although AD/HD and stress-induced PFC dysfunction have much in common, a more likely etiology for AD/HD is insufficient catecholamine stimulation in PFC. PFC function is markedly weakened by insufficient NE stimulation of α2A-adrenoceptors in the PFC. Findings reviewed below suggest that moderate levels of NE release during non-stressful conditions normally engage α2A-adrenoceptors and strengthen working memory. Thus, unlike α1 and β blockers which have no effect when infused in PFC under non-stressful conditions, infusion of the α2 antagonist, yohimbine, markedly impairs working memory performance in monkeys (Li and Mei, 1994). Conversely, studies of nonhuman primates have shown that α2A-adrenoceptor agonists such as guanfacine improve working memory (Arnsten et al., 1988; Mao et al., 1999). Remarkably, α2 adrenoceptor agonists have been shown to improve PFC function in mice (Franowicz et al., 2002), rats (Tanila et al., 1996), monkeys (Arnsten et al., 1988; Rama et al., 1996) and humans (Jakala et al., 1999a; Jakala et al., 1999b). The efficacy of these agents to enhance PFC cognitive function can be completely dissociated from the sedating properties of these compounds at higher doses (Arnsten et al., 1988). Much research has focused on elucidating the mechanisms underlying these beneficial actions of PFC cognitive function.

Initial studies demonstrated actions at *post*-synaptic, α2A adrenoceptors. Although original research focused on the role of α2 adrenoceptors at pre-synaptic sites, it is now known that the vast majority of α2 adrenoceptors are localized post-synaptic to NE neurons (MacDonald et al., 1997). In the monkey PFC, α2A adrenoceptors are localized both pre-synaptically on NE terminals, and post-synaptically on the dendritic spines of PFC pyramidal neurons ((Aoki et al., 1998). The post-synaptic α2A adrenoceptors are classically positioned over the post-synaptic density. Pharmacological data indicates that NE acts at these post-synaptic α2 receptors to strengthen PFC cognitive function; i.e. the α2 agonist, clonidine, becomes more potent and more efficacious when the presynaptic NE terminal is destroyed or depleted of NE (Arnsten and Goldman-Rakic, 1985; Cai et al., 1993).

The subtype of α2 receptor underlying the beneficial effects has also been determined. There are 3 subtypes of α2 receptors cloned in humans: the A, B and C subtypes. Both the A and C subtypes can be found in PFC, with the former predominating. Studies in genetically altered mice indicate that the α2A-preferring agonist, guanfacine, loses its enhancing effects when administered to mice with a functional knockout of the α2A-adrenoceptor (Franowicz et al., 2002). Indeed, α2A mutant mice exhibit weaker working memory abilities than wild type mice (ibid). In contrast, mice with a knockout of the α2C receptor show normal cognitive enhancement with α2 agonist treatment (Tanila et al., 1999). Thus, the α2A adrenoceptor subtype mediates the enhancing effects of NE in PFC.

A number of approaches have shown that α2 agonists like guanfacine act directly in the PFC to enhance executive functions. SPECT imaging of monkeys performing a spatial working memory task show that systemic administration of guanfacine increases regional cerebral blood flow in the dorsolateral PFC, the same brain region that is critical for performance of the task (Avery et al., 2000). Furthermore, infusions of guanfacine directly into this same region of the PFC produce a delay-related improvement in working memory performance in young adult monkeys (Mao et al., 1999). Importantly, iontophoresis of the α2 agonist, clonidine, onto PFC neurons in monkeys performing a spatial working memory task increases delay-related firing, the cellular measure of working memory (Li et al., 1999). Systemic administration similarly increases delay-related firing in PFC, and this strengthening of delay-related firing by either systemic or iontophoretic application can be blocked by iontophoresis of the α2 antagonist, yohimbine (ibid). Thus, α2 receptor stimulation in PFC strengthens working memory at the cellular and behavioral levels. Conversely, blockade of α2 receptors with yohimbine infusions directly into PFC dramatically weakens working memory (Li and Mei, 1994) and markedly erodes delay-related activity (Li et al., 1999). As reviewed above, delay-related activity is an important cellular basis for regulating attention, reducing distractibility, and inhibiting prepotent behavioral responses. The finding that α2 receptor stimulation enhances delay-related activity is consistent with findings that agonists such as guanfacine reduce distractibility (Arnsten and Contant, 1992) and enhance behavioral inhibition (Steere and Arnsten, 1997). Most recently, infusion of guanfacine into the ventrolateral PFC in monkeys strengthened associative learning and impulse control (Wang et al., 2004b; Wang et al., 2004a). Thus, α2 receptor stimulation is critical for a variety of PFC cognitive processes at both the behavioral and cellular levels.

Importantly, studies in monkeys indicate that many of the symptoms of AD/HD can be recreated by blocking α2 NE receptors in the PFC (Figure 1). In addition to the weakened working memory and delay-related activity described above, infusions of the α2 antagonist, yohimbine, into PFC increase impulsivity (Ma et al., 2003). Monkeys tested on a go- no go task showed increased errors of commission, but no change in errors of omission, following yohimbine infusion into dorsolateral PFC (Ma et al., 2003). ADHD patients also show errors of commission on the go-no go task, and methylphenidate ameliorates these errors (Trommer et al., 1991).

Most recently, yohimbine infusions into the dorsolateral PFC of rhesus monkeys have been shown to induce locomotor hyperactivity (Ma et al., 2005), reminiscent of the increased activity found with PFC ablations. Infusion of saline was without effect. Many researchers have speculated that striatal DA mechanisms underlie the hyperactivity observed in AD/HD patients.

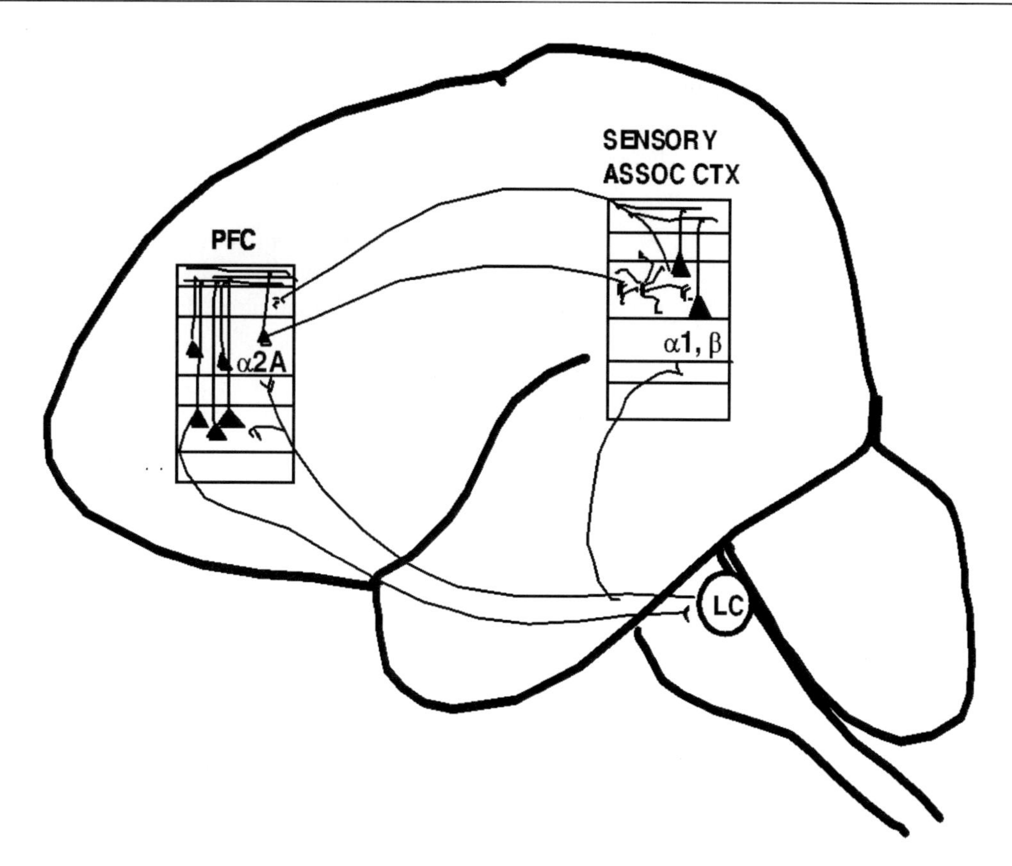

Figure 1. NE cells in the LC project to the PFC, where they improve working memory via α2A receptors, and to the posterior cortices, where they may enhance sensory function via α1 and β receptors. The PFC projects back to the LC and to posterior cortices. Blockade of NE α2A adrenoceptors in the monkey PFC produces behavioral changes similar to AD/HD, including hyperactivity, distractibility, impulsivity and impaired working memory. See text for details.

However, these results in monkeys suggest that PFC dysfunction may contribute to the locomotor hyperactivity, as well as the impulsivity and poor attention regulation/working memory, which form the cardinal symptomatology of AD/HD.

The PFC projects back to posterior sensory cortices (Barbas et al., 2005) and to subcortical structures such as the amygdala (Ghashghaei and Barbas, 2002) in order to gate sensory (Knight et al., 1989; Yamaguchi and Knight, 1990) and affective (Quirk et al., 2003) processing. Thus, NE actions in PFC can indirectly influence posterior and subcortical practices. However, NE also has direct modulatory effects on these brain regions, and these are described below.

NE Influences on Posterior Cortical Functions

NE can influence attention and perception through its actions on sensory and association cortices posterior to the PFC. Thirty years ago, Foote and colleagues first discovered that NE increases the signal/noise properties of neurons in the auditory cortex of monkeys, primarily by suppressing noise (Foote et al., 1975). Since that seminal paper, there have been numerous other examples whereby NE has been shown to enhance signal/noise response, e.g. in hippocampus (Segal and Bloom, 1976), cerebellum (Moises et al., 1981), and in other sensory cortices (Waterhouse et al., 1980, 1981). Waterhouse and colleagues have continued to perform elegant studies of NE actions in somatosensory cortex, demonstrating that NE can enhance inhibitory actions via β adrenergic receptors, while enhancing excitatory actions via α1 adrenergic receptors coupled to protein kinase C signaling (Waterhouse et al., 1980, 1981; Mouradian et al., 1991; Devilbiss and Waterhouse, 2000). For example, subthreshold somatosensory stimulation will come to evoke a response in somatosensory cortical neurons if it is coupled with phasic activation of noradrenergic cells in locus coeruleus the (LC) (Waterhouse et al., 1998). These electrophysiological findings suggest that NE should facilitate sensory perception, although this hypothesis has not been specifically tested at the behavioral level.

In contrast to these in depth studies of NE actions in primary sensory cortex, there has been relatively little research on NE modulation of higher association areas in posterior cortex, perhaps because these studies must be done in primates. For example, the posterior parietal association cortex is essential for paying attention- allocating attentional resources in time and space (Coull and Nobre, 1998). This cortical area receives a dense NE innervation (Morrison and Foote, 1986), but we know little about NE's modulatory actions in this cortical area. Pharmacological and imaging studies in humans and monkeys indicate that α2 adrenoceptor stimulation may impair the attention orienting abilities of posterior parietal cortex (Clark et al., 1987; Witte and Marrocco, 1997; Coull et al., 2001). However, as all these studies involved systemic administration of drug, it is not known if the compounds had actions in parietal cortex, nor whether the drugs acted by decreasing NE release and/or via post-synaptic receptor mechanisms. Even less is known about NE modulation of the ventral stream, e.g. areas V4 and inferior temporal areas which are critically involved in feature extraction and objection perception. These will be important arenas for future research.

NE Influences on Subcortical Functions

NE prominently innervates the hippocampus and amygdala, and facilitates memory functions in both structures. These structures are likely not of critical relevance to AD/HD, but illustrate the general actions of NE in posterior cortical and subcortical areas. The hippocampus is critical for the consolidation of memory, and long term potentiation (LTP) is an electrophysiological measure of increased synaptic strength commonly associated with this process. NE strengthens LTP at mossy fiber synapses in hippocampus through actions at β adrenergic receptors (Hopkins and Johnston, 1984; Hopkins and Johnston, 1988). The consolidation of emotionally relevant information is enhanced by interactions with the amygdala. NE is released in the amygdala during emotional events (including stressors) and is critical to this process (Cahill and McGaugh, 1996). Both β and α1 adrenergic receptor stimulation facilitate memory consolidation in the amygdala (Ferry et al., 1999) while α2 receptor stimulation impairs plasticity (e.g. LTP) in the amygdala (DeBock et al., 2003). Thus, these subcortical structures are modulated in a manner similar to somatosensory cortex, thriving under conditions of high NE release.

NE and the Orchestration of Brain Response

NE may act as a chemical switch to determine which brain regions predominate in the control of behavior under varying environmental conditions. This orchestration of brain response may be accomplished by NE's differing affinities for its receptors: moderate levels of NE release engaging high affinity α2 receptors and a modest number of lower affinity α1 and β receptors under non-stressful conditions, and higher levels of NE release engaging large numbers of α1 and β receptors during stress (reviewed in (Arnsten, 2000a).

The PFC and amygdala both have widespread projections throughout the brain, and are positioned to play the role of "conductor". Both structures are also intimately connected with the LC, and levels of NE release may determine which of the two structures predominates in the control of brain function. There are reciprocal connections between the LC and PFC (Arnsten and Goldman-Rakic, 1984; Jodo et al., 1998), the LC and the amygdala (Cedarbaum and Aghajanian, 1978; Price and Amaral, 1981), and the PFC and amygdala (Ghashghaei and Barbas, 2002). These connections may create a flip-flop switch between 1) PFC regulation of brain and behavior under conditions of normal waking (phasic LC firing, moderate NE release, α2 receptor stimulation predominating, thus favoring PFC function and weakening amygdala), vs. 2) amygdala regulation of brain and behavior under conditions of stress (high tonic LC firing, high levels of NE release, α1 and β receptor stimulation predominating, strengthening amygdala and weakening PFC). While PFC is positioned to either excite or inhibit a variety of structures, including the amygdala, the amygdala may indirectly shut off PFC regulation through its projections to the LC and to the dopamine neurons in midbrain. Biochemical studies have shown that the amygdala initiates high levels of NE and DA release in the PFC during psychological stress (Goldstein et al., 1996), which impair PFC function (Arnsten, 2000b). The amygdala may have a general and powerful excitatory effect on LC via CRF projections which are activated under stressful conditions (Curtis et al., 2002). The PFC is also positioned to regulate LC firing, and may provide one of the few "intelligent" inputs to sculpt appropriate firing to relevant vs. irrelevant environmental stimuli. The PFC projects to the region of the LC (Arnsten and Goldman-Rakic, 1984; Jodo et al., 1998), especially in the dendritic zone medial to the LC cell bodies where there are GABAergic "interneurons" (Aston-Jones et al., 2004). Thus, the PFC is in the position to either excite (Jodo et al., 1998) or inhibit LC responses (Sara and Herve-Minvielle, 1995). This more refined regulation of LC firing may occur during non-stressed waking when the PFC is in its optimal neurochemical state. Inappropriate LC firing patterns could contribute to symptoms of AD/HD (Aston-Jones et al., 2000).

The Role of NE in Medications that Treat AD/HD

It should be evident even from this review that NE mechanisms and thus NE medications would be highly relevant to disorders of attention and behavioral regulation. Biederman and colleagues have long appreciated the importance of NE and NE compounds in AD/HD (Biederman and Spencer, 1999; Biederman and Spencer, 2000). For example, they found that NE reuptake blockers such as desipramine can be highly efficacious in treating AD/HD symptoms, however, cardiac side effects have limited their utility in a pediatric population

(Spencer et al., 2002b). The recent success of the safer, NE transporter blocker, atomoxetine (Strattera), in treating AD/HD has refocused the interest of the entire field on the role of NE in this disorder (Spencer et al., 2002a). It is now being appreciated that stimulant medications commonly used to treat AD/HD facilitate NE as well as DA transmission: methylphenidate (Ritalin) blocks both DA and NE transporters, amphetamines (e.g. Adderall) block DA and NE transporters and increase catecholamine release, atomoxetine (Strattera) blocks NE transporters (which take up DA in PFC, thus effectively increasing the concentration of both catecholamines in the synapse and extrasynaptic space (Bymaster et al., 2002)), and guanfacine mimics NE at α2A adrenoceptors. As reviewed above, oral, low dose treatment with stimulants can enhance PFC cognitive function in both normal college students (Mehta et al., 2000) and in AD/HD patients e.g. (Aron et al., 2003). Guanfacine has also been shown to enhance working memory or other PFC tasks (e.g. planning, Stroop interference, CPT), in normal adults (Jakala et al., 1999a) and patients with AD/HD (Scahill et al., 2001; Taylor and Russo, 2001). Neuropsychological assessment of atomoxetine is in progress, and it will be interesting to observe whether this compound also improves performance of PFC tasks.

The important effects of NE in stimulant actions have been relatively neglected, especially as many previous studies in animals have used high doses that have prominent effects on DA. However, recent biochemical evidence from studies in rats indicates that low, oral doses of methylphenidate, that produce plasma levels comparable to those observed in AD/HD patients, have more effect on NE than on DA in subcortical structures (Kuczenski and Segal, 2002),. These same low, oral doses *reduce* locomotor activity in rats, highlighting their relevance to AD/HD (ibid). Low doses of stimulant increase NE and DA in PFC, consistent with the greater sensitivity of the PFC than striatal DA systems (Berridge and Stalnaker, 2002). Identification of the appropriate dose regimen for methylphenidate administration in rats has allowed research to be performed of immediate relevance to AD/HD. Our lab has recently found that low, oral doses of methylphenidate improve performance of PFC tasks in rats and mice, and that this improvement results from both α2A adrenoceptor and DA D1 stimulation (Arnsten and Dudley, 2005). Thus, it is highly likely that the therapeutic actions of stimulants such as methylphenidate and amphetamine involve increased NE stimulation of α2A adrenoceptors in PFC, which strengthen PFC executive functions.

Optimizing catecholamine influences in PFC may remediate AD/HD symptomatology irrespective of the cause of AD/HD. It is obvious that such treatments would be helpful if the AD/HD symptomatology was caused by inadequate catecholamine transmission in PFC, as suggested by the Ernst findings (Ernst et al., 1998). However, it is likely to be helpful under other conditions as well- e.g. slowed maturation of PFC circuits, fewer PFC cells (i.e. smaller PFC volume)- as long as there are normal PFC cells available as a substrate for catecholamine actions.

The finding that high levels of catecholamine release impair PFC function is also informative for the treatment and even diagnosis of AD/HD. For example, high doses of stimulants may impair PFC cognitive abilities and induce perseverative or restricted thinking via excessive catecholamine actions in PFC. We have recently observed this in rats given higher doses of methylphenidate (e.g. 2-3 mg/kg, p.o.) who show impaired spatial working memory performance due to perseverative errors (Arnsten and Dudley, 2005). Exposure to uncontrollable stress (e.g. families going through divorce) may also induce excessive catecholamine release in PFC, impairing PFC function and mimicking AD/HD. In such cases stimulant medications would not be indicated; treatment should focus on remediating the stress and if necessary, use of agents (e.g. prazosin, guanfacine) that can protect PFC from stress in animals (Birnbaum et al., 1999; Birnbaum et al., 2000) and humans (Horrigan, 1996; Raskind et al., 2003).

In summary, the PFC mediates executive abilities such as working memory, attention regulation, behavioral inhibition, planning and organization that are problematic in subjects with AD/HD. Research in animals has shown that NE has vital, beneficial effects on PFC function through actions at post-synaptic α2A adrenoceptors in PFC, strengthening working memory, attention regulation, planning and behavioral inhibition. Many of the symptoms of AD/HD, including impaired working memory, impulsivity and hyperactivity, can be recreated by blocking NE α2A adrenoceptors in the monkey PFC. Conversely, medications that treat AD/HD directly or indirectly stimulate α2A adrenoceptors. Thus, there we are beginning to see a strong correspondence between knowledge of NE mechanisms in PFC and therapeutic drug actions in AD/HD. NE may also be important to attentional processes mediated by posterior cortices, although there has been less basic research in this field. These posterior abilities may be weakened in children with symptoms of inattention, i.e. the inability to allocate attentional processing resources at earlier stages of attention regulation. Our ability to differentiate and treat symptoms of AD/HD will evolve as we develop a more refined understanding of higher cortical mechanisms and their dependence on NE actions.

Acknowledgments

This work was supported by AG06036 and MH066393.

References

Aoki, C., Venkatesan, C., Go, C.-G., Forman, R. and Kurose, H. (1998) Cellular and the subcellular sites for noradrenergic action in the monkey dorsolateral prefrontal cortex as revealed by the immunocytochemical localization of noradrenergic receptors and axons. *Cereb. Cortex,* 8, 269-277.

Arnsten, A.F.T. (2000a) Through the looking glass: Differential noradrenergic modulation of prefrontal cortical function. *Neural Plasticity,* 7, 133-146.

Arnsten, A.F.T. (2000b) Stress impairs PFC function in rats and monkeys: Role of dopamine D1 and norepinephrine alpha-1 receptor mechanisms. *Prog. Brain Res.,* 126, 183-192.

Arnsten, A.F.T. and Goldman-Rakic, P.S. (1984) Selective prefrontal cortical projections to the region of the locus coeruleus and raphe nuclei in the rhesus monkey. *Brain Res.,* 306, 9-18.

Arnsten, A.F.T. and Dudley, A.G. (2005) Methylphenidate improves prefrontal cortical cognitive function through α2 adrenoceptor and dopamine D1 receptor actions: Relevance to therapeutic effects in Attention Deficit Hyperactivity Disorder. *Behav Brain Funct.* 1, 2. <http://www.behavioralandbrainfunctions.com/content/1/1/2>.

Arnsten, A.F.T. and Goldman-Rakic, P.S. (1985) Alpha-2 adrenergic mechanisms in prefrontal cortex associated with cognitive decline in aged nonhuman primates. *Science,* 230, 1273-1276.

Arnsten, A.F.T. and Goldman-Rakic, P.S. (1990) Stress impairs prefrontal cortex cognitive function in monkeys: role of dopamine. *Soc. Neurosci. Abstr.* 16, 164.

Arnsten, A.F.T. and Contant, T.A. (1992) Alpha-2 adrenergic agonists decrease distractibility in aged monkeys performing a delayed response task. *Psychopharmacology,* 108, 159-169.

Arnsten, A.F.T. and Goldman-Rakic, P.S. (1998) Noise stress impairs prefrontal cortical cognitive function in monkeys: Evidence for a hyperdopaminergic mechanism. *Arch. Gen. Psychiatry,* 55, 362-369.

Arnsten, A.F.T., Cai, J.X. and Goldman-Rakic, P.S. (1988) The alpha-2 adrenergic agonist guanfacine improves memory in aged monkeys without sedative or hypotensive side effects. *J. Neurosci.,* 8, 4287-4298.

Arnsten, A.F.T., Steere, J.C. and Hunt, R.D. (1996) The contribution of alpha-2 noradrenergic mechanisms to prefrontal cortical cognitive function: potential significance to Attention Deficit Hyperactivity Disorder. *Arch. Gen. Psychiatry,* 53, 448-455.

Arnsten, A.F.T., Mathew, R., Ubriani, R., Taylor, J.R. and Li, B.-M. (1999) Alpha-1 noradrenergic receptor stimulation impairs prefrontal cortical cognitive function. *Biol. Psychiatry,* 45, 26-31.

Aron, A.R., Robbins, T.W. and Poldrack, R.A. (2004) Inhibition and the right inferior frontal cortex. *Trends Cogn. Sci.,* 8, 170-177.

Aron, A.R., Dowson, J.H., Sahakian, B.J. and Robbins, T.W. (2003) Methylphenidate improves response inhibition in adults with attention-deficit/hyperactivity disorder. *Biol P.ychiatry,* 54, 1465-1468.

Aston-Jones, G., Rajkowski, J. and Cohen, J. (1999) Role of locus coeruleus in attention and behavioral flexibility. *Biol. Psychiatry,* 46, 1309-1320.

Aston-Jones, G., Rajkowski, J. and Cohen, J. (2000) Locus coeruleus and regulation of behavioral flexibility and attention. *Prog. Brain Res.,* 126, 165-182.

Aston-Jones, G., Zhu, Y. and Card, J.P. (2004) Numerous GABAergic afferents to locus ceruleus in the pericerulear dendritic zone: possible interneuronal pool. *J. Neurosci.,* 24, 2313-2321.

Avery, R.A., Franowicz, J.S., Studholme, C., van Dyck, C.H. and Arnsten, A.F.T. (2000) The alpha-2A-adenoceptor agonist, guanfacine, increases regional cerebral blood flow in dorsolateral prefrontal cortex of monkeys performing a spatial working memory task. *Neuropsychopharmacology,* 23, 240-249.

Barbas, H., Medalla, M., Alade, O., Suski, J., Zikopoulos, B. and Lera, P. (2005) Relationship of prefrontal connections to inhibitory systems in superior temporal areas in the rhesus monkey. *Cereb. Cortex,* Jan 5; Epub ahead of print.

Barkley, R.A. (1997) *ADHD and the nature of self-control.* New York: Guilford Press.

Bartus, R.T. and Levere, T.E. (1977) Frontal decortication in rhesus monkeys: A test of the interference hypothesis. *Brain Res.,* 119, 233-248.

Bedard, A.C., Ickowicz, A., Logan, G.D., Hogg-Johnson, S., Schachar, R. and Tannock, R. (2003) Selective inhibition in children with attention-deficit hyperactivity disorder off and on stimulant medication. *J. Abnorm. Child Psychol.,* 31, 315-327.

Berridge, C.W. and España, R.A. (2000) Synergistic actions of b- and a1-noradrenergic receptors in the maintenance of alert waking. *Neurosci.,* 99, 495-505.

Berridge, C.W. and Stalnaker, T.A. (2002) Relationship between low-dose amphetamine-induced arousal and extracellular norepinephrine and dopamine levels within prefrontal cortex. *Synapse,* 46, 140-149.

Berridge, C.W. and Waterhouse, B.D. (2003) The locus coeruleus-noradrenergic system: modulation of behavioral state and state-dependent cognitive processes. *Brain Res Brain Res. Rev.,* 42, 33-84.

Biederman, J. and Spencer, T. (1999) Attention-deficit/hyperactivity disorder (ADHD) as a noradrenergic disorder. *Biol. Psychiatry,* 46, 1234-1242.

Biederman, J. and Spencer, T.J. (2000) Genetics of childhood disorders: XIX. ADHD, Part 3: Is ADHD a noradrenergic disorder? *J. Am. Acad. Child Adolesc. Psychiatry,* 39, 1330-1333.

Birnbaum, S.G., Podell, D.M. and Arnsten, A.F.T. (2000) Noradrenergic alpha-2 receptor agonists reverse working memory deficits induced by the anxiogenic drug, FG7142, in rats. *Pharmacol. Biochem. Behav.,* 67, 397-403.

Birnbaum, S.G., Gobeske, K.T., Auerbach, J., Taylor, J.R. and Arnsten, A.F.T. (1999) A role for norepinephrine in stress-induced cognitive deficits: Alpha-1-adrenoceptor mediation in prefrontal cortex. *Biol. Psychiatry,* 46, 1266-1274.

Birnbaum, S.G., Yuan, P., Bloom, A., Davis, D., Gobeske, K., Sweatt, D., Manji, H.K. and Arnsten, A.F.T. (2004) Protein kinase C overactivity impairs prefrontal cortical regulation of working memory. *Science,* 306, 882-884.

Brozoski, T., Brown, R.M., Rosvold, H.E. and Goldman, P.S. (1979) Cognitive deficit caused by regional depletion of dopamine in prefrontal cortex of rhesus monkey. *Science,* 205, 929-931.

Buzsaki, G., Kennedy, B., Solt, V.B. and Ziegler, M. (1991) Noradrenergic control of thalamic oscillation: The role of alpha-2 receptors. *Eur J. Neurosci.,* 3, 222-229.

Bymaster, F.P., Katner, J.S., Nelson, D.L., Hemrick-Luecke, S.K., Threlkeld, P.G., Heiligenstein, J.H., Morin, S.M., Gehlert, D.R. and Perry, K.W. (2002) Atomoxetine increases extracellular levels of norepinephrine and dopamine in prefrontal cortex of rat: a potential mechanism for efficacy in attention deficit/hyperactivity disorder. *Neuropsychopharmacology,* 27, 699-711.

Cahill, L. and McGaugh, J.L. (1996) Modulation of memory storage. *Curr. Opin. Neurobiol.,* 6, 237-242.

Cai, J.X., Ma, Y., Xu, L. and Hu, X. (1993) Reserpine impairs spatial working memory performance in monkeys: Reversal by the alpha-2 adrenergic agonist clonidine. *Brain Res.,* 614, 191-196.

Casey, B.J., Castellanos, F.X., Giedd, J.N., Marsh, W.L., Hamburger, S.D., Schubert, A.B., Vauss, Y.C., Vaituzis, A.C., Dickstein, D.P., Sarfatti, S.E. and Rapoport, J.L. (1997) Implication of right frontostriatal circuitry in response inhibition and attention-deficit/hyperactivity disorder. *J Am. Acad. Child Adolesc. Psychiatry,* 36, 374-383.

Castellanos, F.X., Giedd, J.N., Marsh, W.L., Hamburger, S.D., Vaituzis, A.C., Dickstein, D.P., Sarfatti, S.E., Vauss, Y.C., Snell, J.W., Lange, N., Kaysen, D., Krain, A.L., Ritchhie, G.F., Rajapakse, J.C. and Rapoport, J.L. (1996) Quantitative brain magnetic resonance imaging in attention deficit/hyperactivity disorder. *Arch. Gen. Psychiatry,* 53, 607-616.

Cedarbaum, J.M. and Aghajanian, G.K. (1978) Afferent projections to the rat locus coeruleus as determined by a retrograde tracing technique. *J. Comp. Neurol.,* 178, 1-16.

Clark, C.R., Geffen, G.M. and Geffen, L.B. (1987) Catecholamines and attention II: Pharmacological studies in normal humans. *Neurosci. Biobehav. Rev.,* 11, 353-364.

Comings, D.E., Gade-Andavolu, R., Gonzalez, N. and MacMurray, J.P. (1999) Additive effect of three noradrenergic genes (ADRA2A, ADRA2C, DBH) on attention-deficit hyperactivity disorder and learning disabilities in Tourette syndrome subjects. *Clin Genet.,* 55, 160-172.

Coull, J.T. and Nobre, A.C. (1998) Where and when to pay attention: The neural systems for directing attention to spatial locations and to time intervals as revealed by both PET and fMRI. *J. Neurosci.,* 18, 7426-7435.

Coull, J.T., Nobre, A.C. and Frith, C.D. (2001) The noradrenergic alpha2 agonist clonidine modulates behavioural and neuroanatomical correlates of human attentional orienting and alerting. *Cereb. Cortex,* 11, 73-84.

Curtis, A.L., Bello, N.T., Connolly, K.R. and Valentino, R.J. (2002) Corticotropin-releasing factor neurones of the central nucleus of the amygdala mediate locus coeruleus activation by cardiovascular stress. *J. Neuroendocrinol.,* 14, 667-682.

Daly, G., Hawi, Z., Fitzgerald, M. and Gill, M. (1999) Mapping susceptibility loci in attention deficit hyperactivity disorder: preferential transmission of parental alleles at DAT1, DBH and DRD5 to affected children. *Mol. Psychiatry,* 4, 192-196.

DeBock, F., Kurz, J., Azad, S.C., Parsons, C.G., Hapfelmeier, G., Zieglgansberger, W. and Rammes, G. (2003) Alpha2-adrenoreceptor activation inhibits LTP and LTD in the basolateral amygdala: involvement of Gi/o-protein-mediated modulation of Ca2+-channels and inwardly rectifying K+-channels in LTD. *Eur. J. Neurosci.,* 17, 1411-1424.

Devilbiss, D.M. and Waterhouse, B.D. (2000) Norepinephrine exhibits two distinct profiles of action on sensory cortical neuron responses to excitatory synaptic stimuli. *Synapse,* 37, 273-282.

Dias, R., Roberts, A. and Robbins, T.W. (1996) Dissociation in prefrontal cortex of affective and attentional shifts. *Nature,* 380, 69-72.

Egan, M.F., Goldberg, T.E., Kolachana, B.S., Callicott, J.H., Mazzanti, C.M., Straub, R.E., Goldman, D. and Weinberger, D.R. (2001) Effect of COMT Val108/158 Met genotype on frontal lobe function and risk for schizophrenia. *Proc. Natl. Acad. Sci. USA,* 98, 6917-6922.

Ernst, M., Zametkin, A.J., Matochik, J.A., Jons, P.H. and Cohen, R.M. (1998) DOPA decarboxylase activity in attention deficit disorder adults. A [fluorine-18]fluorodopa positron emission tomographic study. *J. Neurosci.,* 18, 5901-5907.

Ferry, B., Roozendaal, B. and McGaugh, J.L. (1999) Involvement of alpha-1-adrenoceptors in the basolateral amygdala in modulation of memory storage. *Eur. J. Pharmacol.,* 372, 9-16.

Filipek, P.A., Semrud-Clikeman, M., Steingard, R.J., Renshaw, P.F., Kennedy, D.N. and Biederman, J. (1997) Volumetric MRI analysis comparing subjects having attention-deficit hyperactivity disorder with normal controls. *Neurology,* 48, 589-601.

Foote, S.L., Freedman, F.E. and Oliver, A.P. (1975) Effects of putative neurotransmitters on neuronal activity in monkey auditory cortex. *Brain Res.,* 86, 229-242.

Foote, S.L., Aston-Jones, G. and Bloom, F.E. (1980) Impulse activity of locus coeruleus neurons in awake rats and monkeys is a function of sensory stimulation and arousal. . *Proc. Natl. Acad. Sci. USA,* 77, 3033-3037.

Franowicz, J.S., Kessler, L., Dailey-Borja, C.M., Kobilka, B.K., Limbird, L.E. and Arnsten, A.F.T. (2002) Mutation of the alpha2A-adrenoceptor impairs working memory performance and annuls cognitive enhancement by guanfacine. *J. Neurosci.,* 22, 8771-8777.

French, G.M. (1959) Locomotor effects of regional ablation of frontal cortex in rhesus monkeys. *J. Comp. Physiol. Psychol.,* 52, 18-24.

Funahashi, S., Chafee, M.V. and Goldman-Rakic, P.S. (1993) Prefrontal neuronal activity in rhesus monkeys performing a delayed anti-saccade task. *Nature,* 365, 753-756.

Ghashghaei, H.T. and Barbas, H. (2002) Pathways for emotion: interactions of prefrontal and anterior temporal pathways in the amygdala of the rhesus monkey. *Neurosci.,* 115, 1261-1279.

Giedd, J.N., Blumenthal, J., Molloy, E. and Castellanos, F.X. (2001) Brain imaging of attention deficit/hyperactivity disorder. *Ann. NY Acad. Sci.,* 931, 33-49.

Godefroy, O. and Rousseaux, M. (1996) Divided and focused attention in patients with lesion of the prefrontal cortex. *Brain Cogn.,* 30, 155-174.

Goldman-Rakic, P.S. (1995) Cellular basis of working memory. *Neuron,* 14, 477-485.

Goldstein, L.E., Rasmusson, A.M., Bunney, S.B. and Roth, R.H. (1996) Role of the amygdala in the coordination of behavioral, neuroendocrine and prefrontal cortical monoamine responses to psychological stress in the rat. *J Neurosci,* 16, 4787-4798.

Gross, C.G. (1963) Locomotor activity following lateral frontal lesions in rhesus monkeys. *J. Comp. Physiol. Psychol.,* 56, 232-236.

Gross, C.G. and Weiskrantz, L. (1964) Some changes in behavior produced by lateral frontal lesions in the macaque. In: The frontal granular cortex and behavior (Warren JM and Akert K, eds), pp 74-101. New York: McGraw-Hill Book Co.

Hill, D.E., Yeo, R.A., Campbell, R.A., Hart, B., Vigil, J. and Brooks, W. (2003) Magnetic resonance imaging correlates of attention-deficit/hyperactivity disorder in children. *Neuropsychology,* 17, 496-506.

Hopkins, W.F. and Johnston, D. (1984) Frequency-dependent noradrenergic modulation of long-term potentiation in the hippocampus. *Science,* 226, 350-352.

Hopkins, W.F. and Johnston, D. (1988) Noradrenergic enhancement of long-term potentiation at mossy fiber synapses in the hippocampus. *J. Neurophysiol.,* 59, 667-687.

Horrigan, J.P. (1996) Guanfacine for PTSD nightmares. *J. Am. Acad. Child Adolesc. Psychiatry,* 35, 975-976.

Itami, S. and Uno, H. (2002) Orbitofrontal cortex dysfunction in attention-deficit hyperactivity disorder revealed by reversal and extinction tasks. *Neuroreport,* 13, 2453-2457.

Jakala, P., Riekkinen, M., Sirvio, J., Koivisto, E., Kejonen, K., Vanhanen, M. and Riekkinen, P.J. (1999a) Guanfacine, but not clonidine, improves planning and working memory performance in humans. *Neuropsychopharmacol.,* 20, 460-470.

Jakala, P., Sirvio, J., Riekkinen, M., Koivisto, E., Kejonen, K., Vanhanen, M. and Riekkinen, P.J. (1999b) Guanfacine and clonidine, alpha-2 agonists, improve paired associates learning, but not delayed matching to sample, in humans. *Neuropsychopharmacol.,* 20, 119-130.

Jodo, E., Chiang, C. and Aston-Jones, G. (1998) Potent excitatory influence of prefrontal cortex activity on noradrenergic locus coeruleus neurons. *Neurosci.,* 83, 63-79.

Kates, W.R., Frederikse, M., Mostofsky, S.H., Folley, B.S., Cooper, K., Mazur-Hopkins, P., Kofman, O., Singer, H.S., Denckla, M.B., Pearlson, G.D. and Kaufmann, W.E. (2002) MRI parcellation of the frontal lobe in boys with attention deficit hyperactivity disorder or Tourette syndrome. *Psychiat. Res.,* 116, 63-81.

Kennard, M.A., Spencer, S. and Fountain, G. (1941) Hyperactivity in monkeys following lesions of the frontal lobes. *J. Neurophysiol.,* 4, 512-524.

Kimberg, D.Y., D'Esposito, M. and Farah, M.J. (1997) Effects of bromocriptine on human subjects depend on working memory capacity. *NeuroReport,* 8, 3581-3585.

Knight, R.T., Scabini, D. and Woods, D.L. (1989) Prefrontal cortex gating of auditory transmission in humans. *Brain Res.,* 504, 338-342.

Konishi, S., Nakajima, K., Uchida, I., Kikyo, H., Kameyama, M. and Miyashita, Y. (1999) Common inhibitory mechanism in human inferior prefrontal cortex revealed by event-related functional MRI. *Brain,* 122, 981-991.

Kuczenski, R. and Segal, D.S. (2002) Exposure of adolescent rats to oral methylphenidate: preferential effects on extracellular norepinephrine and absence of sensitization and cross-sensitization to methamphetamine. *J. Neurosci.,* 22, 7264-7271.

Li, B.-M. and Mei, Z.-T. (1994) Delayed response deficit induced by local injection of the alpha-2 adrenergic antagonist yohimbine into the dorsolateral prefrontal cortex in young adult monkeys. *Behav. Neur. Biol.,* 62, 134-139.

Li, B.-M., Mao, Z.-M., Wang, M. and Mei, Z.-T. (1999) Alpha-2 adrenergic modulation of prefrontal cortical neuronal activity related to spatial working memory in monkeys. *Neuropsychopharmacol,* 21, 601-610.

Lidow, M.S., Koh, P.-O. and Arnsten, A.F.T. (2002) D1 dopamine receptors in the mouse prefrontal cortex: Immunocytochemical and cognitive neuropharmacological analyses. *Synapse,* 47, 101-108.

Ma, C.-L., Arnsten, A.F.T. and Li, B.-M. (2005) Locomotor hyperactivity induced by blockade of prefrontal cortical alpha-2-adrenoceptors in monkeys. *Biol. Psychiatry,*57, 192-195.

Ma, C.-L., Qi, X.-L., Peng, J.-Y. and Li, B.-M. (2003) Selective deficit in no-go performance induced by blockade of prefrontal cortical alpha 2-adrenoceptors in monkeys. *NeuroReport,* 14, 1013-1016.

MacDonald, E., Kobilka, B.K. and Scheinin, M. (1997) Gene targeting- Homing in on alpha-2-adrenoceptor subtype function. *Trends Pharmacol. Sci.,* 18, 211-219.

Malmo, R.B. (1942) Interference factors in delayed response in monkeys after removal of frontal lobes. *Neurophysiol.,* 5, 295-308.

Manes, F., Sahakian, B.J., Clark, L., Rogers, R., Antoun, N., Aitken, M. and Robbins, T. (2002) Decision-making processes following damage to the prefrontal cortex. *Brain,* 125, 624-639.

Mao, Z.-M., Arnsten, A.F.T. and Li, B.-M. (1999) Local infusion of alpha-1 adrenergic agonist into the prefrontal cortex impairs spatial working memory performance in monkeys. *Biol. Psychiatry,* 46, 1259-1265.

McCormick, D.A., Pape, H.C. and Williamson, A. (1991) Actions of norepinephrine in the cerebral cortex and thalamus: implications for function of the central noradrenergic system. *Prog. Brain Res.,* 88, 293-305.

McLean, A., Dowson, J., Toone, B., Young, S., Bazanis, E., Robbins, T.W. and Sahakian, B.J. (2004) Characteristic neurocognitive profile associated with adult attention-deficit/hyperactivity disorder. *Psychol. Med.,* 34, 681-692.

Mehta, M.A., Goodyer, I.M. and Sahakian, B.J. (2004) Methylphenidate improves working memory and set-shifting in AD/HD: relationships to baseline memory capacity. *J. Child Psychol. Psychiat.,* 45, 293-305.

Mehta, M.A., Owen, A.M., Sahakian, B.J., Mavaddat, N., Pickard, J.D. and Robbins, T.W. (2000) Methylphenidate enhances working memory by modulating discrete frontal and parietal lobe regions in the human brain. *J. Neurosci.,* 20, RC651-656.

Miller, E.K., Li, L. and Desimone, R. (1993) Activity of neurons in anterior inferior temporal cortex during a short-term memory task. *J. Neurosc.,* 13, 1460-1478.

Moises, H.C., Waterhouse, B.D. and Woodward, D.J. (1981) Locus coeruleus stimulation potentiates Purkinje cell responses to afferent input: the climbing fiber system. *Brain Res.,* 222, 43-64.

Morita, M., Nakahara, K. and Hayashi, T. (2004) A rapid presentation event-related functional magnetic resonance imaging study of response inhibition in macaque monkeys. *Neurosci. Lett.,* 356, 203-206.

Morrison, J.H. and Foote, S.L. (1986) Noradrenergic and serotonergic innervation of cortical, thalamic, and tectal visual structures in Old and New World monkeys. *J. Comp. Neurol.,* 243, 117-138.

Mouradian, R.D., Seller, F.M. and Waterhouse, B.D. (1991) Noradrenergic potentiation of excitatory transmitter action in cerebrocortical slices: evidence of mediation by an alpha1-receptor-linked second messenger pathway. *Brain Res.,* 546, 83-95.

Muir, J.L., Everitt, B.J. and Robbins, T.W. (1996) The cerebral cortex of thc rat and visual attentional function: Dissociable effects of mediofrontal, cingulate, anterior dorsolateral, and parietal cortex lesions on a five-choice serial reaction time task. *Cereb. Cortex,* 6, 470-481.

Petrides, M. (1986) The effect of periarcuate lesions in the monkey on the performance of symmetrically and asymmetrically reinforced visual and auditory go, no-go tasks. *J. Neurosci.,* 6, 2054-2063.

Price, J.L. and Amaral, D.G. (1981) An autoradiographic study of the projections of the central nucleus of the monkey amygdala. *J Neurosci.,* 1, 1242-1259.

Quirk, G.J., Likhtik, E., Pelletier, J.G. and Pare, D. (2003) Stimulation of medial prefrontal cortex decreases the responsiveness of central amygdala output neurons. *J. Neurosci.,* 23, 8800-8807.

Rajkowski, J., Kubiak, P., Inanova, S. and Aston-Jones, G. (1998) State-related activity, reactivity of locus coeruleus neurons in behaving monkeys. *Adv. Pharmacol.,* 42, 740-744.

Rajkowski, J., Majczynski, H., Clayton, E. and Aston-Jones, G. (2004) Activation of monkey locus coeruleus neurons varies with difficulty and performance in a target detection task. *J. Neurophysiol.,* 92, 361-371.

Rama, P., Linnankoski, I., Tanila, H., Pertovaara, A. and Carlson, S. (1996) Medetomidine, atipamezole, and guanfacine in delayed response performance of aged monkeys. *Pharmacol. Biochem. Behav.,* 54, 1-7.

Raskind, M.A., Peskind, E.R., Kanter, E.D., Petrie, E.C., Radant, A., Thompson, C., Dobie, D.J., Hoff, D., Rein, R.J., Straits-Troster, K., Thomas, R. and McFall, M.M. (2003)

Prazosin reduces nightmares and other PTSD symptoms in combat veterans: A placebo-controlled study. *Am. J. Psychiatry,* 160, 371-373.

Roman, T., Schmitz, M., Polanczyk, G.V., Eizirik, M., Rohde, L.A. and Hutz, M.H. (2002) Further evidence for the association between attention-deficit/hyperactivity disorder and the dopamine-beta-hydroxylase gene. *Am. J. Med. Genet.,* 114, 154-158.

Rubia, K., Smith, A.B., Brammer, M.J. and Taylor, E. (2003) Right inferior prefrontal cortex mediates response inhibition while mesial prefrontal cortex is responsible for error detection. *Neuroimage,* 20, 351-358.

Rubia, K., Overmeyer, S., Taylor, E., Brammer, M., Williams, S.C.R., Simmons, A. and Bullmore, E.T. (1999) Hypofrontality in Attention Deficit Hyperactivity Disorder during higher-order motor control: A study with functional MRI. *Am. J. Psychiatry,* 156, 891-896.

Sara, S.J. and Herve-Minvielle, A. (1995) Inhibitory influence of frontal cortex on locus coeruleus. *Proc. Natl. Acad. Sci. USA,* 92, 6032-6036.

Sawaguchi, T. and Goldman-Rakic, P.S. (1994) The role of D1-dopamine receptors in working memory: local injections of dopamine antagonists into the prefrontal cortex of rhesus monkeys performing an oculomotor delayed response task. *J. Neurophysiol.,* 71, 515-528.

Scahill, L., Chappell, P.B., Kim, Y.S., Schultz, R.T., Katsovich, L., Shepherd, E., Arnsten, A.F.T., Cohen, D.J. and Leckman, J.F. (2001) Guanfacine in the treatment of children with tic disorders and ADHD: A placebo-controlled study. *Am. J. Psychiatry,* 158, 1067-1074.

Segal, M. and Bloom, F.E. (1976) The action of norepinephrine in the rat hippocampus. IV. The effects of locus coeruleus stimulation on evoked hippocampal unit activity. *Brain Res.,* 107, 513-525.

Sowell, E.R., Thompson, P.M., Welcome, S.E., Henkenius, A.L., Toga, A.W. and Peterson, B.S. (2003) Cortical abnormalities in children and adolescents with attention-deficit hyperactivity disorder. *Lancet,* 362, 1699-1707.

Spencer, T., Heiligenstein, J.H., Biederman, J., Faries, D.E., Kratochvil, C.J., Conners, C.K. and Potter, W.Z. (2002a) Results from 2 proof-of-concept, placebo-controlled studies of atomoxetine in children with attention-deficit/hyperactivity disorder. *J. Clin. Psychiatry,* 63, 1140-1147.

Spencer, T., Biederman, J., Coffey, B., Geller, D., Crawford, M., Bearman, S.K., Tarazi, R. and Faraone, S. (2002b) A double-blind comparison of desipramine and placebo in children and adolescents with chronic tic disorder and comorbid attention-deficit/hyperactivity disorder. *Arch. Gen. Psychiatry,* 59, 649-656.

Steere, J.C. and Arnsten, A.F.T. (1997) The alpha-2A noradrenergic agonist, guanfacine, improves visual object discrimination reversal performance in rhesus monkeys. *Behav. Neurosci.,* 111, 1-9.

Swanson, J.M., Posner, M., Potkin, S., Bonforte, S., Youpa, D., Fiore, C., Cantwell, D. and Crinella, F. (1991) Activating tasks for the study of visual-spatial attention in ADHD children: a cognitive anatomic approach. *J. Child Neurol.,* 6, S119-S127.

Tanila, H., Rama, P. and Carlson, S. (1996) The effects of prefrontal intracortical microinjections of an alpha-2 agonist, alpha-2 antagonist and lidocaine on the delayed alternation performance of aged rats. *Brain Res. Bull.,* 40, 117-119.

Tanila, H., Mustonen, K., Sallinen, J., Scheinin, M. and Riekkinen, P. (1999) Role of alpha-2C-adrenoceptor subtype in spatial working memory as revealed by mice with targeted disruption of the alpha-2C-adrenoceptor gene. *Eur J. Neurosci.,* 11, 599-603.

Taylor, F.B. and Russo, J. (2001) Comparing guanfacine and dextroamphetamine for the treatment of adult Attention Deficit-Hyperactivity Disorder. *J. Clin. Psychopharmacol.,* 21, 223-228.

Thompson-Schill, S.L., Jonides, J., Marshuetz, C., Smith, E.E., D'Esposito, M., Kan, I.P., Knight, R.T. and Swick, D. (2002) Effects of frontal lobe damage on interference effects in working memory. *Cogn. Affect Behav. Neurosci.,* 2, 109-120.

Trommer, B., Hoeppner, J. and Zecker, S. (1991) The go-no go test in attention deficit disorder is sensitive to methylphenidate. *J. Child Neurol.,* 6 Suppl., S128-131.

Turner, D.C., Blackwell, A.D., Dowson, J.H., McLean, A. and Sahakian, B.J. (2004) Neurocognitive effects of methylphenidate in adult attention-deficit/hyperactivity disorder. *Psychopharmacol.,* 178, 286-295.

Usher, M., Cohen, J.D., Servan-Schreiber, D., Rajkowski, J. and Aston-Jones, G. (1999) The role of locus coeruleus in the regulation of cognitive performance. *Science,* 283, 549-554.

Vaidya, C.J., Austin, G., Kirkorian, G., Ridlehuber, H.W., Desmond, J.E., Glover, G.H. and Gabrieli, J.D.E. (1998) Selective effects of methylphenidate in attention deficit hyperactivity disorder: A functional magnetic resonance study. *Proc. Natl. Acad. Sci. USA,* 95, 14494-14499.

Van Tol, H.H.M., Bunzow, J.R., Guan, H.-C., Sunahara, R.K., Seeman, P., Niznik, H.B. and Civelli, O. (1991) Cloning of the gene for a human dopamine D4 receptor with high affinity for the antipsychotic clozapine. *Nature,* 350, 610-614.

Wallis, J.D., Anderson, K.C. and Miller, E.K. (2001) Single neurons in prcfrontal cortex encode abstract rules. *Nature,* 411, 953-956.

Wang, M., Tang, Z.X. and Li, B.M. (2004a) Enhanced visuomotor associative learning following stimulation of alpha 2A-adrenoceptors in the ventral prefrontal cortex in monkeys. *Brain Res.,* 1024, 176-182.

Wang, M., Ji, J.Z. and Li, B.M. (2004b) The alpha(2A)-adrenergic agonist guanfacine improves visuomotor associative learning in monkeys. *Neuropsychopharmacology,* 29, 86-92.

Watanabe, M. (1986) Prefrontal unit activity during delayed conditional go/no-go discrimination in the monkey I. Relation to the stimulus. *Brain Res. Rev.,* 382, 1-14.

Waterhouse, B.D., Moises, H.C. and Woodward, D.J. (1980) Noradrenergic modulation of somatosensory cortical neuronal responses to iontophoretically applied putative transmitters. *Exp. Neurol.,* 69, 30-49.

Waterhouse, B.D., Moises, H.C. and Woodward, D.J. (1981) Alpha-receptor-mediated facilitation of somatosensory cortical neuronal responses to excitatory synaptic inputs and iontophoretically applied acetylcholine. *Neuropharmacology,* 20, 907-920.

Waterhouse, B.D., Moises, H.C. and Woodward, D.J. (1998) Phasic activation of the locus coeruleus enhances responses of primary sensory cortical neurons to peripheral receptive field stimulation. *Brain Res.,* 790, 33-44.

Wilkins, A.J., Shallice, T. and McCarthy, R. (1987) Frontal lesions and sustained attention. *Neuropsychologia,* 25, 359-365.

Witte, E.A. and Marrocco, R.T. (1997) Alterations of brain noradrenergic activity in rhesus monkeys affects the alerting component of covert orienting. *Psychopharmacology,* 132, 315-323.

Woods, D.L. and Knight, R.T. (1986) Electrophysiological evidence of increased distractability after dorsolateralll prefrontal lesions. *Neurology,* 36, 212-216.

Xu, C., Schachar, R., Tannock, R., Roberts, W., Malone, M., Kennedy, J.L. and Barr, C.L. (2001) Linkage study of the alpha2A adrenergic receptor in attention-deficit hyperactivity disorder families. *Am. J. Med. Genet.,* 105, 159-162.

Yamaguchi, S. and Knight, R.T. (1990) Gating of somatosensory input by human prefrontal cortex. *Brain Res.,* 521, 281-288.

Yeo, R.A., Hill, D., Campbell, R., Vigil, J. and Brooks, W.M. (2000) Developmental instability and working memory ability in children: a magnetic resonance spectroscopy investigation. *Dev. Neuropsychol.,* 17, 143-159.

Zahrt, J., Taylor, J.R., Mathew, R.G. and Arnsten, A.F.T. (1997) Supranormal stimulation of dopamine D1 receptors in the rodent prefrontal cortex impairs spatial working memory performance. *J. Neurosci.,* 17, 8528-8535.

In: Encyclopedia of Neuroscience Research
Editors: Eileen J. Sampson and Donald R. Glevins
ISBN 978-1-61324-861-4

Chapter IX

Prefrontal Cortex: Brodmann and Cajal Revisited

Guy N. Elston1 and Laurence J. Garey2
[1]Centre for Cognitive Neuroscience,
Sunshine Coast, Qld 4562, Australia
[2]Centre for Psychiatric Neuroscience
CH-1008 Lausanne, Switzerland

Abstract

Modern imaging techniques allow the mapping of gross brain structure to an exacting standard; however, these methodologies have been applied inappropriately in recent times to the study of the prefrontal cortex – the structure identified by Brodmann as being central to intelligence. Currently, these methodologies cannot detect prefrontal cortex, nor be used to quantify it. Here we review Brodmann's data on comparative aspects of the gross structure of the cerebral cortex, some of which have been largely ignored in recent times. We believe this still remains the definitive data set for prefrontal cortex. In addition we compare Brodmann's observations with those of his contemporary Santiago Ramon y Cajal. Cajal's studies of the fine structure of the cerebral cortex led him to the conclusion that human intelligence is largely attributable to the complexity of the pyramidal (psychic) cell. These observations are compared and contrasted with modern data on prefrontal cortex, essentially substantiating the conclusions of these two great founding fathers of modern neuroscience.

Introduction

Brodmann on Prefrontal Cortex

Apparently mostly unbeknown to neuroscientists, Korbinian Brodmann published an unprecedented and unsurpassed set of data almost 100 years ago (Brodmann, 1912, 1913; translated by Elston and Garey, 2004) in which he quantified, among other things, the size of the frontal lobe and "prefrontal cortex" in a variety of primate and non-primate species. Brodmann's definition of prefrontal cortex is clear: granular cortex (that containing an identifiable layer IV) in the frontal lobe. Because this term has become confused in modern studies we use the term granular prefrontal cortex (gPFC) here in accordance with Brodmann's definition. We highlight in italics examples of where the term has been misused. Brodmann's observations of gPFC were derived from his Nissl preparations of the brains of a large number of species, both primate and non-primate (Figure 1). Brodmann's data on the frontal lobe and gPFC (Table 1) reveal several interesting observations which predate modern findings. Moreover, statements made in modern studies are often inconsistent with the Brodmann data. In particular Brodmann's data reveal that the size of the frontal lobe of humans is smaller than that predicted from the non-human primate data (Figure 2A). In addition, his data reveal that the proportion of the frontal lobe occupied by the gPFC varies considerably between species (Figure 2B).

Table 1. Brodmann's (1913) original data on the area of various cortices in the frontal lobe, motor cortex, granular prefrontal cortex, and the total cerebrum in a variety of mammalian species

Species	Frontal	Motor	Prefrontal	Total
Man	49120	9833	39287	135470
Chimpanzee	12108	5389	6719	39572
Gibbon	3490	1651	1839	16301
Mandrill	4362	2194	2168	21321
Baboon (*Papio hamadryas*)	4865	2898	1967	20594
Baboon (*Papio cynocephalus*)	4311	2200	2111	20376
Macaque	3550	1817	1733	15308
Guenon	3601	1976	1625	14641
Capuchin monkey	3082	1822	1260	13682
Marmoset	315	167	148	1649
Black lemur	790	453	337	4054
Dwarf lemur	164	94	70	921
Fruit bat	99	73	26	1097
Dog (fox terrier)	1940	1283	657	9527
Cat	564	412	152	4474
Rabbit	184	148	36	1627
Hedgehog	24	24	0	575
Armadillo	93	93	0	2010
Opossum	51	51	0	804

Notably, the proportion of the frontal lobe occupied by gPFC is considerably higher in human than in all other species studied (Figure 2C). His data also show that the proportion of the cortical mantle occupied by gPFC in man is greater, as a function of either the frontal lobe or the total cortical surface area, than that in all other species studied (Figure 2D). Because of Brodmann's observations it is widely believed that the disproportionate increase in the size of the gPFC, not the frontal lobe, is related to increasing intelligence during hominine evolution.

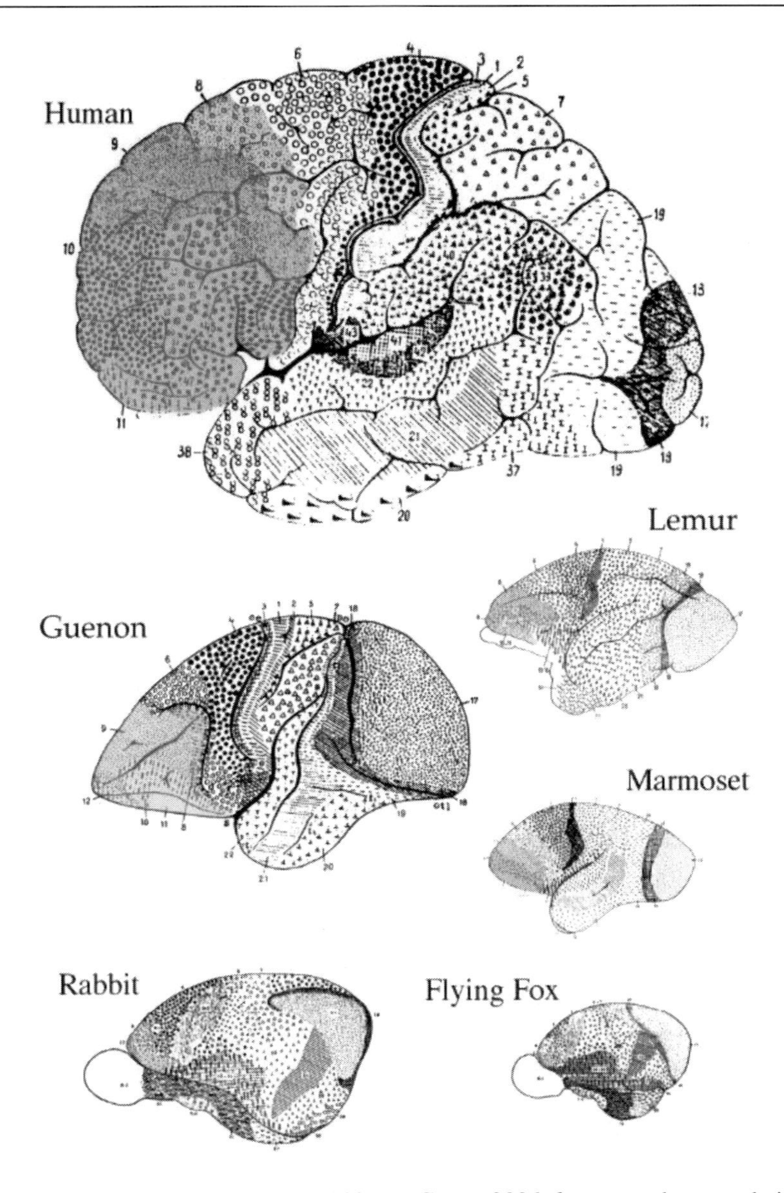

Figure 1. Drawings, based on Brodmann (1909; see Garey, 2006, for a complete translation) illustrating cortical areas identified in the frontal lobe as having a well-developed layer IV ("inner granular layer"). As per Brodmann's definition (1913) these areas (blue) represent the extent of *prefrontal* cortex. Note that the number and size of cortical areas included as granular prefrontal cortex, as well as their topography, differ between species. The most differentiated prefrontal cortex is found in humans, including areas 8, 9, 10, 11, 44, 45, 46 and 47. The number and location of cortical areas in granular prefrontal cortex have since been modified with the advent of new methodologies used to distinguish cortical areas. However, changes to nomenclature do not adversely affect the original Brodmann data (1913) regarding relative differences in the size of the gPFC in different species. Figure modified from Elston, 2003a. Not to scale.

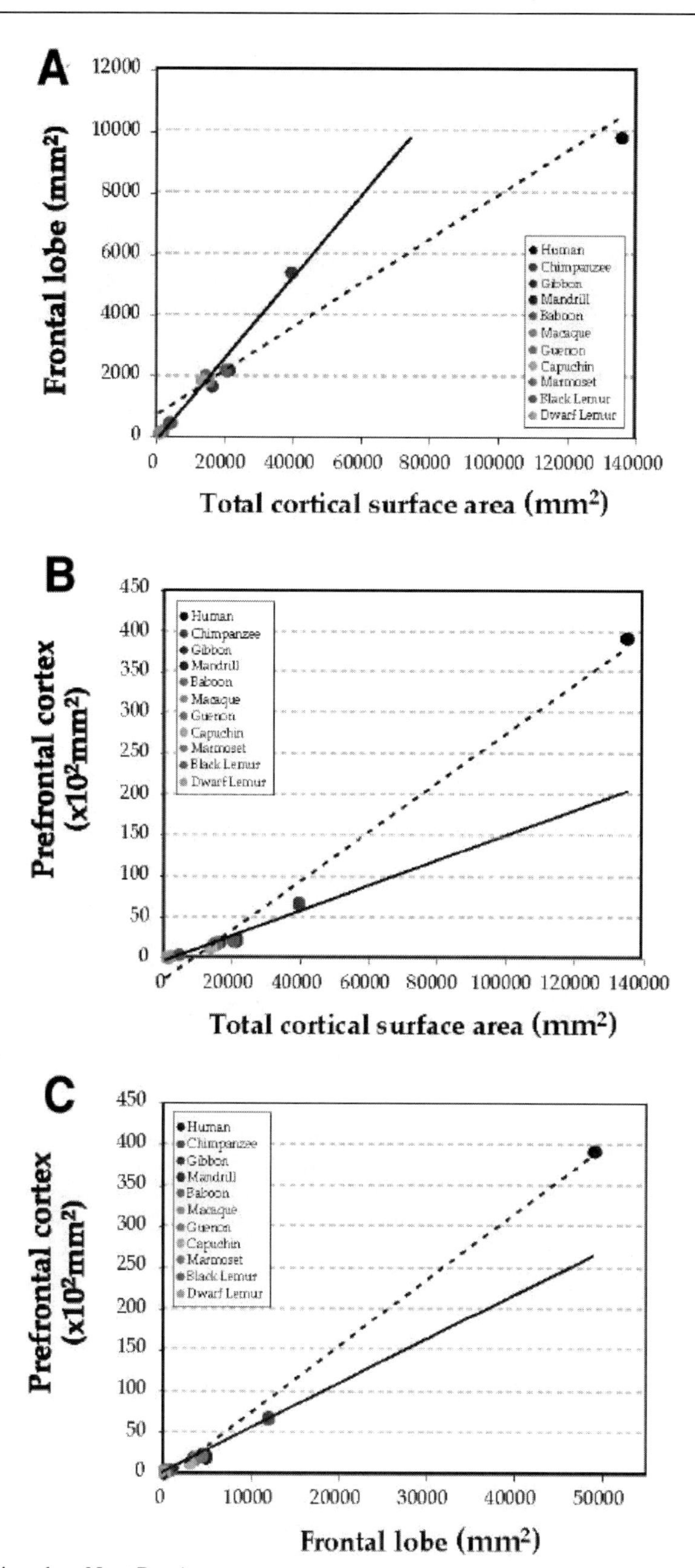

Figure 2. (Continued on Next Page)

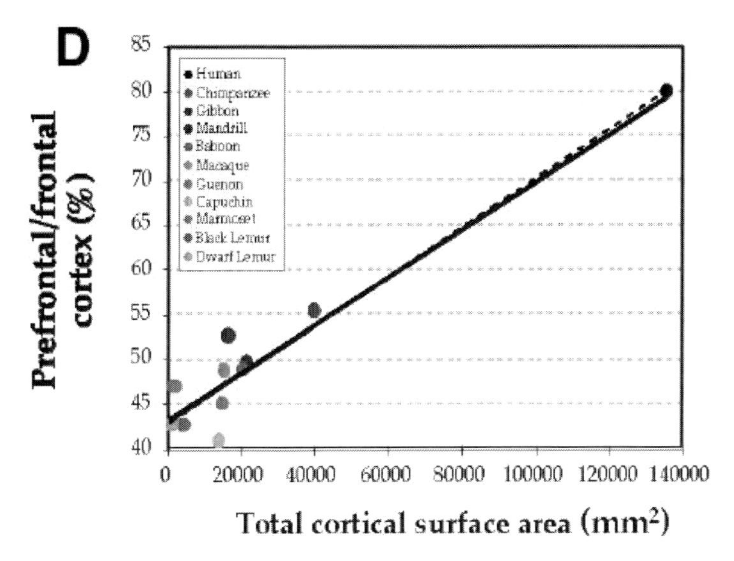

Figure 2. Plots of original data from Brodmann (1913) illustrating (A) the area of the frontal lobe vs. total cortical surface area, (B) area of the granular prefrontal cortex vs. total cortical surface area, (C) area of the granular prefrontal cortex vs. frontal lobe and (D) the proportion of the frontal lobe occupied by granular prefrontal cortex vs. total cortical surface area. Linear regressions of the entire data set are illustrated in dashed black lines; those excluding the human data are illustrated by solid black lines. In A, exclusion of the human data increased the r^2 coefficient. Exclusion of the human data in B-D decreased the r^2 coefficient. Note that according to these data (A) the frontal lobe of the human brain is smaller than that expected from the plot of the non-human primate data. Note also that the proportion of the frontal lobe occupied by granular prefrontal cortex varies quite dramatically between primate species (D). Thus, in some primates, there has been a disproportionate increase in the size of prefrontal cortex during the expansion of the frontal lobe. This is highlighted in C and D; prefrontal cortex occupies a much higher proportion of both the cerebrum and the frontal lobe in humans compared to non-human primates. In addition to these relative values, it should be kept in mind that the absolute size of human prefrontal cortex is considerably larger than that in other species.

Prefrontal Cortex

Modern Imaging Studies

Unfortunately, the field has become confused in recent times as modern imaging technologies have in some cases been applied inappropriately to test Brodmann's conclusion related to the size of the *prefrontal* cortex and intelligence. The confusion arises largely out of a misunderstanding of what *prefrontal cortex* is or because something that is not *prefrontal* cortex has been used to test this relationship. These studies fail to acknowledge the existence of the Brodmann data, which have been reviewed previously (Passingham, 1973). For example, Semendeferi and colleagues (2002) quantified the size of the frontal lobes of great apes and monkeys and concluded that "Human frontal cortices were not disproportionately large in comparison to those of the great apes" (as a function of total cerebral cortical volume). Although this study included a larger number of species than Brodmann's, the conclusion was no different from that made by Brodmann (1913) some 90 years earlier (see Figure 2A). However, unlike the modern investigators, Brodmann further extended his line of investigation to the gPFC, leading him to conclude that it was this region of the cerebral cortex, not the frontal lobe, that was larger in man than in other species (both absolutely and relatively). Indeed, it is difficult to understand the justification presented by Semendeferi and colleagues (2002) for studying size relationships of the frontal lobe and intelligence. Studying the Brodmann data (1913) reveals that the frontal lobe of humans *and great apes* is considerably smaller than what would be expected in a primate of similar brain size when extrapolating the non-human/non-great ape data. Taking these data together with those of Brodmann (who also studied the chimpanzee), the logical conclusion is that the frontal lobe of humans *and great apes* is considerably smaller than what would be expected in a primate of similar brain size when extrapolating the non-human/non-great ape data. In essence, this is just what Bush and Allman (2004) demonstrated: frontal cortex hyperscaling is greater in lemurs and lorises compared with higher primates. While new and important data have been fleshed out, these modern studies represent no advance over Brodmann's original conclusions (1913): the relative size of the frontal lobe compared to total cortical volume reveals little about cognitive abilities.

As mentioned above, the field has become confused due to a misunderstanding of *what prefrontal cortex is* in estimating its size. Schoenemann and colleagues (2005) concluded that there is a disproportionately larger volume of white matter in human *prefrontal* cortex as compared to that in other primates, and suggested that this reflects a substrate for human cognitive abilities. However, these authors used what they called a "proxy" for the *prefrontal* cortex, which neither reflects granular prefrontal cortex nor is consistent across species. It is likely that white matter in this study was sampled from a different selection of areas in the gPFC of the different species studied due to differences in the size and topography of prefrontal areas (see Brodmann, 1909, translated by Garey, 2006 for pictorial illustration). The study was too ambitious. Moreover, reanalyses of their data led Sherwood and colleagues (2005) to conclude that the white matter volume in the human brain is less that expected from the great ape data. Clearly, there is no agreement among these studies. Moreover, conspicuous in all of these studies is that the methodologies used do not allow identification of the structure in question – gPFC.

Why the Confusion?

Most of the recent confusion arises because inappropriate methodology has been applied to test the relationship between the size of the gPFC and "intelligence". No doubt this last statement will be cause for concern for some, perhaps rightly so. Despite the technological advances available to us today, the only way to quantify prefrontal cortex is to identify granular cortex in the frontal lobe. The only way to identify granular cortex is to employ age-old techniques that generally do not attract funding. Herein lies the conundrum faced by some of today's neuroscientists interested in the study of the gPFC: submit a scientifically sound application that may not attract funding or submit an application based on an "attractive methodology" that will likely get funded. In the latter case, it is necessary to state why the applicant chooses the modern technology over the appropriate technology. In essence, the argument becomes that the modern technique is superior to the old technique and data collected by the old technique are then usually discredited. For these reasons, some discount the Brodmann data: the brains were collected over 100 years ago, methods of preservation may have caused differential shrinkage and it is unclear how many brains were included for study for a given species (see Passingham, 1973 for a review). However, Brodmann published these data at the end of his career after having the benefit of studying the brains of many different species over a period of decades. It is difficult to believe that he was incapable of distinguishing granular from non-granular cortex. In addition, the evidence suggests that Brodmann processed all the brains himself, making it likely that a standard methodology was applied in all studies. Furthermore, at least for the human data, we know that Brodmann studied many brains. Human brain weights were determined from fourteen cases and the size of human granular prefrontal cortex was reported in the brains obtained from two normal and three pathological cases (Brodmann, 1913; Elston and Garey, 2004). While modern imaging techniques allow the sampling of larger numbers of individuals, and reveal important data related to interindividual variation in brain structure (Sherwood et al., 2004), these methodologies still cannot be used to identify and quantify the gPFC - the structure identified by Brodmann as being central to intelligence. This goal remains off-limits to imaging technology until such time as granular cortex can be distinguished from non-granular cortex.

A Glimpse of Circuit Specialisations in Prefrontal Cortex

It is not all bad news. Considerable effort has been focussed on the determination and quantification of patterns of connections between cortical areas in the gPFC - at least, that of the macaque monkey (Felleman and Van Essen, 1991; Young, 1993; Goldman-Rakic, 1996; Hilgetag et al., 1996; Barbas, 2000; Lewis and Van Essen, 2000; Petrides, 2000; Passingham et al., 2002). Moreover, considerable advances have been made in quantifying patterns of intrinsic connections in the gPFC of the macaque (Lund et al., 1993; Melchitzky et al., 1998, 2001; Barbas et al., 1999; González-Burgos et al., 2000; Rempel-Clower and Barbas, 2000; Soloway et al., 2002; Germuska et al., 2006). However, it is remarkable how little has been done to quantify connectivity in the gPFC of other primate species. To the best of our knowledge there exists not a single study in which patterns of connectivity have been quantified systematically in homologous cortical areas of the *granular* prefrontal cortex of primates by a standardised methodology (that allows axon projections to be visualised and quantified). Two studies have addressed this question qualitatively in non-human primates, both from the laboratory of the late Patricia Goldman-Rakic (Bugbee and Goldman-Rakic, 1983; Preuss and Goldman-Rakic, 1991). Petrides and colleagues have utilised patterns of connectivity to determine the internal organisation and homology of cortical areas of the *granular* prefrontal cortex between the macaque monkey and man (Petrides and Pandya, 2001); however, no quantitative comparisons were made between species. This is particularly striking when we consider that methodologies used in these connectional studies were developed over 30 years ago. Perhaps the lack of application of these techniques to comparative analyses results from the fact that these methodologies were developed at a time when it was generally believed that the cortex was uniform in structure. At that time it was widely believed that all functional differences in the cerebral cortex were solely attributed to the origin of inputs (e.g. Szentagothai and Arbib, 1974; Creutzfeldt, 1977; Zeki, 1978). We now know that this very definitely is not the case.

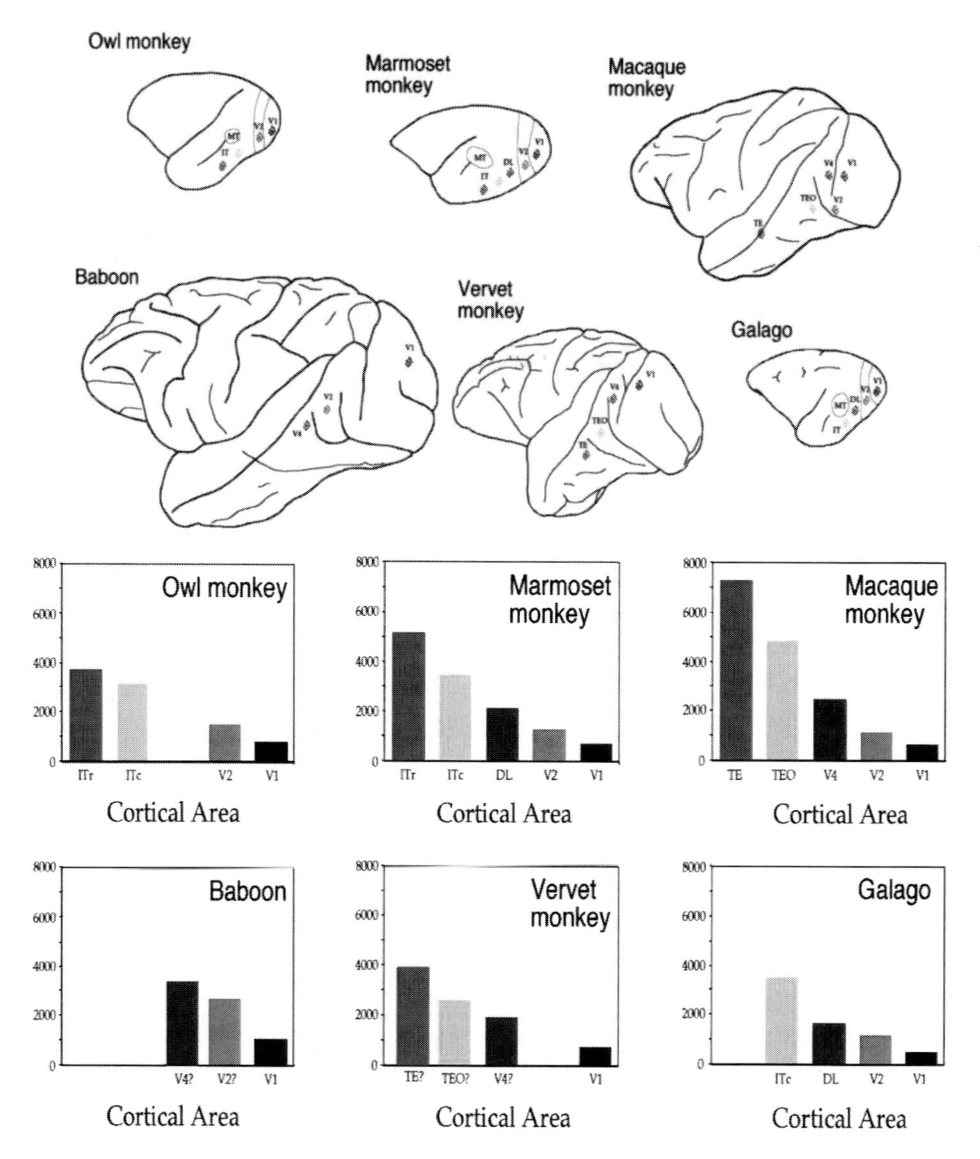

Figure 3. Differences in the number of dendritic spines (sites of excitatory synapses) on the dendritic trees of pyramidal neurons in cortical areas of the ventral visual pathway of the galago, owl monkey, marmoset monkey, macaque monkey, vervet monkey and baboon (dots illustrate regions where neurons were sampled; X-axis = total number of spines in the basal dendritic tree calculated by summing the product of the number of branches by spine density along concentric annuli [25mm] from the cell body and extending to the distal tips). Note there is a clear and consistent trend for neurons to have more spines with anterior progression through cortical areas associated with this pathway. See also Figures 4 and 5. The different brains are not drawn to scale. Modified from Elston (2007).

Cajal's Psychic Cell

Recent studies of the morphology of the pyramidal cell, the archetypal ubiquitous neuron in the cerebral cortex, have revealed extensive phenotypic variation in its structure across the cortical mantle. The dendritic trees of pyramidal cells differ in their size and branching structure between cortical areas. Significant variations have also been demonstrated in the density of spines, the major site for excitatory synaptic input to pyramidal neurons, along the dendrites. Our estimates of the total number of spines in the dendritic trees (calculated by summing the product of the number of branches by spine density along concentric annuli [25mm] from the cell body and extending to the distal tips) reveal systematic differences among cortical areas associated with different cortical pathways (Figures 3-6).

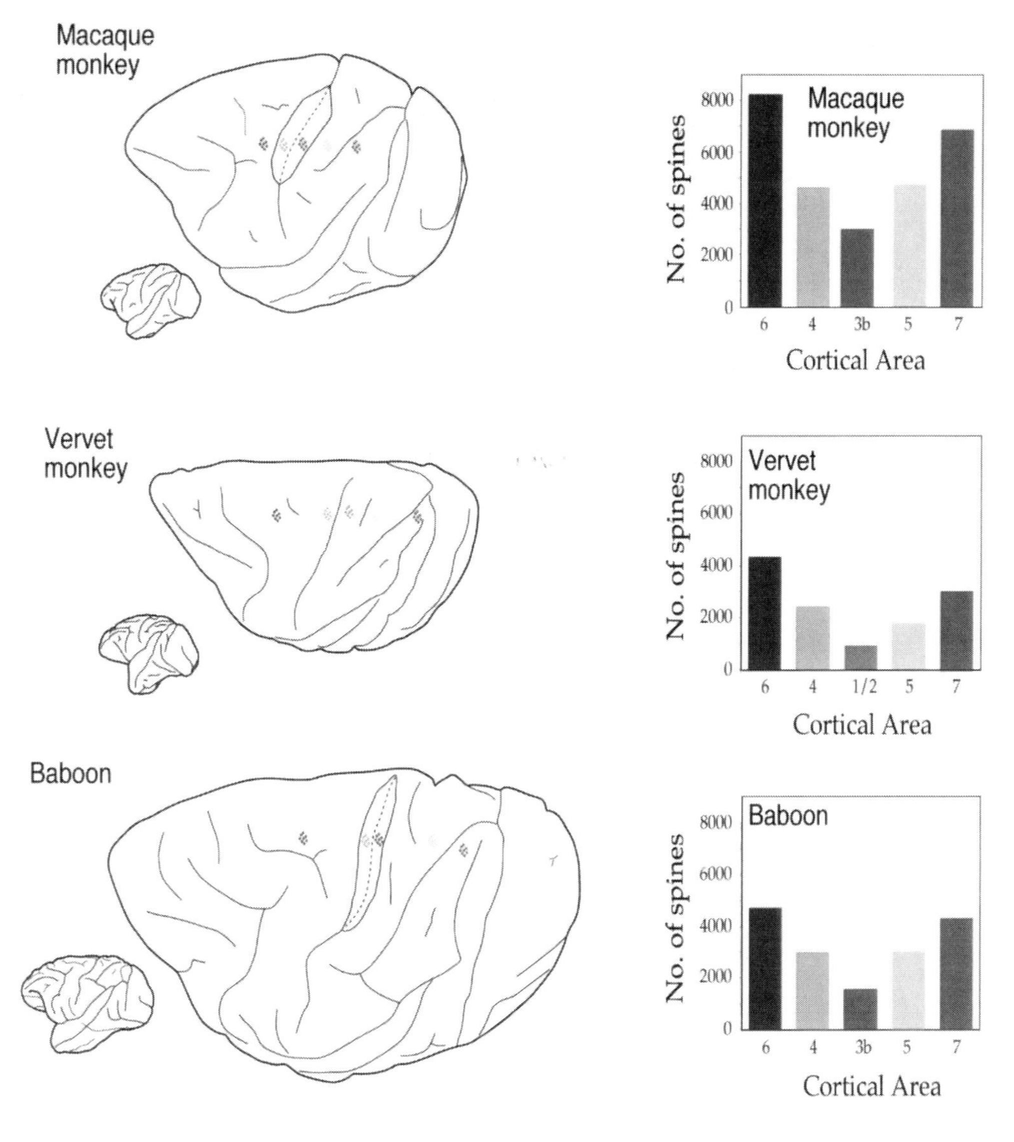

Figure 4. Differences in the number of dendritic spines on pyramidal neurons in cortical areas of the somatosensory-motor cortex of the macaque monkey, vervet monkey and baboon (dots illustrate regions where neurons were sampled; X-axis = total number of spines, calculated as in Figure 3). Note the consistent trend for neurons to have more spines with anterior progression from the primary motor area (area 4) to premotor cortex (area 6) in all three species. Note also the consistent trend for progressively more spinous cells with posterior progression through somatosensory areas from the central sulcus to the angular gyrus. Modified from Elston, 2007.

Moreover, there is a systematic increase in pyramidal cell complexity from primary sensory to sensory association, limbic and granular prefrontal cortex (Lund et al., 1993; Elston et al., 1999a,b, 2001, 2005a-c; Elston, 2000, 2003b; Elston and Rosa, 1998, 2000; Jacobs et al., 2001; Elston and Rockland, 2002). This systematic increase in complexity has a dramatic functional parallel: neurons with simple structure in primary sensory cortex are characterised by phasic discharge, those in sensory association cortex are characterised by tonic discharge that can be interrupted by distractors and those most complex cells in the granular prefrontal cortex are characterised by tonic discharge that is resistant to distractors (see Fuster, 1997; Miller and Cohen, 2001; Elston, 2003a for reviews). Moreover, comparison of pyramidal cell structure in the gPFC of different primates reveals a dramatic variation in the pyramidal cell phenotype among species (Elston et al., 2001, 2007). There is a trend for increasingly more complex pyramidal cells in species with a progressively larger gPFC. Thus, not only is the human *granular* prefrontal cortex larger than that in other species (in absolute terms), but it is composed of neurons of greater functional capacity (see Elston, 2007 for a review).

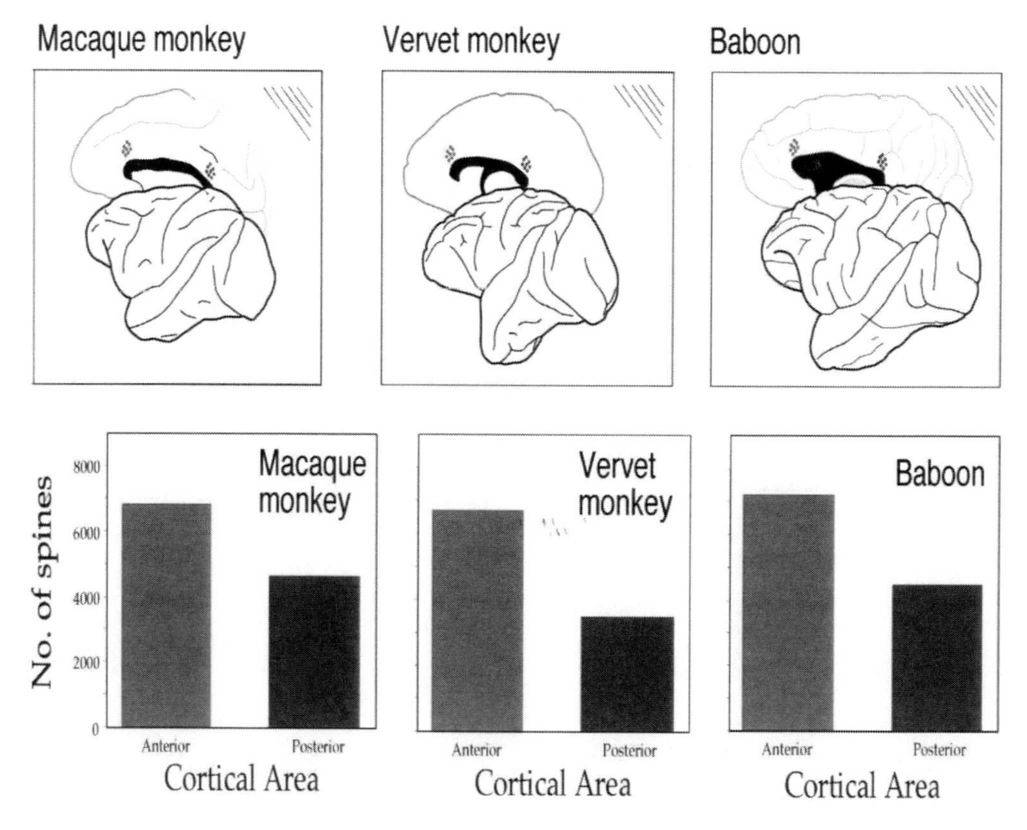

Figure 5. Differences in the number of dendritic spines (sites of excitatory synapses) on the dendritic trees of pyramidal cells in cortical areas of the cingulate gyrus (Brodmann's areas 23 and 24) in macaque monkey, vervet monkey and baboon. (Dots illustrate regions where neurons were sampled; X-axis = total number of spines, calculated as in Figure 3). In all three species there was a consistent trend for pyramidal neurons to have more spines in the anterior cingulate gyrus than in the posterior. The different brains not drawn to scale. Modified from Elston, 2007.

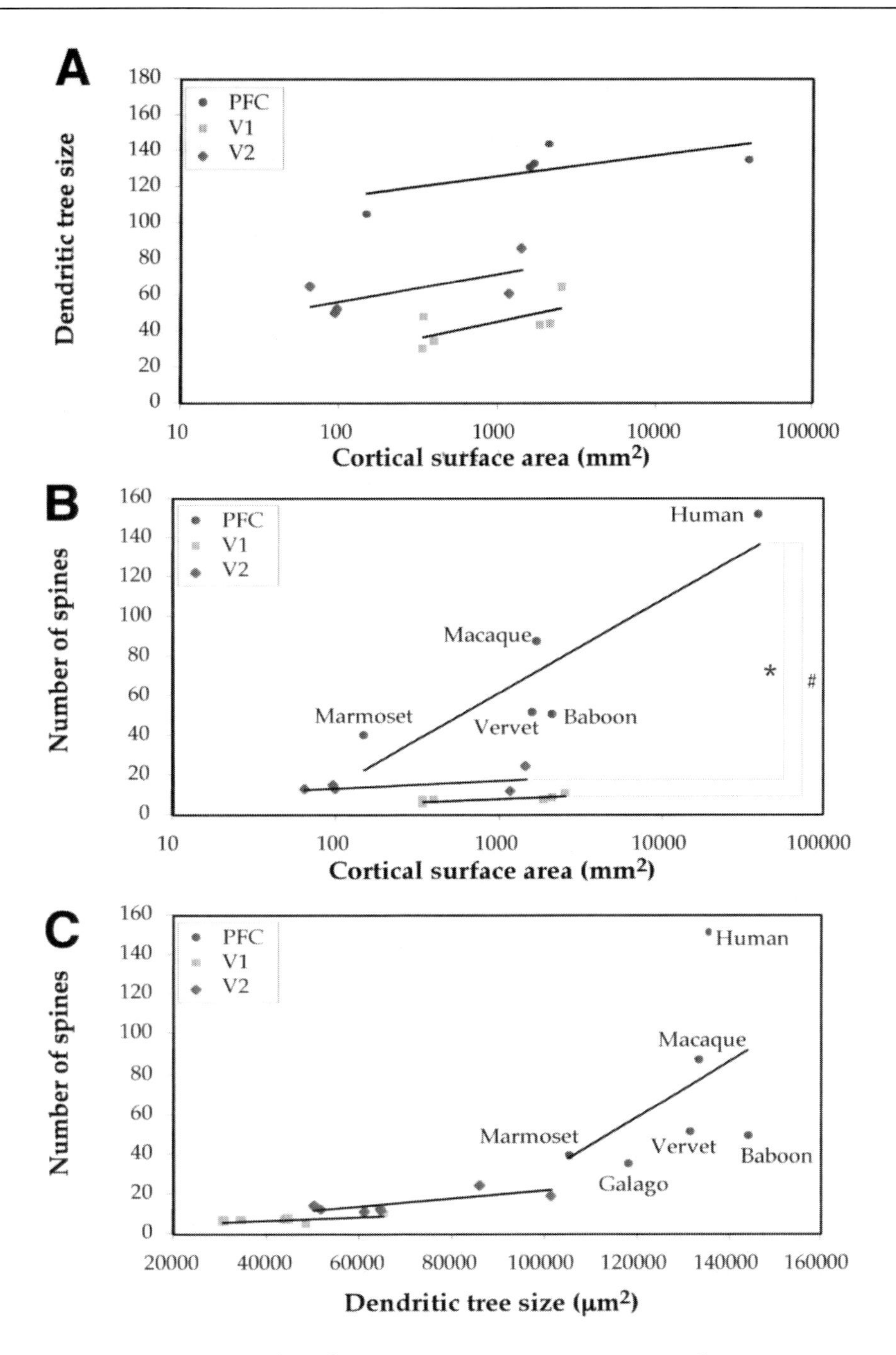

Figure 6. Plots of (A) the area (μm^2 x10^3), and (B) the number of spines (x10^2) on the basal dendritic trees of pyramidal cells in the granular prefrontal cortex (gPFC), primary (V1) and secondary (V2) visual cortex vs. the tangential surface area of each cortical area. (C) Plot of the number of spines (x10^2) vs. the size of the basal dendritic trees of pyramidal cells in the gPFC, primary (V1) and secondary (V2) visual cortex. Moderated multiple regression revealed a significant difference between the slopes of regression lines of gPFC and V2 for comparisons between total number of spines on the dendritic arbors of pyramidal cells vs. cortical surface area (F_{change} (1,6) = 6.19, $p < 0.05$). Significance was approached for the comparison between the total number of spines in the dendritic arbors of pyramidal cells vs. cortical surface area for gPFC and V1 (F_{change} (1,7) = 5.20, $p < 0.057$). Reproduced from Elston et al. 2006.

Conclusion

Brodmann's data reveal that the frontal lobe of modern man is smaller than what might be expected if extrapolating from data obtained from other primate species. However, the size of human granular prefrontal cortex is larger than in other primate species, in both absolute and relative terms. Because of this, it is often stated that the granular prefrontal cortex is associated with intelligence or cognitive ability. Recent studies demonstrate that not only is the human granular prefrontal cortex larger than in other primate species, but it is composed of neurons of more complex structure. The highly complex neuronal structure observed in the gPFC of humans endows a higher degree of connectivity and functional capacity than those in the gPFC of other species. Santiago Ramon y Cajal said it beautifully in 1893: "as one ascends the animal scale the psychic (pyramidal) cell becomes larger and more complex; it is natural to attribute this progressive morphological complexity, in part at least, to its progressive functional state…it can thus be considered probable that the psychic cell performs its activity more amply and usefully the larger the number of somatic and collateral dendrites that it offers".

Just after the one hundredth anniversary of the award of the Nobel Prize to Ramon y Cajal it might be appropriate for modern neuroscientists to embrace this specialisation and further probe the microstructure of intelligence in a series of systematic comparative studies (see also DeFelipe, 2006).

References

Barbas H. Connections underlying the synthesis of cognition, memory, and emotion in primate prefrontal cortices. *Brain Research Bulletin* 2000 52:319-330.

Barbas H, Ghashghaei H, Dombrowski SM, Rempel-Clower NL. Medial prefrontal cortices are unified by common connections with superior temporal cortices and distinguished by input from memory-related areas in the rhesus monkey. *Journal of Comparative Neurology* 1999 410:343-367.

Brodmann K. Vergleichende Lokalisationslehre der Grosshirnrinde in ihren Prinzipien dargestellt auf Grund des Zellenbaues. Barth, Leipzig 1909 (for translation see Garey, 2006, below).

Brodmann K. Neue Ergebnisse über die vergleichende histologische Lokalisation der Grosshirnrinde mit besonderer Berücksichtigung des Stirnhirns. *Anatomischer Anzeiger* 1912 41:157-216.

Brodmann K. Neue Forschungsergebnisse der Grosshirnrindenanatomie mit besonderer Berücksichtigung anthropologischer Fragen. Verhandlungen der Gesellschaft deutscher *Naturforscher und Ärzte* 1913 85:200-240.

Bugbee NM, Goldman-Rakic PS. Columnar organization of corticocortical projections in squirrel and rhesus monkeys: similarity of column width in species differing in cortical volume. *Journal of Comparative Neurology* 1983 220:355-364.

Bush EC, Allman JM. The scaling of frontal cortex in primates and carnivores. *Proceedings of the National Academy of Sciences USA* 2004 101:3962-3966.

Cajal S R y. 1893. Neuvo concepto de la histologia de los centros nerviosos. *Revista de ciencias medicas*, Barcelona 18:21–40.

Creutzfeldt OD. Generality of the functional structure of the neocortex. *Naturwissenschaften* 1977 64:507-517.

DeFelipe J. Brain plasticity and mental processes: Cajal again. *Nature Reviews Neuroscience* 2006 7:811-817.

Elston GN. Pyramidal cells of the frontal lobe: all the more spinous to think with. *Journal of Neuroscience* 2000 20:RC95(1-4).

Elston GN. Cortex, cognition and the cell: new insights into the pyramidal neuron and prefrontal function. *Cerebral Cortex* 2003a 13:1124-1138.

Elston GN. The pyramidal neuron in occipital, temporal and prefrontal cortex of the owl monkey (*Aotus trivirgatus*): regional specialization in cell structure. *European Journal of Neuroscience* 2003b 17:1313-1318.

Elston GN. 2007. Specializations in pyramidal cell structure during primate evolution. In: Kaas JH, Preuss TM (Eds.) *Evolution of Nervous Systems*. Elsevier Oxford. pp 191-242

Elston GN, Benavides-Piccione R, DeFelipe J. The pyramidal cell in cognition: a comparative study in human and monkey. *Journal of Neuroscience* 2001 21:RC163(1-5).

Elston GN, Benavides-Piccione R, Elston A, DeFelipe J, Manger P. Specialization in pyramidal cell structure in the cingulate cortex of the Chacma baboon (*Papio ursinus*): an intracellular injection study of the posterior and anterior cingulate gyrus with comparative notes on the macaque and vervet monkeys. *Neuroscience Letters* 2005a 387:130-135.

Elston GN, Benavides-Piccione R, Elston A, Manger P, DeFelipe J. Specialization in pyramidal cell structure in the sensory-motor cortex of the Chacma baboon (*Papio ursinus*) with comparative notes on macaque and vervet monkeys. *Anatomical Record* 2005b 286A:854-865.

Elston GN, Benavides-Piccione R, Elston A, Zietsch B, DeFelipe J, Manger P, Casagrande V, Kaas JH. Specializations of the granular prefrontal cortex of primates: implications for cognitive processing. *Anatomical Record* 2006 288A:26-35.

Elston GN, Elston A, Casagrande VA, Kaas JH. Pyramidal neurons of granular prefrontal cortex of the galago: complexity in evolution of the psychic cell in primates. *Anatomical Record* 2005c 285A:610-618.

Elston GN, Garey LJ. New research findings on the anatomy of the cerebral cortex of special relevance to anthropological questions. 2004: University of Queensland Printery. Brisbane

Elston GN, Rockland KS. The pyramidal cell of the sensorimotor cortex of the macaque monkey: phenotypic variation. *Cerebral Cortex* 2002 12:1071-1078.

Elston GN, Rosa MG. Morphological variation of layer III pyramidal neurones in the occipitotemporal pathway of the macaque monkey visual cortex. *Cerebral Cortex* 1998 8:278-294.

Elston GN, Rosa MGP. Pyramidal cells, patches, and cortical columns: a comparative study of infragranular neurons in TEO, TE, and the superior temporal polysensory area of the macaque monkey. *Journal of Neuroscience* 2000 20:RC117(1-5).

Elston GN, Tweedale R, Rosa MGP. Cortical integration in the visual system of the macaque monkey: large-scale morphological differences in the pyramidal neurons in the occipital, parietal and temporal lobes. *Proceedings of the Royal Society*, London, Series B 1999a 266:1367-1374.

Elston GN, Tweedale R, Rosa MGP. Cellular heterogeneity in cerebral cortex. A study of the morphology of pyramidal neurones in visual areas of the marmoset monkey. *Journal of Comparative Neurology* 1999b 415:33-51.

Felleman DJ, Van Essen DC. Distributed hierarchical processing in the primate cerebral cortex. *Cerebral Cortex* 1991 1:1-47.

Fuster JM. 1997. *The Prefrontal Cortex: Anatomy, Physiology, and Neuropsychology of the Frontal Lobe.* Lippincott-Raven: Philadelphia.

Garey LJ. 2006. *Brodmann's Localisation in the Cerebral Cortex.* Springer: New York.

Germuska M, Saha S, Fiala J, Barbas H. Synaptic distinction of laminar-specific prefrontal-temporal pathways in primates. *Cerebral Cortex* 2006 16:865-875.

Goldman-Rakic PS. The prefrontal landscape: implications of functional architecture for understanding human mentation and the central executive. Philosophical Transactions of the Royal Society, London, Series B 1996 351:1445-1453.

González-Burgos G, Barrionuevo G, Lewis DA. Horizontal synaptic connections in monkey prefrontal cortex: an in vitro electrophysiological study. *Cerebral Cortex* 2000 10:82-92.

Hilgetag CC, O'Neill MA, Young MP. Indeterminate organization of the visual system. *Science* 1996 271:776-777.

Jacobs B, Schall M, Prather M, Kapler L, Driscoll L, Baca S, Jacobs J, Ford K, Wainwright M, Treml M. Regional dendritic and spine variation in human cerebral cortex: a quantitative study. *Cerebral Cortex* 2001 11:558-571.

Lewis JW, van Essen DC. Corticocortical connections of visual, sensorimotor, and multimodal processing areas in the parietal lobe of the macaque monkey. *Journal of Comparative Neurology* 2000 428:112-137.

Lund JS, Yoshioka T, Levitt JB. Comparison of intrinsic connectivity in different areas of macaque monkey cerebral cortex. *Cerebral Cortex* 1993 3:148-162.

Melchitzky DS, González-Burgos G, Barrionuevo G, Lewis DA. Synaptic targets of the intrinsic axon collaterals of supragranular pyramidal neurons in monkey prefrontal cortex. *Journal of Comparative Neurology* 2001 430:209-221.

Melchitzky DS, Sesack SR, Pucak ML, Lewis DA. Synaptic targets of pyramidal neurons providing intrinsic horizontal connections in monkey prefrontal cortex. *Journal of Comparative Neurology* 1998 390:211-224.

Miller EK, Cohen JD. An integrative theory of prefrontal cortex function. *Annual Reviews of Neuroscience* 2001 24:167-202.

Passingham R. Anatomical differences between the neocortex of man and other primates. *Brain Behavior and Evolution* 1973 7:337-359.

Passingham R, Stephan KE, Kötter R. The anatomical basis of functional localization in the cortex. *Nature Reviews Neuroscience* 2002 3:606-616.

Petrides M. 2000. Mapping prefrontal cortical systems for the control of cognition. In: Toga AW, Mazziotta JC, Frackowiak RS (Eds.) *Brain Mapping: The Systems.* Academic: San Dieg:. pp 159-176.

Petrides M, Pandya DN. Comparative cytoarchitectonic analysis of the human and the macaque ventrolateral prefrontal cortex and corticocortical connection patterns in the monkey. *European Journal of Neuroscience* 2001 16:291-310.

Preuss TM, Goldman-Rakic PS. Ipsilateral cortical connections of granular frontal cortex in the strepsirhine primate Galago, with comparative comments on anthropoid primates. *Journal of Comparative Neurology* 1991 310:507-549.

Rempel-Clower NL, Barbas H. The laminar pattern of connections between prefrontal and anterior temporal cortices in the rhesus monkey is related to cortical structure and function. *Cerebral Cortex* 2000 10:851-865.

Schoenemann PT, Sheehan MJ, Glotzer LD. Prefrontal white matter volume is disproportionately larger in humans than in other primates. *Nature Neuroscience* 2005 8:242-252.

Semendeferi K, Lu A, Schenker N, Damasio H. Humans and great apes share a large frontal cortex. *Nature Neuroscience* 2002 5:272-276.

Sherwood CC, Cranfield MR, Mehlman PT, Lilly AA, Garbe JA, Whittier CA, Nutter FB, Rein TR, Bruner HJ, Holloway RL, Tang CY, Naidich TP, Delman BN, Steklis HD, Erwin JM, Hof PR. Brain structure variation in great apes, with attention to the mountain gorilla (*Gorilla beringei beringei*). *American Journal of Primatology* 2004 63:149-164.

Sherwood CC, Holloway RL, Semendeferi K, Hof PR. Is prefrontal white matter enlargement a human evolutionary specialization? *Nature Neuroscience* 2005 8:537-538.

Soloway AS, Pucak ML, Melchitzky DS, Lewis DA. Dendritic morphology of callosal and ipsilateral projection neurons in monkey prefrontal cortex. *Neuroscience* 2002 109:461-471.

Szentagothai J, Arbib MA. Conceptual models of neural organization. *Neuroscience Research Programme Bulletin* 1974 12:305-510.

Young MP. The organization of neural systems in the primate cerebral cortex. *Proceedings of the Royal Society,* London, Series B 1993 252:13-18.

Zeki SM. Uniformity and diversity of structure and function in rhesus monkey prestriate visual cortex. *Journal of Physiology* (London) 1978 277:273-290.

In: Encyclopedia of Neuroscience Research
Editors: Eileen J. Sampson and Donald R. Glevins
ISBN 978-1-61324-861-4

Chapter X

Developmental Characteristics in Category Generation Reflects Differential Prefrontal Cortex Maturation

Julio Cesar Flores Lázaro1, 2 and Feggy Ostrosky-Solís1*
[1]Universidad Nacional Autónoma de México, Psychology Faculty,
Laboratory of Neuropsychology and Psychophysiology México, D.F.
[2]Universidad Juárez Autónoma de Tabasco,
división académica de ciencias de la salud,
Villahermosa Tabasco, México

Abstract

The developmental characteristics of semantic category generation were studied in a sample of 200 neurologically intact subjects from 6 to 30 years old. Results indicate that diverse brain areas support different the category criteria used: subjects in the 6-8 age range used a concrete criteria (posterior brain areas), in the 9-11 age range a functional criteria is dominant (dorsolateral prefrontal cortex), while abstract criteria becomes dominant in the 12-14 age-range (anterior prefrontal cortex) and continuous to develop in the 16 to 30 age range. It is proposed that different prefrontal areas participate during development, shaping cognitive approach to categorisation during childhood and adolescence.

Keywords: categorisation, prefrontal cortex, development

* Correspondence: Feggy Ostrosky-Solís, Universidad Nacional Autónoma de México, facultad de Psicología, Laboratorio de Neuropsicología y Psicofisiología. Rivera de Cupia 110-71. Lomas de Reforma. México D.F. C.P. 11900. E-mail address: feggy@servidor.unam.mx.

Introduction

The ability to categorize is fundamental for cognitive development. The evolution from concrete to abstract thinking goes trough different developmental stages that shape cognitive performance in children (Gelaes and Thibaut, 2006).

Young children frequently compare and categorize objects using perceptual features, like for example the shape of an object. Progressively when children learn new words that represent semantic labels, they prefer to use the category such as "animals" or "furniture", rather than a perceptual criteria (Gentner and Namy, 2000). While performing comparisons between objects, children systematically introduce cognitive changes and variations in their semantic representations. These changes allow the transformation and construction of experiences into categories, achieved by continuously delineating each category's characteristics and its boundaries, thus creating a conceptual structure for each representation (Borodistky, 2007).

The processes of comparison are mainly based on similarity, a basic component of cognition. As Markman and Gentner (1996) point out, humans tend to categorize objects based mainly on similarities. Similar objects are grouped into the same categories and serve to support inductive inferences between each other. These categories also form a structure to further acquire and learn new information.

The development of categorization transforms experience into well-defined categories by constructing conceptual structures that goes beyond concrete experience (Boroditsky 2007).

When young children have to include objects within categories they prefer to use perceptual features (Gelaes and Thibaut, 2006). Children continuously compare objects, and integrate them into categories formed by structurally related items. Throughout children's development, perceptual comparison leads to the continuously dynamic identification of similarities and differences, but most importantly to conceptual category development that progressively evolves from concrete to abstract features (Gelaes and Thibaut, 2006).

It has already been found that children are able to know and differentiate a significant number of complex categories showed by adults. However, this can only happen when these categories are shown by the adult, and not necessarily when children have generate them spontaneously (Gentner and Namy, 2000).

Frontal Lobes and Categorization

The capacity to generate abstract classification criteria and to maintain an attitude for an adequate abstract thinking is a very important property of the frontal lobes (Lezak, 2004; Luria, 1986). Normal adults (with average education) tend to analyze information in an abstract form, even if instructed not to do so (Noppeney et al., 2005). This phenomenon has been named *abstract attitude* (Lezak, 2004).

It is a common characteristic that damage to the frontal lobes frequently alters the capacity to identify abstract classification criteria in both spontaneous and active form (Delis et al., 1992; Kertesz, 1994; Luria, 1986). Often patients with frontal lobe damage don't lose the ability for abstract processing, but prefer to analyze objects, events and situations in a concrete form. This phenomenon has been named *concretism* (Lezak, 2004). A severe bilateral frontal lobe damage affects the ability for abstract processing in a significant way (Kertsez, 1994; Luria, 1989).

Recent neuroimaging studies have found that during abstract categorization in adults, the prefrontal cortex and mainly the anterior prefrontal cortex (APFC) are significantly activated (Green et al., 2006). This can be seen particularly during visual object categorization that requires a dynamic scan of elements and an active features comparison, where the right prefrontal cortex, mainly areas 10 and 47, are the most active (Bright, Most and Tyler, 2004; Pernet et al., 2004; Reber et al., 1998). In contrast, the comparison and retrieval of concrete and perceptual object features activates mainly the posterior cortex (occipito-temporal cortex) (Rossion et al., 2000).

Within neuropsychology, there is a conceptual division between semantic representations and their active utilization, where semantic representation activates posterior-cortical areas (Rossion et al., 2000), while the active processing occurs mainly in the prefrontal cortex (Delis et al., 1992). This dissociation represents the neural basis for the division between knowledge and its adequate use, named *know/do* (Bright et al., 2004; Noppeney et al., 2005 Peranni et al., 1999).

As a result of diverse neuroimaging studies, a functional division between the anterior prefrontal cortex (APFC) and the dorsolateral prefrontal cortex (DLPFC) has been found, in which the latter region participates during executive categorization: updating specific information and directs element feature comparison (Rossion et al., 2000). In contrast, the APFC participates during semantic comparison (Cools, Clark and Robbins, 2004; Kroger et al., 2002; Noppeney et al., 2002; Ricci et al., 1999).

Classification tests regularly used in neuropsychological evaluations depend on the capacity to generate abstract classification criteria, and normal performance in these tests requires prefrontal cortex participation

In neuropsychological testing as well as in most developmental studies, evaluators or researchers ask the child to intentionally produce a specific category or they provide a prompt-cue to influence the subject to generate this category, however in this testing situation the examiner is given the structure thus is acting as the "child's frontal lobes". Here, the question of whether the child possesses the conceptual level of knowledge for the category or if he can recognize it is answered, however, the question regarding what type of category the child chooses spontaneously can't be answered by these evaluation procedures. Therefore, the main objective was to study the developmental characteristics for the capacity of self-regulated category generation in subjects between 6 and 30 years old. Specific questions were 1) How do children spontaneously use their semantic knowledge in a production category task? 2) At what moment during development is present the "abstract attitude present" and finally 3) at what age does abstract category generation reach maximum performance?

By adopting a cognitive neuropsychological perspective, we tried to answer each one of these different questions by using a particular evaluation form that is sensitive to different brain network involvement during development, mainly anterior vs. posterior network divisions and their functional dissociations.

Method

A transversal and comparative study was performed, using the self-regulated category generation test, extracted from the *Executive Functions Battery* (Flores Lázaro and Ostrosky-Solis, 2005). This battery contains several tests that evaluate diverse executive functions, which depend on different prefrontal cortical areas. The sample included 200 normal subjects between 6 to 30 years of age. All subjects were free from neurological, neuropsychological or psychiatric background. All had average school performance. The sample was divided into five age groups: 6-8, 9-11,12-14, 15-17 and 18-30 years old.

Material: The instrument is a category-generation test based on one A4 size foil including 25 animal figures extracted from Snodgrass and Vandertwart´s (1980). Each of these animals can be classified into several semantic categories, such as mammals, aquatic animals, etc. It also gives the opportunity to generate other types of classification, named *concrete*: size, shape, feathers, and other physical attributes; *functional:* they fly; they are fast, etc. (see figure 1).

Procedure: subjects were given the following instruction: you must generate as many categories as you can, in a maximum time of five minutes. For children between ages 6-8, instructions were: you must make as many animals groups as you can, I'll give you five minutes to do so. Test application was individual, and it was part of a larger research on the development of executive functions.

Statistical Analysis: Results were registered in a database with SPSS 11. Statistical analysis included one way ANOVA. Contrast analysis was made considering the age range as the independent variable and curvilinear estimation was made considering year by year as a lineal as the independent variable.

Figure 1. Examples of animals included in the test.

Results

Results are divided by age ranges and presented by category type. Table 1 shows the category type percentages (in round numbers). It can be noticed (outlined in bold) that *concrete* category type is the preferred classification criteria for children between 6-8 years of age; functional characteristics are the preferred classification criteria for 9-11 year old children, abstract criteria is the preferred classification criteria for the 12-14 year old subject group, in the 15-17 and 18-30 year old groups, abstract categories represent the dominant semantic criteria for category generation. Category type averages are shown in table 2. It can be seen that abstract category average increases as a function of age, whereas concrete category average decreases, although it never completely disappears (not even in university students). Number of functional categories increases between ages 6-8 and 9-11, but later decreases.

Table 1. Category type percentage according to age range groups

AGE RANGE GROUPS	CONCRETE	FUNCTIONAL	ABSTRACT
1 (6-8)	58 %	34 %	8 %
2 (9-11)	20 %	47 %	33 %
3 (12-14)	11 %	26 %	63 %
4 (15-17)	10 %	19 %	71 %
5 (18-30)	8 %	16 %	76 %

Table 2. Category type average produced by age range groups

AGE RANGE GROUPS	CONCRETE	FUNCTIONAL	ABSTRACT
1 (6-8)	3.84	2.23	0.61
2 (9-11)	1.38	3.18	2.22
3 (12-14)	0.67	1.76	4.24
4 (15-17)	0.73	1.33	5.03
5 (18-30)	0.63	1.33	6.36

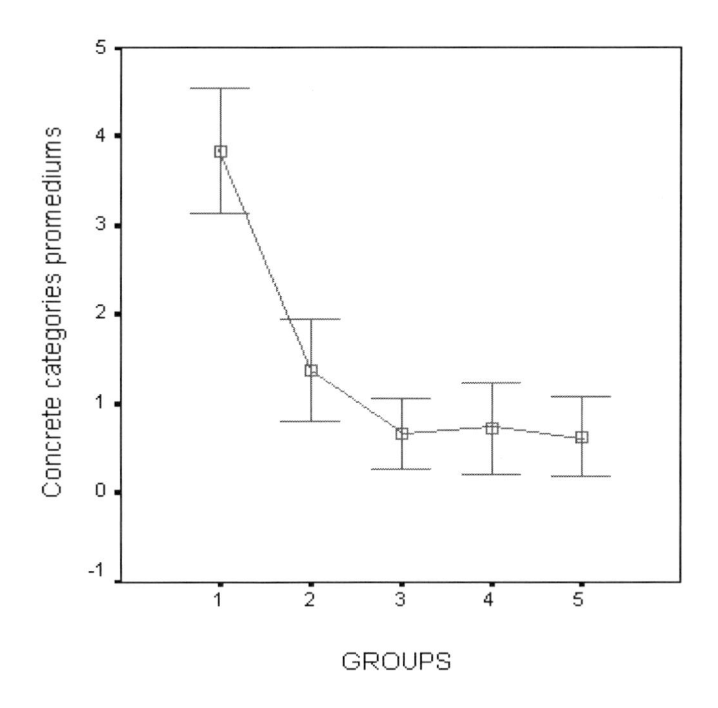

Figure 2. Average of concrete categories by age range . Contrast analysis values present a significant inverse relationship (F= 75.587 p= .000). Curvilinear analysis detects a main inverse lineal effect (R^2= .407, F= 117.92 p=. 000) Age range of group 1=(6-8), 2=(9-11) 3= (12-14), 4=(15-17).

Figure 2 presents a graphical representation of the data, where it can be observed that concrete categories fall dramatically between 6-8 and 9-11, presenting statistical significance among these age ranges (table 3). After this, decrement is very slow and it is not statistically significant among any group, from the 9-11 age range and later age groups. Contrast analysis reveals a main linear relationship, while linear analysis reveals an inverse effect (see figure 2).

Table 3. Statistical differences between averages in number of Concrete categories, produced by age range

Age Range Groups	9-11	12-14	15-17	18-30
1 (6-8)	2.46 (p= .000)	3.17 (p= .000)	3.11 (p= .000)	3.21 (p= .000)
2 (9-11)		0.71 (p= .239)	0.65 (p= .430)	0.75 (p= .240)
3 (12-14)			- 0.06 (p= 1.000)	0.03 (p=1 .000)
4 (15-17)				0.09 (p=. 999)

One-way ANOVA between groups showing mean differences and statistical significance (F(4,181)= 27.739 p= .000).

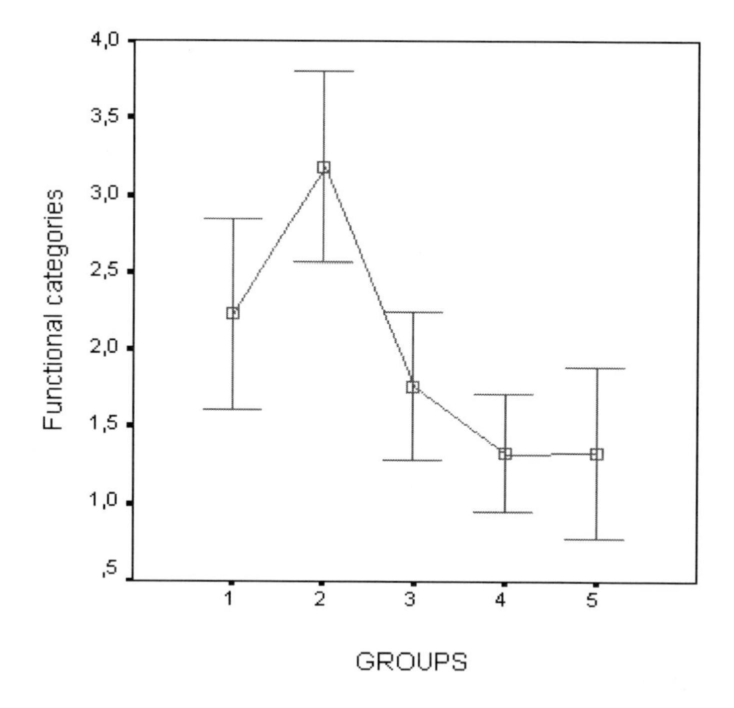

Figure 3. Average of functional categories by age range. Contrast analysis presents a significant lineal relationship (F= 15.833 p= .000). Curvilinear analysis detects a main discreet lineal effect (F=14.34 p=. 000); although quadratic model was significant, its value was inferior to lineal (F= 8.18 p= .000). Age range of group 1=(6-8), 2=(9-11) 3= (12-14), 4=(15-17).

The generation of functional categories presents the most interesting behavior during development (see figure 3), increasing between 6-8 and 9-11 years of age and decrementing through the 12-14 age range (see figure 2). Statistical significance is presented between 9-11 year old group and the subsequent groups. No statistical differences are presented between 6-8 and 9-11 year old group. Abstract category production presents a linear type of behavior during development. All groups were statistically different except group 12-14 and 15-17 (see table 5 and figure 4).

Table 4. Statistical differences between averages in number of Functional categories, produced by age range

Age Range Groups	9-11	12-14	15-17	18-30
1 (6-8)	-0.96 (p= .187)	0.47 (p= .752)	0.89 (p= .106)	0.89 (p= .199)
2 (9-11)		1.42 (p= .004)	1.85 (p= .000)	1.86 (p= .000)
3 (12-14)			0.43 (p= .622)	0.43 (p= .760)
4 (15-17)				0.00 (p= 1.000)

One-way ANOVA between groups showing mean differences and statistical significance (F(4,182)= 7.677 p= .000).

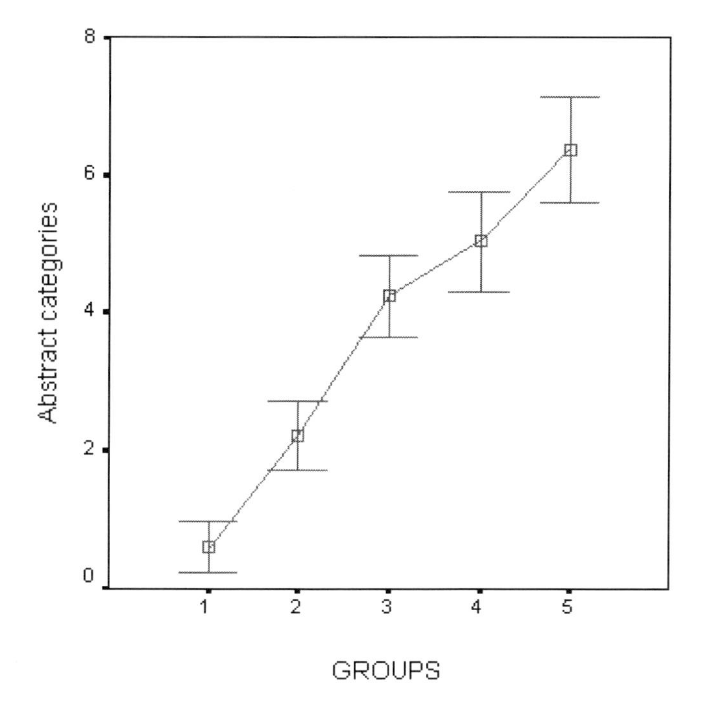

Figure 4. Average of abstract categories by age range . Contrast analysis values present a significant lineal relationship (F=238.946 p= .000). Curvilinear analysis detects a main inverse lineal effect (R2= .592, F=271.31 p=. 000). Age range of group 1=(6-8), 2=(9-11) 3= (12-14), 4=(15-17).

Considering that although some category types are the main preferred category, a second category is also present, shaping cognitive development in a more complex form. In table 6 it can be seen that in the 6-8 age group, concrete categories represent the dominant cognitive approach, accompanied by the functional features. An inverse relation occurs in the next age range (9-11). In the 12-14 age range, abstract classification is the dominant cognitive approach accompanied by the functional criteria. The 15-17 age range it can be noticed that abstract classification is the dominant cognitive criteria for spontaneous semantic classification (see table 6).

Table 5. Statistical differences in average number of Abstract categories, produced by age range

Age Range Groups	9-11	Comparison 12-14	15-17	18-30
1 (6-8)	-1.61 (p= .000)	- 3.63 (p= .000)	- 4.42 (p= .000)	- 5.75 (p= .000)
2 (9-11)		- 2.02 (p= .000)	- 2.81 (p= .000)	- 4.14 (p= .000)
3 (12-14)			0.79 (p= .437)	- 2.12 (p= .000)
4 (15-17)				-1.33 (p= .089)

One-way ANOVA between groups showing mean differences and statistical significance ($F_{(4,184)}$= 60.973 p= .000).

Table 6. Category combination reflecting spontaneous cognitive approach to classification

Age Range Groups	Category's Combination	Percentage relationship
1 (6-8)	Concrete-functional	58-34 %
2 (9-11)	Functional-abstract	47-33 %
3 (12-14)	Abstract-Functional	63-26 %
4 (15-17)	Abstract	71 %
5 (18-30)	Abstract	77 %

Discussion

If we consider that concrete classification criteria depend mainly on posterior cortical networks (or the lack of involvement of APFC) (Noppeney et al., 2002), that functional classification criteria depend mainly on prefrontal cortex -fronto-parietal network- (Piatt et al., 1999; Wood et al., 2004) and that abstract classification criteria depend mainly on APFC (Cools, Clark and Robbins, 2004; Reber et al., 1998), then results show different prefrontal cortex development stages and their influence on cognitive development processes, like categorization.

During the 6-8 age range, concrete features are the preferred criteria for spontaneous classification of objects, indicating that until these ages, children do not present a sufficiently active (or dominant) APFC-network. For this reason, they prefer to spontaneously categorize objects by concrete features (posterior cortical mode of processing). Recent neuroimaging studies have found that the anterior temporal lobe participates in the perceptual analysis of concrete objects' features (Noppeney et al., 2002). The combination of classification criteria reflects that DLPFC is already active and also present during spontaneous classification in this age range. A concrete-functional form of processing indicates a dominant posterior network, combined with an emergent and secondary DLPFC network (fronto-parietal) participation.

In the 9-11 age range, functional features are the dominant criteria for spontaneous classification; this may reflect a very important shift from posterior cortical to prefrontal processing. Action naming is closely related to dorsolateral prefrontal cortex functioning in adults and children (Holland et al., 2001; Piatt et al., 1999; Weiss et al., 2003; Wood et al.,

2004). These results show that from ages 9-11, the dominant networks for spontaneous classification depend mainly on DLPFC. Other studies have outlined the acceleration in cognitive performance in tests that depend on DLPFC during this same period, suggesting that these brain regions present the most important growth during late childhood (Diamond, 2002). A secondary abstract criterion is already present and shapes categorization during this phase. These results indicate that DLPFC (fronto-parietal) is the dominant network and that APFC begins to appear as a secondary cortical network during this age range.

During the 12-14 age range the abstract categorization criteria becomes dominant and the functional criteria becomes secondary, reflecting an interesting within-frontal network shifting stage from DLPFC to APFC networks. The last phase occurs from 15-17 years old in which a clear dominant abstract criteria for categorization is present, with a slow but linear increment in the next age range (18-30).

According to our results, four developmental stages have been determined during spontaneous classification: Stage 1 (6-8 years old): concrete-functional. Stage 2 (9-11years old): functional-abstract. Stage 3 (12-14 years old): abstract-functional, Stage 4 (15-17 and 18-30 years old): abstract.

Contrast and curvilinear analysis show different lineal age-effects for each categorization criteria: an inverse effect for concrete criteria, a lineal and discrete quadratic effect for functional criteria, and a clear lineal effect for abstract criteria. The cognitive resources (indeed prefrontal cortex networks) slowly but linearly increase spontaneous attention to abstract processing, while rapidly decreasing the concrete form. Spontaneously functional analysis presents an increment in a very specific age-range (9-11), to later decrease slowly but continuously.

Results suggest that during cognitive development, the prefrontal cortex provide the active use of abstract processing, transforming concrete and functional cognition into abstract spontaneous processing. The development of the prefrontal cortex restructures cognition from concrete to functional, and later on to abstract. Results also suggest a developmental functional dissociation within the prefrontal cortex for the 9-11 age range dorsolateral prefrontal cortex may be the particular dominant prefrontal cortex network, while anterior prefrontal cortex networks are already dominant at the beginning of adolescence, reaching almost exclusively cognitive dominance at the end of adolescence. Regarding these findings, we consider that abstract categorization is more closely related to AFPC than DLPFC, and is clearly present at the beginning of adolescence. Therefore, the operational phase described by Piaget seem to depend not on “the prefrontal cortex” as a whole but mainly on APFC development.

In conclusion, the ability to spontaneously categorize into semantic fields passes through several cognitive stages that in turn depend on developmental characteristics of the prefrontal cortex. Gradual transition occurs from concrete categorization in early childhood to functional criteria utilization in late childhood. At the beginning of adolescence, abstract features become the main preferred criteria for categorization, whereas inn later age ranges including young adulthood, abstract criteria (indeed abstract attitude) develops further and becomes the more dominant cognitive analysis mode. A posterior to frontal cortical network shift occurs during childhood. During adolescence a shift occurs from DLPFC (functional categorization criteria) to APFC (abstract categorization criteria) within the prefrontal cortex. Results indicate how prefrontal cortex development shapes cognitive development from childhood to youthood.

References

Borodistky, L. (2007). Comparison and the development of knowledge. *Cognition,* 102, 118-128.

Bright, P., Most, H., and Tyler, L.K. (2004). Unitary vs multiple semantics: PET studies of word and picture processing. *Brain and Language,* 89, 17-32.

Cools, R., Clark L., and Robbins, T.W. (2004). Differential responses in human striatum and prefrontal cortex to changes in object and rule relevance. *Journal of Neuroscience*, 4, 1129-35.

Delis, D.C., Squire, L.R., Bihrle, A., and Massman, P. (1992). Componential analysis of problem-solving ability: performance of patients with frontal lobe damage and amnesic patients on a new sorting test. *Neuropsychologia*, 30, 683-97.

Diamond, A. (2002) Normal development of prefrontal cortex from birth to young adulthood, cognitive functions, anatony and biochemistry. In D. T. Stuss and R. T. Knight, (Eds.), *Principles of frontal lobes function* (pp. 466-503). New York: Oxford University Press.

Flores Lázaro, J. C., and Ostrosky-Solís, F. (2005). Batería de funciones ejecutivas. *Instituto nacional de derechos de Autor.* Registro: 03-2005-03141105400-01. México, D. F.

Gelaes, S., and Thibaut, J.P. (2006). The role of the structure of parts and of the overall object shape in children's generalization of novel object names. *Cognitive development,* 21, 369-376.

Gentner, D., and Namy, L.L. (2000). Comparison in the development of categories. *Cognitive development*, 14,487-513.

Green, A.E., Fugelsang, J.A., Kraemer, D.J., Shamosh, N.A., Dunbar, K.N. (2006). Frontopolar cortex mediates abstract integration in analogy. *Brain Research*. 22,125-37.

Holland, S.K., Plante, E,, Byars, W.B. (2001). Normal fMRI brain activation patterns in children performing a verb generation task. *Neuroimage*, 14 837-843.

Kertesz, A. (1994). *Localization and Neuroimaging in Neuropsychology*. San Diego: Academic Press.

Kroger, J.K., Sabb, F.W., Fales, C.L., Bookheimer, S.Y., Cohen, S., and Holyoak, K.J. (2002). Recruitment of anterior dorsolateral prefrontal cortex in human reasoning: a parametric study of relational complexity.*Cerebral Cortex*. 12 (5), 477-85.

Lezak, M.D. (2004). *Neuropsychological Evaluation*. 4th Edition. New York: Oxford University Press.

Luria, A.R. (1986). *Las Funciones Corticales Superiores del Hombre*. México: Fontamara.

Luria, A.R. (1989). *El Cerebro en Acción*. España: Roca.

Markman, A.B. and Gentner, D., (1996) Commonalities and differences in similarity comparisons. *Memory and Cognition*. 24, 235-49.

Noppeney, U., Price, C.J., Penny, W.D., and Friston, K.J. (2005). Two Distinct Neural Mechanisms for Category-selective Responses. *Cerebral Cortex*. 16(3), 437-45.

Perani, D., Schnur, T., Tettamanti, M., Gorno-Tempini, M., Cappa, S.F., and Fazio, F. (1999). Word and picture matching: a PET study of semantic category effects. *Neuropsychologia*. 37(3), 293-306.

Pernet, C,, Franceries, X,, Basan, S., Cassol, E., Demonet, J.F., and Celsis, P. (2004). Anatomy and time course of discrimination and categorization processes in vision: an fMRI study. *Neuroimage,* 22 (4), 1563-77.

Piatt, A., Fields, J., Paolo, A.M., Troster, A.I. (1999). Action (verb naming) fluency as an executive function measure: convergent and divergent evidence of validity. *Neuropsychologia*, 37, 1499-1503.

Reber, P.J., Stark, C.E., and Squire, L.R. (1998). Cortical areas supporting category learning identified using functional MRI. *Proceedings of the National Academy of Science U. S. A*. 20 (2), 747-50.

Ricci, P.T., Zelkowicz, B.J., Nebes, R.D., Meltzer, C.C., Mintun, M.A., Becker, J.T. (1999). Functional neuroanatomy of semantic memory: recognition of semantic associations. *Neuroimage*, 9, 88-96.

Rossion, B., Bodart, J.M., Pourtois, G., Thioux, M., Bol, A., Cosnard, G., Benoit, G., Michel, C., and De Volder, A. (2000). Functional imaging of visual semantic processing in the human brain. *Cortex*. 36(4), 579-91.

Snodgrass, J. G., and Vanderwart, M. (1980). A standardized set of 260 pictures: Norms for name agreement, familiarity and visual complexity. *Journal of Experimental Psychology: Human Learning and Memory*, 6, 174-215.

Weiss, E.M., Siedentopf, C., Hofer, A., and Deisenhammer, E.A. (2003). Brain activation patterns during a verbal fluency test in healthy male and female volunteers: a functional magnetic imaging study, *Neuroscience letters*, 352,191-194.

Wood, A.G., Harvey, A.S., Wellard, R.M., Abbott, D.F., Anderson, V., Kean, M., Saling, M.M., and Jackson, G.D. (2004). Language cortex activation in normal children. *Neurology*. 28, (6), 1035-44.

In: Encyclopedia of Neuroscience Research
Editors: Eileen J. Sampson and Donald R. Glevins
ISBN 978-1-61324-861-4

Chapter XI

Common Questions and Answers to Deep Brain Stimulation Surgery

Fernando Seijo[1]*, Marco Alvarez-Vega[1], Beatriz Lozano[2], Fernando Fernández-González[2], Elena Santamarta[3] and Antonio Saíz[3]

[1]Functional Neurosurgery Unit, Neurosurgery Department
Hospital Universitario Central deAsturias, Oviedo, Spain
[2]Clinical Neurophysiology Department. Hospital Universitario
Central deAsturias, Oviedo, Spain
[3]Radiology Department. Hospital Universitario Central deAsturias, Oviedo, Spain.

Introduction

When a new technique is introduced, a certain period of time is required for it to be improved and adjusted. The success of functional stereotactic procedures depends on a variety of factors: patient selection, methodology of choice and localization of the target, and the experience of the neurosurgery team. The learning curve resulting from the neurosurgery team's acquired knowledge is also notable.

After 12 years of working with deep brain stimulation (DBS), operating on over 200 patients and implanting more than 380 DBS electrodes for different illnesses, Parkinson´s disease, dystonia, tremor and Horton's headache, we believe that our experience may be useful to others who are starting out in functional stereotactic procedures.

* Servicio de Neurocirugía,Hospital Universitario Central Hospital de Asturias,33006, Oviedo, Spain. Tfno: 00-34-98510800- Ext. 38164 Fax: 00-34-985 273657,E-mail: fseijo@uniovi.es

Question One: Patient Selection

Many patients are referred to surgical centres to be considered for DBS. It is essential to establish criteria that will effectively predict the outcome of DBS before the therapy is applied. It is then necessary to use a rigorous selection process that will successfully identify candidates who will obtain the greatest benefit from this therapy and who will maintain this benefit for long enough to justify the time and expense involved. Candidates must also be in a suitable physical, cognitive and emotional condition to tolerate surgery [24].

The following protocol is used for DBS surgery candidates: a) a clinical diagnosis of idiopathic Parkinson´s disease (PD) with Core Assessment Program for Intracerebral Transplantations (CAPIT) [25], Unified PD Rating Scale (UPDRS, part III, version 3) [9], > 30% improvement after administration of levodopa or apomorphine, OFF phase Hoehn & Yahr stage 3 or above, severe motor fluctuations and/or dyskinesias secondary to medication; absence of severe disease that could interfere with the surgery or concomitant medication that is incompatible with surgery;, stabilised antiparkinsonian treatment 6 months before surgery; dystonia [8], tremor (essential tremor, rubral tremor, PD´s tremor) [16, 34], cluster headache [29]; b) less than 70 years of age; c) absence of dementia, psychosis, severe depression and drug dependence; d) brain magnetic resonance imaging within 6 months of surgery without severe atrophy or any anomalies that could prevent surgery; e) written informed consent.

Depending on the clinical diagnosis, patients underwent the following neurological, neurophysiological and psychological battery before surgery: 1) pharmacological tests and UPDRS scale in ON and OFF states, 2) chronometric tests consisting of right and left hand tapping and gait, 3) video recording, 4) electroencephalogram, 5) mapping of somatosensory, visual and cognitive evoked potentials, (P-300), 6) cognitive evaluation with the Mattis dementia scale, Hanoi tower, Rey oral auditive learning test, Benton visual retention test (VRT form 1), words and colours test (Stroop test), letters and numbers test (Wais-III), oral fluency test, Benton line orientation test, 7) behavioural test with the ISRA anxiety test and the Montgomery Asberg depression scale, 8) functional evaluation with the PDQ-39 Quality of Life questionnaire.

A patient with a third ventricle width of more than 9 mm is not a candidate for DBS surgery in our opinion because of the shift that occurs in the brain from the loss of cerebrospinal fluid and pneumoencephalus. Cortical atrophy is also a major handicap to surgery because of the risk of subdural haematoma.

Also, although it has been suggested in the literature that patients with pacemakers are not candidates for this type of surgery because of incompatibility with magnetic resonance imaging (MRI), we had a patient with a pacemaker due to bundle branch block, and we obtained the stereotactic coordinates with computed tomography (CT) imaging alone. The patient did not have any problems peri- or post-surgery when the brain neurostimulator was implanted, because of the distance from the pacemaker. Likewise, a history of myocardial infarction or diabetes increases morbidity/mortality but are not exclusion criteria for intervention. We have operated on three patients with a history of myocardial infarction as well as patients with previous controlled diabetes, with good results.

Exclusion criteria are: serious systemic disease, carotid stenosis, cerebral atrophy, dementia or major psychological disorder, lack of collaboration on the patient's part, or unsigned informed consent.

Question Two: There Are Different Stereotactic Devices with Different Properties, Depending on the Type of Surgery in Question. Which Device Is the Most Suitable for Deep Brain Stimulation Surgery?

Stereotactic surgery is a surgical method that provides access to any part of the brain with the utmost accuracy. The safety of these devices is ± 1 mm for the X and Y coordinates, and 1.5 mm for the Z coordinate [7, 12]. To achieve their objective, stereotactic devices are based on two different systems: the arch-centre system and the polar coordinate system.

The arch-centre system means that any given point of the brain expressed in Cartesian coordinates (X, Y, Z) can be reached from anywhere in the cranial area. In other words, there may be multiple entries on the skull for a particular site in the brain.

These coordinates are bound by three planes. First, a midsagittal plane to the brain passes through the septum pellucidum and commissure anterior-commissure posterior line (Y coordinate). Second, a plane that is perpendicular to the previous basal plane and passes through the upper edge of the external auditory canal-edge lower orbits (X coordinate). And third, another plane perpendicular to the midsagittal plane and the basal plane called the coronal plane (Z coordinate).

This system is used in stereotactic devices such as the Leksell apparatus (Elekta Instruments, Stockholm, Sweden) and the Cosman-Roberts-Wells stereotactic apparatus (CRW) (Radionics, Burlington, MA, USA).

The coordinate system or spherical polar point target is defined by a distance (R) and two angles (Δ, Ω) from a reference point known as 0. This is the system that is used in astronomy. R is the distance between point 0 and the target, the Δ angle is the angle on the sagittal plane, and the Ω angle is on the horizontal plane. This system enables any point in the brain to be reached through a previously selected burr-hole. The mathematical calculations are complex because of its operability-specifying trigonometric calculations. This system is used less than the arch-centre system.

An example of stereotaxy apparatus using this system is the Riechert-Mundinger apparatus. It has an accuracy of ± 0.5 mm [39].

During the first few years we used the Riechert-Mundinger apparatus, and we then went on to use the CRW-FN frame (CRW improved) and the G model Leksell frame. Despite the safety of using the polar coordinate methodology in the Riechert Mundiger-frame, its practical complexity led us to replace it with appliances that use the simpler arch-centre system.

The first of these devices we used was the CRW frame that can only be used with CT. When a CRW frame was used in MRI, unpredictable geometric shifts were observed. However, images obtained using a Leksell stereotactic frame was virtually free of these geometric shifts [15]. Then we used the CRW-FN frame and the Leksell G frame. These two models have the advantage that they can be used equally with CT and MRI alike. We also observed that there is greater distortion in MRI studies with the CRW FN-frame than with the Leksell frame. Using the Philips MRI scanner (Philips Medical Systems) and Siemens MRI scanner (Siemens Medical System) with the Leksell frame the mean values of the maximum errors for the X, Y, and Z axes were 0.9-1.0 mm, 0.2-0.4 mm, and 1.6-3.8 mm respectively [56]. In our experience using MRI (General Electric, 1.5 teslas) the largest distortion was found around the periphery (2-3 mm) and the least distortion was present in the middle (1-2 mm). A number of factors cause the spatial distortion, heterogeneities in the scanner's constant static magnetic field, nonlinearity of the gradient field, linear scale error, instrument imperfections, magnetic susceptibility artefacts, effects of local magnetization [42], and the magnetic properties of the stereotactic frame [5]. Magnetic susceptibility perturbations introduced by human head anatomy and by stereotactic frames can produce geometric shifts of up to several mm [5].

The head ring of the CRW-FN frame and Leksell frame have earplugs to help stabilise the frame and prevent significant tilt ("roll" in the coronal plane) or rotation ("yaw" in the axial plane) of the frame with respect to the brain [22], but they do not affect the anterior-posterior axis. To minimise the sagittal pitch angle, the axial plane is aligned with the anterior commissure-posterior commissure line (AC-PC line) by angling the base ring parallel to an imaginary line between the external auditory meatus (EAM) and the orbital floor.

We have avoided this problem by developing a frame for the CRW-FN that is fixed from the head ring on the CRW-FN and has two earplugs and a nose bar. With this mechanism, the CRW-FN´s head ring is fixed to the head by three points and it can therefore define an axial plane parallel to the anterior commissure-posterior commissure plane (AC-PC plane). In the case of Leksell frame we fixed the ring head to the head with three points, on a parallel to the AC-PC plane, by adding another earplugs at the front of the head ring, so that the three earplugs on the Leksell frame could set the ring head on a parallel to AC-PC plane (figure 1).

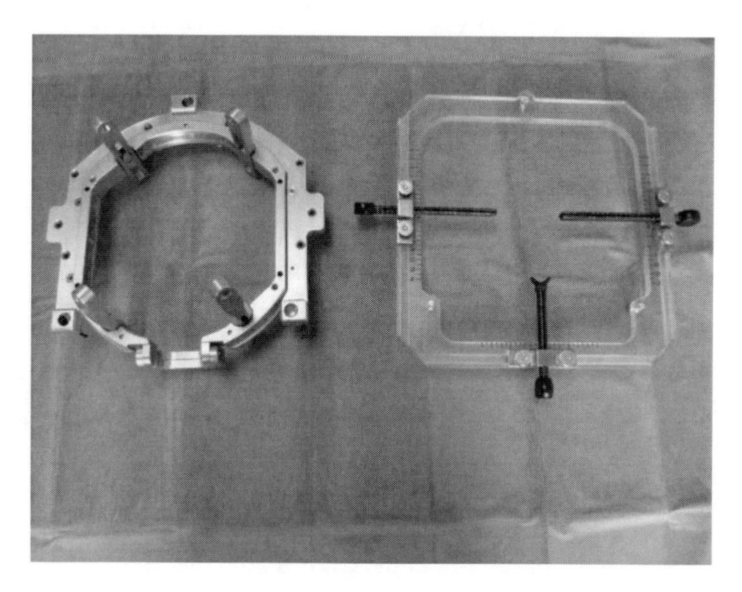

Figure 1. A) CRW head ring with adapter to earplugs and nose bars. B) Leksell head ring with earplugs.

Question Three: What Is the Best Method for Placing the Stereotactic Ring in the Correct Position?

Good identification of the AC-PC line in the CT or MRI requires correct frame placement. In theory the AC-PC line is parallel to a line connecting the EAM with the inferior orbital wall but not in all cases. Some authors believe that the AC-PC line draw an angle of -10º with the orbit-meatal line, which is the vertex angle of the line joining the EAM [57]. It is essential to place the frame with its axes orthogonally to the standard anatomic planes of the brain, to ensure that preoperative images as well as the intraoperative microelectrode-derived brain maps are interpretable in terms of familiar anatomy corresponding to standard brain atlases [48]. Then, the frame must be placed on the head on a parallel with the AC-PC plane.

The stereotactic frames have earplugs to help stabilise the frame and prevent any "roll" or "yaw" with respect to the brain, but they do not affect the anterior-posterior axis. It is also possible that the EAMs may not be symmetrical. To minimise the sagittal pitch angle, the axial plane is angled on the base ring parallel to an imaginary line between the EAM and the orbital floor. In theory, the plane of frame is parallel to the AC-PC plane.

In order to place the stereotactic ring as parallel as possible to the AC-PC plane we use a methodology that can be performed directly on the MRI console or MRI images. Days before surgery on a sagittal image MRI slide taken beforehand, we mark the EAM. Then, this point (EAM) is moved to midsagittal image MRI slide where it shows the AC-PC line. Once this point (EAM) is marked on the midsagittal image MRI slide where the AC-PC line is displayed, we draw a line that starts at the point of the EAM that is parallel to the AC-PC line and runs through the patient's nose. Then, we measure the distance between the nasion and the point where the parallel line cuts the patient's nose. The distance between the nasion-intersection of the line parallel to the AC-PC line and the patient's nose gives us the point on the nose where we put the head ring nose bar. A mark is done in this point.

Once this is done, on the morning of surgery, we fix the head ring to the patient's head so that the earplugs are fixed on the EAM, and the nose bar is fixed to the mark that has been made on the patient's nose. Thus, the head ring forms an axial parallel plane to the AC-PC of the patient's head and the frame is parallel to the axis of scanning (Figure 2).

Example (Figure 3). The first image in Figure 3 shows the EAM mark and determines its position on the sagittal MRI images display using the radiological scale. In the second image in Figure 3, having moved the point representing the EAM, we draw a parallel line to the AC-PC line. This line starts at the point where the EAM crosses the patient's nose. Then, we measure the distance from the nasion-parallel line intersection point to the AC-PC line on the patient's nose. This distance measures 18 mm. Then, we mark this point on the patient's nose and the head ring is placed on the patient's head. The earplugs are placed in the EAM and the nose bar on the mark of the nose. Thus the head ring is parallel to the AC-PC plane. In this patient the AC was lower than PC (Figure 4).

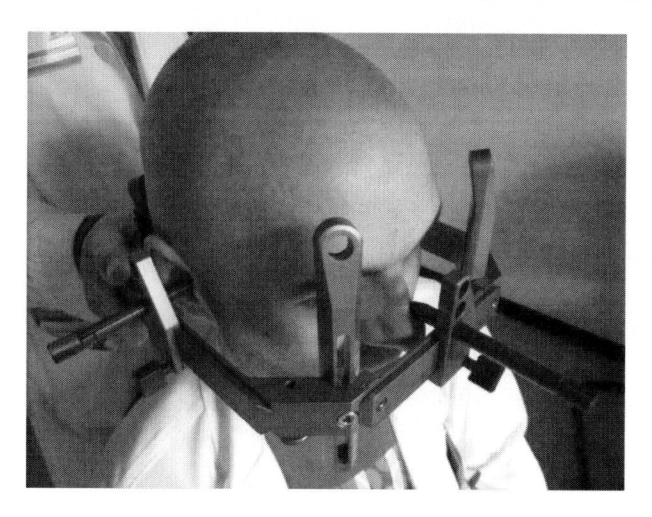

Figure 2. A) CRW head ring with adapter to earplugs and nose bar. B) Leksell head ring with earplugs. The head ring forms a parallel plane to the anterior commissure-posterior commissure plane.

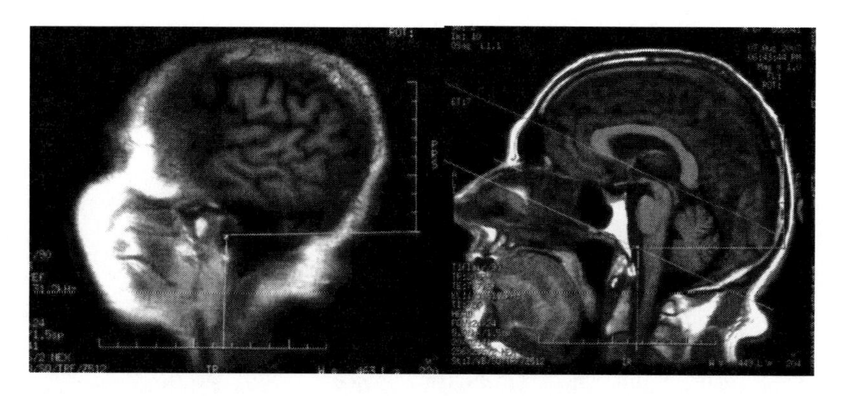

Figure 3. A) Position of the external auditory canal. B) Line parallel to anterior commisure-posterior commisure line and its intersection with the nose.

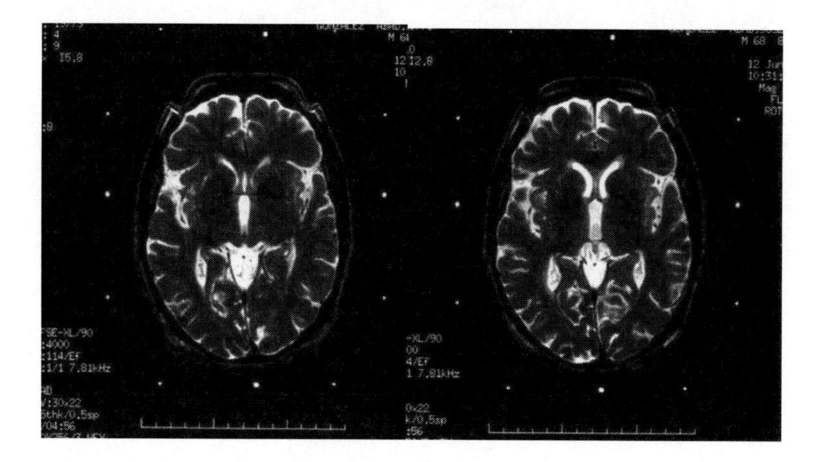

Figure 4. A) The first slide shows the anterior commissure alone. B) The second slide shows the posterior commissure. The two slides are separated by 3 mm.

Therefore, the angle formed between the head ring and AC-PC plane was 2.5°. During surgery, the subthalamus nucleus was identified in the first track. The neurophysiological identification was observed on the AC-PC plane rather than 2 mm below the level of the AC-PC line which is more commonly reported [2].

Having performed this method in over 180 patients with PD, we have obtained a half deviation angle between the AC-PC plane and the ring plane of 1° ± 3°. Others authors have reported a mean pitch angle of 7.5° [58].

Question Four: What Is the Best Radiological Procedure for Fixing the Target, CT, MRI or a Fusion of Images? How Are the Coordinates Calculated? Is There a Specific Computer Program or a Direct Method?

Accurate target localisation is of prime importance in functional stereotactic neurosurgery. Classically, localization of subcortical nuclei (targets) has been performed indirectly, based on fixed distance from an internal reference that can be visualized on a stereotactically-acquired brain imaging study. The internal reference used to obtain the stereotactic coordinates of target is the AC-PC line.

Normally the targets are obtained from different stereotactic atlases. Human brain atlases are used to measure the distance in each dimension [47]. The atlases have been written from the anatomical examination of a limited number of post-mortem brains. They are of limited statistical value and artefacts due to fixation and sections are difficult to assess and correct. Adaptations have to be made to an individual patient's dimensions.

In the Schaltenbrand and Wahren atlas the AC-PC line is 23 mm [43], in Tailarach and Tournoux atlas it is 24 mm [49], by Van Buren and Maccubbin it is 25 mm [54], by Riechert it is 20.9-27.3 mm [40] and by Taveras it is 23-28 mm [51]. Schaltenbrand et al [43] and Talairach et al [49] are the most human atlases used. However, if we use this method, we make an intrinsic error of calculation because the length of the AC-PC line differs between post-mortem brains (atlas) and brains of living persons (patients) [43, 49]. On the other hand, in the Schaltelbrand et al atlas [43], the AC-PC line is measured between the upper border of AC (length 2.5 mm) [55] and the lower border of PC. CT and MRI images of the AC-PC line are measured between the posterior edge of the AC and the anterior border of the PC. To standardise measurements, the length of the AC-PC line is established as 24 mm.

Finally, another error factor is the wide diversity of coordinates for the same target [23].

The aim of radiological studios is able to calculate coordinates of stereotactic frame from the AC-PC line. First of all, the stereotactic frame coordinates of the patient's reference AC-PC line must be determined. The X axis is defined as a perpendicular line to the AC-PC line on an axial plane, the Y axis is defined as a line crossing the midpoint of the AC-PC line in the anteroposterior direction on an axial plane, and the Z axis is defined as a perpendicular line to the AC-PC line on a midsagittal plane. Then, the target coordinates that have been determined in the patient´s AC-PC line must be transcribed into the frame stereotactic coordinate system.

To display the AC-PC line, ventriculography [5] was initially used, then CT [26] became more popular and recently MRI is used [36]. Although ventriculography is still used at several medical centres and is considered the gold standard for localisation in stereotactic coordinates calculation [3], it has since been replaced by CT and MRI. The advantages of ventriculography are that it provides accurate and rapid localization of the commissures and is free of image distortion. Its disadvantages are: there can be inaccurate imaging if the geometrical radiological set-up is not perfect, leading to potential morbidity [44]. Finally, there is a displacement of structural fiducial markers and an enlarged width of the third ventricle [14].

At present, CT and MRI are the most commonly-used methods in functional stereotactic neurosurgery, but there is no universal consensus as to which is the most advantageous.

CT provides spatially accurate imaging of the AC and PC, but does not provide the necessary resolution of anatomic detail for direct observation of the target, and final target coordinates must be derived from atlas predictions based on the AC-PC line. On the other hand, MRI can provide neuroanatomic detail and could be used to directly target desired structures, but MRI images are subject to spatial distortion [4, 7, 15]. Moreover, the average difference in level between the axial CT and MRI is about 3.29 mm, 3.52 mm in the coronal plane and 4.28 mm in the sagittal plane [12, 19]. The maximum error distance between CT and MRI is between 1.2 mm in the mediolateral dimension to 2 mm in the anteroposterior dimension [58].

On the other hand, there are significant differences in coordinates obtained by MRI performed concomitantly with ventriculography (anteroposterior coordinates, mean error: 2.6 mm, range 0.5-6.8 mm) [33]. Therefore, the use of MRI alone for the determination of stereotactic targets has been questioned.

To summarise the advantages of ventriculography are good AC-PC display, without distortion. Its disadvantages are that it is an invasive method, involves radiation exposure, does not show the surrounding structures of AC-PC and it distorts the shape and size of the third ventricle.

The advantages of CT are that it provides a good AC-PC display, does not produce distortion, does not distort the third ventricle, and it is not an invasive method. Its disadvantages are that it involves radiation exposure, and does not show the surrounding structures of AC-PC.

The advantages of MRI are that is provides a good AC-PC display, it is non-invasive, does not involve radiation, and it displays the surrounding structures of AC-PC well. Its disadvantages MRI images are subject to spatial distortion

In order to eliminate the disadvantages of these studies and only retain their advantages, software has been designed to merge the images of CT and MRI [27, 41]. These programs have two disadvantages which lie in the economic cost and that the fact that MRI images are not in real time.

The philosophy of our surgical method is to calculate stereotactic coordinates in real time. An MRI and CT are performed on the day of the operation to fix the stereotactic frame to the patient's head. Then, we calculate the stereotactic coordinates on CT and MRI. The great advantage of this method is that the stereotactic coordinates in the CT and MRI are calculated in real time and can be made on the same slide.

On the other hand, with the aim of avoiding this error several authors recommend direct localization of target [6, 18]. This method seems the most logical and correct, but, it also entails mistakes. The targets are difficult to identify in MRI and it has the problem of distortion [7, 28, 47].

In the actuality, we believe that the current method which provides a more operational, fast and safe way to calculate the stereotactic coordinates of different targets is the indirect method based on real time images obtained with CT and MRI.

Then we choose MRI scanning and CT as the routine imaging procedures to guide the implanted DBS electrode. We compare the two imaging methods used frequently in stereotactic neurosurgery for anatomical localization of the target: CT and MRI. Once the patient is selected for DBS procedure we perform a standard MRI to disclose the presence of anatomical variations or unexpected pathology. On the MRI monitor and using the middle sagittal T1 MRI as a reference, we find the AC-PC line and translate it to the EAM level. We project this line to the nasal bridge. In this way, (so) we obtain the superficial nasal reference and calculate the nasion-nasal point distance (about 2 cm). In this plane, the neurosurgeon will fit the patient with the stereotactic frame.

Under local anaesthesia, this device will be temporarily affixed to the skull by four shallow screws so that the patient's head will not move during diagnostic imaging, or the actual DBS procedure. Following placement of the stereotactic frame, we performed a CT scan (Toshiba Aquilion 16) on all our patients prior to the MRI. After fixing the frame onto the CT table serial CT scans were obtained at a 1 mm interval and a scan thickness of 2 mm, with the CT gantry parallel to the stereotactic frame. We obtained an AC-PC plane where the AC and PC were identified.

After CT scan, four MR image sets were obtained using a GE MR 1.5 T. A stereotactic T2 magnetic resonance sequence was performed on the three spatial planes (first sagittal, then coronal and finally the axial plane) with the following parameters: (TR/TE) repetition time msec/echo time msec, 4000/113.8; number of sections acquired 15; section thickness, 2.5 mm; spacing 0.5; NEX 2; matrix, 320 x 256; field of view, 30x30; and receive bandwidth, 7.8 KHz. In addition to the images above, imaging with a three-dimensional data set for the purpose of anatomic localization was performed using 3-dimensional T-1-weighted 2.6 mm thick gradient-echo images to obtain 110 contiguous axial slices (MPRAGE) 3D/FGR (magnetization-prepared rapid acquisition gradient-echo transverse acquisition). Parameters were as follows: 11.1/4.2/500 msec TR/TE/TI; flip angle 15°; number of sections acquired 112; spacing 1.3; matrix 256x192; field of view, 28x21; and receive bandwidth, 15.6 KHz. This 3D sequence optimizes the surgical approach, avoiding traversing sulci, cortical veins, dural venous lakes and lateral ventricles. Images were reconstructed in a 3-D GE-workstation (Advantage Windows, Version 4.0, software (GEMS, Milwaukee, WI. USA).

To account for a potential shift in the AC-PC coordinates and the ICP of the patient's head with respect to the stereotactic frame, we calculated the following measurements on the 2D T2-weighted images. In the axial AC-PC plane on T2-WI, we measured AC-PC distance and ICP. We also measured the width of the third ventricle. To find the geometrical centre of the stereotactic frame, we intersected lines from peripheral reference points on the stereotactic frame, thus, fixing the coordinate axis (X, Y). To calculate Z axis of ICP we used the measurement of the lateral oblique fiducial markers on the stereotactic frame as a reference. After this, we compared the patient coordinates to the real centre of the stereotactic frame *(Scanner localization guidelines with Leksell stereotactic frame. Elekta Instrument. Stockholm, Sweden)*. We then verified the AC-PC distance on the reformatted sagittal plane of the volumetric MPRAGE images. We calculate α and ß angles on the coronal and sagittal reformatted planes, respectively (Figure 5).

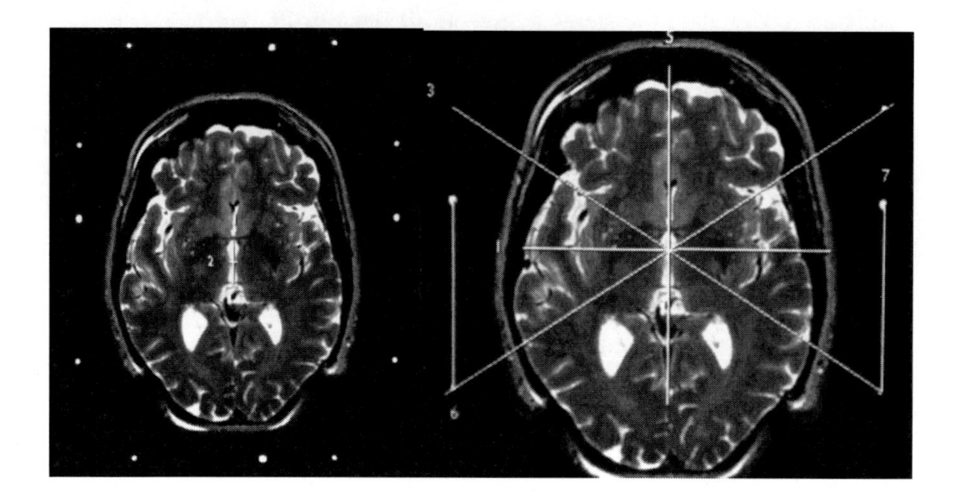

Figure 5. Axial AC-PC plane on T2 magnetic resonance sequence. A) The first image shows the width of the third ventricle. B) The second image shows target localization with Leksell stereotactic frame.

Question Five: Which Is the Best Anesthesia for Functional Neurosurgery?

DBS is a long operation that usually requires the collaboration of the patient. Moreover, the patient's personality, the uncomfortable position on the operating table and the anxiety it produces throughout the long surgical process leads to discomfort and less collaboration with the patient's medical team. Therefore, anaesthesia in functional neurosurgery must meet three strategic objectives, to keep the patient awake, cooperative and comfortable (no pain, no anxiety, etc.). Thus, the surgeon can assess the patient throughout the operation as well as his/her response to neurostimulation. In exceptional cases, such as generalized dystonia, general anaesthesia will be essential.

We have operated on more than 200 patients and only 4 patients needed general anaesthesia; three with generalized primary dystonia and one 51 year-old patient with PD who had painful dystonia that prevented CT and MRI-imaging from being performed. Generally, we use the following anaesthesia protocol: Antiparkinsonian drugs are discontinued 12 hours before surgery.The patient is placed in a semi-seated position to minimize cerebrospinal fluid loss and facilitate neurological examination. Monitoring included electrocardiography, pulse oximetry, arterial pressure (non invasive), end-tidal CO_2, urinary catheter and entropy (it informs us of the degree of sedation of the patient). Control of fluids and electrolyte imbalances. Analgesics that do not interfere with coagulation (paracetamol).

Opioid analgesics (fentanyl) administered in IV bolus. We do not use remifentanyl because of the potential side effect of respiratory depression. This side effect is difficult to resolve by the patient's position and his head placement and fixation in the stereotactic frame.

Benzodiazepines (potassium clorazepate) are administered intravenously. It keeps the patient sedated but able to give verbal responses.

Hypnotic sedatives derived from benzodiazepines (midazolam) administered in IV bolus. The potent effect of anxiolytics enables us to keep the patient asleep at certain times.

The long duration of the surgery and fixing the head is difficult to endure for some patients.

We try not to use propofol because it interfered with parkinsonian symptoms even when used in low amounts. It also elicited abnormal movements in individual patients [21]; the most adverse event is its interference with neurophysiological monitoring (personal experience).

Once the surgery is finished, a CT control is done and the patient is transferred to the intensive care unit where he is hospitalized 24 hours.

Question Six: Neurophysiological Recordings in the Surgical Procedure. What Are the Advantages and Disadvantages of the Microelectrode and the Semi-Microelectrode?

In any functional neurosurgery scenario, the neurophysiological targeting is still controversial and is identified as a luxury, research or a risk. In our experience, stereotactic imaging is used only to define the general location of the target then intra operative neurophysiological monitoring (IONM), to avoid neurologic impairment, is currently used for surgical decision-making. The average distance error between the final physiological targets and the magnetic resonance imaging-derived targets is 2.6 ± 1.3 mm [46] and IONM improved the initial accuracy from 64% to 100% [53]. The identification, localization and preservation of anatomically ambiguous nervous tissue and the continuous "on-line" assessment of functional integrity of neural pathways are the IONM's neurological primary objectives.

In this context, IONM is a cooperative process between anaesthetist, neurophysiologist and neurosurgeon in the operating theatre. The classical reference to Functional Neurosurgery as Applied Neurophysiology is representative of the neurophysiologist-neurosurgeon teamwork outside and/or inside the operating room. This cooperative expertise optimises the final decision. Moreover, today, surgical processes require candid communication, checking time-tables and quality assurance of the implemented procedures. Hence, the accurate placement of the electrode in specific brain nuclei for ablative procedures or DBS require IONM to identify, to register and to document each trajectory's incidences and to correlate the trajectories with the relevant post-operative neurophysiological assessment. Furthermore, the physical dilemma of accuracy versus sharpness is present during target neuroimaging-based planning. The reconstruction from MRI and CT needed to measure the coordinates of the direct or indirect theoretical target, would be regarded as a technical approach to the localization of the ambiguous tissue. In this context, the role of IONM is related to the improvement of the detected planning inaccuracies since they are in relationship to millimeters.

To summarize, neuroimaging techniques for stereotactic planning, neurophysiological mapping to confirm the target and the exploration of the target's adjacent structures before placement of a DBS lead are essential to improve the accuracy of the process and to reduce side effects. This is the reason we think the recording techniques are mandatory in order to accurately localize the optimal target site.

From Albe-Fessard and Pendfield, as early as 1920-1970 [1, 37], a set of neurophysiological procedures, from microelectrode recording to evoked potentials, have been used for localizing cortical and subcortical targets for stereotactic brain surgery. Years ago, to cover these technical procedures, a set of sophisticated laboratory equipment was required in the operating room, but at the present time, standard IOM equipment with any facility can be used. In our opinion, requirements such as easy application and easy interpretation, unimpaired and unimpeded surgery, uninterrupted and continuous monitoring, no risk and low rate of failure, affordable costs, high sensitivity and specificity should be kept in mind when doing IONM planning [10].

The neurophysiological methodologies used in the operating theatre require trained and skilled personnel, not only to collect but also to interpret the data and to cooperate with the neurosurgeon in decision-making. Different IONM protocols for subcortical targets are available in the medical literature.

The present opinion is based on our monitoring, assessment and research experience at the Hospital Universitario Central de Asturias, Spain, from 1996 to July 2008, with more than 400 hemispheric procedures. We implemented IONM's protocol based on standard IOM criteria plus specific software tools. Somatosensory evoked potentials (SSEP), visual evoked potentials (VEP), integrated-interference pattern and turn-amplitude analysis in electromyography (EMG), motor evoked potential (MEP), spectral filters, threshold and delay facilities, etc., available as standard hardware and software in IONM equipment, are used for monitoring and mapping an optimal and impeded target. Special software tools could be implemented to improve the standard IOM equipment performance. In general, these tools are related to neuronal single-unit and/or multi-unit microelectrode recording (MER) on-line analysis and the data displayed on a MER/atlas background.

Neuroimaging techniques for stereotactic planning, neurophysiological mapping to confirm the target and the exploration of the target's adjacent structures before placement of a DBS electrode are essential to improve the accuracy of the process and to reduce side effects. However, opinions differ not only in regards to the inaccuracies of the CT-MR imaging fusion techniques for direct or indirect planning, but also which neurophysiological methods should be considered for target mapping and DBS electrode securing after identifying an appropriate track and target.

Our procedure has four steps: Pre-surgical, intra-operative, inter-operative and post-surgical:

1. Pre-surgical: Patients considered for surgery are also selected for pre-surgical neurophysiological assessment. A standard EEG, afferent evoked potentials -PEV for the internal globus pallidus (GPi) target; SSEP for the ventralis intermedius nucleus (Vim) and subthalamic nucleus (STN)-, P30 and P300 event related potentials and motor evoked potentials (MEP magnetic stimulation) are performed. Patients should be willing and able to undergo a conscious, long and noisy surgical procedure in

which the exploration of patient behavioural responses are empirically assessed and recorded.

2. Intra-operative: This concerns monitoring, mapping and securing the DBS electrode at the selected target site. Because the differences between the type of anatomic planning and the functional information needed to locate the optimal target vary from patient to patient, free and evoked deep brain macro- and micro-electrode recordings, EMG polygraph recordings, SSEP or VEP and MEP are performed. In regards to equipment, over the course of 12 years we have changed the IOM equipment from an initial low cost four-channel Premier (MEDELEC, England) to a moderately priced Keypoint plus Lead-Point facilities (DANTEC-MEDTRONIC, Denmark) and/or a 10Ch. XLTECH (Canada). The MER continuous on-line recording is analyzed and displayed on an auxiliary laptop with NDRS (CIREN, Cuba) software. We use an Ohye's electrode (Unique Medical, Japan) for multi-unit MER [30] and two types of DBS tetraelectrodes (MEDTRONIC, USA), Model 3389 for STN and hypothalamus surgery and Model 3387 for Gpi and thalamus ones for macro-electrode recording and stimulation. We also used to include mapping performed by MER of extracellular single-unit activities (neuronal activities), but only for research purposes. An ISO-80 (WPI, England) isolated differential amplifier is used as a preamplifier to secure deep brain intrusion and to optimize the high impedance noise from microelectrodes [10].

Neurophysiological measurements in a hostile operating theatre environment, with a conscious patient, face strong limitations when compared to the freedom we can find in a physiological laboratory. Thus, MER-IONM requires not only surgical time but skills, knowledge and neurophysiological expertise. In order to answer the neurosurgeon's questions and to make decisions in the operating room, the neurophysiologist needs to select, record and compare all the neurons recorded so far in a track, in order to detect the transition between the various structures crossed by the microelectrode and to identify the abnormal pattern of discharge and synchronization of the neuronal activities of different nuclei. Not all centres use MER-IONM, and not all centres that use MER, have a physiologist-surgeon team. As a result of our experience, we can confirm that the interplay between neurosurgeon and neurophysiologist optimizes safety, accuracy, constraints, knowledge and final outcomes. In our opinion, the use of the different mapping methodologies depends upon the circumstances and characteristics of the medical team involved in the neurosurgery, such as their level of expertise-training, skills-knowledge, research-clinical background and medical specialization, as well as the degree of their relationship. In our experience, if an IOM criterion is applied, in order to facilitate the neurophysiological intraoperative assessment after frame placement, image acquisition, theoretical target and initial trajectory planning, a set of electrodes is distributed with a 10-20 IS criterion over the patient's scalp and respectively, on the extensor digitorum communis and tibialis anterior muscles in the operating room. Ocular movement and facial expression video-recording are performed for monitoring purposes.

The extracellular single-unit MER is a well-designed mapping technique whose analysis includes a minimum of three steps: A subjective characterization of the neuronal discharge pattern, an empirical amplitude threshold selection and an objective quantification of the selected single-unit. This technique uses sharp 5-30 μm diameter, high impedance (> 1MΩ) micro-electrode tips. By recording on-line single-unit activity, we can identify the free and evoked activities from different nuclei and characterize the superior and inferior borders of the ambiguous nervous tissue. The sporadic or rhythmic neuronal discharge patterns are auditory discriminated and quantitatively analyzed. Usually an amplifier 500 Hz filter is used for single-unit MER. It is a time-consuming technique if properly performed (30-40 min by track), because the microelectrode must advance along the recording track by small steps (<0.5 mm) and frequently not all steps show a useful signal.

However it is very efficient when the limitations of this methodology are known by the neurosurgeon and the technique is mastered with experience and patience by a neurophysiologist.

Another IONM point of view or physiological mapping alternative is the multi-unit MER analysis, a methodology we normally use. This technique uses a concentric bipolar sharp microelectrode that has a 0.2 mm^2 active area with a 5 μm tip, referred to as a *semi-microelectrode.* This active area is separated from the reference area by 1 mm to 0.8 mm^2 (Ohye's electrode). This MER detects the activities of 10 to 40 perceptible neurons localized in a small volume (100-200 μm^3) around the sharp and facilitates the bipolar recording of local-field evoked potentials. The EMG standard software is useful for multi-unit MER analysis. Our procedure resembles that of EMG pattern analysis applied to assess recruitment and synchronization/interference of motor unit potentials. Therefore, amplitude, time and frequency domain analysis are considered for on-line identification and characterization of ambiguous nervous tissue.

Compared to single-unit MER, multi-unit MER has lower spatial resolution and cannot specify the pattern of neuronal discharge. Conversely, multi-unit MER better estimates the number of active neurons within certain areas and enables easy and reliable quantification of neuronal activity with a few sources of technical pitfalls [11]. The amplitude change or "*integrated*" multi-unit values of the recorded interference pattern significantly distinguishes between inside and outside the nucleus. The use of other standard on-line automatic EMGs such as turn-amplitude measurements of the interferential multi-unit MER pattern show specific features. Usually multi-unit MER is a timeless technique that consumes 10 to 15 min. per track.

For our mapping proposes we implement both multi-unit and local-field potential using MER, but in special research cases, a single-unit study was performed. The standard amplifier settings are carried out usually through a 100 Hz-5 KHz band pass for multi-unit MER and 10 Hz-5 KHz for local-field potential MER or beta-band MER signal analysis. The intraoperative MER is performed with the patient awake, but occasionally low doses of fentanyl are required and, if necessary, a reduced infusion of propofol. In our experience, both sleep and propofol modify multi-unit neuronal activity pattern. Once the duramatter has been opened, the Ohye's electrode is inserted through the burr hole and is then advanced by 1 mm steps using a microdrive device. In order to identify and localize ambiguous tissue, mapping for deep brain multi-unit MER is performed from the striatum to the inferior border of the expected nuclei. After two seconds, using the new step-impact, two seconds of free of artifacts multi-unit patterns are recorded and then the two-second "*integrated*" multi-unit value is graphically displayed as a step-related histogram on the theoretical outlined atlas-plane background (Figure 6).

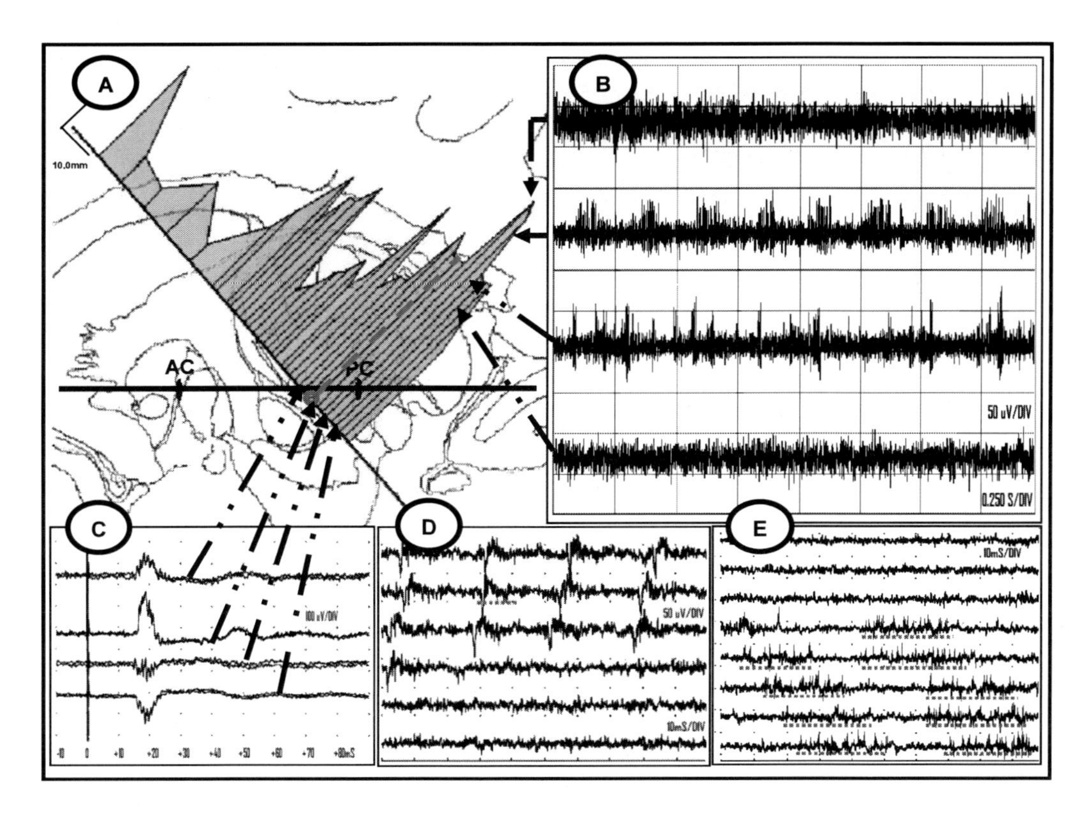

Figure 6. Mapping of the thalamic ventral intermedius nucleus (Vim). A) Display of the processed micro-**electrode** "*step by step*" **multi**-unit recording histograms and anterior commisure (AC) – posterior commisure (PC) plane over a 12 mm sagital plate of Schaltebrand and Wahren Atlas. B) Four traces of two seconds free run multi-unit recording from thalamic track showing tremor related multi-unit recruitment at a frequency of 4 Hz. C) Vim-ventroposterolateral nuclei (VPL) transition identify by median nerve somatosensory evoked potentials polarity reversal. D) Scrolling of three underlines consecutive traces of tapping event related somatosensory evoked responses. E) Scrolling of five underlined consecutive traces of rushting event related responses.

In a downward direction, the abrupt amplitude, frequency and pattern changes of multi-unit MER significantly distinguish between the inside (top) and outside (bottom) of the nucleus or nuclei transitions, and the length of the track in the nucleus measurement. This methodology makes the spatial abstraction of the three-dimensional track easier. The 3D displayed neurophysiological trajectory is a user-friendly and useful tool for decision-making by the neurosurgeon-neurophysiologist .

Usually, in 78.3% of the explored hemispheres in DBS tetraelectrode implantation, we perform no more than five MER tracks plus the final DBS tetraelectrode track (5.93 tracks ±1.82). It can be useful to perform local-field SSEP and VEP-MER when, for example, the STN nucleus can not be located or we need to know the inferior border of the GPi. In this context, if the first STN track shows a thalamic trajectory and the nuclei were not detected, then, we perform a median nerve SSEP-MER. In the case of clearly identified SSEP-MER responses, we do a rectification for the next track, 6 mm anterior and 2 mm lateral from the previous one. In our experience, the theoretical initial target result of the neuroimaging-based planning needs to be corrected. For example, in STN mapping, when the correction of inaccuracies were related to a necessary rostro-dorso-lateral displacement (75% of explored hemispheres) of the final track, the corrections were in the order of 1.91 mm±0.79 (median 2.0 mm) lateral, 3.24 mm±1.64 (median 2.5 mm) rostral and 1.04 mm±0.66 (median 1.0mm) dorsal. Conversely, when we need a caudo-ventro-medial displacement (25% of explored hemispheres) to localize the best final target, the corrections were in the order of 1.62 mm±0.60 (median 2 mm) medial, 2.96 mm±1.6 posterior (median 2.25 mm) and 1.0 mm±0.65 ventral (median 1 mm). In this context, the precise and effective functional localization of the nucleus is only possible using the mapping of the single-unit or multi-unit MER neuronal activity. However, the definitive implant of the DBS tetraelectrode can deviate significantly from the selected neurophysiological target due to the variability introduced by the manual neurosurgery method. An integrative methodology is needed to allow the correction of the inaccuracies between the target identified and the final implant of the DBS tetraelectrode. In order to optimize this objective, a bipolar macro-recording from the DBS tetraelectrode is performed to check the correct final position of the implant. To achieve this goal, the recordings from 0-1, 1-2 and 2-3 tetraelectrode contact, when the bottom contact is named as 0 and the top one as 3, were analyzed and displayed respectively on three channels. The "*integrated*" macro-recording signal from 0-1 contact derivation was correlated with the data of the final-track

Although many MER systems allow the surgical team to switch between recording and stimulation modes, we do not perform micro-stimulation to identify motor bundles or motor nuclei during IONM. In our experience, the "integrated" multi-unit value change identifies nucleus limits without requiring this assurance technique. Secondly, as Shils et al [46] points out, "*the volume of tissue that can be affected with microelectrode stimulation is so small that gross clinical changes are rarely observed with this technique*".

3. Inter-operative: This period refers to the 6 days between DBS tetralectrode implantation and the surgical act for definitive plug-in to the impulse generator stimulator (IGS). After 72 hrs with the DBS tetraelectrode and during the 3 days before the IGS implants, we perform a set of neurophysiological techniques to find the best contact and any possible inaccuracies. The relative distance between the different DBS tetraelectrode contacts and the efferent motor fibres is assessed by low rate monopolar DBS tetraelectrode contact stimulation and EMG recording. From a 0.5 mA intensity, 5 Hz frequency and 200 μS duration initial pulse, the threshold MEP from masseter, extensor digitorum communis and tibialis anterior muscles is identified and the intensity/duration characteristics of the stimuli measured. If the necessary stimulus to induce a response is low, an abnormal placement of the electrode is confirmed. The anatomo/functional relationship between DBS tetraelectrode and the lemniscus medialis and ventro-postero-lateral (VPL) thalamus is assessed by SSEP for Vim and STN targets [11]. In this way, the median nerve and/or tibialis anterior nerve stimulation evoked event-related potentials can be recorded as far-field potentials from the DBS tetraelectrode contacts. From the 0-1, 1-2, 2-3 bipolar derivations, we measure the amplitude, polarity and configuration responses. This data is correlated with the MEP. Macro-recordings from these bipolar derivations in addition to a standard polysomnography (PSG) are performed and the polarity of the beta rhythms and the sharp waves recorded during the REM sleep are measured to make a final decision. When the target is the GPi to assess the

position of the DBS tetraelectrode with respect to the optic bundle, VEP are performed. If the results from SSEP or VEP, MEP and STN-PSG indicate an adequate 3D-location of the DBS tetraelectrode on the subthalamic volume, the IGS is implanted.

4. Post-operative: The initial contact for therapeutic DBS is selected based on interoperative neurophysiological data previously described. After one year, the outcome is evaluated by an adequate clinical assessment [20] and a scalp-recorded STN, Vim or GPi evoked potentials by bipolar DBS tetraelectrode stimulation. Short latency and cortical mapping can be evaluated and correlated with the clinical improvement.

Question Seven: How Do You Implant the DBS? Is an Introductory Cannula Necessary? Is a Radiological Control Necessary? How Can We Check the Correct Positioning of the DBS in the Operating Room?

Currently, there are only two DBS leads models available, Model 3387 and 3389 (Medtronic Inc, Minneapolis, MN, USA). The Model 3387 lead has four electrodes, of which each one is 1.5 mm long, separated by 1.5 mm, thus spanning a linear distance of 10.5 mm. The Model 3389 lead has the same four electrodes separated by only 0.5 mm, spanning a total of 7.5 mm. In both cases, the leads are 1.27 mm in diameter.

Even though it seems logical to think that the higher number of trajectories would increase the risk of bleeding, in our experience, no statistical significance was found between the number of sites and bleeding [45]. An increased number of tracks have a higher risk of brain edema that spreads along the longitudinal direction of the axons and generates a shift in midline structures. In our series, such edema did not produce morbidity (Figure 8).

Once semi-microelectrode recording confirms the target, we checked its position with control X-ray films. After the semi-microelectrode and its guiding cannula are withdrawn and replaced by the guiding cannula for the DBS lead in the same place. This cannula is inserted to the brain slowly and gently until 10 mm above the target. Then, the DBS lead is deployed and DBS lead electrode recording is performed to confirm the physiology of the neural activity of the target. Its position is checked by X-ray films and compared with the position of the semi-microelectrode. If DBS lead´ electrode recording, radiological position and the macrostimulation test are correct, we implant the DBS lead.

We always implant the DBS lead with a guiding cannula because the DBS lead stylet that facilitates stereotaxic placement of the lead is weak, thus, there is a potential for a breach of procedure. In the beginning, after getting the final trajectory with semi-microelectrode, a lateral X-ray film was taken for attitude control of the implant. Then the semi-microelectrode is withdrawn and a lesion probe with a 1.1x3 mm exposed tip (Radionics, Burlington, Mass, USA) is introduced into the same site. We take another X-ray film to check that the lesion probe remained in the same position as the semi-microelectrode. Afterwards, the lesion probe is removed and the DBS lead, Model 3387 or Model 3389 is introduced with its stylet, tracing the path undertaken by the lesion probe. Finally, an X-ray film was taken to verify that the DBS was in the same position as the semi-microelectrode (Figure 10).

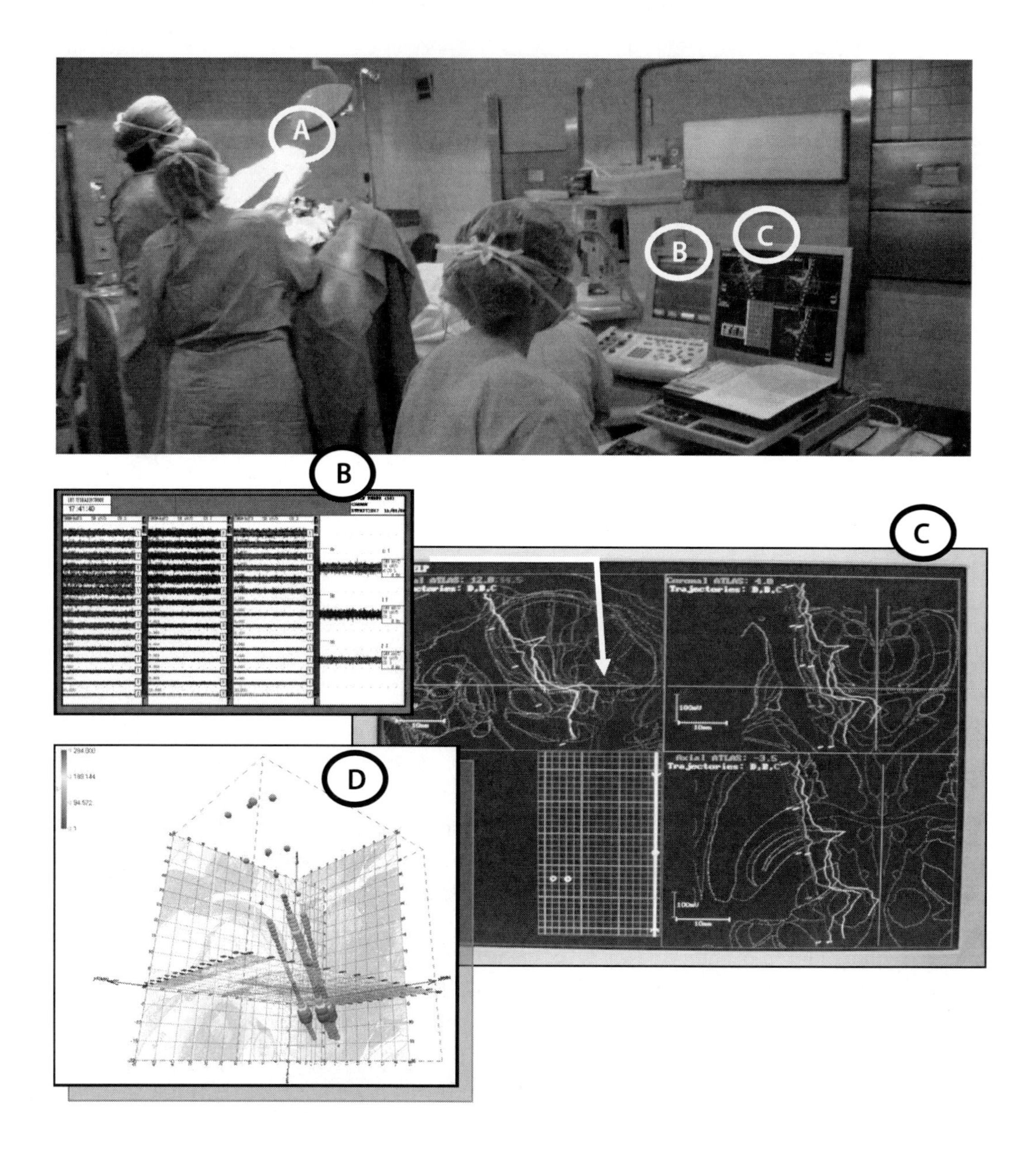

Figure 7. Operating room. A) Micro- tetra-electrode tower. B) Identification and recording of subthalamic nucleus (STN) activity and bipolar tetraelectrode activity display recording from EMG/PE equipment. C) On-line sagital. Coronal and axial display of the processed micro- and tetra-electrode "*step by step*" recording histograms superimposition from the STN multi-unit recording (red and blue traces) and the tetraelectrode 0-1 bipolar recoding (red trace) to identified final DBS tetraelectrode position. D) 3D-display mapping of three tracks. Color and size indicate "*integrated*" multi-unit MER values where red color points correspond to STN.

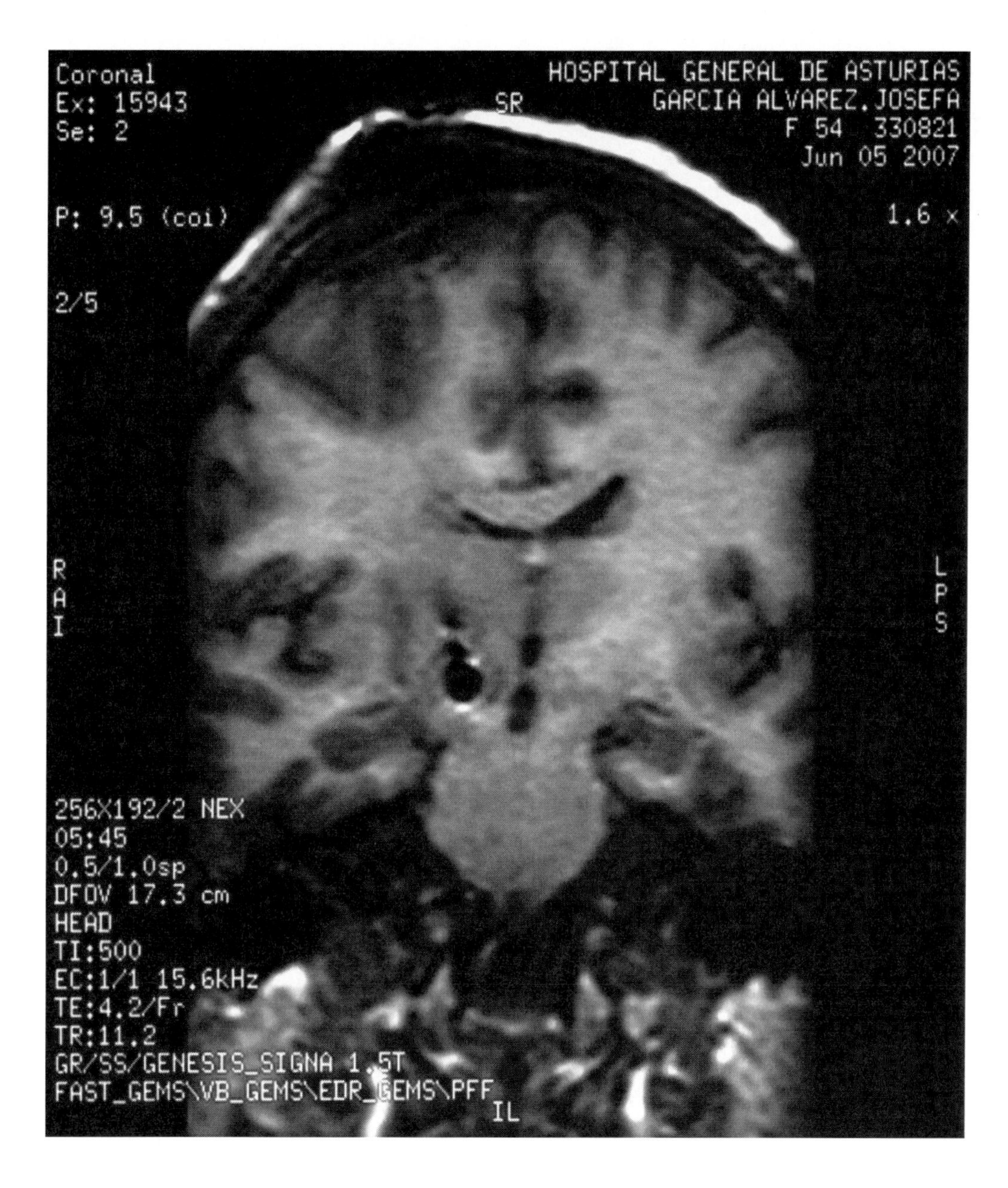

Figure 8. Shift by cerebral brain after DBS lead implanted

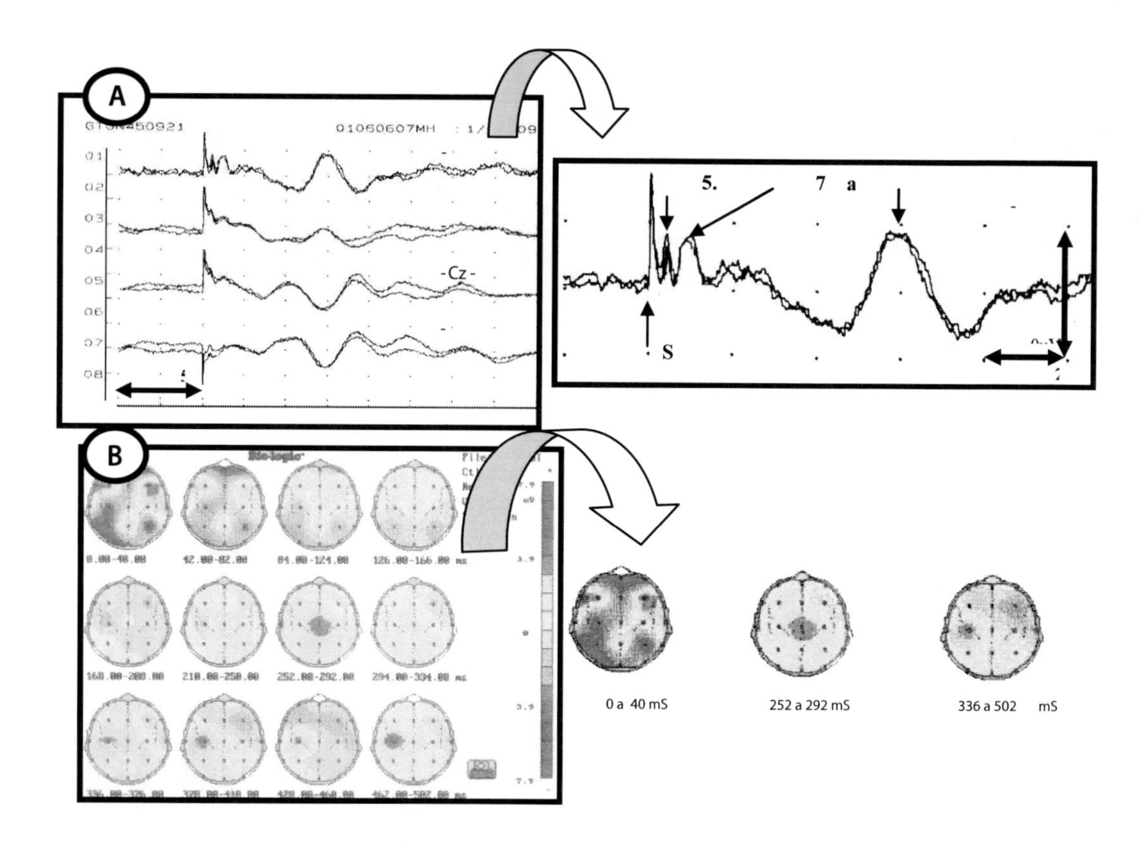

Figure 9. Scalp evoked responses after low frequency subthalamic nucleus (STN) stimulation of the therapeutically effective DBS tetraelectrode contact. A) Short and middle latency responses showing frontal predominance of the 5 to 12 ms potentials and polarity change over Cz 10-20 IS electrode. B) Display of twelve consecutive relative amplitude/fourteen second maps of the scalp recorded event related responses showing as red color higher amplitude.

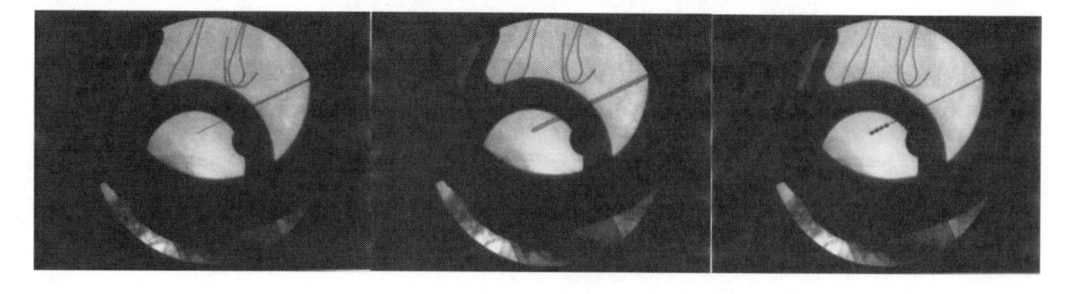

Figure 10. X-ray films. A) Semi-microelectrode. B) Lesion probe. C) DBS lead.

After verifying that this technique can be performed parallel paths of 1-3 mm between the semi-microelectrode and DBS lead, we chose to directly enter the DBS through a cannula (Figure 11). Since we do this method we have not seen any diversion of DBS lead over the semi-microelectrode.

Finally in the last 115 patients we devised a method to verify that DBS electrodes are well established. In the implantation of DBS lead we recorded neuronal activity through own

DBS electrodes. This confirmed the good final position of DBS electrodes. The distal contact is located where the recording of cell activities and beneficial effects induced by stimulation are lost.

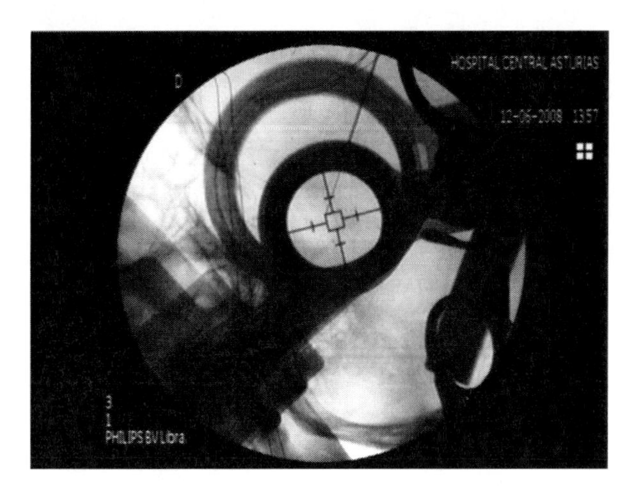

Figure 11. X-ray films. A) Semi-microelectrode. B) DBS lead with cannula.

Question Eight: When is the Correct Moment for Radiological Control?

Immediately after surgery we performed a standard CT scan to exclude possible brain lesions (pneumoencephalus, haemorrhage).

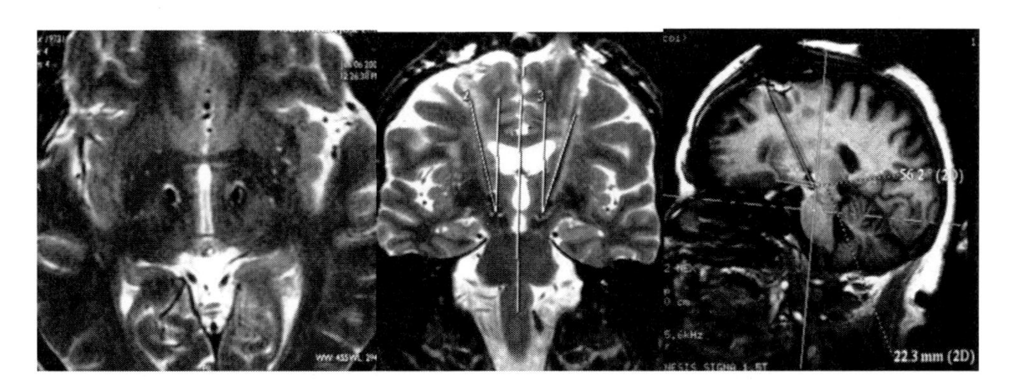

Figure 12. A) Axial T2-Weighted images. Measure X coordinate of electrodes with respect to third ventricle midline and Y coordinate with respect to intercommisural point (ICP). B) Coronal T2-Weighted images. Mesures β angle. C) Sagittal T2-Weighted images. Mesures α angle.

Seventy-two hours after the operation, a new radiological control with a three-dimensional volumetric magnetic resonance was performed, with special focus on the CA-CP line and the position of the implanted DBS lead from that line because the final target position after semi-microelectrode recording must be deviated from the original anatomic coordinates [29]. We think that MRI is superior to CT in evaluating the possible shifting of electrodes during surgical placement because of the multiplanar capacities and because pins create significant artifactual distortion of CT imaging which can create a source of targeting error (Figure 12).

Although subcortical structure localization, using MRI or CT and a stereotactic head frame can have an accuracy of approximately 1 mm, discrepancies between the initial selected target and the DBS electrode position (electrophysiology based) are described [32, 56]. These discrepancies must be due to several factors including imaging distortion, mechanical inaccuracy of the stereotactic frame and brain shift. Many factors may contribute to cerebral shift, including gravitational force, loss of cerebral fluid, changes in pressure due to skull opening during the surgery, pneumocephalus, and forces due to insertion of the DBS lead. The number of microelectrode tracks may be a significant factor that can increase brain edema and increase the shift [13, 17]. Finally another factor that can affect the assessment of brain shift is the difference between the completely supine-positioned patient in the MRI scanner and the semi-seated position during surgery.

Normally, brain shift and pneumocephalus disappeared 72 hours after surgery [17]. Furthermore, the permanent DBS lead may vary ±1 mm, compared to the best localization of the semi-microelectrode due to potential movement during stylet removal [30, 31].

Since 2005, the numerical calculations of the specific absorption rate (SAR) [52] were taken into account and have always been below the regulatory limits in accordance with the Safety Guidelines (T2 SAR: 0.0056 W/kg; MPRAGE: 0.0034 W/kg.). Then we selected a specific MRI protocol with limited slices (five T2 slices), for post-operative electrode placement control. We followed the post-operative MRI image protocol below:

(1) Volumetric 3D MPRAGE in 1.5 mm sagittal slices was performed. The parameters were: 11.2/4,2/500 TR/TE/TI msec; flip angle 15°; 2 NEX; field of view, 28 x 21; section thickness, 2 mm; spacing 1.0; 116 number of sections, matrix, 256x192; and receive bandwidth, 7.8 KHz;
(2) Two sets of T2-weighted FR- FSE T2 images were performed at coronal and axial planes with the following parameters: 6000/114,2 msec TR/TE; number of sections acquired 5; section thickness, 3.0 mm; spacing 0.5; NEX 2; matrix, 320x256; field of view, 26x19,5; and receive bandwidth, 7.8 KHz.

On axial T2-Weighted images, at the AC-PC plane, we measured the X coordinate of the electrodes with respect to the third ventricle midline and the Y coordinate with respect to ICP. The Z coordinate is calculated at the plane crossing trough of the electrode tip, usually 1 or 2 slices inferior to the AC-PC plane (3-6 mm). On 3D MPRAGE reformatted oblique images, we measured the α angle with respect to the mid-coronal plane passing through the third ventricle and the β angle with respect to the AC-PC plane.

Before that appeared Specific Absorption rate recommendations (SAR) [52] we perform a 1.5-tesla MRI control on all operated patients (147 patients). All subjects were able to successfully complete the MR imaging examination. None the subjects complained of any abnormal sensations or discomfort caused by MR imaging with DBS electrodes in place.

Question Nine: What Are the Short-Term Complications in Deep Brain Stimulation Surgery?

Between June 1996 and December 2007, we operated on 207 patients; 183 PD, 16 tremors, 4 dystonias and 4 cluster headaches. A total of 387 DBS electrodes were implanted in the thalamus, internal pallidus, subthalamic nucleus (STN) and the hypothalamus.

We reviewed the complications in our Hospital between March 1998 and December 2004 [45]. One hundred and five consecutive PD patients underwent 223 STN DBS procedures. One hundred received bilateral and five unilateral STN DBS. The total number of implanted leads was 203. All DBS were implanted by the same surgeon (F.S.). Sixty one patients did not present any complications (58% of the patients), thirty seven presented one (35% of the patients) and seven patients presented one or more (7% of the patients).

Of the seven patients who presented two or more complications, two of them presented 3 complications and five of them presented 2 complications. In the two cases with 3 complications, one patient experienced bleeding, convulsions and a confusional state and the other patient presented convulsions, a confusional state and urinary tract infection. In the five patients who presented two complications, three presented bleeding and a confusional state and the other two presented an aborted procedure due to non identification of the STN with two additional procedures because of a misplaced lead, and an aborted procedure due to non identification of the STN and cardiac problems (this patient had a cardiac pacemaker), respectively.

The complications reviewed were aborted procedures, misplaced leads, intracranial haemorrhage, seizures, hardware complications and other complications.

An aborted procedure is a procedure in which the DBS lead could not be implanted in the STN. There were a total of 14 aborted procedures, which means 6.2% of all the procedures. Of the 14 aborted procedures, the STN could not be found in 11 of them; in one case, the stereotactic frame was disconnected during the surgery after implanting the first lead, the neurophysiologic recording system failed in one patient, disabling the left STN and the second lead implant, and finally, clinical changes were observed during the first procedure in another patient, which made us suspect the possibility of bleeding for which we cancelled surgery. After performing a CT, we observed a mild haematoma in the right thalamus. DBS leads were implanted afterwards in a second surgery in 12 of the 14 patients in which the lead could not be implanted in the first surgery. In spite of the sophisticated neuroimaging techniques, a difference of around 35% between the initial theoretical difference estimated by radiological sites and the final target localization was observed [50].

For us, a misplaced lead is any situation that required us to reposition the leads, due to bad positioning in the control MRI and a clinical response lower than 30% in the UPDRS (Part III) 6 months after surgery. The clinical improvement was below 30% in the UPDRS (Part III) in a total of 5 patients and the MRI indicated that the leads were not in the correct position. Four patients had unilateral misplacement and one patient had bilateral misplacement. Following repositioning of the lead, all patients presented a UPDRS (Part III) clinical improvement above 30% and the MRI indicated correct positioning of the lead. These cases represented 2.69% of all the procedures. The misplaced lead may be due to brain shift during the brain lead implant. When the brain regains its normal position after some time, the DBS electrode remains deeper in the brain due to its semi rigidity. This cerebral brain shift can be demonstrated during surgery when comparing the amount of physiological solution we can introduce into the cranial cavity in the first tract to the amount introduced when placing the lead.

Intracranial haemorrhage was present in 7 patients, representing 3.13% of the procedures. Two patients had cortical venous infarct with convulsions. One patient had a mild intraventricular haemorrhage without clinical symptoms. One patient had a mild thalamic haemorrhage associated with convulsions. Another had a small right frontal intracerebral haematoma with convulsions and a third patient developed an intracerebral haematoma throughout the lead trajectory which required emergency surgery in order to evacuate the haematoma, but which did not require lead explantation. As a sequela, a mild residual hemiparesis appeared. Another patient had a transitory intraoperative consciousness deterioration which forced us to cancel the procedure. CT was performed and a thalamic haematoma was seen. This case also appears with those described as aborted procedures.

Seizures affected a total of 11 patients out of 105, representing 4.9% of the procedures. Four patients presented bleeding and 8 did not have radiological findings in the post-operative routine CT scan. All seizures occurred within the first 24 hours after STN lead implant surgery and they improved with intravenous anticonvulsive treatment (phenytoin). The seizures did not reappear, and therefore, except for the patient who developed an intracerebral haematoma throughout the lead trajectory requiring emergency surgery, no patient required anticonvulsive treatment after discharge from the hospital. In the case of this last patient, despite the fact that he never had seizures, anticonvulsive treatment was not discontinued until 6 months after hospital discharge. This being due to the fact that up until the third month after discharge the electroencephalograms presented an irritable focus.

A total of 5 patients presented direct hardware-related complications, affecting 4.8% of all patients and 2.2% of the procedures. One month after surgery, a patient presented scarring and infection at the battery site, forcing explantation and antibiotic treatment. Three months later the battery was implanted without complications. Three months after DBS NST surgery, another patient presented scarring and infection at the lead/wire connection site, leading to removal of the whole DBS system and repositioning it 6 months later. The evaluation of the infectious episodes was performed based on clinical criteria (pain, heat and local tumefaction) and analytical values (leukocytosis and increased GSR). Another patient presented scarring at the wire trajectory one month after implantation, which resolved itself without surgical removal. Fourteen months after surgery, one patient had a lead fracture requiring replacement. Finally, during immediate post-operative lead placement, a patient had a cerebrospinal fluid leak through the burr hole ring and cap, requiring surgery.

A total of 13 patients presented complications different from those already mentioned, representing 5.4% of the procedures. These complications can be divided into minor complications (11 patients) and major complications such as worsening of cardiopathy in the patient with the cardiac pacemaker which made a new procedure in order to implant the second lead impossible, and a lower extremity deep venous thrombosis which required anticoagulant treatment for 6 months due to severe respiratory problems. The most frequent complications found in the abovementioned section on complications were confusional episodes. Even though we did not find a co- relationship between the number of sites, age and confusional episodes, it is likely that this symptomology, during immediate postoperative, may have a multifactor origin. This is a minor complication that disappears with time, and requires only vigilance during the first hours. On the other hand, all patients undergo in-dwelling urinary catheterization at the beginning of the intervention, in order to control their diuresis, and this could have influenced our two reported confusional episodes. Antibiotic treatment was started in the presence of clinical symptoms and pathological urinary sediment, without waiting for the result of the urine culture. In both cases the empirical treatment was proven to be sufficient for the treatment of confusion. We believe that the patient's in-dwelling urinary catheter is needed in order to maintain a correct and constant hydric balance during the long period of the surgery, despite the fact that surgery is performed with conscious patients.

References

[1] Albe-Fessard, DG, Arfel G. and Guiot G. Activités électriques characteristiques de quelques structures cérébrales chez l'homme. *Ann. Chir* 1963; 17:1185-1214.

[2] Benabid AL, Krack P, Benazzouz A, Limousin P, Koudsie A, Pollak P. Deep brain stimulation of the subthalamic nucleus for Parkinson´s disease: methodologic aspects and clinical criteria. *Neurology* 2000; 55(suppl 6): S40-S44.

[3] Benabid AL, Pollak P, Gervason C, Hoffmamm D, Gao DM, Hommel M, Perret JE, de Rougemont J. Long-term suppression of tremor by chronic stimulation of the ventral intermediate thalamic nucleus. *Lancet,* 1991; 337:403-406

[4] Burchiel K, Nguyen T, Coombs B, Suzumoski J. MRI distortion and stereotactic neurosurgery using the Cosman-Roberts-Wells and Leksell frames. Stereotac Funct *Neurosurg* 1996; 66:123-136.

[5] Campbell RL, Campbell JA, Heimburger R, Klasbeck JE, Mealey J. Ventriculography and myelography with absorbable radiopaque medium. *Radiology* 1964; 82:286-289.

[6] Cuny E, Guehl D, Burbaud P, Gross C, Dousset V, Rougier A. Lack of agreement between direct magnetic resonance imaging and statistical determination of a subthalamic target: role of electrophysiological guidance. *J Neurosurg* 2002; 97:591-597.

[7] Derosier C, Delegue G, Munier T, Pharaboz C, Cosnard G. IRM distortion géometrique de l´image et stéréotaxie. *J Radiol* 1991; 72:349-35.

[8] Fahn S. Concept and classification of dystonia. *Adv Neurol* 1988; 50:1-8.

[9] Fahn S, Elton RL, and the members of the Unified Parkinson's Disease Rating Scale. In: Fahn S, Marsden CD, Calne CB, Goldstein M, eds. *Recent Developments in Parkinson's Disease.* Vol. 2. Florham Park, NJ: MacMillan Healthcare Information; 1987. p 153-163, 293-304.

[10] Fernández-González F, Seijo F, Menéndez-Guisasola L, Salvador C, Roger RL, González-García FJ, Fernández-Martínez JM, Bulla B, Fernández-García C, González-González S, Galindo A. Identificación de las dianas esterotáxicas en la cirugía de la enfermedad de Parkinson. *Rev Neurol* 1999; 28:600-608.

[11] Fernández-González F, Seijo F, Salvador C, Menéndez-Guisasola, Lozano B, Valles C, Galindo A. Neurofisiología aplicada en el tratamiento con estimulación cerebral profunda del temblor severo de la esclerosis múltiple. *Rev Neurol* 2001; 32:559-567.

[12] Guerdes J, Hitchon P, Neerangun W, Torner J. Computed tomography versus magnetic resonance imaging in stereotactic localization. *Stereotact Funct Neurosurg* 1994; 63:124-129.

[13] Halpern CH, Danish SF, Baltuch GH, Jaggi JL. Brain shift during deep brain stimulation surgery for Parkinson´s disease. *Stereotact Funct Neurosurg* 2008;86:37-43.

[14] Hariz MI, Bergenheim T, Fodstard H. Comparison between ventriculographic and CT-guided determination of brain targets in functional stereotaxis. *Stereotact Funct Neurosurg (Abstract)* 1990; 54/55:240-241.

[15] Holtzheimer III PE, Roberts DW, Darcey TM. Magnetic resonance imaging versus computed tomography for target localization in functional stereotactic neurosurgery. *Neurosurgery* 1999; 45:290-298.

[16] Hubble JP, Busenbark KL, Wilkinson S, Pahwa R, Paulson GW, Lyons K, Koller WC: Effects of thalamic deep brain stimulation based on tremor type and diagnosis. *Mov Disord* 1997; 12:337-341.

[17] Khan MF, Mewes K, Gross R, Skrinjar O. Assessment of brain shift related to deep brain stimulation surgery. *Stereotac Funct Neurosurg* 2008; 86:44-53.

[18] Koike Y, Shima F, Nakamizo A, Miyagi Y. Direct localization of subthalamic nucleus supplemented by single-track electrophysiological guidance in deep brain stimulation lead implantation: techniques and clinical results. *Stereotact Funct Neurosurg* 2008; 86:173-178.

[19] Kondziolka D, Dempsey PK, Lunsford LD, Kostle JR, Dolan EJ, Kanal E, Tasker JJl. A comparison between magnetic resonante imaging and computed tomography for stereotactic coordinate determination. *Neurosurgery* 1992; 30:402-407.

[20] Krack P, Fraix V, Mendes A, Benabid AL and Pollack P. Postoperative Management of Subthalamic Nucleus Stimulation for Parkinson's disease Mov Disord 2002; 17:S188-197.

[21] Krauss JK, Akeyson EW, Giam P, Jankovic J. Propofol-induced dyskinesias in Parkinson´s disease. *Anesth Analg* 1996; 83:420-422.

[22] Krauss JK, King DE, Grossman RG: Alignment correction algorithm for transfomation of anterior comissure/posterior comissure based coordinates into frame coordinates in image-guided functional neurosurgery. *Neurosurgery* 1998; 42:806-811.

[23] Laitinen LV. Brain targets in surgery for Parkinson´s disease. *J Neurosurg* 1985; 62:349-351.

[24] Lang AE, Widner H. Deep brain stimulation for Parkinson´s disease: patient selection and evaluation. *Mov Disord* 2002; S3:S94-S101.

[25] Langston JW, Widner H, Gotees CG, Brooks D, Fahn S, Freeman T, Watts R. Core assessment program for intracerebral transplantations *Mov Disord* 1992; 7:2-13.

[26] Latchaw R, Lunsford L, Kennedy W. Reformatted imaging to define the intercommisural line for CT-guided stereotaxic functional neurosurgery. *A.J.N.R.* 1985; 6:429-433.

[27] Lee MW,De Salles AA, Frighetto L, Torres R, Behnke E, Bronstein JM. Deep brain stimulation in intraoperative MRI environment-Comparison of imaging techniques and electrode fixation methods. *Minim Invas Neurosurg* 2005; 48:1-6.

[28] Lehman R, Mezrich R, Sage J, Goldbe L. Peri and postoperative magnetic resonance imaging localization of pallidotomy. *Stereotact Funct Neurosurg* 1994; 62:61-70.

[29] Leone M, May A, Franzini A, Broggi G, Dodick D, Rapaport A, Goadsby P, Schoenen J, Bonavita V, Bussone G. Deep brain stimulation for intractable chronic cluster headache: proposals for patient selection. *Cephalalgia* 2004; 24:934-937.

[30] Litt B, Cranstoun. EEG and the anterior thalamic nucleus. *In:Deep brain stimulation and epilepsy*. Pp.171-185. Lüders HO Ed. Thomson Publishing Services. Andover, UK. 2004.

[31] Lozano AM, Hutchinson W, KissA, Tasker R, Dostrovsky J. Methods for microelectrode-guided posteroventral pallidotomy. *J Neurosurg* 1996; 84:194-202.

[32] Maciunas RJ, Galloway RL, Latimer JW. The application accuracy of stereotactic frames. *Neurosurg* Online 1994;35:682-695.

[33] Mandydur G, Morenski J, Kuniyoshi S, Iacomo R. Comparison of MRI and ventriculographic target acquisition for posteroventral pallidotomy. *Stereotact Funct Neurosurg* 1995; 65:54-59.

[34] Obwegeser A, Uitti RJ, Witte RJ, Lucas JA, Turk MF, Wharen RE. Quantitative and qualitative outcome measures after thalamic deep brain stimulation to treat disabling tremors. *Neurosurgery* 2001; 48:274-284.

[35] Ohye CH. and Narabayashi H. Activity of thalamic neurons and their receptive fields in different functional states in man. *In: Somjen GG editor Neurophysiology studies in man.* Amsterdam. Excerpta Med 1972; 23:79-84.

[36] Olivier A, Peters T, Bertrand G. Stereotactic system and apparatus for use with MRI, CT and DSA. *Appl Neurophysiol* 1985; 48:94-96.

[37] Pendfield W and Rasmussen T. *The cerebral cortex of man: A clinical study of localization and functions* The Macmillian Co., New York. 1957.

[38] Pralong E, Ghika J, Temperli P, Pollo C, Vingerhoets F and Villemure JG. Electrophysiological localization of the subthalamic nucleus in parkinsonian patients. *Neurosci Lett* (2002) 325:144-146.

[39] Riechert T. Development of human stereotactic surgery. *Confin Neurol* 1975; 37:399-409.

[40] Riechert T. *Stereotactic Brain Operations*. Bern; Hans Huber Publishers, 1980.

[41] Saint-Cyr JA, Hoque T, Pereira LC, Dostrosvsky JO, Hutchison WD, Mikulis DJ, Abosch A, Sime E, Lang AE, Lozano AM. Localization of clinically effective stimulating electrodes in the human subthalamic nucleus on magnetic resonance imaging. *J Neurosurg* 2002; 97:1152-1166.

[42] Schad L, Lott S, Schmitt F, Sturm V, Lorenz W. Correction of spatial distorsion in MR imaging: a prerequisite for accurate stereotaxy. *J Compt Assist Tomogr* 1987; 11:499-505.

[43] Schaltenbrand G, Wahren W. *Atlas for Stereotaxy of the Human Brain*. Stuttgart; Thieme, 1977.

[44] Schuurman PR, de Bie RM, Majoie CB, Speelman JD, Bosch DA. A prospective comparison between three-dimensional magnetic resonance imaging and ventriculography for target-coordinate determination in frame-based functional stereotactic neurosurgery. *J Neurosurg* 1999; 91:911-914.

[45] Seijo FJ, Alvarez-Vega MA, Gutiérrez JC, Fdez-Glez F, Lozano B. Complications in subthalamic nucleus stimulation surgery for treatment of Parkinson disease. Review of 272 procedures. *Acta Neurchirurgica*; 149:867-876.

[46] Shils JL, Taqliati M, Alterman RL. Intraoperative microelectrode recording equipment. What features are necessary? *Stereotact Funct Neurosurg* 2001; 77:101-107.

[47] Starr PA, Vitek JL, DeLong M, Bakay RA. Magnetic resonance imaging-based stereotactic localization of the globus pallidus and subthalamic nucleus. *Neurosurgery* 1999; 44:303-314.

[48] Starr PA, Vitek JL, DeLong M, Mewes K, Bakay AE. Pallidoty: theory and technique. *In: Techniques in Neurosurgery*, vol 5, nº 1:31-45. 1999. Lippincott Willians and Wilkins, Inc. Philadelphia.

[49] Talairach J, Tournoux P. *Co-planar Stereotaxic Atlas of the Human Brain*. Stuttgart; Thieme, 1988.

[50] Tasker RR. Deep brain stimulation is preferable to thalamotomy for tremor suppresision. *Surg Neurol* 1998; 49:145-154.

[51] Taveras JM, Word E. *Diagnostic Neuroradiology*. Vol 1. Baltimore; Williams & Wilkins 1976.

[52] Tremmel J (ed): Urgent device correction-change of safe limits for MRI procedures used with Medtronic Activa deep brain stimulation systems. *Medtronic*, 2005.

[53] Tsao K, Wilkinson S, Overman J, Koller WC, Batnitzky S, Gordon MA. Pallidotomy lesion locations: significance of microelectrode refinement. *Neurosurg* 1998; 43:506-513.

[54] Van Buren JM, Maccubbin DA. An outline atlas of the human basal ganglio with stimulation of anatomical variants. *J Neurosurg* 1962; 19:811-839.

[55] Williams D, Warwick R. *Functional Neuroanatomy of Man*. London; Churchill Livingstone, 1975.

[56] Yu C, Apuzzo M, Zee C, Petrovich Z. A phantom study of the geometric accuracy of computed tomographic and magnetic resonance imaging stereotactic localization with the Leksell stereotactic system. *Neurosurgery* 2001; 48:1092-1099.

[57] Yuan S, Zhang J, Gu M, Xu Y,Chen L, Yao Q, He Q. A new method to localize brain nuclei for surgery in extrapyramidal disease. *Stereotact Funct Neurosurg* 1995; 65:47-53.

[58] Zonenshanyn M, Rezai AR, Mogilner AY, Beric A, Sterio D, Nelly PJ. Comparison of anatomic and neurophysiological methods for subthalamic nucleus targeting. *Neurosurgery* 2000; 47:282-294.

In: Encyclopedia of Neuroscience Research
Editors: Eileen J. Sampson and Donald R. Glevins
ISBN 978-1-61324-861-4

Chapter XII

Deep Brain Stimulation and Cortical Stimulation Methods: A Commentary on Established Applications and Expected Developments

Damianos E. Sakas* * and Ioannis G. Panourias
Department of Neurosurgery, University of Athens Medical School,
Evangelismos Hospital, Athens, Greece

Abstract

Over the last twenty years, deep brain stimulation (DBS) has been increasingly accepted as an effective alternative treatment to the selective lesioning methods previously used in stereotactic and functional neurosurgery. This method, based on its reversibility and adaptability, has currently been used for the alleviation of medically-refractory cases of movement disorders, epilepsy, pain, and psychiatric diseases. In the present commentary, data comparing DBS with other brain stimulation methods such as electrical cortical stimulation (CS) and transcranial magnetic stimulation (TMS) are analyzed and the perspectives of the field are outlined. Progress in the related fields of neuromodulation, functional neuroprosthetic surgery, neuroinformatics, neurocomputation, and developments in neuroengineering that are expected to refine, enhance and widen the therapeutic applications of DBS are also discussed.

Introduction

Electricity and some of its various effects on the human body have been known since ancient times. The use of electrical stimulation, however, as a therapeutic tool has been limited. Traditionally, medicine has been applied by administration of pharmaceuticals and

*Damianos E. Sakas, MD. Department of Neurosurgery, Evangelismos Hospital, 4 Marasli Street, 10676 Athens, Greece. Tel: +30 210 7201704 -5. Fax: +30 210 7249986. Email: sakasde@med.uoa.gr.

surgical interventions. Over the last 50 years, great progress in neurophysiology and neurosurgery made it possible to investigate the therapeutic potential of electrical brain stimulation. During the last two decades, the introduction of deep brain stimulation (DBS) and, to a lesser extent, of chronic electrical cortical stimulation (CS) transformed functional neurosurgery and neuroscience, in general. DBS offers superior clinical efficacy with fewer complications compared to the selective lesional procedures of conventional functional neurosurgery. More importantly, the success of DBS in improving outcome has signified the role of electricity in neurological disorders and the great benefits that can be derived by its modulation, thus creating a new field in medicine this of *selective focal electrotherapy* of the nervous system.

Therapeutic brain stimulation can be applied in networks which are located: a) in cortical or subcortical layers, b) in deep nuclei groups (relay nodes) or c) in a combination of both. DBS belongs to the domain of *operative neuromodulation* and particularly, to the field of *neural networks surgery* [45]; this field studies and applies advancements in neural networks research, digitised stereotactic brain imaging and implantable electrical or electronic devices in order to alter electrically the signal transmission in the nervous system, modulate neural networks and produce therapeutic effects.

In this article, we comment on the current state and future prospects of DBS of the human brain for the treatment of chronic medically-refractory neurological conditions such as Parkinson's disease, dystonia, tremor, epilepsy, pain, and psychiatric disorders. In certain specific applications, DBS is compared with CS with respect to their indications, benefits, advantages, and limitations as therapeutic tools. In addition, we comment on current limitations and future prospects of DBS and other brain stimulation methods and on expected scientific and technological advances and their implications for future therapeutic applications of DBS.

Parkinson's Disease

DBS has been acknowledged worldwide as the surgical treatment of choice for improving the incapacitating symptoms and the quality of life in patients with Parkinson's disease (PD). The efficacy of *subthalamic nucleus (STN)* DBS in alleviating the off-medication state of PD patients has been validated in prospective, randomized trials and meta-analysis studies [17, 23, 71, 75]. STN has gained general acceptance over *globus pallidus internal (GPi)* as the preferred target [16, 21] mainly because of easier targeting on MRI, longer effect of its stimulation, resulting to bigger reduction of required medication, and longer battery life due to lower stimulation settings required for the smaller in size STN compared to the GPi. Despite the undeniable benefit of DBS in PD, almost half of the patients are excluded from DBS therapy due to advanced age (>70 years), dementia, psychiatric comorbidity, poor response to levodopa, low score (<30-40) in the UPDRS part III or other contraindications.

The application of motor cortical stimulation (MCS) in PD has been limited but this may prove to be an efficacious alternative treatment modality, particularly, in patients presenting with marked laterality of symptoms and age >70 years. This number of patients will steadily increase, in the years to come, and will require, certainly, an alternative to DBS therapy. Nevertheless, there are patient categories that, currently, cannot be helped by either DBS or

CS. These include patients with: a) unresponsive non-motor or levodopa-resistant symptoms, or b) parkinsonism due to multisystem atrophy or progressive supranuclear palsy. It is promising that TMS carries positive predictive value for MCS efficacy [4]. New predictive tests need to be developed that will be individualized for the specific method, i.e. DBS or CS, the age, and the stage of the disease. Future research in PD should explore the significance of new targets such as the peduculopontine nucleus (PPN), as well as the neuroprotective effect of DBS [51].

Dystonia

DBS of the *GPi* is currently the most effective treatment for medically-refractory primary dystonia and, particularly, cervical dystonia; improvements in mobility have been proven by randomized, sham stimulation-controlled trials and can approach the 70% mark [22, 24, 26, 32]. Recently, GPi DBS was also proved effective in treating truncal dystonia such as camptocephalia and camptocormia [47, 48]. Various investigators have tried other targets, such as the *thalamus* and *STN*; the reported results, however, although encouraging, are still scarce and contradictory [54]. The above suggests that GPi DBS will remain the mainstay of therapeutic electrical stimulation in dystonia in the foreseeable future. Following DBS, the phasic dystonic movements tend to improve soon after surgery but the response of tonic movements usually is delayed [25]. This difference should be studied further while taking seriously into consideration the observation that fixed dystonia responds much better to CS rather than to DBS [43]. Bidirectional interconnectivity between motor cortex and basal ganglia could explain why certain types of dystonia respond to electrical stimulation at either deep or cortical areas. CS may induce distant neuromodulatory activity either at the cortical and/or subcortical levels through orthodromic or antidromic effects [61]. It is very important to investigate further the therapeutic potential of modulating this interconnectivity. In addition, research in the future will likely continue to explore the mechanism of DBS action, the most appropriate programming settings for each specific type of dystonia, and alternative deep or cortical brain targets which, when stimulated, might have greater efficacy and fewer side effects compared to GPi DBS.

Tremor

DBS of the *ventral intermediate thalamic (VIM)* nucleus has resulted to great improvements in the control of essential tremor (ET), limb functionality, quality of life, and need for medication [27, 57]. VIM DBS has also controlled successfully multiple sclerosis-related tremor [73], Holmes tremor [42] and tremor associated with inherited cerebellar ataxia [50]. Unilateral VIM DBS controls effectively contralateral idiopathic and secondary tremor [10], but head, trunk and voice tremor are better suppressed by bilateral VIM DBS [59]. DBS of the subthalamic posteromedial white matter, i.e. *zona incerta (ZI)* has also suppressed markedly proximal tremor [33]. STN DBS has suppressed ET as well in patients with coexisting movement disorders such as PD [52]. It seems that distal limb tremor responds well to VIM DBS, whereas proximal limb tremor is better controlled by ZI DBS [34].

There are only few reports of successful application of CS in non-parkinsonian post-stroke distal rest and/or action tremor [4, 20, 35]. Noninvasive CS such as TMS has also not provided any clinical benefits in non-parkinsonian tremor [15]. Hence, it is quite doubtful that CS could represent an alternative to DBS treatment for ET. This may be attributed to the fact that the underlying pathophysiological dysfunction of ET affects the central portion of the corticospinal tracts rather than the cortex *per se* [9]. In addition, there is sufficient evidence on the involvement of the somatosensory thalamus and VIM nucleus in the integration of cortico-basal ganglia-spinal networks and the cerebello-thalamocortical system [62]. Currently, tremor appears to be the best indication of DBS and VIM DBS the "gold standard" in the treatment of medically-refractory tremor. Future progress may depend on understanding better the significance and the role of other nuclei such as ZI or STN.

Epilepsy

The promising results of DBS in animal epilepsy [67] and the experience gained in humans on DBS in movement disorders and, particularly, in dystonia, which frequently has a paroxysmal character similar to epilepsy, provided a solid scientific basis for conducting clinical trials of DBS in epilepsy. After the first reports of efficacy [63], several deep brain targets have been stimulated. Thalamus is a pivotal structure in epileptogenesis and two nuclei have been targeted for controlling seizures: the centromedian nucleus of the thalamus (CMT) and the anterior nucleus of the thalamus (ANT). In addition, STN and caudate nucleus have also been investigated. DBS of the aforementioned targets aims to modulate the abnormal cortical function by activating or inhibiting relay nodes which, although remote to the epileptogenic area, are critical epilepsy "gating mechanisms". Theoretically, DBS of the CMT or STN, which are relay nodes with extensive cortical afferent and efferent projections, should control generalized tonic/clonic seizures. Conversely, DBS of the ANT or hippocampus, which are integrated to the limbic system, is expected to control complex partial seizures [40]. SNr and STN are parts of the nigral control of the epilepsy system (NCES) [8]. High-frequency stimulation of the STN interrupts the normally exerted by STN inhibition on SNr; at this state, the SNr ceases inhibiting the Dorsal Midbrain Antiepileptic Zone (DMAZ) and the latter can exert its antieptileptic effects [5, 66].

Other areas that have been targeted by electrical stimulation are the cerebellum and hippocampus. Cerebellum, due to its thalamic projections, was first electrically stimulated in the 1970s [6], and more recently, a mean reduction of approximately 30% in motor seizures was reported [65]. However, the mumber of treated patients remains very small. A different approach involves the direct stimulation of the *epileptogenic zone per se* such as the hippocampus. Hippocampus has a pivotal role in the generation and propagation of temporal lobe seizures [37, 56], and its stimulation provided good or excellent control of temporal lobe seizures [64, 69].

Undoubtedly, epilepsy, particularly of the drug-resistant type, is a complex, multivariate, and global rather than localized phenomenon in the brain. It is, therefore, unlikely that DBS, in its current form to become ultimately the only solution to the problem of epilepsy. Alternative types of electrical stimulation will certainly play their role in modulating the mechanisms of epileptogenesis and the numerous epilepsy phenotypes. CS and TMS have evolved as alternative adjunctive therapies for epilepsies originating primarily at the cortical level. TMS decays with the square of the distance; therefore, TMS or CS, in general, are unlikely to have an impact on seizures originating in structures deeper than the cortical layers [70]. CS or TMS methods may have to be used in combination with DBS providing thus, a big hope for sufferers and a challenging field for further research and clinical work.

Current methods deliver stimulation in an "open-loop" arrangement, i.e. according to pre-programmed electrical parameters regardless of the patient's condition. A great leap forward is the development of intelligent "closed-loop" systems of "responsive stimulation" with integrated combined use of DBS and CS. These devices have incorporated software with seizure detection algorithms and deliver stimulation to the epileptogenic focus "on demand", i.e. once premonitory electroencephalographic (EEG) signs of imminent ictal activity are detected in order to suppress seizures before their clinical manifestation. To date, two such fully implantable systems, the RNS (Neuropace, Mountain View, CA) [13] and the Intercept^{TM} (Medtronic, Minneapolis, US), which is activated by the patient, whenever he feels that a seizure is likely to start (Medtronic's website, www.medtronics.com), have offered reductions in seizure frequency and are under clinical evaluation [55].

DBS or CS are promising methods for controlling intractable epilepsy. However, before either of them becomes an effective therapy, a few important issues that are described below should be addressed. Given the lack of homogeneity in the epileptic population, *patient selection* for DBS is a factor that can determine the outcome of treatment. DBS should be offered only in patients with well defined clinical, neuroimaging, anatomical, and genetic characteristics, and similar epileptic EEG activity. This will allow us to: a) evaluate objectively the efficacy of DBS or CS for each epileptic syndrome and b) identify the best candidates for DBS or CS. The determination of the *optimal stimulation parameters* for each epileptic syndrome will maximize the clinical benefits. This would prolong battery life and reduce the time spent in programming sessions. The *volume of stimulated cerebral tissue* around the electrode can affect the overall efficacy of either DBS or CS; this should become a subject of further intensive study. For instance, in the SNr, there are two different cell groups located very densely together, which have both reciprocal GABA projections [62]. Depending on which of the two is stimulated the effect of SNr stimulation on the DMAZ could be quite different. The *interaction of antiepileptic medication* with the electrical stimulation is another important issue that has not been investigated sufficiently. Notably, in most series, patients continue their drugs during DBS therapy. In addition, it has been found that CS, in the form of low-frequency TMS, combined with valproate acid induces epileptic activity. Obviously, well-controlled studies are needed to evaluate the antiepileptic effect of DBS or CS as monotherapy in order to validate DBS alone or combined with CS as powerful therapies for epilepsy.

Pain

The use of DBS in the management of intractable pain has been rather limited although pain was the first neurological condition on which DBS was applied. Wider application of DBS in pain management has been hindered by important limitations such as the heterogeneous nature of pain and the uncertainty regarding mechanisms and appropriate targets. Similarly to epilepsy, good patient selection is mandatory for the success of DBS. The existence of a known neuroanatomical basis of the pain is a factor that favors the success of DBS and the failure or resistance to conservative treatments and both spinal cord and peripheral nerve stimulation are important prerequisites. Compared to CS, another great limitation of DBS has been the lack of a noninvasive predictive test of efficacy. It is difficult to envisage how predictive tests of successful use of DBS in chronic pain could be developed in the foreseeable future since the DBS targets are small in size and deeply located in the brain and, at present, they can be investigated only by stereotactic approaches. In the coming years, all these factors are likely to restrict DBS for pain only in highly-specialized centers [7].

Compared to DBS, CS is likely to expand its role in pain. CS seems to have better prospects to become a more generalized method in the management of intractable pain, particularly in patients who have: a) profound laterality of painful symptoms, b) extensive representation of the painful area in the lateral cortical surface, or c) advanced age. Patients with trigeminopathic pain or arm deafferentation pain are more suitable for CS as compared to DBS [28]. Another important advantage of CS over DBS is the availability of a noninvasive highly positive predictor of potential efficacy of CS such as the TMS. Progress in DBS application in pain may come from a better understanding of the underlying mechanisms and advancements in brain functional imaging, and EEG source-localization of deep nuclei which become activated during the exacerbation of pain. Such advances may also help us in identifying more precisely the suitable candidates and the critical deep brain targets. Finally, in both DBS and CS, it is necessary to make the postoperative titration easier and the follow-up sessions less time-consuming for both practitioners and patients. Undoubtedly, this will be facilitated by the development of software-guided consensus stimulation protocols.

Psychiatric Disorders

Modern research indicates that the disorders of cognition or emotion can be understood as dysfunction of specific neuronal networks or circuits, which operate upon defined anatomical pathways and convey messages by neurotransmitters or sequences of electrical stimuli. The recognition that thought and emotion are rooted in the biology of the brain and that electrochemical abnormalities in the brain either genetic or acquired can give rise to cognitive or psychiatric symptoms has become a universally accepted notion in modern neuroscience. This fundamental concept leads to the conclusion that highly selective stimulation of the brain should be further investigated as an alternative therapy of intractable psychiatric disorders. In spite of this, neuromodulatory interventions in psychiatric disorders have remained in latency due to poor understanding of pathogenetic mechanisms, fear of potential irreversible side effects, debates over long-term efficacy, but, particularly, because of social criticism and hesitancy of the medical community based on the extensive, not well-founded and indiscriminate use of crude lesional methods in the 1950s [46].

Undoubtedly, the stigmatized past of lesional psychosurgery has influenced negatively patients and carers' attitude towards DBS applications in psychiatry. In spite of this, DBS practice has been guided by prior relevant lesioning experience and has shown promising results in treatment-refractory depression (TRD), in obsessive-compulsive disorder (OCD), and in Tourette's syndrome (TS). In addition, new targets for DBS in psychiatry have been discovered as a result of serendipity. Such an example is DBS of the ventral capsule/ventral striatum for OCD which provided pronounced antidepressant benefits [14]. With respect to depression, strong evidence on the potential role of DBS in the treatment of TRD came inadvertently from PD patients who developed severe depressive behavioural changes while undergoing DBS [60]. Remarkable improvement in mood and symptoms of anxiety and depression in patients who undergo STN DBS has also been reported [74]. This was attributed to inadvertent stimulation of areas such as ZI or SNr, which are parts of the limbic system, and has supported the assumption that DBS of relay nodes in the limbic system may prove beneficial in TRD. Various other targets have been investigated such as the inferior thalamic peduncle [19], ventral internal capsule/nucleus accumbens [49], and white matter in the subgenual cingulate gyrus (Cg25 area) with the latter target showning the best clinical results [30]. Brodman's area 24a (BA 24a) has also been proposed as a putative target for DBS in TRD [44]. Many investigators consider depression as the next major therapeutic application of DBS. Given that the population of depressed patients is far greater compared to that of patients suffering from movement disorders, the impact of DBS in depression is expected to be far greater than that in movement disorders.

In OCD, neuroimaging studies have demonstrated decreased volume or increased grey matter density in cortico-striatal-thalamic-cortical circuits, increased activity in orbitofrontal cortex, cingulate, and striatum at rest, as well as temporal dysfunction and amygdala involvement [41, 58, 76]. The targeted areas by DBS that have offered considerable reductions in the intensity of symptoms and substantial decreases in required medication include the anterior limb of the internal capsule and the ventral striatum/nucleus accumbens [14, 36, 53]. Apart from the IPG battery's short life expectancy and hypomania, DBS therapy has been well tolerated.

With respect to *TS,* imaging and physiological studies suggest that the disorder involves dysfunction in both the limbic and cortical-basal gaglia-thalamocortical circuitry, explaining thus the prevailing motor and non-motor symptoms of the disease. DBS of the medial intralaminar thalamic [68] and centromedian parafascicular nuclei [18], anterior limb of internal capsule [12] and GPi [1, 11] have shown high efficacy (>70%) in suppressing motor and vocal tics. Given that TS is a borderline condition among the traditionally divided domains of neurology and psychiatry, it is likely that DBS treatment will be accepted more easily for TS compared to OCD or depression. In our opinion, TS may prove to be the condition that will facilitate the acceptance of DBS for psychiatric disorders.

As research moves forward, certain important issues should be addressed before DBS becomes an effective therapeutic option for psychiatric disorders. Few of these issues are discussed below.

Neuropsychiatric disorders are diverse with regard to their clinical profile; symptoms of different psychiatric conditions may be present in the same patient, thus making diagnosis difficult. Pre- and postoperative assessments should be done by common evaluation rating scales in order that *patient selection* would be the best possible for DBS. This would ensure that data, produced by different groups, would be comparable and strong conclusions could be inferred. Other important issues are the *optimization of stimulation parameters* and the *delineation* of the *stimulation zone*. It is well known that DBS cannot stimulate a large volume of brain tissue. In current DBS practice in psychiatry, adjustments of stimulation parameters are largely guided by experience gained in movement disorders and short battery life has been a commonly reported problem. This implies either that high-intensity stimulation is needed in order to affect a large tissue volume and produce a clinical benefit or, alternatively, that our current delineation and targeting of the appropriate stimulation zones is suboptimal. These observations indicate that it is important to standardize and individualize the stimulation settings for each psychiatric illness. This will maximize clinical benefits, increase cost-effectiveness, and improve acceptance of DBS as a therapeutic tool by the referring physicians and psychiatrists, patient's relatives, and the patients themselves.

The *DBS-induced psychiatric effects* should be analyzed and studied. In PD patients, DBS has been associated with symptoms of anxiety, fear [38] and increased rates of suicide attempts [2]. Such observations, however, should not be extrapolated to DBS procedures for psychiatric disorders. The *interaction of DBS with medications* has not been studied adequately. Currently, only seriously debilitated and pharmacologically-resistant psychiatric patients are considered for DBS. DBS efficacy is usually evaluated without the medication being discontinued. Hence, the blindness of studies is doubtful and placebo effects are possible. Undeniably, it is important to evaluate the efficacy of DBS in earlier stages of intractable neuropsychiatric disorders and also the potential beneficial effects of DBS on illness progress. The *lack of animal models of psychiatric disorders* is a great hindrance to new developments in the field. In movement disorders, progress has been facilitated by reliable animal models where efficacy of DBS ensured a smooth and safe transition to human studies. This reality brings, again, to the spotlight the fact that a great limitation of DBS is the lack of a noninvasive predictor of its potential efficacy. Conversely, before any new potential application of CS, we can have some predicting evidence of possible efficacy by TMS or transcranial direct current stimulation (tDCS). This consideration suggests that a significant expansion of DBS therapy in neuropsychiatric disorders will be difficult in the foreseeable future. On the other hand, application of other methods of brain stimulation may be accepted more easily.

The best prospect for DBS in psychiatry may be found in progress in sophisticated structural and functional imaging methods. Advanced imaging may correlate disease-related brain activity and clinical improvement with effective DBS stimulation patterns, provide clarification of mechanisms of action and indicate new potential targets for DBS. The combination of advanced electrophysiological mapping with high-resolution imaging studies is expected to detect accurately deep targets or white matter bundles which may be critical structures in the treatment of psychiatric disorders. In conclusion, DBS is a viable and safe experimental treatment for intractable neuropsychiatric disorders and holds a great hope for the severe psychiatric patients. Its efficacy in OCD, TS and depression derives from reports of single cases or small series and it should be rigorously validated and replicated in big randomized controlled trials. Only further extensive and serious research and clinical work will prove that DBS is a reliable and powerful therapy for neuropsychiatric disorders. Such therapeutic developments should be transparent in their design and evolve only within a framework of serious scientific criteria. Finally, as we move forward, it is critical to safeguard human rights and keep high *ethical standards* in every step of exploring DBS in psychiatry.

Current General Limitations of DBS

The characteristics of minimal invasiveness, reversibility, and adjustability have rendered DBS to a safe and effective treatment for neurological and psychiatric disorders which are characterized by altered electrical conductivity. In order to foresee the future prospects of DBS, one has to review its established and emerging therapeutic uses, as well as to consider the technical problems and difficulties which, currently, hinder the potential future applications of DBS and are described below.

With respect to the *required resources*, DBS practice is based on expensive high-technology equipment including frame-based stereotaxis, sophisticated software for planning and monitoring, and, particularly, multidisciplinary personel teams consisting of neuro-surgeons, neurophysiologists, movement disorder neurologists, pain therapists, psychiatrists, neuropsychologists, experienced technicians and specially-trained nurses. A wider application across the world cannot happen unless every interested hospital or health authority manages to fund the above resources and assemble the right team of specialists.

The *patient's preoperative clinical condition and evaluation* are other important variables. The patient should be mentally and physically able to undergo the preoperative clinical assessments and to attend the postoperative follow-up visits for titration of the stimulation. A significant proportion of sufferers are excluded because of advanced age (more than 70 years), medical co-morbidities, brain atrophy or cognitive and psychiatric impairment. There is a small risk (1-3%) of surgery-related complications (intracerebral hemorrhage, seizures, infection, neuropsychiatric and congitive effects). With respect to preoperative imaging evaluation, DBS is currently restricted by the imaging technology which is subject to spatial and geometrical distortions and the lack of reproducibility and uniformity among different imaging systems and devices. Thus, high accuracy becomes difficult and the comparison of data from different groups problematic.

It is generally accepted that the current DBS device, i.e. lead, wires and IPG, is susceptible and interferes with strong environmental magnetic fields. With respect to *electrode-tissue interactions*, our current understanding is limited, particularly regarding the long-term induced changes in the brain surrounding the electrode, the potential development of gliosis and how this can affect electrical conductivity and the efficacy of DBS, particularly in the long-term. Furthermore, in the coming years, it would be useful to work on the production of DBS electrodes with a high number of poles which could offer more somatotopically refined stimulation.

Although DBS has gained a protagonist role in the management of PD and other neurological disorders a comparison with CS leads to the conclusion that, currently, CS presents, in certain conditions, significant advantages over DBS. CS is less invasive in the cerebral tissue, carries a minimal perioperative risk of intracerebral hemorrhage and has a noninvasive preoperative predictive test in the form of either TMS or tDCS. Furthermore, CS does not require a stereotactic frame and, as being much less invasive, can be more acceptable by patients and referring physicians. Undoubtedly, with respect to both DBS and CS, the unsettled technical issues represent limitations; the process of elucidating these issues, however, will be associated with countless opportunities to gain new knowledge and offer great benefits to patients.

Scientific and Technological Advances: Implications for Future Prospects of Therapeutic DBS

The future of DBS will be influenced by developments in many important relevant areas. Some of them are described below.

1. Sophisticated batteries and electrodes

   In the future, IPGs will be smaller in size, miniaturized and contour-shaped to apply to the cranium or be inserted in a burr hole. Advanced chemistry will offer rechargeable batteries with more power in smaller size, extended life, and wireless connectivity to the electrodes. These devices may prove particularly helpful in patients with epilepsy or dystonia where electrode damage due to abrupt or violent movements is commonly encountered. The incorporation into the DBS electrodes of *microactuators* will enable precise permanent implantation after the initial temporary insertion, reducing thus considerably the operating time, as fine adjustments in the precise position of the electrode could be done more comfortably in the ward rather than in the operating theatre. The construction of ''*smart stimulators*'' with the capability for dynamic internal adjustments will also refine DBS practice [39].

2. Sophisticated graphics and programming software

   The electrode implantation will be assisted by software programmes that combine magnetic resonance imaging, computed tomography, three-dimensional brain atlases and neurophysiological microelectrode recording data with the volume of brain tissue activated by DBS; such programmes have been evaluated in preliminary studies [31]. Developments in programming software are likely to optimize the pole selection after permanent implantation, individualize the stimulation parameters and maximize clinical efficacy [3]. The patient's dependence from the reference center and the need for follow-up visits will be minimized by programming the stimulation devices through remote access, telephone lines or via the internet.

3. Computational neuromodulation and computer models of neural networks.

   New applications of DBS will be influenced by progress in computational neuromodulation which is a field of computational biology dedicated to the study of the biophysical and mathematical characteristics of the electrochemical modulation in the nervous system. All types of networks and neurons (motor, sensory, and interneurons) are subject to modulation. An area of expected developments that would have a great impact on DBS is the development of computer models of the bioelectrical and statistical aspects of neural recording and stimulation-induced recruitment. Neural computation and coding can be used to model, decipher and simulate the neural coding of biological neural systems. We could potentially alter the intrinsic properties of neurons, change their time-course, voltage-dependence and synaptic conductance and the strength of synaptic connections and, thus, reconfigure an anatomically defined network into a different functional circuit by altering the intrinsic properties or the synaptic strength of the neurons within the network. Hence, such intelligent implanted DBS devices could mimick the natural neuromodulation.

4. Microtechnology and nanotechnology systems

Our current implantable devices will improve and incorporate complex microsystems and sophisticated software. These will enable them to integrate information exchange between the device and the patient's brain. Microsystems technologies, microelectronics, and nanotechnologies promise us to develop cost-effective *advanced sensor and stimulation systems* and expand the therapeutic potential of both DBS and CS. The neurological disorders currently treated by DBS have as a common feature the altered electrical conductivity in certain areas of the cerebral tissue. These conditions could potentially be helped by nanomaterials such as *carbon nanofibers array* [72]. These can act as minimally traumatic CNS electrodes with increased accuracy of stimulation. Currently, such nanoelectrode arrays utilising aligned carbon nanofibers are under development at the NASA Ames Research Center. This technology holds a great promise that our ability to offer complex but precise patterns of stimulation would be enhanced making feasible to perform not only electrical microrecording but also electrochemical recording and stimulation [29]. These expected developments in micro- and nanotechnologies are likely to make feasible to link bi-directionally DBS with the human brain by "closed-loop" integrated monitoring and stimulating implantable systems, i.e. electrodes that monitor neuronal activity from multiple regions, generate ''network level'' representations of the brain and simultaneously stimulate. Promising work in epilepsy paves the road for Parkinson's disease. For instance, in PD patients, stimulation could be triggered by EMG signals elicited by patient's movement and detected by an advanced sensor being placed either in contact with the skin or embedded in muscles.

Conclusion

The methods of applying DBS and other electrical stimulation therapies are likely to improve and the indications of DBS are going to expand. The future of electrical brain stimulation will be influenced greatly by accumulating convincing evidence regarding efficacy, improved patient selection, increased awareness and understanding of current and new indications, and by developing reliable methods for assessing outcome. A more mature understanding of mechanisms of action will allow us to perform more effective and precise stimulation for each neurological or psychiatric condition, and predict more accurately the degree and duration of expected clinical benefits. Great advancements are likely to take place in the field of responsive DBS such as a closed-loop stimulation, triggered by the electrical changes that precede a seizure or an abnormal movement. Another great prospect is the integration of DBS with CS. Hence, investigators and clinical practitioners in this field should adopt a judicious but bold approach to those potential new applications of DBS and ensure a smooth transition from experimental evidence to clinical efficacy.

DBS, but similarly CS, are high-technology dependent fields that will be largely influenced by a shift away from the current dependence on pharmacological treatment and by the forthcoming advances in neuroengineering. Therapeutic electrical brain stimulation is an area of intersection, exchange and cross-fertilization of ideas from many disciplines. In this respect, a great breakthrough will be the integration of DBS and/or CS with processes of neuroplasticity, neural repair, and neuroprotection; this could have far-reaching consequences in modern neuroscience and the biomedical world, in general.

References

[1] Ackermans L., Temel Y, Cath D, et al. Deep brain stimulation in Tourette's syndrome: two targets? *Mov Disord* 2006;21:709-13.

[2] Burkhard PR, Vingerhoets FJ, Berney A, et al. Suicide after successful deep brain stimulation for movement disorders. *Neurology* 2004;63:2170-2.

[3] Butson CR, Noecker AM, Maks CB, McIntyre CC. StimExplorer: deep brain stimulation parameter selection software system. *Acta Neurochir Suppl* 2007;97(2):569-74.

[4] Canavero S, Bonicalzi V. Extradural cortical stimulation for movement disorders. *Acta Neurochir Suppl* 2007;97(2):223-32.

[5] Chabardes S, Kahane P, Minotti L, et al. Deep brain stimulation in epilepsy with particular reference to the subthalamic nucleus. *Epileptic Disord* 2002;4(suppl 3):S83–S93.

[6] Cooper IS, Amin I, Gilman S. The effect of chronic cerebellar stimulation upon epilepsy in man. *Trans Am Neurol Assoc* 1973;98:192-6.

[7] Cruccu G, Aziz TZ, Garcia-Larrea L, Hansson P, Jensena TS, Lefaucheur J-P, Simpson, Taylori RS. EFNS guidelines on neurostimulation therapy for neuropathic pain. *Eur J Neurol* 2007;14:952-70.

[8] Depaulis A, Moshe SL, "The basal ganglia and the epilepsies: translating experimental concepts to new therapies"*Epileptic Disord* (2002);4(Suppl. 3): pp. S7–8.

[9] Deuschl G, Elble RJ. The pathophysiology of essential tremor. *Neurology* 2000;54(11 Suppl 4):S14-S20.

[10] Deuschl G, Bain P. Deep brain stimulation for tremor [correction of trauma]: patient selection and evaluation. *Mov Disord* 2002;17(Suppl 3):S102-11.

[11] Diederich NJ, Kalteis K, Stamenkovic M, et al. Efficient internal pallidal stimulation in Gilles de la Tourette syndrome: a case report. *Mov Disord* 2005;20:1496-9.

[12] Flaherty AW, Willimas ZM, Amimovin R, et al. Deep brain stimulation of the anterior limb of internal capsule for the treatment of Tourette's syndrome: technical case report. *Neurosurgery* 2005;57(Suppl 4):E403.

[13] Fountas KN, Smith JR, Murro AM, Politsky J, Park YD, Jenkins PD. Implantation of a closed-loop stimulation in the management of medically refractory focal epilepsy: a technical note. *Stereotact Funct Neurosurg* 2005;83:153-8.

[14] Greenberg BD, Malone DA, Friehs GM, et al. Three-year outcomes in deep brain stimulation for highly resistant obsessive-compulsive disorder. *Neuropsychopharmacol* 2006;31:2384-2393.

[15] Hallett M. Transcranial Magnetic Stimulation: a primer. *Neuron* 2007;55:187-98.

[16] Halpern C, Urtig H, Jaggi J, Grossman M, Won M, Baltuch G. Deep brain stimulation in neurologic disorders. *Parkinsonims Relat Disord* 2007;13:1-16.

[17] Hamani C. Richter E, Schwalb JM, Lozano AM. Bilateral subthalmic nucleus stimulation for Parkinson's disease: a systematic review of clinical literature. *Neurosurgery* 2005;56:1313-21.

[18] Houeto JL, Karachi C, Mallet L, Pillon B, Yelnik J, Mesnage V et al. Tourette's syndrome and deep brain stimulation. *J Neurol Neurosurg* Psychiatr 2005;76:992-5.

[19] Jimenez F, Velasco F, Salin-Pascual R, Hernandez JA, Velasco M, Criales JL, Nicolini H. A patient with a resistant major depression disorder treated with deep brain stimulation in the inferior thalamic peduncle. *Neurosurgery* 2005;57:585-93.

[20] Katayama Y, Oshima H, Fukaya C, Kawamata T, Yamamoto T. Congrol of post-stroke movement disorders using chronic motor cortex stimulation. *Acta Neurochir Suppl* 2002;79:89-92.

[21] Kern DS, Kumar R. Deep brain stimulation. The Neurologist 2007;13:237-252.

[22] Kiss ZH, Doig-Beyaert K, Eliasziw M, et al. The Canadian multicentre study of deep brain stimulation for cervical dystonia. *Brain* 2007;130:2879-86.

[23] Kleiner-Fisman G. Herzog J, Fisman DN, et al. Subthalamic nucleus deep brain stimulation: summary and meta-analysis of outcomes. *Mov Disord* 2006;21(Suppl /14):S290-S304.

[24] Krauss JK, Pohle T, Weber S, Ozdoba C, Burgunder JM. Bilateral stimulation of globus pallidus internus for treatment of cervical dystonia. *Lancet* 1999;354:837-8.

[25] Krauss JK. Deep brain stimulation for dystonia in adults. Overview and developments. *Stereotact Funct Neurosurg*. 2002;78:168-82.

[26] Krauss JK, Loher TJ, Weigel R, Capelle HH, Weber S, Burgunder JM. Chronic stimulation of the globus pallidus internus for treatment of non-dYT1 generalized dystonia and choreoathetosis: 2-year follow up. *J Neurosurg*. 2003;98:785-92.

[27] Kumar R, Lozano AM, Sime E, Lang AE. Long-term follow-up of thalamic deep brain stimulation for essential and parkinsonian tremor. *Neurology* 2003;61:1601-04.

[28] Lazorthes Y, Sol JC, Roux FE, Verdie JC. Motor cortex stimulation for neuropathic pain. *Acta Neurochir Suppl* 2007;97(2):37-44.

[29] Li J, Andrews RJ. Trimodal nanoelectrode array for precise deep brain stimulation: prospects of a new technology based on carbon nanofiber arrays. *Acta Neurochir Suppl* 2007;97(2):537-46.

[30] Mayberg HS, Lozano AM, Voon V, et al. Deep brain stimulation for treatment-resistant depression. *Neuron* 2005;45:651-60.

[31] Miocinovic S, Noecker AM, Maks CB, Butson CR, McIntyre CC. Cicerone: stereotactic neurophysiological recording and deep brain stimulation electrode placement software system. *Acta Neurochir Suppl* 2007;97(2):569-74.

[32] Mueller J, Skogseid IM, Benecke R, , et al. Pallidal deep brain stimulation improves quality of life in segmental and generalized dystonia: results from a prospective, randomized sham-controlled trial. *Mov Disord.* 2008;23:131-4.

[33] Murata J, Kitagawa M, Uesugi H, Saitoh H, Kikuchi S, Tashiro K, Sawamura Y. Electrical stimulation of the posterior subthalamic area for the treatment of intractable proximal tremor. *J Neurosurg* 2003;99:708-15.

[34] Nandi D, Chir M, Liu X, et al. Electrophysiological confirmation of the zona incerta as a target for surgical treatment of disabling involuntary arm movements in multiple sclerosis: use of local field potentials. *J Clin Neurosci* 2002;9:64-8.

[35] Nguyen JP, Lefaucher JP, Le Guerinerl C, Eizenbaum JF, Nakano N, Carpentier et al. Motor cortex stimulation in the treatment of central and neuropathic pain. *Arch Med Res* 2000;31:263-5.

[36] Nuttin BJ, Gabriels LA, Cosyns PR. Long-term electrical capsular stimulation in patients with obsessive-compulsive disorder. *Neurosurg* 2003;52:1263-72.

[37] Oikawa H, Sasaki M, Tamakawa Y, Kamei A. The circuit of Papez in mesial temporal sclerosis: MRI. *Neuroradiology* 2001;43:205-10.

[38] Okun MS, Mann G, Foote KD, et al. Deep brain stimulation in the internal capsule and nucleus accumbens region: responses observed during active and sham programming. *J Neurol Neurosurg Psychiatry* 2007;78:310-4.

[39] Pancrazio JJ, Chen D, Fertig SJ, et al. Towards neurotechnology innovation: report from the 2005 Neural Interfaces Workshop. An NIH-sponsored event. *Neuromodulation* 2006;9:1-7.

[40] Pollo C, Villemure, J-G. Rationale, mechanisms of efficacy, anatomical targets and future prospects of electrical deep brain stimulation for epilepsy. *Acta Neurochir Suppl* 2007;97(2):311-20.

[41] Rauch SL, Baxter LR Jr. Neuroimaging in obsessive-compulsive disorder and related disorders. *In: Jenicke MA, Baer L, Minichiello WE, eds. Obsessive-compulsive disorders: practical management*, 3rd edn. St Louis, MI, USA: Mosby, 1998.

[42] Romanelli P, Bronte-Stewart H, Courtney T, Heit G. Possible necessity for deep brain stimulation of both the ventralis intermedius and subthalamic nuclei to resolve Holmes tremor. Case report. *J Neurosurg* 2003;99:566-71.

[43] Romito LM, Franzini A, Perani D, et al. Fixed dystonia unresponsive to pallidal stimulation improved by motor cortex stimulation. *Neurology* 2007;68:875-6.

[44] Sakas DE, Panourias IG. Rostral cingulate gyrus: A putative target for deep brain stimulation in treatment-refractory depression. *Med Hypothes* 2006;66:491-4.

[45] Sakas DE, Panourias IG, Simpson BA. An introduction to neural networks surgery, a field of neuromodulation which is based on advances in neural networks science and digitised brain imaging. *Acta Neurochir Suppl* 2007;97(2):3-13.

[46] Sakas DE, Panourias IG, Singounas E, Simpson BA. Neurosurgery for psychiatric disorders: from the excision of brain tissue to the chronic electrical stimulation of neural networks. *In: Sakas DE, Simpson BA, editors. Operative Neuromodulation: Neural networks surgery*. New York: Springer-Verlag; 2007. p. 365-74.

[47] Sakas DE, Panourias IG, Boviatsis ES et al. *Treatment of primary camptocormia* (dystonic bent spine or bent neck) by electrical stimulation of the globus pallidus internus: report of 3 cases. Abstract book, The 8th International Neuromodulation Society and the 11th Annual Meeting of the North American Neuromodulation Society, Acapulco, 2007, December 7-12.

[48] Sakas DE, Panourias IG, Boviatsis E, Themistocleous M, Stavrinou L, Gatzonis S. Treatment of idiopathic head-drop syndrome (camptocephalia) by bilateral chronic electrical stimulation of the globus pallidum internus. *J Neurosurg* 2008b [in press].
[49] Schlaepfer TE, Cohen MX, Frick C,et al. Deep brain stimulation to reward circuitry alleviates anhedonia in refractory major depression. *Neuropsychopharmacology* 2008;33:368-77.
[50] Schramm P, Scheihing M, Rasche D, Tronnier VM. Behr syndrome variant with tremor treated by VIM stimulation. *Acta Neurochir* (Wien) 2005;147:679-83.
[51] Simpson BA. The role of neruostimulation: the neurosurgical perspective. *J Pain Symptom Manage* 2006;31:S3-S5.
[52] Stover NP, Okun MS, Evatt ML, Raju DV, Bakay RA, Vitek JL. Stimulation of the subthalamic nucleus in a patient with Parkinson's disease and essential tremor. *Arch Neurol* 2005;62:141-3.
[53] Sturm V, Lenartz D, Koulousakis A, et al. The nucleus accumbens: a target for deep brain stimulation in obsessive-compulsive and anxiety disorders. *J Chem Neuroanat* 2003;26:293-9.
[54] Sun B, Chen S, Zhan S, Le W, Krahl SE. Subthalamic nucleus stimulation for primary dystonia and tardive dystonia. *Acta Neurochir Suppl* 2007;97:207-14.
[55] Sun FT, Morrell MJ, Wharen RE Jr. Responsive cortical stimulation for the treatment of epilepsy. *Neurotherapeutics* 2008;5:68-74.
[56] Swanson TH. The pathophysiology of human mesial temporal lobe epilepsy. *J Clin Neurophysiol* 1995;12:2-22.
[57] Sydow O, Thobois S, Alesch F, Speelman JD. Multicenter European study of thalamic stimulation in essential tremor: a six year follow-up. *J Neurol Neurosurg Psychiatr* 2003;74:1387-91.
[58] Szeszko PR, Robinson D, Alvir JMJ, et al. Orbital frontal and amygdala volume reductions in obsessive-compulsive disorder. *Arch Gen Psychiatry* 1999; 56**:** 913–19.
[59] Taha JM, Jansen MA, Favre J. Thalamic deep brain stimulation for the treatment of head, voice, and bilateral limb tremor. *J Neurosurg* 1999;91:68-72.
[60] Temel Y, Kessels A, Tan S, Topdag A, Boon P, Visser-Vandewalle V. Behavioural changes after bilateral subthalamic nucleus stimulation in advanced Parkinson's disease: A systematic review. *Parkinsonism Relat Disord* 2006;12:265-72.
[61] Tisch S, Rothwell JC, Limousin P, Hariz MI, Corcos DM. The physiological effects of pallidal deep brain stimulation in dystonia. *IEEE Trans Neural Syst Rehabil Eng* 2007;15:166-72.
[62] Utter AA, Basso MA. The basal ganglia: An overview of circuits and function. *Neurosci Biobehavior Rev 2008*;32:333–42.
[63] Velasco F, Velasco M, Jimenez F, et al. Predictors in the treatment of difficult-to-control seizures b electrical stimulation of the centromedian thalamic nucleus. *Neurosurgery* 2000;47:295-304.
[64] Velasco AL, Velasco M, Velasco F, et al. Sucabute and chronic electrical stimulation of the hippocampus on intractable temporal lobe seizures: preliminary report. *Arch Med Res* 2000;31:316-28.
[65] Velasco F, Carrillo-Ruiz JD, Brito F, Velasco M, Velasco AL, Marquez I, Davis R. Double-blind, randomized controlled pilot study of bilateral cerebellar stimulation for treatment of intractable motor seizures. *Epilepsia* 2005;46:1071-81.

[66] Veliskova J, Velsek L, Moshe SL. Subthalamic nucleus: a new anticonvulsant site in the brain. *Neuroreport* 1996;7:1786–1788.

[67] Vercueil L, Benazzouz A, Deransart C, Bressand K, Marescaux C, DepauLis A, Benabid A-L. High frequency stimulation of the subthalamic nucleus suppresses absence seizures in the rat: comparison with neurotoxic lesions. *Epilepsy Res* 1998;28:158-61.

[68] Visser-Vandewalle V, Temel Y, Boon P, et al. Chronic bilateral thalamic stimulation: a new therapeutic approach in intractable Tourette's syndrome. Report of three cases. *J Neurosurg* 2003;99:1094-1100.

[69] Vonck K, Boon P, Claeys P, Dedeurwaerdere S, Achten R, Van Roost D. Long-term deep brain stimulation for refractory lobe epilepsy. *Ann Neurol* 2005;52:556-65.

[70] Wagner T, Gangitano M, Romero R et al. Intracranial measurement of current densities induced by transcranial magnetic stimulation in the human brain. *Neurosci Lett* 2004;354:91-4.

[71] Weaver F. Follett K, Hur K, Ippolito D, Stern M. Deep brain stimulation in Parkinson's disease: a meta-analysis of patient outcomes. *J Neurosurg* 2005;103:956-67.

[72] Webster TJ, Waid MC, McKenzie JL, Price RL, Ejiofor JU. Nano-biotechnology: carbon nanofibers as improved nerual and orthopaedic implants. *Nanotechnology* 2004;15:48-54.

[73] Wishart HA, Roberts DW, Roth RM, et al. Chronic deep brain stimulation for the treatment of tremor in multiple sclerosis: review and case reports. *J Neurol Neurosurg Psychiatr* 2003;74:1392-7.

[74] Woods SP, Fields JA, Troster AI. Neuropsychological sequelae of subthalamic nucleus deep brain stimulation in Parkinson's disease: a critical review. *Neuropsychol Rev* 2002;12:111–6.

[75] Yu H, Neimat J. The treatment of movement disorders by deep brain stimulation. *Neurotherapeutics* 2008;5:26-36.

[76] Zungu-Dirwayi N, Hugo F, van Heerden BB, Stein DJ. Are musical obsessions a temporal lobe phenomenon? *J Neuropsychiatry Clin Neurosci* 1999; 11: 398–400.

In: Encyclopedia of Neuroscience Research
Editors: Eileen J. Sampson and Donald R. Glevins

ISBN 978-1-61324-861-4

Chapter XIII

Cortical Stimulation versus Deep Brain Stimulation in Neurological and Psychiatric Disorders: Current State and Future Prospects [*]

Damianos E. Sakas*† *and Ioannis G. Panourias

Department of Neurosurgery, University of Athens Medical School, Evangelismos Hospital, and P.S. Kokkalis Hellenic Center for Neurosurgical Research, Athens, Greece

Abstract

Therapeutic brain stimulation can be applied: a) in networks which are located in cortical and subcortical layers, b) in deep nuclei groups (relay nodes) or c) in combination. CS and DBS belong to the domain of *operative neuromodulation* and particularly to the field of *neural networks surgery* (Sakas et al 2007a), defined as the field that studies and applies advancements in neural networks research, digitized stereotactic brain imaging and implantable electrical or electronic devices in order to alter electrically the signal transmission in the nervous system, modulate neural networks and produce therapeutic effects.

Therapeutic brain stimulation methods are currently distinguished into two major categories: a) noninvasive and b) invasive. The *noninvasive* methods (rTMS, tDCS) stimulate electrically the brain cortex after applying electrical currents on the human scalp. The *invasive* methods require the surgical implantation of specially designed electrodes either in contact with the cortex of the brain, i.e., CS, or in deeply located selected targets, i.e., DBS. Instead of "*invasive*", the terms *implantable* or *by implantable devices*" are much more appropriate, as they do not carry the negative association of a traumatic intervention that is implied by the term *invasive CS*. In this chapter, we

[*] A version of this chapter was also published in *Text of Therapentic Cortical Stimulation*, edited by Sergio Canavero published by Nova Science Publishers, Inc. It was submitted for appropriate modification in an effort to encourage wider dissemination of research.

† Correspondence concerning this article should be addressed to Dr. Damianos E. Sakas, MD. E-mail: sakasde@med.uoa.gr.

describe and compare the current state and future prospects of DBS and CS in the fields of neurology and psychiatry.

Current Applications

1. Pain

DBS: The rationale for the use of DBS in chronic pain dates back to the 1950s and, to date, more than 1,300 patients have been offered such a treatment. However, many factors limited its use: insufficient understanding of the neural networks that convey the perception of pain, uncertainties about optimal brain targets, inconsistent efficacy, and the success of spinal cord stimulation and opioid infusion pumps. The US Food and Drug Administration, based on the results of two multicenter trials, withdrew permission to use DBS as a treatment for pain (Burchiel 2001). The method, however, still remains in practice in Canada, Europe and Asia, but the number of reported cases over the last decade is limited (Bartsch et al, 2008; Owen et al., 2006a and b; Rasche et al., 2006). Traditionally, the two most targeted areas for treatment of pain are: a) the sensory thalamus (Vc, ventrocaudalis) and b) the periaqueductal and periventricular gray (PAG and PVG). Most retrospective studies support the view that thalamic DBS has greater efficacy in neuropathic pain with 50-60% success rate (Levy et al., 1987; Bendok and Levy, 1998); mesencephalic DBS has proved more efficacious in nociceptive pain (including osteoarthritis and cancer (Rezai and Lozano, 2002) and the mixed-pain failed back surgery syndrome (Young and Rinaldi, 1997; Kumar et al., 1997) with 70% pain relief. However, DBS is poorly effective for central pain (Canavero and Bonicalzi 2007a,b), and contradictory results exist for phantom limb pain and postherpetic neuralgia (Owen et al 2006b; Rasche et al 2006). Recently, DBS of the posterior hypothalamic region has emerged as an effective treatment for otherwise medically-refractory cluster headache (CH) (Bartsch et al., 2008; Owen et al., 2007; Schoenen J et al., 2005; Starr et al., 2007), with half of the reported patients experiencing fair-to-excellent long-term relief in terms of frequency of pain attacks and severity of pain. The risk of intracerebral hemorrhage when bilaterally targeting such deeply-located structures remains a pervasive concern.

CS: MI ECS was introduced for the management of pain following the recognition that DBS of the sensory thalamic nuclei had offered disappointing long-term results. Initially, MI ECS was proposed as a treatment for central thalamic pain (chapter 9) and facial neurogenic pain (trigeminal neuropathy) (chapter 10). MI ECS has proved most effective for trigeminal neuropathic pain with about 60% of the sufferers reporting a fair to excellent long-term pain relief (chapter 10). MI ECS is also effective in about half of all reported patients with CPSP (chapter 9). In phantom limb and brachial plexus avulsion pain, MI ECS offers an average success rate of slightly above 40% (chapter 10). Sensory cortex stimulation may also afford analgesia (chapter 9) and should be explored further. The application of either motor or sensory CS by implantable systems is moving along with that of noninvasive cortical stimulation such as TMS or tDCS (chapter 8). Despite modest effects, analgesia is usually transient, and even with protracted sessions, it lasts no more than 1 month, at best. However, they may predict a positive outcome to ECS (Canavero and Bonicalzi, 2007a and b; chapter 8). The apparent superiority of tDCS over rTMS may pave the way to clinical application.

CS vs. DBS: Considerations, Limitations and Future Prospects

The small number of patients treated in randomized placebo-controlled clinical trials, the heterogeneous nature of pain disorders, the uncertainty regarding the appropriate target for each specific pain syndrome, the lack of universal stimulation protocols and systematic follow-up data, and DBS-associated complications (diplopia, seizures, nausea, paresthesia, and, above all, intracerebral hemorrhage) are all factors that are likely to hinder an expansion of DBS in pain therapy for the foreseeable future and restrict it only to a few highly-specialized centers (Cruccu et al., 2007).

Apparently, implantable CS devices offer the best alternative for the surgical treatment of intractable pain conditions. Apart from the aforementioned factors, other important criteria in selecting CS over DBS inlude: a) the profound laterality of painful symptoms, b) the extensive representation of the painful area in the lateral cortical surface, and c) advanced patient's age. Bilateral pain is not a contraindication for ECS: bilateral effects from unilateral stimulation have been reported. Yet, bilateral surgery should be offered with a word of caution. Another problem arises when the pain afflicts a body region, such as the leg, which is represented in the medial interhemispheric cortical surface in several patients. In this case, only the much more invasive subdural CS would be offered. This problem may become more complex if leg pain is bilateral; in such cases, the electrodes should be inserted in both aspects of the interhemispheric fissure which may not be an easy procedure in many patients. Undoubtedly, it is much easier to stimulate the lateral aspect of the cortex unilaterally rather than the interhemispheric surface bilaterally. Fortunately, there is evidence that unilateral CS can have bilatateral beneficial effects (Canavero et al., 2007b). Patients with trigeminal or arm neuropathic pains are especially suitable for ECS as compared to patients with bilateral leg pain. ECS appears to be a viable therapeutic option for drug-resistant central and trigeminal neuropathic pain. This is important, because neuropathic pain is a major concern for pain therapists due to its unresponsiveness to available drug and conservative treatments. Notably, the cut-off rate for success in pain relief is arbitrarily set at >50%; in the case of NP, however, pain improvement of 40% or even 30% has been sufficient to consider ECS effective. At this time, TMS or tDCS can be proposed as a preoperative test for selecting candidates for implantable CS procedures and possibly to alleviate painful conditions for a short period before a scheduled ECS procedure (chapter 8). A meta-analysis concluded that MI ECS appears to be more effective in alleviating chronic pain syndromes as compared to either MI rTMS or tDCS (Lima and Fregni, 2008). The availability of a noninvasive screening procedure for CS represents an important advantage of implantable (invasive) CS over DBS. Conversely, no drug-based or other test method has been developed that could predict the efficacy of DBS and, therefore, help us in selecting the right patients. DBS targets are small in size and deeply located in the brain and, at present, they can be investigated only by stereotactic approaches.

CS is likely to expand its role in pain management. Patient selection will be optimized by developing drug dissection protocols in order to identify "neurochemical signatures" portending a successful outcome and selecting stimulation targets on the basis of anomalies as seen in neuroimaging studies (Canavero and Bonicalzi 2007, chapter 9). For instance, SI cortex has been first targeted on the basis of theory and metabolic findings (Canavero and Bonicalzi 1995). The combination of anatomical, functional imaging, clinical, and intraoperative neurophysiological data (by keeping the patient awake during the procedure) will further improve target localization, together with confirmation by means of noninvasive stimulation. Securing the electrode in the predetermined position may reduce possible subsequent migration, with loss of effect. Software-guided stimulation protocols must be developed to facilitate postoperative titration and make the follow-up sessions less time-consuming and more comfortable for both practitioners and patients; moreover, it will probably eliminate the need of "intensive reprogramming" or surgical repositioning of the electrode in case of tolerance-like phenomena or loss of ECS efficacy over time. On the basis of the available published studies and evidence from noninvasive brain stimulation, CS by implantable devices will likely remain at the forefront of pain therapies in the foreseeable future, particularly as the mechanisms of action and clinical efficacy will be addressed and evaluated in large-scale randomized controlled trials.

2. Parkinson's Disesae

DBS: DBS has metamorphosed the treatment of PD by alleviating its troublesome symptoms and improving substantially the quality of life of sufferers. The reversibility, adjustability and programmability of DBS have favored it over the lesional procedures of the past. A change of rigidity is usually seen within 20-30 seconds with a maximum within 1 minute. Tremor can improve over weeks or months. Bradykinesia may respond after several hours or days and returns to baseline within up to 24 hours. The induction of dyskinesias with DBS in the short term predicts a favorable long-term outcome. Postoperatively, dopaminergic drugs may be reduced (about 50%) due to the additive effect; however, too rapid or too drastic a decrease in dopaminergic drugs carries the risk of unmasking apathy or depression or mania or psychosis or akinetic crisis. After optimization of parameters, levodopa may be gradually replaced with long-acting dopamine agonists. Postoperative dyskinesias or temporary worsening of PD due to drug reduction are possible. The thalamic *ventral intermediate (Vim) nucleus* was the first DBS target in the 1980s, with almost 90% of PD patients experiencing years-long control of their tremor (Benabid et al., 1996). The minimal effect of VIM DBS on rigidity and bradykinesia spurred researchers to explore alternative nuclei as potential targets. Animal studies showed that the *subthalamic nucleus (STN)* is an integral relay node in the cortical-basal ganglia-thalamocortical pathways. The efficacy of STN DBS has now been documented by numerous groups and validated in prospective, randomized trials. Long-term follow-up and meta-analysis studies have demonstrated sustained improvement in the Unified PD Rating Scale (UPDRS) motor subscores of tremor, rigidity, and akinesia in the off-medication state, but minimal improvement in the on-medication state (Kleiner-Fisman et al., 2006; Yu and Neimat, 2008). In prospective open-label trials, STN DBS offered a significant benefit on the quality of life, particularly in patients younger than 65 years (Derost et al., 2007; Deuschl et al., 2006; Krack et al., 2003). Bilateral STN DBS is considered the standard surgical therapy for medically-refractory PD; unilateral STN DBS is efficacious in highly asymmetric parkinsonism (Kern and Kumar, 2007). However, concerns regarding continuous ipsilateral deterioration of PD symptoms over time restrict unilateral STN DBS only to highly selected patients. Based on reports of efficacy of pallidotomy for PD, *globus pallidus internus (GPi)* emerged as a target for DBS. High-frequency stimulation of the GPi alleviates rigidity, tremor and bradykinesia in the off-medication state, substantially decreases disabling levodopa-induced dyskinesias in the on- and off-medication states and improves the quality of life (Deuschl et al., 2006; Wichmann and DeLong, 2006; Yu and Neimat, 2008). In contrast to patients who received STN DBS, those with GPi DBS are not able to substantially reduce their medications (Rodriguez-Oroz et al., 2005). *Peduculopontine nucleus (PPN)*, a relay node in the network connecting GPi and STN with the basal ganglia and the peripheral nervous system, has recently been proposed as a new target for DBS in PD. Clinical studies support its efficacy, particularly, for postural instability and akinesia in parkinsonism (Nandi et al., 2002a; Plaha and Gill, 2005). Similarly, DBS of the caudal *zona incerta (ZI)* has shown beneficial effects on parkinsonian tremor, rigidity and akinesia (Kitagawa et al., 2005; Plaha et al., 2005).

CS: Functional neuroimaging studies have shown that cortical areas such as the primary motor cortex and the supplementary motor area are hypo- or hyperactive in both early and late stages of PD (chapter 12). Canavero and collaborators first reported on the clinical benefits obtained from unilateral extradural MI ECS in three PD patients (Canavero and Paolotti, 2000; Canavero et al., 2002). UPDRS motor subscores (Part III) were decreased by 48% and the need for levodopa by 80%. Afterwards, the benefits of unilateral MI ECS in improving all parkinsonian symptoms (tremor, rigidity, bradykinesia, posture and gait disturbances, freezing) and reducing the required daily medications were confirmed in a large multicenter study (chapter 13 and 13b). The neurophysiologically documented prolonged modulatory activity of TMS on pathologically hyperactive or hypoactive areas of the human brain cortex has supported the assumption that TMS may also improve the motor components of PD (chapter 12).

CS vs. DBS: Considerations, Limitations and Future Prospects

DBS has been acknowledged worldwide as the surgical treatment of choice for improving the incapacitating symptoms and the quality of life in PD patients: in 2002, it received FDA's approval as a treatment option for PD. Bilateral STN or GPi DBS is effective in patients with severe medication-refractory tremor-dominant PD. DBS of the Vim is no longer recommended for PD (Kumar et al., 1999; Kern and Kumar, 2007; Krack et al., 2003). In spite of the absence of published blinded, randomized, controlled studies comparing the efficacy of DBS of the STN or GPi, the STN has gained general acceptance as the preferred target (Halpern et al., 2007; Kern and Kumar, 2007). The reasons for the supposed superiority of the STN over GPi DBS in PD include: easier targeting on MRI, longer stimulation effects of STN stimulation, larger reductions in medication, longer IPG battery life due to the lower intensity required for stimulating the smaller (as compared to GPi). Although longer follow-up data from randomized, controlled studies are needed to correlate the improvement induced by DBS with that due to medication, there is little doubt that DBS represents a breakthrough in the history of the treatment of PD. However, almost half of all PD sufferers are excluded from DBS mainly due to advanced age (>70 years), significant dementia or psychiatric comorbidity, poor response to levodopa, low scores (<30-40) in the UPDRS part III or other general contraindication to DBS. Non-motor and non-dopaminergic symptoms of PD respond poorly to DBS, but newer targets may boost DBS potential. DBS has been associated with neuropsychiatric untoward effects (e.g., dementia, apathy, anxiety, suicidal attempts) and cognitive undesired sequelae (e.g., a decline in lexical fluency and executive functions: Temel et al., 2006). There also exists a considerable risk of surgery- and hardware-related complications estimated to be 1-3% and up to 25%, respectively (Breit et al., 2004; Oh et al., 2002). A hint that DBS may slow the progression of Parkinson's disease remains unproven and, if DBS is not applied early in the course of PD, such evidence will not become available (Simpson, 2006a).

Due to the limited number of PD sufferers that have been treated by MI ECS so far, it is difficult to compare it to DBS. However, it seems that a considerable number of patients remain desperate and unrelieved. MI ECS appears to have some advantages over DBS in the treatment of PD: 1) it can alleviate all cardinal symptoms, but particularly the axial ones, 2) unilateral MI ECS induces beneficial effects bilaterally, 3) MI ECS is more suitable in elderly PD cases due to minimal surgery-related side effects and lower complication rates; moreover, the bilateral beneficial effects induced by unilateral MI ECS may delay the progressive deterioration of PD in the non-stimulated side, 4) MI ECS is more cost-effective because it does not require stereotactic equipment, and 5) MI ECS promises to potentially benefit almost half of the PD patients who have been excluded from DBS. Given that the number of PD sufferers aged >70 years will steadily increase, a considerable part of this patient population will require an alternative to DBS therapy; currently, CS appears the most likely option. A few caveats are in order: a) patients with levodopa-resistant parkinsonian symptoms are unlikely to respond to MI ECS, b) parkinsonism related to multisystem atrophy or progressive supranuclear palsy hardly benefits from MI ECS, and c) TMS-induced improvement of PD symptoms carries positive predictive value for MI ECS efficacy (Canavero 2007a). Nevertheless, a negative rTMS test does not predict failure of MI ECS (Cioni, et al., 2007). Future efforts should focus on exploring the potential efficacy of stimulation of premotor area and the neuroprotective potential of MI ECS (Canavero and Bonicalzi, 2007a).

3. Tremor

DBS: Tremor affects up to 2% of the general population worldwide, and can be essential (idiopathic) or associated with various conditions such as cerebellar dysfunction, multiple sclerosis, lesions of the brainstem (Holmes tremor) or head injury. Unilateral DBS has proved to be particularly effective in controlling contralateral idiopathic and secondary forms of tremor (Deuschl and Bain, 2002). Bilateral DBS has proved beneficial in head, trunk and voice tremor (Taha et al., 1999). Traditionally, based on the existing experience of lesional surgery, the *Vim thalamic nucleus* has been targeted. Long-term follow-up data of unilateral Vim-DBS have documented the great improvement in essential tremor (ET) suppression, limb functionality, quality of life, and reduced need for medication (Kumar et al., 2003; Sydow et al., 2003). In 8 patients with ET, unilateral DBS of the posteromedial subthalamic white matter, i.e., *zona incerta* (ZI) suppressed contralateral proximal tremor by 81% for a mean follow-up of 22 months (Murata et al., 2003). STN DBS has suppressed ET in patients with coexisting movement disorders, such as PD (Stover et al 2005). It is generally accepted that distal limb tremor responds well to Vim DBS, whereas proximal limb tremor is better controlled by ZI DBS (Nandi et al, 2002b). In a systematic review of 78 patients with tremor secondary to multiple sclerosis, unilateral Vim DBS offered postoperative benefits in motor function and daily activities in 88% and 76% of sufferers, respectively (Wishart et al 2003). Vim DBS has also successfully controlled tremor associated with unusual pathological conditions such as Holmes tremor (Romanelli et al 2003) or inherited cerebellar ataxia (Schramm et al 2005).

CS: There are only few reports of successful CS application in non-parkinsonian tremor. MI ECS has shown the best efficacy in post-stroke distal rest and/or action tremor (Canavero and Bonicalzi, 2007a; Katayama et al., 2002). In most cases, it reduced tremor significantly when it was offered to alleviate coexisting pain or movement disorders. Otherwise, the results of MI ECS in the management of ET have been disappointing (chapter 13). Similarly, noninvasive cortical stimulation has not provided clinical benefits in non-parkinsonian tremor (Hallett, 2007).

CS vs. DBS: Considerations, Limitations and Future Prospects

Tremor has been part of the repertoire of stereotactic and functional surgery over the last fifty years. VIM DBS is the "gold standard" in the treatment of medically-refractory non-parkinsonian tremor with an efficacy up to 80% and long-lasting benefit for more than 7 years. MI ECS has not proved particularly effective in non-parkinsonian tremor, so far. This may be due to the pathophysiologic substrate of ET. Although recent reports support a cortical involvement (Raethjen et al., 2007), the role of the cortex in the generation of ET has not been elucidated. On the other hand, there is sufficient demonstration of the key role of the somatosensory thalamus, and the Vim nucleus in particular, in the integration of neural loops within both the cortico-basal ganglia-spinal networks and, also, the cerebello-thalamocortical system. These networks are essential in the pathogenesis of tremor and, therefore, have offered convincing support that Vim is the optimal target for alleviating tremulous conditions, regardless of the underlying cause. Yet, given MI ECS efficacy in relieving parkinsonian tremor (although less so than akinesia and rigidity), it is possible that its limited efficacy in non-parkinsonian tremor may be due to wrong targeting (other areas rather than MI should be stimulated), suboptimal stimulation protocols or electrode-related shortcomings. The development of new electrode arrays with a larger number of poles supported by multiple programming IPGs may enhance its efficacy in the future.

4. Dystonia

DBS: DBS is currently the most effective treatment for medically refractive dystonia. The main target used in primary dystonia (DYT1 gene positive) is the *Globus Pallidus internus (GPi)* and its efficacy has been shown in generalized dystonia, segmental dystonia, and complex cervical dystonia with movement scores typically improving by 75% (Kiss et al., 2007; Mueller et al., 2008; Krauss et al., 1999, 2003). Certain types of truncal dystonia (camptocephalia and camptocormia) are emerging as good candidates for pallidal DBS (Sakas et al., 2008b; Sakas et al., 2007c). DBS maintains marked long-term symptomatic and functional improvement in the majority of patients with dystonia. (Loher et al., 2008). Upon activating DBS, phasic dystonic movements tend to improve early after surgery, but the response of tonic movements to chronic stimulation may be delayed (Krauss et al., 2002). Other targets, such as the *thalamus* and *STN* have shown encouraging results, but the available data is still scarce and sometimes contradictory (Sun et al., 2007). In sum, GPi DBS is an established treatment for idiopathic dystonia, either generalized or segmental, and its efficacy has been proven by extensive experience and by randomized, sham stimulation-controlled trials. Future research will likely continue to address the most appropriate programming settings for various subpopulations of dystonia, the mechanism by which DBS affects this ailment, and the possibility of alternative brain targets that might have fewer associated side effects or greater efficacy than the GPi.

CS: CS has been reported to be effective in certain forms of fixed dystonia (chapter 13). The reports are only a few, but preliminary data implies that the mechanism of fixed dystonia is more corticalized than in primary torsion dystonia. Some task-specific forms of dystonia, such as writer's cramp or musician's cramp, may be related to maladaptive cortical plasticity as a result of overuse or improper use of the hand (Byl et al., 1996; Classen et al., 2003). Again, those focal forms of dystonia may be suitable candidates for CS. The optimal stimulation site remains to be elucidated. Based on results from rTMS on various forms of the disease, stimulation over the premotor cortex seems to yield better results than stimulation over the motor cortex (Murase et al., 2005). This may reflect both local and remote effects on distant, but connected sites (e.g., MI, the posterior parietal cortex and the basal ganglia). The assumption that stimulation over the premotor cortex may affect both the deep relay-nodes in the basal ganglia and the cortical sensory areas is in agreement with the known pathophysiology of dystonia, which involves the basal ganglia, MI and the somatosensory system in a diffuse sensory-motor processing dysfunction. Despite reports of invasive and non-invasive MI stimulation for fixed dystonia (chapter 13; Murase et al., 2005), it would be interesting to compare those effects with premotor cortex stimulation effects as well. Hence, even though MI ECS appears to have beneficial effects, it may be postulated that CS over the premotor cortex might be more effective.

CS vs. DBS: Considerations, Limitations and Future Prospects

Bidirectional interconnectivity between motor cortical areas and deep-seated structures, such as the basal ganglia, could explain why dystonia and other movement disorders respond to neuromodulation at multiple sites. CS may induce distant neuromodulatory effects either at the cortical and/or subcortical levels through orthodromic or antidromic effects, similarly to what is postulated as a possible mechanism of action of DBS (Tisch et al., 2007). The proven efficacy of GPi DBS and, to a lesser extent, STN DBS in treating a wide spectrum of dystonic phenotypes suggests that DBS will remain the mainstay of therapeutic neuromodulation of dystonia in the near future. CS should however be better assessed by, e.g., comparing unilateral versus bilateral stimulation of MI and non-primary motor areas in fixed rather than in torsion dystonia, and even combined with DBS. Initial experimental and clinical data suggest that the potential efficacy of CS should be explored in cases of fixed or unilateral focal segmental dystonia involving one arm and neck or part of the trunk or leg, in cases of unilateral secondary dystonia of a multisegmental character, especially with a normal MRI scan, where DBS is likely to be less effective. rTMS may provide a useful predictor of clinical outcome and help triaging those patients who are likely to respond to CS.

5. Epilepsy

DBS: 30% of epileptics are drug-resistant, due to either failure to respond to available antiepileptic drugs or medication-related adverse effects (Silanpää and Schmidt, 2006; Kwan and Brodie, 2000). Resective brain surgery of the epileptic focus can be effective in judiciously selected patients; unfortunately, up to 40% of sufferers are not eligible for such surgical treatment. Moreover, post-operatively, anticonvulsant agents cannot be discontinued, seizure remissions may occur and undesired neuropsychiatric sequelae often lessen the surgical success rate. Several brain areas have been targeted by DBS (Velasco et al., 2000a). The thalamus has been long considered a pivotal brain structure in the epileptogenic process and two thalamic nuclei have been targeted for suppressing intractable seizures: the *centromedian nucleus of the thalamus* (CMT) and the *anterior nucleus of the thalamus* (ANT). In addition, four other areas have been investigated: STN, cerebellum, caudate nucleus and hippocampus. The CMT has been considered a suitable DBS target on the assumption that it has an integrating role in the propagation of generalized seizures through its participation in an ascending brainstem-diencephalon-subcortical loop. CMT DBS has been successful in controlling mainly generalized tonic-clonic seizures and atypical absences (Velasco et al., 1993a; Velasco et al., 1993b; Velasco et al., 1995; Velasco et al., 2001) and, to a lesser degree, partial complex seizures (Fisher et al., 1992). The role of CMT DBS, although promising, remains moot and should be further evaluated by large scale randomized-controlled studies (Andrade et al., 2006; Fisher et al., 1992). The ANT, with its relatively small size, safe distance from vascular structures and documented anterograde projections to cortical, subcortical and limbic regions (Halpern et al., 2008) has been targeted for DBS. Bilateral ANT DBS reduced the frequency of partial complex seizures by approximately 50% in a small series and is being submitted to a controlled trial (Andrade et al., 2006; Hodaie et al., 2002; Kerrigan et al., 2004; Halpern et al., 2008). The STN has been also evaluated as a potential DBS target in drug-resistant epilepsy. In animal models, inhibition of the GABAergic neurons of the substantia nigra pars reticulata (SNr) reduces partial and generalized epileptic seizures (Iadarola and Gale 1982). Based on its excitatory effect on the nigral system, STN has been targeted in small, open-label prospective studies; notably, STN DBS reduced seizure frequency by 60-80% (Benabid et al., 2001; Chabardes et al., 2002; Handforth et al., 2006), but large trials are needed to confirm these initial results. In the early 1970s, the cerebellum, due to its thalamic projections, was electrically stimulated for suppressing seizures (Cooper et al., 1973) and Velasco et al., (2005) reported a mean reduction of approximately 30% in motor seizures in five patients. At present, less than 20 patients have been treated by cerebellar stimulation in controlled studies and less than five experienced a significant reduction in their generalized tonic/clonic seizures. Chronic electrical stimulation of the head of the caudate nucleus (HCN) is now being evaluated (Chkhenkeli and Chkhenkeli, 1997; Chkhenkeli et al., 2004). The reported suppression of seizures has been attributed to the hyperpolarization of cortical neurons, induced by activation of the HCN by low-frequency stimulation (4-8 Hz). DBS of the aforementioned targets aims to modulate the abnormal cortical function by activating or inhibiting relay nodes which, although distant from the epileptogenic area, are critical epilepsy “gating mechanisms”. A different approach involves the direct stimulation of the epileptic zone per se. There is growing evidence that, in the years to come, stimulation of the amygdalohippocampal region will receive special attention. This technique may prove particularly helpful when the epileptogenic focus either involves eloquent brain areas or is not suitable for a resective procedure. The hippocampus, an integral component of Papez’s circuit, plays a pivotal role in both the generation and propagation of temporal lobe seizures (Swanson, 1995; Oikawa et al 1994). Hippocampal stimulation provided excellent control of temporal lobe seizures in seven of 10 patients (Velasco et al., 2000b). Vonck et al. (2005) reported satisfactory suppression of seizures by bilateral amygdalohippocampal DBS; at 14-month follow-up, one of seven patients was free of complex partial seizures, five enjoyed a 20-50% reduction in seizure frequency and only one did not improve. Undeniably, results are still unconvincing for this target.

CS: Due to their intrinsic characteristics, CS and TMS are more likely to suppress seizures originating primarily in the cortex. Thus, if the epilepticogenic zone lies over the motor or speech area and a surgical excision is risky, CS could be an alternative method. In contrast, CS is unlikely to prove effective in cases with a deeply-located epileptogenic focus. The application of TMS to testing the mode of action and responsiveness to antiepileptic drugs (Macdonell et al 2002; Ziemann et al 1996) constituted the basis for the clinical evaluation of rTMS in epilepsy. Repetitive TMS provided promising, but unvalidated, results in a small number of patients suffering from complex partial seizures and temporal lobe epilepsy, focal cortical dysplasia or cortical myoclonus (chapter 18). Clearly, more work is needed. One issue with all commercially available DBS and CS devices is that the electrical stimulation is delivered in a pre-programmed manner with defined stimulation parameters (intensity, pulse width, frequency), mode of action (continuous versus cycling) and duration of stimulation, regardless of the subjects condition (so called *open-loop* stimulation). An important leap forward is the development of *closed-loop* stimulation systems. These devices have incorporated software with seizure detection algorithms and deliver stimulation to the epileptogenic focus "on demand", i.e., once premonitory electroencephalographic (EEG) signs of imminent ictal activity are detected, well before clinical manifestation. CS combined with depth monitoring of epileptic activity is the first closed-loop system to have entered a controlled trial (chapter 19). Another controlled trial is also underway in US to test ANT stimulation in partial-onset drug-resistant epilepsy; the patient can activate the neuro-stimulator (Intercept™, Medtronic, Minneapolis, US) when he/she feels the onset of seizure.

CS vs. DBS: Considerations, Limitations and Future Prospects

Complex partial seizures, atypical absences and generalized tonic/clonic seizures seem to be responsive to DBS. Theoretically, DBS of the CMT or STN, which are relay nodes with extensive cortical afferent and efferent projections, is likely to alleviate generalized seizures. In contrast, DBS of the ANT or hippocampus, which are functionally integrated into the limbic system, is expected to suppress partial seizures. Nevertheless, several critical issues should be addressed before either CS or DBS become effective therapeutic options for intractable epilepsy, such as a lack of homogeneity in the epileptic population. In clinical trials of CS or DBS, it is mandatory to include patients who have well defined anatomical, clinical, radiological and genetic characteristics, similar EEG activity in terms of intensity and frequency of spikes and also similar time elapsed since their last seizure. Standardized selection protocols based on proper phenotypical classification through video-EEG and neuroimaging studies should be developed, in order to help detect the appropriate candidates for CS or DBS, either as monotherapy or as an adjunct to other surgical treatments. At present, it remains unclear whether it is preferable to target the epileptogenic zone or the epileptogenic focus per se or whether subcortical epilepsy is best treated with DBS rather than CS. An important issue, therefore, is the definition of the optimal deep node or cortical zone to be targeted in each epileptic syndrome. H-coil TMS for depth (6 cm) stimulation may benefit some patients. Also, both high- (50 Hz) and low- (1 Hz) frequency CS of the epileptogenic zone can suppress epileptic activity and the impact of different stimulation parameters must be investigated (which applies to DBS too). In most series, patients continue their drugs during DBS therapy. It is important to evaluate the effect of DBS separately, but also compare it with that of the antiepileptic drugs. The synergistic or antagonistic actions of CS with antiepileptic drugs should also be studied. For instance, low-frequency TMS combined with valproate acid induces epileptic activity. Again, only well-conducted studies will answer these and other questions.

6. Psychiatric Disorders

DBS: Early neuropsychiatric DBS practice has been largely guided by prior relevant lesioning experience (Sakas et al., 2007b). Based on results of anterior capsulotomy and fMRI studies (Rauch et al 2006), DBS of the anterior limb of the internal capsule and the ventral striatum/nucleus accumbens (Nuttin et al., 2003; Sturm et al., 2003) has been brought to bear on treatment-refractory obsessive compulsive disorder (OCD) patients (30% of all cases: Picinelli et al., 1995). In one study, four of eight patients treated by DBS enjoyed a greater than 35% reduction in the intensity of their symptoms and a substantial decrease in medication (Greenberg et al 2006). Apart from the IPG battery's short life expectancy and hypomania, DBS therapy has been well tolerated. DBS of the medial intralaminar thalamic (Visser-Vandewalle et al., 2003) and centromedian parafascicular nuclei (Houeto et al., 2005), anterior limb of internal capsule (Flaherty et al., 2005) and GPi (Ackermans et al., 2006; Diederich et al., 2005) has also shown high rates of efficacy (>70%) in suppressing drug-refractory motor and vocal tics in Tourette's syndrome, which are driven by dysfunctional limbic and cortico-basal gaglia-thalamo-cortical circuits. Patient selection guidelines (Mink et al., 2006) have been published. Severe behavioral changes may be induced in PD patients undergoing DBS (Temel et al., 2006) due, most probably, to inadvertent stimulation of parts of the limbic system such as ZI or SNr; this data supports the assumption that DBS of relay nodes of the limbic system may prove beneficial in treatment-resistant depression (TRD) (Fava, 2003). At present, DBS of the inferior thalamic peduncle (Jimenez et al., 2005), ventral internal capsule/nucleus accumbens (Schlaepfer et al., 2008), and particularly the white matter in the subgenual cingulate gyrus (Cg25 area) (Mayberg et al., 2005) have shown efficacy in depressed patients, with the latter target being now submitted to a controlled trial. Brodman's area 24a has also been proposed as a putative target for DBS in TRD (Sakas and Panourias, 2006). The superiority of Cg25 over the other targets may be due to its extensive connections to brain regions such as the brainstem, hypothalamus, insula, orbitofrontal, and cingulate cortex, which are all clearly implicated in the pathophysiology of depression (Marangell et al., 2007). Despite all these challenging data, DBS remains highly experimental for intractable neuropsychiatric disorders. These efficacy data need validation and replication in large randomized controlled trials in order to spare it the fate of ablative psychosurgery in the past.

CS: Noninvasive CS has been extensively tested and CS by implanted electrodes is now becoming reality. Over the last decade, the number of rTMS trials in psychiatric patients has steadily increased due to the high number of intractable or drug-intolerant psychiatric patients and the need to find alternatives to the rather brute-force approach of electroconvulsive therapy. The early promising results obtained from TMS of the dorsolateral prefrontal cortex (DLPFC) in intractable OCD were not replicated in other studies (chapter 17 and Sachdev et al 2007). Interestingly, in ten patients with comorbid OCD and TS, low-frequency rTMS of the supplementary motor area alleviated the symptoms and normalized overactive motor cortical regions (Mantovani et al., 2006; see also Munchau et al., 2002). Similar beneficial effect was shown in two other patients with TS and comorbid ADHD and MDD with an average improvement of 52%; this is highly comparable to that of approved behavioral or pharmacological treatments for TS (Mantovani et al 2007a). While low-frequency rTMS of the temporoparietal cortex has reduced hallucinations in schizophrenic patients (Aleman et al., 2007), high frequency rTMS of the left prefrontal (or bilateral prefrontal) cortices has provided equivocal results (Prikryl et al., 2007; chapter 17). DLPFC has been the most implicated area in the dysregulation of mood and the pathophysiology of depression. Antidepressant efficacy of high frequency rTMS of the left DLPFC has been repeatedly documented in randomized trials and meta-analyses (Fitzgerald, 2008; O'Reardon et al., 2007). Low frequency TMS of the right DLPFC has also been effective in TRD when offered either as a single TMS treatment (Isenberg et al., 2005) or combined with high-frequency rTMS of the left DLPFC (Fitzgerald et al., 2006). Efficacy of rTMS has also been reported in other comorbid or depressive conditions such as after stroke (Jorge et al., 2004), cerebrovascular disease-associated depression (Fabre et al., 2004), PD (chapter 12) or panic disorder (Mantovani et al., 2007b). tDCS of the left prefrontal region provided clinical benefits compared to sham stimulation in depressed patients (Boggio et al., 2007) and studies are underway to investigate the possible efficacy of tDCS in enhancing cognitive functions such as working memory, decision making and mood (Been et al., 2007). At this time, non-implantable CS should be viewed as a treatment for the acute state rather than a realistic long-term therapy for TRD. However, TMS/tDCS studies opened the way to implantable (invasive) CS. Certainly, this latter cannot be offered as easily as TMS or tDCS. Yet, implantable CS may prove more efficacious, because the electrodes are positioned very closely to the precise cortical area which has been indicated by functional imaging studies as the most appropriate to be stimulated. Given the high level of precision of implantable CS, we need to establish reliable criteria in order to select, among responders to TMS or tDCS, those who can expect the most benefit by implantable CS.

CS vs. DBS: Considerations, Limitations and Future Prospects

In the new era of brain stimulation, the application of CS or DBS in psychiatric disorders has advanced less compared to movement disorders or pain. Undeniably, the unfortunate past of ablative psychosurgery has influenced negatively physicians' attitude towards current brain stimulation efforts. DBS or CS experience in psychiatry is quite limited; therefore, definitive judgments on their clinical efficacy is impossible.

As research moves forward, a few important issues must be raised. The clinical profile of neuropsychiatric disorders is diverse and symptoms of different psychiatric conditions, i.e., TS, OCD and MDD, may coexist making accurate diagnosis debatable. The establishment of consensus guidelines in selecting candidates suitable for DBS therapy is highly recommended and has already become feasible in OCD and TS (OCD-DBS Collaborative Group, 2002; Mink et al., 2006). Additionally, keeping common evaluation rating scales during the patient selection process, plus preoperative and postoperative assessments ensures reliable comparative data from different centers and, hence, inference of strong conclusions. Reported TMS series include small cohorts of medically-refractory psychiatric patients at late stages of the disease. Most of the patients suffered from co-morbid conditions such as OCD and TS. In order to improve patient selection, the clinical benefits should be validated and replicated in patients with a clear psychiatric diagnosis who are enrolled in large-scale double-blind, sham-controlled studies. Hopefully, correlative studies between non-invasive and invasive stimulation will make response prediction feasible for implantable CS, similar to what is seen in chronic pain (chapter 8). This will spur the development of the field in a way that is not – apparently – possible for DBS. Compared to TMS or tDCS, implantable CS offers the opportunity to stimulate on a continuous or cyclical basis, without the patient having to return every month for therapy, and is better suited for sham-controlled studies. The development of the H-coil for TMS will make stimulation of, e.g., the orbitofrontal cortex feasible, while advanced imaging may reveal stimulation patterns of CS or DBS that can be correlated with disease-related activity and clinical improvement. High-resolution imaging needs to be combined with electrophysiological mapping in order to accurately detect cortical targets or white matter bundles which may be critical in mood performance. Advancements in cortical mapping are expected to clarify the precise cortical area that is stimulated and possibly indicate other potential targets for CS therapy. Although DBS cannot stimulate a large area of brain tissue, yet, deep structures such as the STN or thalamic nuclei can be precisely delineated by current neuro-monitoring methods and perhaps are more suitable for further exploration by DBS (Visser-Vandewalle et al 2003). New relay-nodes for DBS have been discovered rather fortuitously, e.g., DBS of the ventral capsule/ventral striatum for OCD provided prominent antidepressant effect (Greenberg et al., 2006). It should become a matter of priority, therefore, for investigation groups to provide detailed clinical and imaging information, as well as observations regarding target coordinates, lead location and clinical effectiveness; this will allow meaningful comparisons of targets and facilitate future developments. Short battery life is commonly reported in DBS studies for psychiatric disorders, implying either suboptimal targeting or that high-intensity stimulation is needed in order to produce a clinical effect. While adjustments of stimulation parameters are largely guided by experience gained in movement disorders, efforts should be directed to identifying stimulation settings individualized for each psychiatric illness in order to maximize the clinical benefits and improve acceptance of DBS as a therapeutic tool from both practitioners and patients. As research moves forward in the field, reported results should be comparable through common sets of stimulation parameters. With respect to TMS, further studies will clarify whether alternative stimulation paradigms such as theta burst stimulation (chapter 4) or priming stimulation (Iyer et al., 2003) will impact psychiatric disorders. Up to now, mostly seriously debilitated and drug-resistant psychiatric patients have been studied with TMS and DBS. In most series, efficacy of DBS is evaluated without discontinuing medications. This makes the "blindness" of studies doubtful and placebo effects possible. In the future, CS and DBS efficacy should be evaluated in the early stages of disease and in non-drug-resistant patients too and compared to patient cohorts treated only by medications. Since TMS may augment or hasten the clinical response of the patients to their medication (Fitzgerald, 2008), it would be interesting to explore possible effects of stimulation therapies on hastening the expected progress of psychiatric symptoms over time. Although DBS has been clearly associated with psychiatric symptoms such as anxiety, fear (Okun et al., 2007) and increased rates of suicide attempts (Burkhard et al., 2004) in movement disorders patients, caution should be exercised when extrapolating the above observations into psychiatry-oriented DBS practice: DBS is a safe and viable therapeutic option and should be further explored in psychiatry.

In sum, it should be clear for both practitioners and patients that invasive and noninvasive brain stimulation methods are still highly experimental and only much more work will establish them as powerful therapies for neuropsychiatric disorders. In this endeavor, it is critical to safeguard human rights and ethical rules as the future of the field is delineated. Therapeutic developments should evolve within a frame of evidence-based scientific knowledge, strict patient selection criteria and transparency in design and evaluation of clinical studies.

7. Stroke Rehabilitation

Stroke is a debilitating disorder: roughly half of all stroke victims have some degree of residual motor disability substantially affecting their daily activities, self-independence and overall quality of life (Hendricks et al., 2002). Functional recovery is mainly expected within a year after stroke; it may occur either spontaneously, possibly from partially or temporarily damaged cortical regions surrounding the stroke area, or by neurological reorganization of adjacent brain areas which take over the function of the irreversibly damaged areas. Physical therapy combined with peripheral nerve electrical stimulation is the only approved treatment for motor disability; unfortunately, many sufferers experience limited benefits at the end of the rehabilitation program.

DBS has been used to ameliorate stroke-induced movement disorders such as tremor; to the best of our knowledge, however, it has not yet been used as a therapeutic tool or an add-on treatment in stroke rehabilitation. DBS is exerted on well-defined deep "relay-nodes"; in the post-stroke state, however, no important deep targets implicated in the regeneration process and suitable for therapeutic stimulation have been recognized yet. The disease per se extensively damages cortical and white matter pathways rather than specific brain nuclei; our limited understanding of the mechanisms of neuroplasticity may account for the investigators' reluctance to offer DBS in stroke patients.

CS has been introduced in stroke rehabilitation on the basis of its efficacy in modulating neuroplasticity and enhancing motor performance (Hummel and Cohen, 2005). Extradural MI ECS has been reported to improve motor recovery after stroke (chapter 15); however, the invasive character of the method, the uncertainty regarding potential MI ECS interferences with the natural recovery processes and the overall high cost remain considerable limitations for its wider use as an adjunctive neurorehabilitation treatment. Moreover, in a phase III study enrolling 164 patients with chronic stroke, MI ECS plus rehabilitation therapy did not provide improved motor status as compared to rehabilitation therapy alone (Levy et al., 2008). Reasons can be adduced to explain such failure (chapter 15B).

CS by implantable devices grew out of TMS(<1Hz/inhibitory and >10Hz/excitatory) and tDCS (anodal/excitatory and cathodal/inhibitory) studies (chapter 14). The concept of "interhemispheric rivalry" in the motor areas between the stroke-injured and the non-injured hemisphere has been a key hypothesis in reported series (chapter 14 and Talleli and Rothwell 2006; but see Lotze et al., 2006). Motor impairments are supposed to be elicited by reduced output of the injured hemisphere and/or increased inhibitory activity of the non-injured hemisphere over the injured hemisphere. Both the stroke-affected and non-affected hemispheres have been stimulated variably in an effort to improve recovery from post-stroke motor impairments. Both TMS and tDCS have been offered in single or multiple sessions in order either to enhance the function of the injured hemisphere or to inhibit the non-injured hemisphere; the functional improvement ranged between 10-20% (Chapter 14 and Talelli and

Rothwell 2006). Notably, TMS-induced improvements were also noted in stroke victims suffering from aphasia or visuospatial neglect (chapter 14).

CS vs. DBS: Considerations, Limitations and Future Prospects

DBS has been offered only in post-stroke tremor or hemiballism (Wichmann and Delong 2006) and thus the discussion can only be limited to CS. To date, TMS treatment has been offered only to small series of patients with subcortical lesions and rather mild motor impairment. To improve recovery by CS, it is important to define selection criteria to identify stroke patients most likely to benefit; several clinical parameters need to be defined including the somatic distribution and severity of the motor impairment and the type of brain lesion (cortical, subcortical or deep). It is mandatory to define, by means of randomized controlled studies, which type of intervention is more suitable for each type of stroke and each type of patient, taking into account the size and location of the lesion and the interval from stroke. The hand area has been the most targeted cortical region. The availability of tests evaluating hand performance, the accessibility of the hand cortical area, and the high incidence of residual hand motor deficits after stroke explain the above preference over other cortical regions. rTMS, however, has also provided moderate to good recovery of gait when the leg motor cortex was targeted (chapter 14). Clarification of whether stimulation of the ipsilesional, contralesional or both cerebral hemispheres provides the best recovery is mandatory. After stroke, functional magnetic resonance imaging (fMRI) can effectively locate the hand motor region or other brain regions with residual function and discriminate between the affected and non-affected brain areas which could offer better functional recovery when stimulated. An interesting feature is the feasibility of combining neuroimaging studies with TMS-induced motor-evoked potentials, in order to determine the degree of functional recovery and guide the stimulation dose (Cramer 2008). TMS is offered within the first year after stroke, but the best interval to stimulate the motor cortex remains unknown. rTMS has been tested in patients after the first year, after the conventional rehabilitation had been completed and the motor impairments were considered permanent; improvements in motor performance of paretic extremities were reported even five years after stroke (Mally and Dinya, 2008). TMS induced neuroregeneration in a Parkinsonian animal model (Arrias-Carrion et al., 2004); however, such a role in stroke remains speculative. It is important, therefore, to explore whether CS or DBS can influence processes such as synapse formation and outgrowth and, hence, could be used in neural regeneration and development. Future studies including larger series of patients with various degrees of severity of symptoms in both early and late stages of the disease are expected to clarify key interactions between CS efficacy and type of lesion, laterality of symptoms, time since stroke, stimulation algorithms, duration of stimulation, and stimulation combined with other forms of neurorehabilition.

One recent development might improve the prospects of this therapy, according to Canavero (personal communication). A recently developed neurochip (Jackson et al., 2006) creates an artificial connection between two sites in MI by using action potentials recorded on one electrode to trigger electrical stimuli delivered to another, in a chronically implanted array, over the long-term. Once configured, it operates autonomously. It can induce changes mediated by hebbian mechanisms by delivering stimuli within 50 ms (20 ms best) of recorded spikes. The linkage is explained by potentiation of horizontal pathways within MI and in cases of partial injury this could strengthen surviving projections between sites connected by the prosthesis.

At present, CS for neurorehabilitation should be considered as a highly experimental, but potentially effective method that holds great promise for functional improvement of otherwise treatment-refractory stroke sufferers. CS has the potential to address the multifaceted clinical profile of stroke that includes motor, sensory, cognitive and psychiatric impairments, and favor the rehabilitation outcome of various components of stroke.

Limits

Compared to other neurosurgical procedures, both CS and DBS, are minimally invasive and associated with low complication rates. The stimulation can be programmed to meet patient's needs or discontinued in case of undesirable effects or ineffectiveness. Yet, there is neither agreement on which targets should be stimulated in order to obtain maximum clinical benefit for each neurological or psychiatric disorder, nor sufficiently standardized stimulation protocols. These procedures have further limitations.

1. Resources and Personnel

Both DBS and CS require expensive high-technology equipment including frame-based stereotaxis or frameless neuronavigation, respectively, and multidisciplinary teams consisting of neurosurgeons, neurophysiologists, movement disorder neurologists, pain therapists, neuropsychologists, experienced technicians and specially-trained nurses. One key difference is that DBS for movement disorders requires the close collaboration of a neurologist-neurophysiologist during surgery, unless it is performed without microrecordings monitoring and intraoperative clinical testings. Conversely, in CS, the intraoperative collaboration of a neurologist is not usually necessary, unless meticulous awake mapping is considered essential for the success of the procedure. Both TMS and tDCS are dependent on high-technology equipment, advanced neuroimaging studies for identifying optimal targets, and experienced teams: although they may evolve into powerful predicting tools of efficacy of implantable CS, their therapeutic potential has not been elucidated yet.

2. Preoperative Evaluation

The application of DBS is restricted by certain technical limitations: a) current imaging technology is still subject to spatial and geometrical distortions and b) the reproducibility and uniformity among different imaging systems and devices varies considerably. This makes high accuracy problematic and the comparison of data from different groups very difficult. In CS, an important limitation is the difficulty in the precise identification of the cortical areas in the single individual, given the huge inter-subject variability. Various techniques have been used to ensure accurate cortical targeting including fMRI, TMS and PET fused with neuronavigation data and combined with intraoperative neurophysiological recordings (sensory and motor evoked potentials, bipolar extradural stimulation). FMRI, though, has limitations: 1) cerebral areas are variably activated depending on task performance and individual thresholding, 2) large draining veins, motion artifacts, and echo planar images compromise accuracy, 3) activation protocols are not yet standardized, 4) cortical activity is hard to detect within the fissures, and 5) often, the examination must be repeated due to lack of patient's cooperation. Intraoperative neurophysiological mapping too requires local anesthesia and a larger craniotomy and may not be well tolerated by chronic pain sufferers or elderly patients.

3. Electrode-tissue Interface

Deeply implanted DBS electrodes are in contact with the brain, i.e., there is a direct interface of the electrode with the neural tissue, which is not the case for extradural CS; however, alterations in the impedance due to the thickness of the underlying dura or cerebrospinal fluid (CSF) can affect the stimulation results (Manola and Holsheimer, 2007). Subdural CS, like DBS, is interfaced with neural tissue, but given its low overall efficacy and association with such complications as cortical lesions, bleeding, CSF leakage, seizures, secondary gliosis and scar tissue formation, it is rarely used. However, the interdural placement of the electrode (Sakas et al., 2008a) in a "patch-like" fashion might prove an effective alternative option to standard extradural and subdural cortical stimulation techniques.

4. Electrode Design and Stimulation Delivery

The stimulation of more extensive cortical zones by implanting multiple electrode arrays may improve clinical results. An important issue is whether the electrodes with a single array of stimulating poles should be replaced by electrodes with a double- or triple-array of stimulating poles. The three-array electrode (Lamitrode, ANS, Plano, Texas, US) may improve efficacy (figure 1), as it is supposed to deliver a better targeted "wedge-shaped" stimulation which, when necessary, can be offered in an alternating "sweeping" mode. It is also important to establish whether it is preferable to perform bipolar versus monopolar stimulation and what are the most effective electrical stimulation algorithms. The stimulation efficacy could also be affected by other parameters, such as the position of the electrode (perpendicular or parallel to the central sulcus) and the lack of precision in positioning or securing the electrode over the cortex.

5. Patient's Clinical Condition

DBS and CS call for mentally and physically able patients to undergo the demanding preoperative clinical assessments and to attend the postoperative follow-up sessions during the period of titration of the stimulation parameters. However, a considerable percentage of sufferers are excluded from DBS due to advanced age (usually more than 70 years), brain atrophy, cognitive impairment, psychiatric symptoms and medical co-morbidities. In DBS, there is a small (1-3%), but considerable risk of hardware- or surgery-related complications (intracerebral hemorrhage, seizures, infection), plus stimulation-related neuropsychiatric effects, such as apathy, hallucinations, cognitive dysfunction, decline in executive functions, depression, and even suicide attempts. CS is better tolerated because it is unilateral, does not involve penetration of brain tissue with all its associated risks (hematoma, contusion, parenchymal infection) and the risk of intracerebral hemorrhage is virtually non-existent; the risk of seizures is related to the stimulation parameters and usually controlled with changes in programming. In DBS, however, the risk of seizures may be related to the intraparenchymal presence of the electrode *per se* and, therefore, not easily corrected by better programming only.

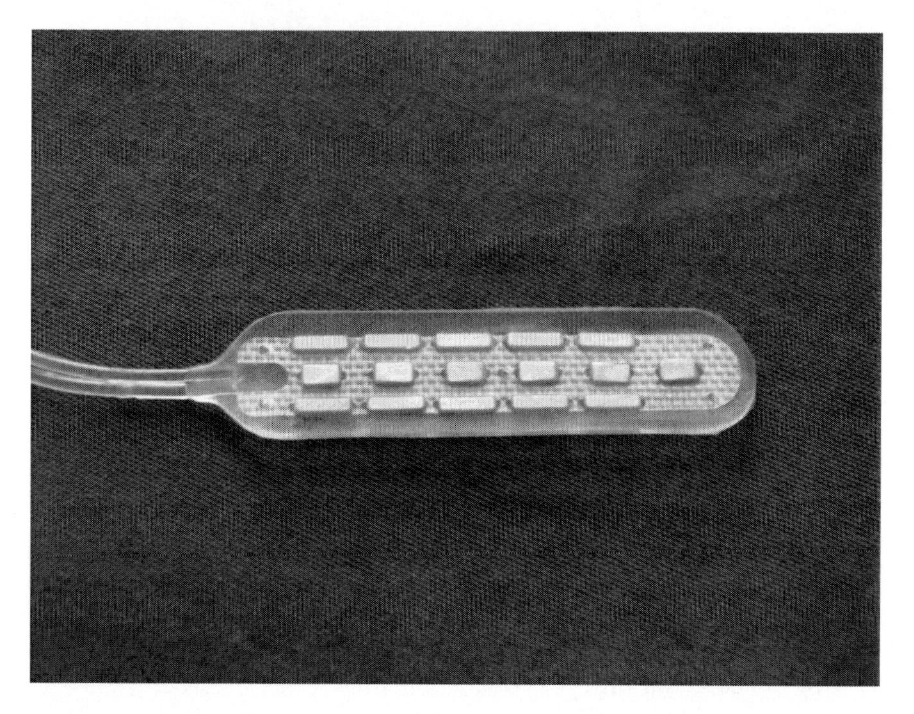

Figure 1. The three-array electrode (Lamitrode, ANS, Plano, Texas, US) offers the option of delivering "wedge-shaped" stimulation in an alternating "sweeping" mode. Such a stimulation type is expected to achieve a better targeted stimulation focus and maximize clinical benefits.

6. Device Features

Given the widespread appreciation of its unfulfilled potential, improvements and refinements in CS electrodes should be anticipated, whereas the more established and widespread use of the electrodes creates a lesser industrial incentive for development and refinement of the current quadripolar systems for DBS. Nevertheless, a higher number of poles would allow more refined stimulation. All current implantable devices for brain stimulation are susceptible to and interfere with magnetic environmental fields and they have not been cleared for safety in high-field (3-Tesla) MRI scanners. This is a problem that must be addressed by the device industry. As concerns the noninvasive technology, major shortcomings are the cost, the lack of portability of most machines that makes its routine use not easy, the patient's discomfort for certain stimulation algorithms and the difficulty to conduct blinded studies. tDCS has promise over tTMS, because the required equipment is less expensive, the technique is easier in application and maintenance, the stimulators can be battery-driven and are portable, the modulatory effect appears to last longer, and sham-stimulation is easily implemented as patients are unable to detect stimulation.

In Conclusion

a. Only well-structured and controlled clinical trials of rTMS and tDCS will verify their therapeutic potential and justify the enthusiasm generated by their introduction in clinical practice. At present, noninvasive brain stimulation methods have a strong potential for predicting a positive response to implantable CS and guide patient selection, timing of intervention and type of stimulation.
b. CS by implantable devices appears to be a promising and effective treatment modality for several neurological conditions. However, the clinical effectiveness of CS may wane over time and the long-term benefits may become less pronounced. There is a large inter-individual variability in terms of response to stimulation (in some cases leading to excitation, in others to inhibition, but a similar final result, e.g. analgesia). Moreover, some individuals on fMRI show pure SI activation during a motor task and in such patients SI must be engaged, not MI. This inconsistency of the effect and unpredictability of the outcome makes surgeons and neurologists hesitant to propose the procedure to patients. It is important to define -and possibly standardize- optimal cortical mapping, ideal targeting and protocols of programming in order for CS to become a first line treatment for neuropsychiatric disorders. Although DBS plays a starring role in the management of PD and other neurological disorders, implantable CS presents certain significant advantages, such as a lack of requirement for a stereotactic frame, close to nil risk of intracerebral hemorrhage, less invasiveness, better acceptability to both patients and neurologists, and the availability of a noninvasive preoperative predictive test.

Future Prospects

An exciting frontier is the development of the capability to artificially relay environmental sensory information by CS implants, i.e., relay of hearing or visual information, directly into the human brain. From a strict anatomical and surgical safety perspective, the cerebral cortex is likely to be a much more attractive implantation site compared to the brainstem, basal ganglia or other deep brain sites. However, one of the most difficult aspects of this work is the encoding of environmental sound or vision into parameters of electrical stimulation applicable to the cortex in order to convey the information. Research has shown that patients with profound auditory loss can discriminate between electrical stimuli based on the differences in the parameters of the stimulation current with the level of the electric current, being correlated to sound loudness and the frequency of the electric current to sound pitch. Following this discovery and taking advantage of the human's discriminatory ability, it became possible to design and build effective speech processors; these devices receive input from an external microphone and then electrically encode this acoustic information in a manner specifically designed to exploit the patient's ability to perceive differences in certain electrical stimulus parameters. On this basis, a *cortical auditory neural prosthesis* becomes possible (Howard et al., 2000). However, one of the most ambitious goals is the transmission of visual perception by cortical stimulation electrodes into the visual cortex of blind volunteers, i.e., a *cortical visual prosthesis* (chapter 22).

It is encouraging that the cerebral cortex seems to have the ability to adapt and to interpret in efficient manner electrical information that is applied in the right sequences and within a proper range of stimulation parameters by a large cortical-intracortical array. A relevant and important finding is that deaf patients do not appear to sustain deafferentation changes that would preclude the ''reactivation'' of normal auditory processing by the prosthetic stimulation device (Howard et al 2000) and a similar phenomenon could be expected to occur in the visual cortex and pathways of blind patients (chapter 22). A significant discovery that resulted from this research is that penetrating electrodes into the cerebral cortex are greatly superior in delivering more precisely the electrical stimulation compared to those electrodes that are placed in contact with the cortical surface. This opens the possibility of another field of therapeutic cortical stimulation namely *deep cortical* or *subcortical stimulation* (via an array of penetrating recording and stimulating electrodes), to be distinguished from both surface cortical stimulation (extradural or subdural) and DBS.

Computational neuromodulation is a field of computational biology dedicated to the study of the biophysical and mathematical characteristics of the electrochemical modulation in the nervous system. It is widely acknowledged that all types of networks and neurons (motor, sensory, and interneurons) are subject to modulation. Modulation may be induced in a circuit either extrinsically or intrinsically by circuit neurons or neural fibers (Fellous and Linster, 1998) or by artifical (manufactured) projecting systems or interfaces, including CS and DBS, with the aim to mimic "natural" neuromodulation. Given the complexity, convergence and divergence of neuromodulation in nature, the computational approach may provide a deeper understanding and the foundation for refined clinical applications. Areas amenable to computational analysis include synaptic drive, synaptic efficacy, and sensory encoding (Hille, 2001). With such approaches, altering the intrinsic properties of neurons, change their time-course, voltage-dependence and synaptic conductance and the strength of synaptic connections may become feasible. A single membrane current could be influenced and, depending on the conductance of the neuron membrane, a neuron brought across the boundaries of different behavior (Hille, 2001; Marder and Thirumalai, 2002). Furthermore, on the basis of current progress in computational biology, it is important to explore whether CS or DBS could be used in order: a) to offer a short synaptic input that can ''jump start'' a circuit, b) to influence the encoding of sensory information in spike trains, or c) to reconfigure an anatomically defined network into a different functional circuit by altering the intrinsic properties or the synaptic strength of the neurons within the network. Extrinsic modulation in particular (either natural or by implanted CS or DBS devices), could potentially be used to tune and configure whole networks and organize ensembles of circuits (Katz, 1995), in order to bias extensive networks or neurological systems into different functional outputs, in much the same way as changing parameters in a network model should bias or modify the output of the network (Marder and Tiramulai, 2002). Computer models of the bioelectrical and statistical aspects of neural recording and stimulation-induced recruitment can be used to model, and decipher neural coding. For instance, hippocampal functions can be replicated with a microchip implementation of the predictive mathematical models. Advances in this field hold the promise that individuals with damaged areas of the brain could be helped by CS based on such computational methods and that the use of neural interfaces could enhance normal or impaired neural function (Berger et al 2005).

CS and DBS may be further refined by developing electrodes that monitor neuronal activity from multiple regions, generate ''network level'' representations of the brain and simultaneously stimulate, i.e., "*closed-loop*" integrated monitoring and stimulating implantable systems. In these systems, the stimulation is activated "on demand" by intrinsic signals such as specific brain activity. Hence, these devices are capable of ''*responsive neurostimulation*''. A ''closed loop'' CS or DBS could be much more effective when it is activated by, i.e. it is responsive to, control signals derived from brain or body signals. Such technologies would be useful not only for epilepsy (chapter 19), but also for, e.g., Parkinson's disease; research has shown that intended and self-timed movements are preceded by increased activity in the parietal cortex and sensorimotor putamen (Pancrazio et al., 2006). Patients will have an advanced sensor either in contact or embedded in muscles and stimulation will be triggered by EMG signals elicited by the patient's movement. The possibility to offer effective intermittent stimulation would provide extended battery life. Responsive direct brain stimulation carries certain advantages over other methods as treatment is provided as needed, when needed, and for the time needed. The undesired effects of chronic stimulation are reduced, habituation is minimized, and, in epilepsy, the time of stimulation is the least necessary to suppress epileptiform activity; hence, the chances of potential failure due to reorganization of the epileptogenic circuitry are minimized. Temporal and spatial specificity, however, appear to play a pivotal role in the success of responsive stimulation. It is a high priority to develop sophisticated software that will allow early detection of epileptiform activity and enhance accuracy in lead or array implantation over the epileptogenic focus or zone. It is also important to develop seizure-prediction algorithms for each sufferer and stimulation devices able to drive several multicontact leads and arrays, and offer numerous combinations of cortical, subcortical or deep brain stimulation. Finally, all this technology should be incorporated into implantable, cosmetically acceptable devices in the most clinically effective fashion.

Numerous exciting developments in *neuroengineering* and *neurotechnology* are likely to influence the future applications of both CS and DBS. Future telemetric IPGs are expected to be smaller in size, more cosmetic, contour-shaped to apply to the cranium convexity or miniaturized to be implanted through a burr hole or small craniotomy. Batteries will be based on advanced chemistry offering more power in smaller size, will be rechargeable with extended life, and capable to be connected to the electrode(s) without extension wires. Such advancements may prove particular helpful in cases of cervical dystonia where hardware-related complications are commonly encountered. A great advance will be the incorporation into DBS electrodes of *microactuators* which will enable precise implantation after the initial insertion. This will reduce substantially the operating time, since fine adjustments for maximizing the clinical effect or minimizing untoward sequelae will be easier to carry out, at a later stage, rather than in the operating theatre. The construction of ''*smart stimulators*'' with the capability for dynamic internal adjustments will refine CS and DBS practice (Pancrazio et al 2006). Sophisticated software with advanced graphics for precise electrode insertion and for programming the stimulation in both CS and DBS procedures is expected to optimize pole selection after permanent implantation, individualize stimulation parameters and maximize clinical efficacy. Notably, such advanced software programs that combine MRI, CT, three-dimensional (3-D) brain atlases and neurophysiological microelectrode recording data with the volume of brain tissue activated by DBS have already been tested in preliminary studies (Miocinovic et al 2007; Butson et al 2007). Finally, the ability to program stimulation devices through remote access, telephone lines or via the Internet may render patients independent from the reference center and lessen the number of required follow-up visits. Sensitive, highly specific sensors which can be applied on the skin and detect physiological biosignals (e.g., changes in the heart rate, skin temperature, respiration rate, eyelid closure, EMG) are in development, and so are implantable chronic recording microelectrode arrays incorporating on board amplification, spike detection and wireless transmission of data and power (Pancrazio et al., 2006).

Finally, brain computer interfaces (BCI), defined as an electronic brain implant that translates the intention "to either communicate or move" into "communication or movement" through computer cursors, robotic devices or into actual movement of paralyzed limbs, represent the next opportunity to integrate CS and DBS into a larger clinical perspective. A BCI can detect changes in the user's brain activity and convert them into commands for a computer application. This is achieved through the application of signal processing techniques to the signals that the patient is still able to control. The key element in a BCI is a decoding algorithm that converts the main electrophysiological signal into an output that is suitable to control an external device. The interfaces rely on the natural adaptive ability of the human brain. The users have to learn to adapt their biological response, i.e., change the amplitude or frequency of the signal monitored. BCI technology will open new possibilities for severely disabled humans and offer systems that will deliver "communication or action" based on data derived from incorporated cortical monitoring and stimulation systems (Angelakis et al., 2007; Warwick et al., 2006). In this regard, the development of miniaturized, multi-functional, chronic neural implants *(BioMicroElectro-Mechanical Systems or Bio MEMS)* aims to obtain and control signals extracted from neuronal activity in order to enhance residual capabilities or perform actions such as moving a prosthetic arm with near natural performance. *Microelectrode arrays* are special types of micro-hardware constructed by using microfabrication and microelectronics by thin film-based planar and 3D-arrays of substrate microelectrodes. *Hybrid neural interfaces* or *cultured neural probes* can be coupled in vitro to populations of cultured neurons (Sanguineti et al 2003), where each electrode is covered and surrounded by a locally confined network of cultured neurons, obtained by chemical patterning of the substrate. These microsystems are designed to make connections and communicate with regenerating neurons and can be intended as hybrid neural information prosthetic transducers for stimulation and/or recording of neural activity in the brain or the spinal cord (Rutten et al., 2007).

Any neurological disorder that has altered electrical conductivity could potentially be helped by nanomaterials (Webster et al., 2004), which interact much more closely with cells than currently available materials. Carbon nanofibers can act as minimally traumatic CNS electrodes and carbon nanofibers arrays are expected to increase the accuracy of cerebral electrical stimulation. With such technology, our ability to offer complex, but precise patterns of stimulation may be enhanced; it will become possible to perform not only electrical microrecording, but also electrochemical recording and stimulation (Li and Andrews, 2007). Neuromimetic or neuromorphic engineering, in which very large scale integration (VLSI) systems containing electronic analog circuits mimic the architecture and design of biological nervous systems in order to replicate their function, e.g., vision or hearing (Silver et al., 2007; Smith and Hamilton, 1999) may one day be integrated into CS or DBS.

Conclusion

Currently, the main indication of CS is central and trigeminal neuropathic pains. In the coming years, PD, dystonia, epilepsy, and psychiatric disorders are likely to become established indications for CS as well. Undoubtedly, CS will also play a starring role in neurorehabilitation, facing the great challenge of inducing movement in paretic extremities and releasing medically-refractory spasticity even years after stroke. The future of CS will be influenced greatly by increased awareness and understanding of existing indications and applications, improved case selection, introduction of new indications, more mature assessment of outcome and convincing evidence regarding efficacy (Simpson, 2006b). A deeper understanding of mechanisms of action will allow us to answer critical questions regarding type of stimulation (extradural, subdural or interdural), mode of stimulation (anodal or cathodal), specific sets of effective stimulation algorithms for each neurological or psychiatric condition, and prediction of degree and duration of expected clinical benefit. Certainly, the ability of unilateral CS to influence brain activity bilaterally, be it pain, PD or stroke, is a major advantage over DBS. Great advancements are likely to take place in the field of responsive CS, which will particularly impact young patients suffering from progressive disabling disease, such as epilepsy or dystonia. CS also holds the potential to enhance cognitive performances (Topper et al., 1998; Olivieri et al., 2001; Boroojerdi et al., 2001; Montes et al., 2002) and should be explored for dementia (Canavero S, personal communication). CS is a high-technology dependent field that will be largely influenced by forthcoming neuroengineering developments and a shift away from the current dependence on pharmacological treatment. A big challenge will be to integrate CS technology with relevant advancements in human neurobiology, neuroplasticity, and neural repair, as well as to explore the potential neuroprotective effects of CS. Current implantable CS devices may need to be redesigned in order to incorporate complex and intelligent miniaturized systems and sophisticated software which will enable them to integrate information exchange between the device and the patient's brain, while remaining cost-effective. Progress in microsystems technologies, microelectronics, nanotechnologies, and computer modelling software are likely to create new opportunities for developing advanced stimulation systems and for expanding the therapeutic potential of CS.

References

Ackermans L Temel Y, Cath D, et al. Deep brain stimulation in Tourette's syndrome: two targets? *Mov Disord* 2006;21:709-13.

Aleman A, Sommer IE, Kahn RS. Efficacy of slow repetitive transcranial magnetic stimulation in the treatment of resistant auditory hallucinations in schizophrenia: a meta-analysis. *J Clin Psychiatr* 2007;68:416-21.

Andrade DM, Zumsteg D, Hamani C, et al. Long-term follow-up of patients with thalamic deep brain stimulation for epilepsy. *Neurology* 2006;66:1571-3.

Angelakis E, Hatzis A, Panourias IG, Sakas DE. Brain computer interfaces: a reciprocal self-regulated neuromodulation. *Acta Neurochir Suppl* 2007;97(2):555-9.

Arrias-Carrion O, Verdugo-Diaz L, Feria-Velasco A. Neurogenesis in the subventricular zone following transcranial magnetic stimulation and nigrostriatal lesions. *J Neurosci Res* 2004;78:16-28.

Bartsch T, Pinsker MO, Rasche D, et al. Hypothalamic deep brain stimulation for cluster headache: experience from a new multicase series. *Cephalalgia* 2008;28:285-95.

Been G, Ngo TT, Miller SM, Fitzgerald PB. The use of t DCS and CVS as methods of noninvasive brain stimulation. *Brain Res Rev* 2007;56:346-61.

Benabid AL, Koudsie A, Benazzouz A, et al. Deep brain stimulation of the corpus luysi (subthalamic nucleus) and other targets in Parkinson's disease: extension to new indications such as dystonia and epilepsy. *J Neurol* 2001;248(Suppl 3):III37-47.

Benabid A-L, Pollak P, Gao D, et al. Chronic electrical stimulation of the ventralis intermedius nucleus of the thalamus as a treatment of movement disorders *J Neurosurg* 1996;84:203-14.

Bendok B, Levy RM. Brain stimulation for persistent pain management. In: Gildenberg PL, Tasker RR, editors. *Textbook of stereotactic and functional neurosurgery*. New York: McGraw Hill; 1998. p.1539.

Berger TW, Ahuja A, Courellis SH, Deadwyler SA, Erinjippurath G, Gerhardt GA et al. Restoring lost cognitive function. *IEEE Engl Med Biol Mag* 2005;24:30-44.

Boggio PS, Rigonatti SP, Ribeiro RB, et al. A randomized, double-blind clinical trial on the efficacy of cortical direct current stimulation for the treatment of major depression. *Int J Neuropsychopharmacol* 2007;11:173-83.

Boroojerdi B, Phipps M, Kopylev L, Wharton CM, Cohen LG, Grafman J. enhancing analogic reasoning with rTMS over the left prefrontal cortex. *Neurology* 2001; 56: 526-528

Breit S, Schultz JB, Benabid A-L. Deep brain stimulation. *Cell Tissue Res* 2004;318:275-8.

Burchiel K. Deep brain stimulation for chronic pain: the results of two multi-center trials and a structured review. *Pain Med* 2001;2:177.

Burkhard PR, Vingerhoets FJ, Berney A, et al. Suicide after successful deep brain stimulation for movement disorders. *Neurology* 2004;63:2170-2.

Butson CR, Noecker AM, Maks CB, McIntyre CC. StimExplorer: deep brain stimulation parameter selection software system. *Acta Neurochir Suppl* 2007;97(2):569-74.

Byl N, Wilson F, Merzenich M, Melnick M, Scott P, Oakes A, McKenzie A. Sensory dysfunction associated with repetitive strain injuries of tendinitis and focal hand dystonia: a comparative study. *J Orthop Sports Phys Ther* 1996;23:234-44.

Canavero S, Bonicalzi V. Cortical stimulation for central pain. *J Neurosurg* 1995;83:1117.

Canavero S and Paolotti R. Extradural motor cortex stimulation for advanced Parkinson's disease: case report. *Mov Disord* 2000;15:169-71.

Canavero S, Paolotti R, Bonicalzi V, et al. Extradural motor cortex stimulation for advanced Parkinson's disease: report of two cases. *J Neurosurg* 2002;97:1208-11.

Canavero S, Bonicalzi V. Extradural cortical stimulation for movement disorders. *Acta Neurochir Suppl* 2007a;97(2):223-32.

Canavero S, Bonicalzi V. Extradural cortical stimulation for central pain. *Acta Neurochir* Suppl 2007b;97(2):27-36

Canavero S, Bonicalzi V. *Central pain sindrome*. New York: Cambridge University Press, 2007

Canavero S, Bonicalzi V. Central pain sindrome: elucidation of genesis and treatment. *Exp Rev Neurother* 2007; 7: 1485-1497

Chabadres S, Kahane P, Minotti L, Koudsie A, Hirsch E, Benabid A-L. Deep brain stimulation in epilepsy with particular reference to the subthalamic nucleus. *Epileptic Disord* 2002;4 (Suppl 3): S83-S93.

Chkhenkeli SA, Chkhenkeli IS. Effects of therapeutic stimulation of nucleus caudatus on epileptic electrical activity of brain in patients with intractable epilepsy. *Stereotact Funct Neurosurg* 1997;69:221-4.

Chkhenkeli SA, Sramka M, Lortkipanidze GS, et al. Electrophysiological effects and clinical results of direct brain stimulation for intractable epilepsy. *Clin Neurol Neurosurg* 2004;106:318-29.

Cioni B, Meglio M, Perotti V, De Bonis P, Montano N. Neurophysiological aspects of chronic motor cortex stimulation. *Clin Neurophysiol* 2007;37:441-7.

Classen J. Focal hand dystonia-a disorder of neuroplasticity? *Brain* 2003;126:2571-2.

Cooper IS, Amin I, Gilman S. The effect of chronic cerebellar stimulation upon epilepsy in man. *Trans Am Neurol Assoc* 1973;98:192-6.

Cramer SC. Repairing the human brain after stroke. II. Restorative therapies. *Ann Neurol* 2008;63:549-60.

Cruccu G, Aziz TZ, Garcia-Larrea L, et al. EFNS guidelines on neurostimulation therapy for neuropathic pain. *Eur J Neurol* 2007;14:952-70.

Derost PP, Ouchchane L, Morand D, et al. Is DBS-STN appropriate to treat severe Parkinson's disease in an elderly population? *Neurology* 2007;68:267-271.

Deuschl G, Bain P. Deep brain stimulation for tremor [correction of trauma]: patient selection and evaluation. *Mov Disord* 2002;17(Suppl 3):S102-11.

Deuschl G, Schade-Brittinger C, Krack P, et al. A randomized trail of deep-brain stimulation for Parkinson's disease. *N Engl J Med* 2006;355:896-908.

Diederich NJ, Kalteis K, Stamenkovic M, et al. Efficient internal pallidal stimulation in Gilles de la Tourette syndrome: a case report. *Mov Disord* 2005;20:1496-9.

Fabre I, Galinowski A, Oppenheim C, et al. Antidepressant efficacy and cognitive effects of repetitive transcranial magnetic stimulation in vascular depression: an open trial. *Int J Ger Psychiatr* 2004;19:833-42.

Fava M. Diagnosis and definition of treatment-resistant depression. *Biol Psychiatry* 2003;53:649-59.

Fellous J-M, Linster C. Computational models of neuromodulation. *Neural Comput* 1998;10:771-85.

Fisher RS, Uematsu S, Krauss GL, et al. Placebo-controlled pilot study of centromedian thalamic stimulation in treatment of intractable seizures. *Epilepsy Res Suppl* 1992;5:217-29.

Fitzgerald PB, Benitez J, Brown T, et al. A double-blind sham controlled trial of repetitive transcranial magnetic stimulation in the treatment of refractory auditory hallucinations. *J Clin Psychopharmacol* 2005;25:358-62.

Fitzgerald PB, Benitez J, de Castella A, Daskalakis ZJ, Brown TL, Kulkarni J. A randomized, controlled trial of sequential bilateral repetitive transcranial magnetic stimulation for treatment-resistant depression. *Am J Psychiatr* 2006;163:88-94.

Fitzgerald P. Brain stimulation techniques for the treatment of depression and other psychiatric disorders. *Australasian Psychiatr* 2008;16:183-90.

Flaherty AW, Willimas ZM, Amimovin R, et al. Deep brain stimulation of the anterior limb of internal capsule for the treatment of Tourette's syndrome: technical case report. *Neurosurgery* 2005;57(Suppl 4):E403.

Greenberg BD, Malone DA, Friehs GM, et al. Three-year outcomes in deep brain stimulation for highly resistant obsessive-compulsive disorder. *Neuropsychopharmacol* 2006;31:2384-2393.

Hallett M. Transcranial Magnetic Stimulation: a primer. *Neuron* 2007;55:187-98.

Halpern C, Urtig H, Jaggi J, Grossman M, Won M, Baltuch G. Deep brain stimulation in neurologic disorders. *Parkinsonims Relat Disord* 2007;13:1-16.

Halpern C, Samadani U, Litt BB, Jaggi JL, Baltuch GH. Deep brain stimulation for epilepsy. *Neurotherapeutics* 2008;5:59-67.

Handforth A, DeSalles AA, Krahl SE. Deep brain stimulation of the subthalamic nucleus as adjunct treatment for refractory epilepsy. *Epilepsia* 2006;47:1239-41.

Hendricks HT, van Limbeek J, Geruts AC, Zwarts MJ. Motor recovery after stroke: a systematic review of the literature. *Arch Phys Med Rehab* 2002;11:1629-1637.

Hille B. *Ion channels of excitable membranes.* 3rd edition. Mass. Sinauer Associates Inc Sunderland, 2001.

Hodaie M, Wennberg RA, Dostrovsky JO, Lozano AM. Chronic anterior thalamus stimulation for intractable epilepsy. *Epilepsia* 2002;43:603-8.

Houeto JL, Karachi C, Mallet L, Pillon B, Yelnik J, Mesnage V et al. Tourette's syndrome and deep brain stimulation. *J Neurol Neurosurg Psychiatr* 2005;76:992-5.

Howard MA, Volkov IO, Noh D, Garell C, Abbas PJ, Rubinstein JT et al. Auditory cortex neural prosthetic devices. In: *Maciunas eds. Neural prostheses. Illinois: American Association of Neurological Surgeons Publications*, Park Ridge. 2000; p. 273-286.

Hummel FC, Cohen LG. Drivers of brain plasticity. *Curr Opin Neurol.* 2005a;18:667-74.

Iadarola MJ, Gale K. Substantia nigra in GABA-mediated anticonvulsant actions. *Adv Neurol* 1982;44:343-64.

Isenberg K, Down D, Pierce K, et al. Low frequency rTMS of the right frontal cortex is as effective as high frequency rTMS of the left frontal cortex for antidepressant-free, treatment-resistant depressed patients. *Ann Clin Psychiatr* 2005;17:153-9.

Iyer MB, Schleper N, Wassermann EM. Priming stimulation enhances the derpessant effect of low-frequency repetitive transcranial magnetic stimulation. *J Neurosci* 2003;23:10867-72.

Jackson A, Mavoori J, Fetz EE. Long-term motor cortex plasticity induced by an electronic neural implant. *Nature* 444 56-60 2006.

Jimenez F, Velasco F, Salin-Pascual R, Hernandez JA, Velasco M, Criales JL, Nicolini H. A patient with a resistant major depression disorder treated with deep brain stimulation in the inferior thalamic peduncle. *Neurosurgery* 2005;57:585-93.

Jorge RE, Robinson RG, Tateno K, et al. Repetitive transcranial magnetic stimulation as treatment of post-stroke depression: a preliminary study. *Biol Psychiatr* 2004;55:398-405.

Katayama Y, Oshima H, Fukaya C, Kawamata T, Yamamoto T. Congrol of post-stroke movement disorders using chronic motor cortex stimulation. *Acta Neurochir Suppl* 2002;79:89-92.

Katz PS. Ion channels of excitable membres. *Curr Opin Neurobiol* 1995;5:799-808.

Kern DS, Kumar R. Deep brain stimulation. *The Neurologist* 2007;13:237-252.

Kerrigan JF, Litt B, Fisher RS, et al. Electrical stimulation of the anterior nucleus of the thalamus for the treatment of intractable epilepsy. *Epilepsia* 2004;45:346-354.

Kiss ZH, Doig-Beyaert K, Eliasziw M, et al. The Canadian multicentre study of deep brain stimulation for cervical dystonia. *Brain* 2007;130:2879-86.

Kitagawa M, Murata J, Uesugi H, et al. Two-year follow-up of chronic stimulation of the posterior subthalamic white matter for tremor-dominant Parkinson's disease. *Neurosurgery* 2005;2005;56:281-9.

Kleiner-Fisman G. Herzog J, Fisman DN, et al. Subthalamic nucleus deep brain stimulation: summary and meta-analysis of outcomes. *Mov Disord* 2006;21(Suppl /14):S290-S304.

Krack P, Batir A, Van Blercom N, et al. Five-year follow-up of bilateral stimulation of the subthalamic nucleus in advanced Parkinson's disease. *N Engl J Med* 2003;349:1925-34.

Krauss JK, Pohle T, Weber S, Ozdoba C, Burgunder JM. Bilateral stimulation of globus pallidus internus for treatment of cervical dystonia. *Lancet* 1999;354:837-8.

Krauss JK. Deep brain stimulation for dystonia in adults. Overview and developments. *Stereotact Funct Neurosurg*. 2002;78:168-82.

Krauss JK, Loher TJ, Weigel R, Capelle HH, Weber S, Burgunder JM. Chronic stimulation of the globus pallidus internus for treatment of non-dYT1 generalized dystonia and choreoathetosis: 2-year follow up. *J Neurosurg*. 2003;98:785-92.

Kumar K, Toth C, Nath RK. Deep brain stimulation for intractable pain: a 15-year experience. *Neurosurgery* 1997;40:736-46.

Kumar R, Lozano AM, Sime E, et al. Comparative effects of unilateral and bilateral subthalamic nucleus deep brain stimulation. *Neurology* 1999;53:561-6.

Kumar R, Lozano AM, Sime E, Lang AE. Long-term follow-up of thalamic deep brain stimulation for essential and parkinsonian tremor. *Neurology* 2003;61:1601-04.

Kwan P, Brodie MJ. Early identification of refractory epilepsy. *N Engl J Med* 2000;342:314-9.

Levy RM, Lamb S, Adams JE. Treatment of chronic pain by deep brain stimulation: long-term follow-up and review of the literature. *Neurosurgery* 1987;21:885-93.

Levy R, Benson R, Winstein C, for the Everest Study Investigators. Cortical stimulation for upper-extremity hemiparesis from ischemic stroke: Everest Study primary endpoint results. *International Stroke Conference*, New Orleans, LA, February 20-22, 2008.

Li J, Andrews RJ. Trimodal nanoelectrode array for precise deep brain stimulation: prospects of a new technology based on carbon nanofiber arrays. *Acta Neurochir* Suppl 2007;97(2):537-46.

Lima MC, Fregni F. Motor cortex stimulation for chronic pain. Systematic review and meta-analysis of the literature. *Neurology* 2008;70:2329-37.

Loher TJ, Capelle HH, Kaelin-Lang A, et al. Deep brain stimulation for dystonia: outcome at long-term follow-up. *J Neurol* 2008; 255: 881-884

Lotze M, Markert J, Sauseng P, et al. The role of multiple contralesional motor areas for complex hand movements after internal capsular lesion. *J Neurosci* 2006;26:6096-6102.

Macdonell RA, Curatolo JM, Berkovic SF. Transcranial magnetic stimulation and epilepsy. *J Clin Neurophysiol* 2002;19:294-306.

Mally J, Dinya E. Recovery of motor disability and spasticity in post-stroke after repetitive transcranial magnetic stimulation (rTMS*). Brain Res Bull* 2008;75:388-95.

Manola L, Holsheimer J: Motor cortex stimulation: role of computer modelling. In: Sakas DE and Simpson B, editors. *Operative Neuromodulation: Neural Networks Surgery* (Vol. II). New York: Springer-Verlag; 2007, 497-503.

Mantovani A, Lisanby SH, Pieraccini F, Ulivelli M, Castrogiovanni P, Rossi S. Repetitive transcranial magnetic stimulation in the treatment of obsessive-compulsive disorder and Tourette's syndrome. *Int J Neuropsychopharmacol* 2006;9:95-100.

Mantovani A, Leckman JF, Grantz H, King RA, Sporn AL, Lisanby SH. Repetitive transcranial magnetic stimulation of the supplementary motor area in the treatment of Tourette's syndrome: report of two cases. *Clin Neurophysiol* 2007a;118:2314-5.

Mantovani A, Lisanby SH, Pieraccini F, Ulivelli M, Castrogiovanni P, Rossi S. Repetitive transcranial magnetic stimulation in the treatment of panic disorder with comorbid major depression. *J Affect Disord* 2007b;102:277-80.

Marangell LB, Martinez M, Jrdi RA, Zboyan H. Neurostimulation therapies in depression: a review of new modalities. *Acta Psychiatr Scand* 2007;116:174-81.

Marder E, Thiramulai V. Cellular, synaptic and network effects of neuromodulation. *Neural Netw* 2002;15:479-93.

Mayberg HS, Lozano AM, Voon V, et al. Deep brain stimulation for treatment-resistant depression. *Neuron* 2005;45:651-60.

Mink JW, Walkup J, Frey KA, et al. Patient selection and assessment recommendations for deep brain stimulation in Tourette syndrome. *Mov Disord* 2006;21:1831-8.

Miocinovic S, Noecker AM, Maks CB, Butson CR, McIntyre CC. Cicerone: stereotactic neurophysiological recording and deep brain stimulation electrode placement software system. *Acta Neurochir* Suppl 2007;97(2):569-74.

Montes C Mertens P Convers P Peyron R Sindou M Laurent B Mauguiere F Garcia-Larrea L Cognitive effects of precentral cortical stimulation for pain control: an ERP study, *Neurophysiol. Clin./Clin. Nurophysiol* 2002 ; 32 : 313-325

Mueller J, Skogseid IM, Benecke R, , et al. Pallidal deep brain stimulation improves quality of life in segmental and generalized dystonia: results from a prospective, randomized sham-controlled trial. *Mov Disord*. 2008;23:131-4.

Munchau A, Bloem BR, Thilo KV, Trimble MR, Rothwell JC, Robertson MM. Repetitive transcranial magnetic stimulation for Tourette syndrome. *Neurology* 2002;59:1789-91.

Murata J, Kitagawa M, Uesugi H, et al. Electrical stimulation of the posterior subthalamic area for the treatment of intractable proximal tremor. *J Neurosurg* 2003;99:708-15.

Murase N, Rothwell JC, Kaji R, et al. Subthreshold low-frequency repetitive transcranial magnetic stimulation over the premotor cortex modulates writer's cramp. *Brain* 2005;128:104-15.

Nandi D, Liu X, Winter JL, Aziz TZ, Stein JF. Deep brain stimulation of the peduculopontine region in the normal non-human primate. *J Clin Neurosci* 2002a;9:170-4.

Nandi D, Chir M, Liu X, et al. Electrophysiological confirmation of the zona incerta as a target for surgical treatment of disabling involuntary arm movements in multiple sclerosis: use of local field potentials. *J Clin Neurosci* 2002b;9:64-8.

Nuttin BJ, Gabriels LA, Cosyns PR. Long-term electrical capsular stimulation in patients with obsessive-compulsive disorder. *Neurosurg* 2003;52:1263-72.

OCD-DBS Collaborative Group. Deep brain stimulation for psychiatric disorders. *Neurosurgery* 2002;51:519.

Oh MY, Abosch A, Kim SH, Lang AE, Lozano AM. Long-term hardware-related complications of deep brain stimulation. *Neurosurgery* 2002;50:1268-74.

Oikawa H, Sasaki M, Tamakawa Y, Kamei A. The circuit of Papez in mesial temporal sclerosis: MRI. *Neuroradiology* 2001;43:205-10.

Okun MS, Mann G, Foote KD, et al. Deep brain stimulation in the internal capsule and nucleus accumbens region: responses observed during active and sham programming. *J Neurol Neurosurg Psychiatry* 2007;78:310-4.

Olivieri M, Turriziani P, Carlesimo Ga, et al. Parieto-frontal interactions in visual-object and visual-spatial working memory: evidence from trancranial magnetic stimulation. *Cereb Cortex* 2001; 11: 606-618

O'Reardon JP, Solvason B, Janicak PG, et al. Efficacy and safety of transcranial magnetic stimulation in the acute treatment of major depression: A multisite randomized controlled trial. *Biol Psychiatry* 2007;62:1208-16.

Owen SL, Green AL, Nandi D, Bittar RG, Wang S, Aziz TZ. Deep brain stimulation for neuropathic pain. *Neuromodulation* 2006a;9:100-6.

Owen SL, Green AL, Stein JF, et al. Deep brain stimulation for the alleviation of post-stroke neuropathic pain. *Pain* 2006b;120:202-6.

Owen SL, Green AL, Davies P, et al. Connectivity of an effective hypothalamic surgical target for cluster headache. *J Clin Neurosci* 2007;14:955-60.

Pancrazio JJ, Chen D, Fertig SJ, et al. Towards neurotechnology innovation: report from the 2005 Neural Interfaces Workshop. An NIH-sponsored event. *Neuromodulation* 2006;9:1-7.

Picinelli M, Pini S, Bellantuono C, Wilkinson G. Efficacy of drug treatment in obsessive compulsive disorder. A meta-analysis review. *Br J Psychiatry* 1995;166:424-43.

Plaha P, Gill SS. Bilateral deep brain stimulation of the pedunculopontine nucleus for Parkinson's disease. *Neuroreport* 2005;16:1883-87.

Prikryl R, Kasparek T, Skotakova S, Ustohal L Kucerova H, Ceskova E. Treatment of negative symptoms of schizophrenia using repetitive transcranial magnetic stimulation in a double-blind, randomized controlled study. *Schizophr Res* 2007;95:151-7.

Raethjen J, Govindan RB, Kopper F, Muthuraman M, Deuschl G. Cortical involvement in the generation of essential tremor. *J Neurophysiol* 2007;97:3219-28.

Rasche D, Rinaldi PC, Young RF, Tronnier VM. Deep brain stimulation for the treatment of various chronic pain syndromes. *Neurosurg Focus* 2006;21:E8

Rauch SL Dougherty DD, Malone D, et al. A functional neuroimaging investigation of deep brain stimulation in patients with obsessive-compulsive disorder. *J Neurosurg* 2006;104:558-65.

Rezai AR, Lozano AM. Deep brain stimulation for chronic pain. In: Burchiel KJ, ed. *Surgical management of pain*. New York: Thieme; 2002. p. 565-76.

Rodriguez-Oroz MC, Obeso JA, et al. Bilateral deep brain stimulation in Parkinson's disease: a multicentre study with 4 years follow-up. *Brain* 2005;128:2240-9.

Romanelli P, Bronte-Stewart H, Courtney T, Heit G. Possible necessity for deep brain stimulation of both the ventralis intermedius and subthalamic nuclei to resolve Holmes tremor. Case report. *J Neurosurg* 2003;99:566-71.

Rutten WLC, Ruardij TG, Marani E, Roelofsen BH. Neural networks on chemically patterned electrode arrays: towards a cultured probe. *Acta Neurochir Suppl* 2007;97(2):547-54.

Sachdev PS, Loo CK, Mitchell PB, McFarquhar TF, Malhi GS. Repetitive transcranial magnetic stimulation for the treatment obsessive-compulsive disorder. A double-blind controlled investigation. *Psychol Med* 2007;37:1645-9.

Sakas DE, Panourias IG. Rostral cingulate gyrus: A putative target for deep brain stimulation in treatment-refractory depression. *Med Hypothes* 2006;66:491-4.

Sakas DE, Panourias IG, Simpson BA. An introduction to neural networks surgery, a field of neuromodulation which is based on advances in neural networks science and digitised brain imaging. *Acta Neurochir* Suppl 2007a;97(2):3-13.

Sakas DE, Panourias IG, Singounas E, Simpson BA. Neurosurgery for psychiatric disorders: from the excision of brain tissue to the chronic electrical stimulation of neural networks. In: Sakas DE, Simpson BA, editors. *Operative Neuromodulation: Neural networks surgery*. New York: Springer-Verlag; 2007b, 365-74.

Sakas DE, Panourias IG, Boviatsis ES et al. Treatment of primary camptocormia (dystonic bent spine or bent neck) by electrical stimulation of the globus pallidus internus: report of 3 cases. Abstract book, *The 8th International Neuromodulation Society and the 11th Annual Meeting of the North American Neuromodulation Society, Acapulco*, 2007, December 7-12.

Sakas DE, Flaskas P, Panourias IG, Georgakoulias N. Cortical electrical stimulation by *interdural* ("patch-like") implantation of the electrodes: a new technique for increased efficacy and safety. *Neurosurgery* 2008a (under review)

Sakas DE, Panourias IG, Boviatsis E, Themistocleous M, Stavrinou L, Gatzonis S. Treatment of idiopathic head-drop syndrome (camptocephalia) by bilateral chronic electrical stimulation of the globus pallidum internus. *J Neurosurg* 2008b [in press].

Sanguineti V, Giugliano M, Grattarola M, Morasso P. Neuroengineering: from neural interfaces to biological computers. In: Riva G, Davide F, editors. *Communications throuhg virtual technology: identity community and technology in the internet age*. Amsterdam, IOS Press; 2003. p. 233-46.

Schlaepfer TE, Cohen MX, Frick C,et al. Deep brain stimulation to reward circuitry alleviates anhedonia in refractory major depression. *Neuropsychopharmacology* 2008 ; 33:368-77.

Schoenen J, Di Clemente L, Vandenheede M, et al. Hypothalamic stimulation in chronic cluster headache: a pilot study of efficacy and mode of action. *Brain* 2005;128:940-7.

Schramm P, Scheihing M, Rasche D, Tronnier VM. Behr syndrome variant with tremor treated by VIM stimulation. *Acta Neurochir* (Wien) 2005;147:679-83.

Silanpää M, Schmidt D. Natural history of treated childhood-onset epilepsy: prospective, long-term population-based study. *Brain* 2006;129:617-24.

Silver R, Boahen K, Grillner S, Kopell N, Olsen KL. Neurotech for neuroscience: unifying concepts, organizing principles, and emerging tools. *J Neurosci* 2007;27:11807-19.

Simpson BA. The role of neruostimulation: the neurosurgical perspective. *J Pain Symptom Manage* 2006a;31:S3-S5.

Simpson BA. Challenges for the 21st century: the future of electrical neuromodulation. *J Pain Symptom Manage* 2006b;31 (Suppl 4):S3-S5.

Smith LS, Hamilton A. Neuromorphic systems. Introduction. *Int J Neural Syst* 1999;9;371-3

Starr PA, Barbaro NM, Raskin NH, Ostrem JL. Chronic stimulation of the posterior hypothalamic region for cluster headache: technique and 1-year results in four patients. *J Neurosurg* 2007;106:999-1005.

Stover NP, Okun MS, Evatt ML, Raju DV, Bakay RA, Vitek JL. Stimulation of the subthalamic nucleus in a patient with Parkinson's disease and essential tremor. *Arch Neurol* 2005;62:141-3.

Sturm V, Lenartz D, Koulousakis A, et al. The nucleus accumbens: a target for deep brain stimulation in obsessive-compulsive and anxiety disorders. *J Chem Neuroanat* 2003;26:293-9.

Sun B, Chen S, Zhan S, Le W, Krahl SE. Subthalamic nucleus stimulation for primary dystonia and tardive dystonia. *Acta Neurochir* Suppl 2007;97:207-14.

Swanson TH. The pathophysiology of human mesial temporal lobe epilepsy. *J Clin Neurophysiol* 1995;12:2-22.

Sydow O, Thobois S, Alesch F, Speelman JD. Multicenter European study of thalamic stimulation in essential tremor: a six year follow-up. *J Neurol Neurosurg Psychiatr* 2003;74:1387-91.

Taha JM, Jansen MA, Favre J. Thalamic deep brain stimulation for the treatment of head, voice, and bilateral limb tremor. *J Neurosurg* 1999;91:68-72.

Talelli P, Rothwell J. Does brain stimulation after stroke have a future? *Curr Opin Neurol* 2006;19:543-50.

Temel Y, Kessels A, Tan S, Topdag A, Boon P, Visser-Vandewalle V. Behavioural changes after bilateral subthalamic nucleus stimulation in advanced Parkinson's disease: A systematic review. *Parkinsonism Relat Disord* 2006;12:265-72.

Tisch S, Rothwell JC, Limousin P, Hariz MI, Corcos DM. The physiological effects of pallidal deep brain stimulation in dystonia. *IEEE Trans Neural Syst Rehabil Eng* 2007;15:166-72.

Topper R, Mottaghy FM, Brugmann M, Noth J, Huber W. facilitation of picture naming by focal transcranial magnetic stimulation of wernicke's area. *Exp Brain Res* 1998; 121: 371-378

Velasco F, Velasco AL, Velasco G, Jimenez F. Effect of chronic electrical stimulation of the centromedian thalamic nuclei on various intractable seizures patterns: I. Clinical seizures and paroxysmal EEG activity. *Epilepsia* 1993a;34:1065-74.

Velasco F, Marquez I, Velasco G. Effect of chronic electrical stimulation of the centromedian thalamic nuclei on various intractable seizure patterns: II. Psychological performance and background EEG activity. *Acta Neurochir* Suppl 1993b;58:201-4.

Velasco M, Velasco AL, Jimenez F et al. Role of the centromedian thalamic nucleus in the genesis, propagation and arrest of epileptic activity. An electrophysiological study in man. *Epilepsia* 1995;36:63-71.

Velasco F, Velasco M, Jimenez F, et al. Predictors in the treatment of difficult-to-control seizures b electrical stimulation of the centromedian thalamic nucleus. *Neurosurgery* 2000a;47:295-304.

Velasco AL, Velasco M, Velasco F, et al. Sucabute and chronic electrical stimulation of the hippocampus on intractable temporal lobe seizures: preliminary report. *Arch Med Res* 2000b;31:316-28.

Velasco M, Velasco AL, Menez D, Rocha L. Centromedian-thalamic and hippocampal electrical stimulation for the control of intractable epileptic seizures. *Stereotactic Funct Neurosurg* 2001;77:223-7.

Velasco F, Carrillo-Ruiz JD, Brito F, et al. Double-blind, randomized controlled pilot study of bilateral cerebellar stimulation for treatment of intractable motor seizures. *Epilepsia* 2005;46:1071-81.

Visser-Vandewalle V, Temel Y, Boon P, et al. Chronic bilateral thalamic stimulation: a new therapeutic approach in intractable Tourette's syndrome. Report of three cases. *J Neurosurg* 2003;99:1094-1100.

Vonck K, Boon P, Claeys P, Dedeurwaerdere S, Achten R, Van Roost D. Long-term deep brain stimulation for refractory lobe epilepsy. *Ann Neurol* 2005;52:556-65.

Warwick K, Gasson MN, Spiers AJ. Therapeutic potential of computer to cerebral cortex implantable devices. *Acta Neurochir* Suppl 2007;97(2):529-35.

Weaver F. Follett K, Hur K, Ippolito D, Stern M. Deep brain stimulation in Parkinson's disease: a meta-analysis of patient outcomes. *J Neurosurg* 2005;103:956-67.

Webster TJ, Waid MC, McKenzie JL, Price RL, Ejiofor JU. Nano-biotechnology: carbon nanofibers as improved nerual and orthopaedic implants. *Nanotechnology* 2004;15:48-54.

Wichmann T, DeLong MR. Deep brain stimulation for neurologic and neuropsychiatric disorders. *Neuron* 2006;52:197-204.

Wishart HA, Roberts DW, Roth RM, et al. Chronic deep brain stimulation for the treatment of tremor in multiple sclerosis: review and case reports. *J Neurol Neurosurg Psychiatr* 2003;74:1392-7.

Young R, Rinaldi PC. Brain stimulation. In: North RB, Levy RM, editors. *Neurosurgical management of pain*. New York: Springer; 1997. p. 283-301.

Yu H, Neimat J. The treatment of movement disorders by deep brain stimulation. *Neurotherapeutics* 2008;5:26-36.

Ziemann U, Lonnecker S, Steinhoff BJ, Paulus W. Effects of antiepileptic drugs on motor cortex excitability in humans: transcranial magnetic stimulation study. *Ann Neurol* 1996;40:367-78.

In: Encyclopedia of Neuroscience Research
Editors: Eileen J. Sampson and Donald R. Glevins
ISBN 978-1-61324-861-4

Chapter XIV

Invasive Cortical Stimulation for Parkinson's Disease and Movement Disorders*

B. Cioni†, A. R. Bentivoglio, C. De Simone, A. Fasano, C. Piano, D. Policicchio, V. Perotti and M. Meglio

Functional Neurosurgery, Anesthesiology, and Neurology, Università Cattolica, Roma, Italy

Abstract

Ever since the observation that motor cortex (MI) stimulation (MI ECS) could relieve some patients suffering post-stroke movement disorders (Katayama et al., 1997, 2002) and, above all, the successful introduction of MI ECS for the treatment of Parkinson's disease by Canavero in 1998 (Canavero and Paolotti 2000), extradural MI ECS has been employed for the treatment of several movement disorders. This chapter reviews this field.

Parkinson's Disease (PD) and Parkinsonism

A. Review of the Literature

A total of approximately 100 patients with Parkinson's disease have been treated by MI ECS.

---

* A version of this chapter was also published in *Text of Therapentic Cortical Stimulation*, edited by B. Cioni published by Nova Science Publishers, Inc. It was submitted for appropriate modification in an effort to encourage wider dissemination of research.

† Correspondence concerning this article should be addressed to: Dr. Beatrice Cioni, MD.E-mail: bcioni@rm.unicatt.it

1. The first report was that by Canavero and Paolotti (2000), who described the case of a 72 year old woman with advanced PD who showed a substantial improvement of all three cardinal parkinsonian symptoms following unilateral extradural MI ECS (electrode strip parallel to Rolandic fissure). In July 1998, a quadripolar electrode was positioned in the extradural space overlying the hand representation area of MI, contralateral to her worst clinical side. Chronic stimulation was delivered subthreshold for any movement or sensation, at 3V, 180 μs, 25Hz, 3+/0- setting, off during sleep. The clinical improvement was bilateral; she could stand without assistance, climb the stairs and walk for a short distance. The UPDRS III in On-Med condition decreased from 44 to 23 after 3 months. She had a moderate to severe PD-associated dementia and this aspect also improved. L-dopa was reduced by 80%. Importantly, after 3 months, the patient developed infection and was explanted. However, she lost benefit only gradually over several weeks, a sign of likely neuroplastic changes induced by stimulation. Upon reimplantation, she improved again. These results were confirmed in two further patients operated on in 2001 and 2002 (Canavero et al., 2002, 2003). Canavero also reported on a patient suffering multiple system atrophy-associated parkinsonism. Bilateral subthalamic deep brain stimulation was ineffective, but moderately responded to bilateral MI stimulation with the following parameters: 25-40 Hz, 90-180μs, 2-2.5V, bipolar stimulation (3+/0- or viceversa), continuously. Motor symptoms improved for about 9 months, while vegetative symptoms remained improved until death almost 3 years later (Canavero et al 2003).

2. Upon Canavero's urge, Pagni et al. (2003, 2005) implanted 5 new cases (in their 2005 publication, patients 5 and 6 were taken from Canavero's series). All of them were submitted to unilateral extradural MI ECS (electrode strip parallel to Rolandic fissure), opposite to the worst clinical side; chronic stimulation was delivered bipolarly at 2.5-6V, 150-180 μs, 25-40Hz, continuously; follow-up ranged from 4 months to 2.5 years. The patients were evaluated only in the On-Med state: global UPDRS decreased by 42-62%, UPDRS III (motor axis) by 32-83%. It was possible to decrease L-dopa by 11-33% in 3 cases and by 70-73% in 2 patients. The improvement was bilateral. One case had to be reimplanted due to misplacement. Interestingly, neuropsychological assessment (MMSE, 15 Rey's Words, Corsi block-tapping test, WAIS with revised digit span subtest and block design subtest, Prose Memory Test, Attention Matrices Test, Verbal Judgement Test, Arithmetic Judgement Test, SPM, Verbal Fluency, Token Test, MADRS, BPRS, QL-Index) of three of these patients at 1 year follow-up revealed stable cognitive functions, stability of mood disorders, disappearance of hallucinations (1 patient) and mildly improved quality of life in 2 of 3 patients (Munno et al 2007).

3. The Toronto's group (Kleiner-Fisman et al., 2003) reported on the efficacy of high-frequency *subdural* MI ECS (bipolar, or monopolar in 1 case) in 5 patients with Parkinsonism due to multiple systemic atrophy (MSA). Three patients reported subjective improvement, although UPDRS scores did not change at 3 and 6 months. Thereafter, they (Strafella et al 2007) submitted 5 patients (age range 70-75)with advanced PD to unilateral subdural MI ECS (hand motor area, left in 3, right in 1). All had 30% motor improvement on levodopa and none met DBS criteria. One patient developed a cortical venous infarct and was excluded from study results. Multiple settings (5-185 Hz, 60-450 μs, 0.5-7V, monopolar and bipolar stimulation) were assessed immediately after surgery and after 1-2 weeks of stimulation. At 6 month follow-up, patients underwent double-blind evaluations randomized to stimulation on or off (for 2 weeks each), with and without medications (acute levodopa challenge); modified

CAPIT and neuropsychological testing were performed and a PDQ-39 questionnaire was administered preoperatively and at 6 months. There were no acute improvements in akinesia, rigidity or gait function with stimulation. Tremor improved acutely in the patient with prominent tremor. The off drug UPDRS motor scores off- and on-stimulation (43 +/- 7.9 versus 39.5 +/-12.5) were not significantly different, nor the on-stim/on-med scores and preoperative on condition (18.7 +/- 8.6 versus 14.2 +/-2.2). Two patients worsened (more off time, less on), one was unchanged, one had some improvement with stimulation. These patients were also assessed with H2O PET at 6 months from surgical implantation following optimization of parameters (3-5V, 60-90 μs). Subjects were tested at rest and while performing a simple motor task with a joystick. Scanning was performed at 10 minutes intervals on stimulation off, 50 and 130 Hz stimulation. Joystick movement compared to rest while MI ECS was turned off augmented rCBF in the sensorimotor cortex (BA 4/3), lateral premotor cortex (BA6) and caudal supplementary motor cortex (SMA) contralateral to hand movement. During 50 Hz and 130 Hz stimulation, the same areas were activated when comparing joystick movement versus rest. However, these changes in rCBF were not significantly different when comparing across different stimulation settings (OFF vs 50 Hz vs 130 Hz); no significant movement-related rCBF changes were detected during MI ECS in other cortical areas such as the rostral SMA and dorsolateral prefrontal cortex. Motor performance during the joystick task did not change significantly across different stimulation frequencies, although a trend towards worsening was observed during 50 Hz stimulation. This study is ill-conceived. MI ECS has no acute clinical effects and the 10 minute intervals are simply inadequate.

4. Pagni, upon retiring in 2003, spearheaded a multicenter study to evaluate the efficacy of MI ECS in advanced PD, on behalf of the Italian Society of Neurosurgery.

The preliminary results of the multicenter study have been published (Bentivoglio et al 2005). Twenty-nine patients were treated. Initially, patients were submitted to a test period (1-2 weeks) before IPG implantation, but in the following cases the whole implant was made in a single session. All symptoms of PD (tremor, rigidity, motor dexterity, bradykinesia, posture and gait, freezing) were improved and L-dopa could be reduced. Best parameters were 2.5-6V, 150-180 (but also 90-120) μs, 25-40 (but also 60-80) Hz. In an "intention to treat analysis", after 6 months of MI ECS, the mean UPDRS III decreased by 21% in the off-med condition and by 34% in the on-med condition. At 1 year, the effect of MI ECS was less evident, with a decrease of UPDRS III of 13% and 21% respectively in the off and on-med conditions. Notably, the stimulation parameters differed in the various centers: monopolar or bipolar stimulation, 2-8V, 60-400 μs, 20-120Hz, continuously or only during day time. Some patients were unresponsive to MI ECS: many of these patients had MRI findings of leucoencephalopathy, white matter ischemic foci, cerebral atrophy, suggesting a diagnosis of Parkinsonism rather than true PD.

The long term results (up to 36 months) have been reported for 41 patients (Pagni et al 2008).

UPDRS evaluation was performed before implantation and after 3, 6,12, 24 and 36 months of MI ECS, both in the on-med and off-med conditions. Analysis of variance for repeated measures and Wilcoxon signed-rank paired samples test were used for statistical analysis. At 12, 24 and 36 months of follow up, significant decreases of global UPDRS off-med (by 21.9%, 14.6%, and 26.4%, respectively), of UPDRS III off-med (by 17.8%, 10.5% and 25.5%) of axial symptoms (subscores 27+28+29: by 19.9%, 16.2% and 21.3%), of

UPDRS IV (by 37.7%, 18.1%, and 21.6%) and of UPDRS II off-med (by 27.6%, 21.3%, and 31.8% , respectively) were apparent. Drug therapy could be decreased by a mean of 20%. No major adverse event was encountered. Thus, MI ECS moderately improves motor symptoms (UPDRS III), decreases complications of therapy (UPDRS IV) and ameliorates activities of daily living (UPDRS II) in advanced PD. The clinical effect can be expected to last at least 3 years.

5. Benvenuti et al. (2006) submitted a 68 year old woman to MI ECS. Motor fluctuations uncontrolled by therapy, dyskinesias, depression and anxiety were severe. During MI ECS, UPDRS scores changed significantly (Activities of Daily Living: from 26 to 17; Motor: from 33 to 22; Complications of Therapy: from 13 to 7; Mentation: from 8 to 6). Dyskinesias and motor fluctuations were reduced and no longer disabling, rigidity was reduced bilaterally; standing, gait and motor performance were also improved. Levodopa was not reduced. Improvement was 35% on the UPDRS scale 8 weeks after surgery.

6. Verhagen et al. (2006) conducted a prospective multicenter study of 9 PD patients (H&Y OFF ≥ III) receiving extradural MI ECS (hand area) for 24 weeks. Stimulation effects were systematically explored at 20-127 Hz with constant pulse width (250 μs). At end of study, the UPDRS III score OFF medication was 42.13 +/- 13.78 compared to 38.78 +/- 8.08 at baseline. Three patients remained on stimulation 1 year later. Their UPDRS III scores at baseline, 24 and 52 weeks were 37, 30, 30; 29,37,20 and 29, 25 and 25. At 24 weeks "On Time Without Troublesome Dyskinesias" was 11.25 +/-2.10 hours, compared to 9.54 +/- 3.89 at baseline. PDQ-39 Total scores were 38.33+/-10.17 at 24 weeks compared to 43.62+/-8.33 at baseline. During programming, there were 2 fits. This study has limitations. The stimulation hardware was different from that used by all other groups, with a different waveform and current output, but, above all, only two contacts available spaced 31 mm apart corresponding to contacts 0 and 3 of the tetrapolar electrode commonly used for ECS. The 10 Hz band was not explored, and the pulse width was fixed at 250 μs. Finally, at least one patient was probably not PD (B Gliner, meeting communication, 2006).

7. Cilia et al. (2007) submitted to extradural MI ECS (hand area; electrode strip parallel to Rolandic fissure) 5 patients with PD who fulfilled CAPSIT criteria for DBS, with the exception of age >70 years (range 71-77). In particular, there were no vascular abnormalities or significant brain atrophy on MRI. Patients were assessed preoperatively and after 6 months of MI ECS of the left hemisphere, on and off medication, with stimulator on and 2 weeks later with stimulator off, by a blinded neurologist. Stimulation was monopolar (0-/3-/case +), 3-4V, 60 (40 in 1 patient) Hz, 180-210 μs, continuous (night and day). MI ECS determined a reduction of daily OFF time (UPDRS item 39) in 3 patients, a mean L-DOPA reduction of 16% and a mean dopamine agonist reduction dosage of 49%, and a reduction of the Abnormal Involuntary Movements Scale (AIMS) of 19%. The off-med UPDRS II and UPDRS III scores did not vary, even if a trend towards improvement was shown in axial symptoms (items 14-28-29). On-Med UPDRS II and III scores values were unmodified. No behavior or mood changes were reported at 6 months. No immediate motor change was reported when stimulation was switched off, except for 1 patient who showed worsening gait. Fifteen days later, 3 patients complained of prolonged daily OFF time and freezing gait, and 2 patients with unchanged medication complained of an increase in dyskinesias as well. Four patients asked spontaneously to have the stimulator turned on again. In sum, "*4 of 5 patients reported subjective improvement involving mainly axial symptoms* (i.e., stability, posture and gait) *as well as a reduction in daily OFF time and dyskinesia*". Subjective clinical benefit was

reported after a variable interval after switching on the stimulator (minutes to days). This study has a few limitations. Four out of 5 patients were monopolarly stimulated. Monopolar stimulation is not the same as bipolar stimulation. In fact, the best responder was the only one to receive bipolar stimulation (0-1-/2+3+): this was also the only akinetic-rigid dominant patient, whereas the remaining 4 were tremor dominant. These authors "*started from a frequency of 30 Hz and then slowly increased up to 60 Hz*". The 10-20 and 70-80 Hz bands were not explored, and these can be effective. In a follow-up paper (Cilia et al., 2008), these same authors reported on the same 5 patients, plus 1 (a 70 year old man), and studied them with Tc-99m cysteinate dimmer bicisate (ECD) SPECT off-medication with stimulator off and on. Patients were assessed preoperatively and 6 months after surgery, with both stimulator on and 2 weeks after switch-off by a blinded neurologist. All medications remained unchanged during the 2-week stim-off period. Parameters were assessed across different configurations (monopolar and bipolar, 0.5-5V, 60-210 μs, 20-120 Hz). Blinded evaluation in individual patients revealed clinical improvement of UPDRS II and III scores during stimulation compared to stim-off condition up to 40 and 20% respectively. UPDRS scores did not change significantly during bipolar stimulation (0-/3+) and 120 Hz. Clinical improvement occurred after several days to 4 weeks after stimulation parameters modifications. Objective motor benefit was observed mainly in axial symptoms (gait, stooped posture and postural instability). Subjective benefit was admitted by 4 of 6 patients, mainly in mobility and walking, associated with an increase in daily on-time (as if stimulation would delay wearing-off phenomenon). In two patients, battery exhausted about 12 months after implant: both experienced slow and progressive clinical deterioration, associated with an objective worsening of UPDRS III scores of 20% and 25% respectively, that reversed after battery replacement. ECD SPECT with a triple head camera and ultra high resolution collimators was conducted with stimulator on and 2 weeks after switching the stimulator off. At a statistical threshold of $p<0.001$, the authors found rCBF decrements in MI (BA4) and lateral premotor cortex (BA6) bilaterally (!), left dorsolateral prefrontal cortex (BA9) and right caudate head. Perfusion decrements occurred in BA19. At $p<0.01$, additional clusters were seen in right thalamus (stimulator was on left MI!), left caudate nucleus and right BA6. Significant perfusion increases were found in the right cerebellar lobe, and, with a less strict threshold, left BA18, left cerebellar lobe and vermis. Thus, effects were widespread and bilateral, paralleling clinical benefits.

8. Arle et al. (2008) applied extradural MI ECS (electrode strip parallel to Rolandic fissure) in 4 patients (bilateral in 3) aged 44-72. Three patients were eligible for DBS. All patients had at least 30% improvement in their UPDRS scores when taking L-Dopa. Programming was started within 24 hours of electrode implantation. Initially all contacts were checked in a monopolar setting at 130 Hz and 210 μs, with voltage slowly increased to 4V. Contacts 1 and 2 were left ON at 3V: if not tolerated, contact 3 or 0 alone was used. In all cases 2 contacts could be activated. Cathodal (not anodal) stimulation was used. If minimal benefit was noted during later sessions, more contacts were added to the monopolar configuration. In case of failure, contact 0 or 3 was placed in cathodal mode and the other 3 contacts set as the anode. Other investigated changes included PW 410 μs, 4V, 80Hz and 185Hz.

In the end, patient 1 was stimulated at 3.2 mA, 240 μs, 130 Hz, 0-/1-/2-/3off/case +; patient 2 at 3.5mA,210 μs, 130 Hz, left sided 0-/1-/2+/3+/case off and right sided 0off/1-/2-/3+/case off; patient 3 at 3.4 mA (left)/3.2 Amp (right), 210 μs, 100 Hz, left sided 0-/1-/2-

/3off/case + and right sided 0-/1-/2-/3off/case +. Patient 4 was not detailed. One patient developed an infection and was explanted at 3 months: he had an improvement of about 50% over his baseline condition in dyskinesias, rigidity and tremor following unilateral implantation, but it significantly regressed upon device explantation. After reimplantation several months later, he achieved the same benefit. Two patients had a 20-30% reduction in medication requirements, but one needed 37% more. One patient had a return of dyskinesias after 3 months due to reduction of amplitude of stimulation following initial benefit: upon reincreasing the amplitude, dyskinesias were again controlled. Globally, there was a trend toward a significant difference in the UPDRS III scores between the baseline score and the scores at 1,3 and 6 months, but not at 12 months, possibly because of fewer data point observations available for analysis. At 1 month, there was a mean decrease in the UPDRS III score of 21(+/-13) points from baseline, at 3 months of 15(+/-11) points, at 6 months of 18(+/-16) points, but only of 9(+/-9) points at 1 year for the 2 patients with sufficient follow-up. More in detail, the UPDRS global scores during on-med/on-stim preoperatively, at 1,3,6 and 12 months in 3 patients (4th not available) were respectively: 46,52,47,51(12th month milestone not yet reached); 57,79,42,94,62; 53.5,23,32,54,37; during off-med and on-stim: 27,17,22,15 (12th month milestone not yet reached); 46,11,31,31,43; 41.5,12,10,5,25; 18,10,9,14 (12th month milestone not yet reached). The UPDRS III (motor) scores in on-med/on-stim in the same 3 patients were respectively: 19,26,25,23 (12th month milestone not yet reached); 32, 47,13,18,35; 24.5,12,18,27,5.

9. Seijo et al. (2009) submitted 6 patients not eligible for DBS to extradural unilateral MI ECS (contralateral to side of clinical onset). Five of six were akinetic-rigid, 1 had left hand tremor. Parameters adjustments were carried out during 1 week after initial implant. Optimal parameters were: *case 1*:3+/0-, 4.5V,450 μs, 30Hz; *case 2*:2+/1-,3V, 400 μs, 10 Hz; *case 3*: 3+/0-, 3.7V, 450 μs, 30 Hz; *case 4*: 3+/0-, 3.6V, 450 μs, 10 Hz; *case 5*: 3+/0-,3V, 330 μs, 25 Hz; *case 6*: 1-/2+, 3.6V,450 μs, 25 Hz. Follow-up was 1 year. In *case 1*, UPDRS III on-med/on-stim was 26 (but 42 off-med/on-stim), Hoehn&Yahr (H&Y) score was 2 and Schwab &England Daily Living Activities Scale (DLA) 90%. In *case 2*, they were 27 (vs 38), 3 and 80%: dyskinesias and freezing while walking were significantly decreased, with freezing restricted only to off-med periods. In *case 3*, they were 17 (vs 55), 2, 70%; levodopa was decreased. In *case 4*, they were: 27 (vs 38), 2 and 90%: dyskinesias were reduced, though tremor was not. In *case 5*, they were 29 (vs 42), 2 and 80%: levodopa was reduced. In *case 6*, they were 26 (vs 42), 2 and 90%: levodopa was decreased. Globally, UPDRS III score improved in 2 patients (cases 1 and 4) comparing preoperative condition to on-med/on-stim conditions, but did not change in cases 2,3 and 5 or slightly worsened in case 6.

B. Personal Experience

In 2003, we started a prospective study to evaluate the efficacy of MI ECS in patients with advanced PD (Cioni et al., 2007; Cioni, 2007). The inclusion criteria were: idiopathic Parkinson disease (PDSBB criteria); at least 5 years disease's length; disease in the advanced state (UPDRS in off >/= 40/180; Hoehn and Yahrs >/= 3; motor complications: fluctuations and disabling dyskinesias); positive response to L-Dopa; DBS not accepted by the patient or contraindicated; patient's ability to give informed consent to the study. The exclusion criteria were: history of epilepsy or EEG epileptic activity; alcohol or drug abuse; mental deterioration; psychiatric symptoms; previous basal ganglia surgery; other major illness. Ten patients met the above mentioned criteria and were submitted to the implant of an epidural plate electrode over the motor cortex contralateral to the worst clinical side in 3 cases, and to a bilateral implant in the remaining 7 cases (Figures 1-2). Therapeutic stimulation during the first year was through the electrode contralateral to the worst clinical side; parameters were: 120μs, 80Hz, 3-6V (subthreshold for movements, and motor or sensory phenomena), delivered continuously through contacts 0 (anode) and 3 (cathode). After 12 months, in cases of bilateral implantation, stimulation was made bilateral (same parameters for the side ipsilateral to the worst clinical side). The clinical assessment before implant and at 1, 3, 6, 12, 18, 24, and 36 months included: UPDRS (Unified Parkinson Disease Rating Scale); finger tapping; walking time; PDQL (Parkinson Disease Quality of Life Scale); neuropsychological evaluation including MMSE (Mini Mental State Evaluation), behavioral assessment of mood and anxiety, tests for verbal short term memory, spatial short term memory, episodic verbal memory, non-verbal abstract reasoning, frontal executive functions and verbal fluency; EEG; oral medications and adverse events. The clinical motor evaluation was performed both in the off and in the on medication state and the motor assessment was videotaped. The "OFF" condition was achieved by withdrawing antiparkinsonian medications as follows: Levodopa for at least 12 hours, pergolide, pramipexole, ropinirole for at least 48 hours, cabergoline for at least 168 hours, apomorphine for at least 3 hours. The "ON" condition was achieved 60 minutes after administering a supra-threshold dose of standard levodopa, according to daily schedule. Cognitive and behavioral assessment were performed preoperatively and at 6, 12 and 18 months, in the on med status.

A statistically (Wilcoxon's test and analysis of variance for repeated measures) significant improvement was present after 12 months of unilateral MI ECS, as of total UPDRS off-med, UPDRS II, UPDRS III off-med, subscore for axial symptoms (UPDRS III: items 27-31), UPDRS IV, PDQL. The effect of unilateral MI ECS was bilateral, with no significant difference between the two sides. It was evident after 1-2 weeks of stimulation, and in a case of accidental switching off of the stimulator the patient became aware of something going wrong after 2-3 weeks. After 1 year of unilateral stimulation, 7 patients needed bilateral MI ECS, the remaining 3 patients continued with unilateral stimulation. The statistical analysis at 24 and 36 months demonstrated a significant improvement in total UPDRS off-med, in UPDRS III off-med (by 20% compared to the preoperative score), in the subscore for axial symptoms, in UPDRS IV and in PDQL. Notably, in all the patients, the UPDRS III off med at 24 and 36 months was always lower than UPDRS III off med at preoperative evaluation. At 2 years, bilateral stimulation showed a trend towards greater benefit compared to unilateral stimulation, but a statistical analysis was not possible due to the small sample. Drug treatment could be decreased by 25%. No complication occurred; no adverse events, particularly no epileptic seizures nor EEG epileptic activity, were encountered. Cognitive assessment in the overall group of patients showed a significant postoperative improvement on the MMSE and on tasks of episodic verbal memory, but this most likely reflects a practice effect (Daniele et al., 2003). No significant postoperative decline was observed on any cognitive task, including those of phonological and semantic verbal fluency, whereas DBS patients show a significant decline on verbal fluency (Contarino et al., 2007). Unilateral stimulation of the left hemisphere showed a statistical trend towards a postoperative improvement of phonological verbal fluency, along with an increase of depressive symptoms. On the contrary, stimulation of the right hemisphere showed a trend towards a decrease of depression, and no effect on verbal fluency. We also assessed our patients with DAT scan (SPECT) in order to evaluate binding to the dopamine transporters before and after 6, 12 and 28 months of MI ECS, and IBZM-SPECT to evaluate activity of striatal postsynaptic D2 receptors, before and after 6 and 12 months of MI ECS. DAT-scans at 6 and 12 months showed an increase in the activity in both putamina, particularly on the side ipsilateral to the stimulation. Postsynaptic receptors activity was unmodified at 6 and 12 months.

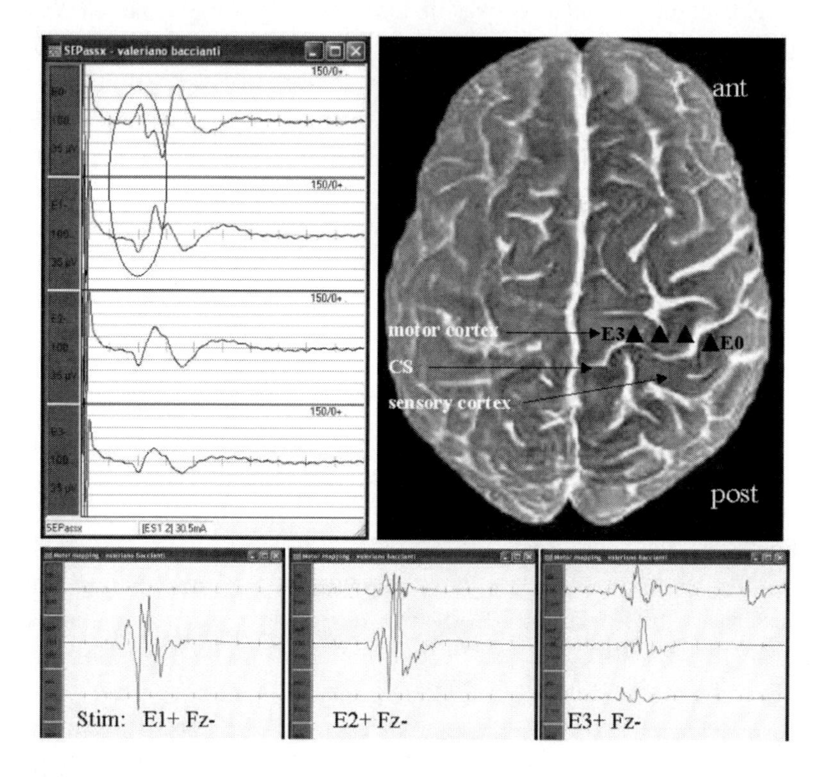

Figure 1. *Top right*: Central sulcus (CS) identification. Cortical median nerve SEPs (recording: E0, E1, E2, E3, referred to Fz) showing phase reversal between traces 1 and 2 (E0 and E1). *Bottom traces*: Motor cortex mapping, Recording from Biceps Brachii, Abductor Pollicis Brevis, Quadriceps after stimulation delivered through electrode E1, E2 and E3 (cathode) and 6 cm in front of Cz (anode). *Top left*: integrated anatomo-functional position of the quadripolar electrode.

Furthermore, we reported (Fasano et al., 2008) the case of a 72-year-old female patient affected by severe PD who underwent bilateral MI ECS. In baseline med-off condition the patient was unable to rise from a chair and to stand without assistance. Stimulation at 3 and 60 Hz failed to provide any improvement, whereas, when stimulating at 130 Hz, axial akinesia and walking improved consistently: the patient, in med-off condition, was able to rise from chair and to walk without assistance. The patient underwent two brain 99mTc-Ethylcysteinate Dimer-SPECT studies, which showed that the regional cerebral perfusion was significantly increased in the supplementary motor area during stimulation at 130 Hz. After five months, the benefit of MI ECS gradually disappeared.

We are now performing a controlled randomized study on the clinical usefulness of MI ECS in PD. The protocol has been published on *www.ClinicalTrials.gov (Identifier #*

NCT00637260). Another controlled study of 10 patients is on file (*Identifier # NCT00159172).*

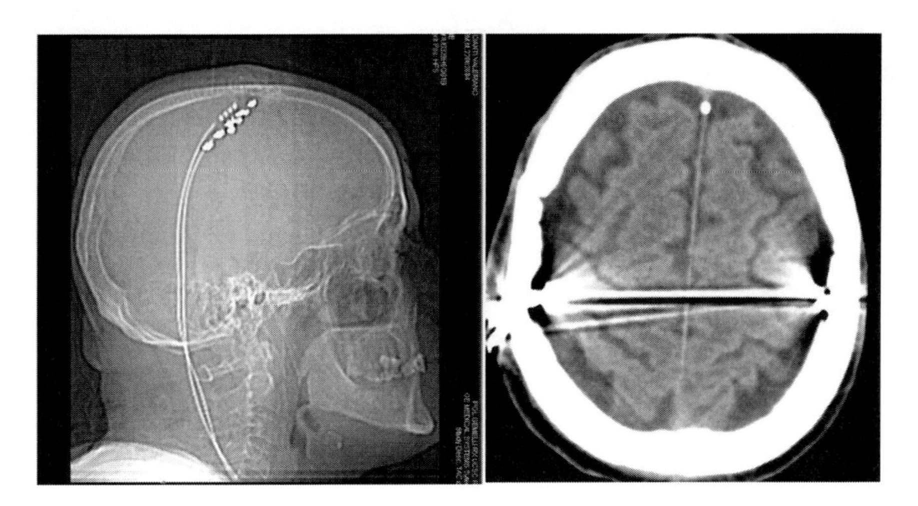

Figure 2. Skull X-rays and CT scan showing the position of the bilateral quadripolar strip electrode.

Pure Akinesia

Tani et al. (2007) presented a case of Levodopa-resistant akinesia, a DBS resistant motor disorder. After a positive trial of 10 Hz MI rTMS (500 pulses) (but negative trial of SMA rTMS), the patient underwent a bilateral implant of subdural electrodes over the motor cortex, close to the sagittal sinus. The best stimulation parameters were: 100Hz, 210μs, 1.8V, 1+/2- electrode setting, 30' on and 120' off (benefit lasted 3 hours). UPDRS decreased form 70 to 41 after 1 year and to 46 after 2 years; walking time (7 m) improved from 84.7 sec to 20sec at 1 year and 15 at 2 years, and so did step size; dysphagia also improved. Blinded switch-off resulted in deterioration within 48 hours. On the fourth day, the stimulator was reactivated and benefit was restored. A PET study showed a significant increase of rCBF in the left SMA (BA6) and right dorso-lateral prefrontal cortex (BA9) after 30' of bilateral MI ECS. Activation of SMA, involved in movement preparation and initiation, was one of the mechanisms hypothesized at the basis of MI ECS effects.

In our series of PD patients the best results were obtained on axial symptoms in patients with severe and unpredictable off periods. We submitted a patient with pure akinesia to bilateral MI ECS. There was a marked improvement of akinesia, posture, and gait; furthermore the patient showed a striking amelioration of verbal fluency.

Essential Tremor

Lyons et al. (2006) submitted to MI ECS (only two contacts available) two patients suffering essential tremor. The two burr-hole technique developed by Canavero was used (see chapter 2). They were evaluated with the Fahn-Tolosa-Marin Tremor Rating Scale at baseline and after 1 and 4 weeks of contralateral (to tremor) MI ECS. Patient 1 (75 year old man) received MI ECS at 30Hz, 3mA, 250 μs. His baseline score was 61, and 1 month after surgery it was 57. Patient 2 (60 year old man)'s stimulation parameters were 50Hz, 5 mA, 250 μs. His score was 47 before surgery and 43 1 month after MI ECS. In sum, MI ECS proved ineffective.

Arle and Shils (2008) reported 2 other patients with essential tremor treated by MI ECS with poor results.

Dystonia

Canavero et al. (2003) described the acute effects of MI ECS in a case of cervical dystonia previously treated with thalamotomy and, after recurrence, cervical rhizotomy, intrathecal baclofen and botulinum toxin. She developed painful dystonic paroxysms involving her right hemibody every 4 hours lasting about 30 min. MI ECS (MI thigh, chest and arm area) at 10 Hz, 450 μs and 1V controlled both pain and dystonia, except for the neck symptoms.

Franzini et al. (2003) reported on the efficacy of MI ECS (130 Hz) for central pain-associated "thalamic hand" in 2 of 3 patients. Two patients suffering intentional myoclonus were also relieved.

Romito et al. (2007) and Albanese et al. (2007) described a case of fixed dystonia unresponsive to pallidal stimulation, but improved by MI ECS. A right handed woman progressively developed a severe segmental dystonia with fixed elevation and anterorotation of the left shoulder, abduction of the upper limb, severe trunk involvement and fixed kyphoscoliosis. Gene mutations were absent. There was no evidence of a complex regional pain syndrome, no preceding trauma, and a complete psychological and psychiatric evaluation ruled out the diagnosis of psychogenic dystonia. The diagnosis was a distinct subtype of primary dystonia – primary fixed dystona. Pharmacological treatments failed as well as botulinum toxin. GPi DBS also failed. Stimulation of the right motor cortex (strip parallel to MI) started the day after implant (3.8V, 60 Hz, 60 μs, 0-/1+/2- setting) was followed by a gradual improvement and the dystonic postures and pain almost resolved after 6 months. The effect was still present at 22 months. A PET study was performed before and after 6 months of MI ECS showing significant hypometabolism in the cerebellum, more pronounced on the right, and no increase in cortical metabolism. Interestingly, PET during Gpi DBS showed significant CBF increases in SI/MI (more on the left) and SMA and ACC bilaterally.

Post-Stroke Movement Disorders

Tsubokawa et al. (1993) and Katayama et al. (1997, 2002) made the original observation that MI ECS can relieve post-stroke movement disorders. Katayama et al. (2002) analyzed the effects of MI ECS on post-stroke movement disorders in 50 patients: the purpose of MI ECS was to control pain in 42 cases and involuntary movements in 8 cases. Good control of involuntary movements was observed in 3 out of 3 patients with hemichoreo-athetosis, in 2 out of 2 cases with distal resting or action tremor and in 1 out of 3 patients with proximal postural tremor. High frequencies (50-125 Hz) seem to be necessary for these effects. Subjective improvement in motor performance was reported by 8 patients who had mild motor weakness; severe motor weakness did not respond. The same group (Katayama et al 2006) developed an on-demand type stimulation system which triggers MI ECS or DBS by detecting intrinsic signals of intention to move. They applied this kind of feed-forward control of involuntary movement in 6 cases in whom motor dysfunction was evident only when patients intended to move their body. On-demand stimulation provided satisfactory feed-forward control in 4 patients with postural tremor and 2 with motor weakness, when the activity of muscles involved in posturing or intention to move was fed into the system.

Nguyen et al (1998) reported a case of severe upper limb action tremor and facial pain following removal of an acoustic schwannoma completely controlled by bipolar MI ECS (1.8V, 60 μs, 50 Hz,cyclical mode) for 32 months.

Arle and Shils (2008) reported the effect of MI ECS in 4 cases of poststroke movement disorders associated with pain. In one case there was no benefit, while the others received varying degrees of benefit.

Our group submitted to MI ECS a patient with hemichorea who had a very good response to rTMS (Di Lazzaro et al., 2006). MI ECS proved inferior to rTMS.

Amyothophic Lateral Sclerosis

Sidoti and Agrillo (2006) published the results obtained by bilateral subdural MI ECS in 4 cases of amyotrophic lateral sclerosis (ALS). The motor hand area was targeted. The stimulating plate was parallel to the central sulcus. Stimulation parameters were 30 Hz, 60 μs, 4.5V, contacts 0 and 4 positive, contacts 3 and 7 negative and the rest off. Clinical evaluation included neurological examination and ALS Functional Rating Scale (ALSFRS). At two year follow up, 2 patients showed an arrest of the natural course of the illness during the first year and a slow worsening in the second year after surgery. In these 2 cases postoperative SPECT disclosed consistent complete recovery of flow disturbances. MI ECS was ineffective in the other two patients. A neuroprotective effect of MI ECS was speculated at least in a subgroup of ASL patients.

Di Lazzaro et al. (2006) investigated whether continuous theta burst TMS could have a beneficial effect on ALS. They performed a double blind placebo-controlled trial in 20 ALS patients. The patients submitted to active stimulation received TMS for five consecutive days every month for 6 consecutive months. The primary outcome was the rate of decline evaluated with the ALSFRS. The active group of patients showed a modest, but significant slowing of the deterioration rate.

On this basis we applied bilateral MI ECS in one patient with ALS. MI ECS was delivered at 3 and at 30 Hz. The rate of disease progression was not significantly affected.

Conclusions

In sum, extradural MI ECS may relieve all three main symptoms of PD (akinesia, rigidity, tremor) on both sides simultaneously, but results vary widely, ranging from positive to negative to mixed. Benefit on the various symptoms can vary too: some are improved, others are not, in the same patient.

Patients' population was not homogeneous in these studies, nor were parameters of stimulation or techniques of implantation (monopolar versus bipolar; low versus high frequency, extradural versus subdural, unilateral versus bilateral, continuous versus daily only, general versus local anaesthesia, burr hole versus craniotomy). Nonetheless, our personal data and published studies suggest that MI ECS can relieve all three major symptoms of PD (akinesia, rigidity, tremor) on both sides simultaneously, as well as other movement disorders. Axial symptoms, gait, akinesia and freezing are particularly improved.

The clinical effect of MI ECS cannot be compared with that of STN DBS due to the different inclusion criteria. DBS is usually contraindicated in patients submitted to MI ECS, because of age limits or the presence of MRI anatomical abnormalities (cerebral atrophy, white matter ischemic foci, etc.). However, STN DBS appears to be more effective on motor symptoms, whereas MI ECS seems to be more effective on axial symptoms (gait, posture). Most dramatically, dyskinesias and painful dystonias are reduced (up to 90%) in almost all patients, well before reducing L-DOPA (a likely direct effect of stimulation). Clinical fluctuations too may be reduced. The complication and adverse events rates are lower for MI ECS, with no risk of intracerebral hemorrhage or infection (extradural approach only); verbal fluency is not impaired by MI ECS, and may even be improved. Hallucinations can occur sporadically and resolve with change of drug therapy and/or stimulation parameters.

Our DAT-Scan data also suggest that MI ECS may have a protective effect on putaminal degeneration, at least in the short term.

The clinical effects of MI ECS seem to decrease with time. This may be due to a placebo effect, and/or to the progressive nature of the disease; or it may reflect a true loss of effectiveness of the stimulation. Changing the parameters of stimulation may reverse this effect: for instance, changing the frequency or the intensity of stimulation according to the impedances (some fibrosis may develop between the electrode surface and the dura) may be useful (neurostimulators delivering impulses in current do not appear superior in this regard). With frequency up to 130Hz, fibers are more likely to be depolarized and excited (Fasano et al 2008). However, if the stimulation is maintained for long periods of time, some synapses may not follow the stimulus train and be blocked, with consequent inhibition. Specific frequencies may be necessary to impose specific patterns of activity, or to suppress abnormal rhythms, or time-protracted processes. Since stimulation is usually delivered continuously, the slight decline in the clinical benefit may be due to cortical habituation. If this is true, alternate stimulation (right side or left side) may be a solution. The clinical effect is long lasting, therefore a cyclic stimulation (only during daytime or even less) is feasible.

Not all the patients respond to MI ECS (7% non-responders in the Italian studies): this may be due to the rather large inclusion criteria, or to the different electrode position and different stimulation parameters. Most patients submitted to MI ECS up to now are older than 70, and cortical atrophy or micro-ischemic lesions may disrupt MI locally in some way in aged people, impairing MI ECS modulation in some cases. Younger patients must be included in future series.

The effect of rTMS may be predictive of the clinical outcome following MI ECS. However, the frequency and the duration of rTMS certainly differ from those used for MI ECS.

As regards the electrode position and stimulation parameters, the number of patients treated with MI ECS is still to low to allow statistical correlation analysis in order to identify prognostic factors.

The need for a bilateral implant has still to be demonstrated. Unilateral MI ECS improves motor performances bilaterally, but bilateral stimulation seems to add to such an improvement.

Finally, there is no place for subdural MI ECS: this can be burdened with fatal brain hemorrhage. If atrophy is found on imaging, the two burr hole technique introduced by Canavero is indicated to close the durocortical gap, even by adding a further silicone layer on the stimulating strip (Canavero S, personal communication).

In sum, MI ECS may be indicated in PD patients with prominent axial symptoms, gait disturbances and therapy complications. It is easy and safe and does away with the cumbersome and time-consuming procedure of DBS. A head-to-head controlled study with DBS is the only way to establish efficacy and safety of both MI ECS and DBS.

References

Albanese A, Romito LM, Piacentini S, Perani D, Carella F, Broggi G. Reply. *Neurology* 2007; 69: 1063

Arle JE, Apetauerova D, Zani J, et al. Motor cortex stimulation for Parkinson disease: 12 months follow up in 4 patients. *J Neurosurg* 2008; 109: 133-139

Arle JE, Shils J. Motor cortex stimulation for pain and movement disorders. *Neurotherapeutics* 2008; 5:37-49

Bentivoglio AR, Cavallo MA, Cioni B, et al. Motor cortex stimulation for movement disorders In: Meglio M (Ed) *Proceedings of the 14th Meeting of the World's Society for Stereotactic and Functional Neurosurgery*. Bologna: Medimond, 2005, pp 89-97

Benvenuti E, Cecchi F, Colombini A, Gori G. Extradural motor cortex stimulation as a method to treat advanced Parkinson's disease:new perspectives in geriatric medicine. *Aging Clin Exp Res* 2006; 18:347-348

Canavero S, Paolotti R. Extradural motor cortex stimulation for advanced Parkinson's disease. *Mov Disord* 2000; 15:169-171

Canavero S, Paolotti R, Bonicalzi V, et al. Extradural motor cortex stimulation for advanced Parkinson's disease: report of two cases. *J Neurosurg* 2002; 97:1208-1211

Canavero S, Bonicalzi V, Paolotti R, et al. Therapeutic extradural cortical stimulation for movement disorders: a review. *Neurol Res* 2003; 25: 118-122

Canavero S, Bonicalzi V. Cortical stimulation for parkinsonism. *Arch Neurol* 2004; 61: 606

Canavero S, Bonicalzi V. Extradural cortical stimulation for movement disorders. In: D Sakas, B Simpson, E Krames, eds. *Operative Neuromodulation*. II Neural Networks Surgery. Wien: Springer-Verlag, 2007, 223-232

Cilia R, Landi A, Vergari F, Sganzerla E, Pezzotti G, Antonini A. Extradural motor cortex stimulation in Parkinson's disease. *Mov Disord* 2007; 22:111-114

Cilia R, Marotta G, Landi A, et al. Cerebral activity modulation by extradural motor cortex stimulation in Parkinson's disease: a perfusion SPECT study. *Eur J Neurol* 2008; 15: 22-28

Cioni B. Motor cortex stimulation for Parkinson's disease. In: D Sakas, B Simpson, E Krames, eds. *Operative Neuromodulation.* II Neural Networks Surgery. Wien: Springer-Verlag, 2007, 233-238

Cioni B, Meglio M, Perotti V, DeBonis P, Montano N. Neurophysiological aspects of chronic motor cortex stimulation. *Clin Neurophysiol* 2007; 37:441-447

Contarino MF, Daniele A, Sibilia AH, et al. Cognitive outcome 5 years after bilateral chronic stimulation of the subthalamic nucleus in patients with Parkinson's disease. *J Neurol Neurosurg Psychiatry* 2007; 78:248-252

Daniele A, Albanese A, Contarino MF, et al. Cognitive and behavioural effects of chronic stimulation of the subthalamic nucleus in patients with Parkinson's disease *J Neurol Neurosurg Psychiatry* 2003; 74:175-182

Fasano A, Piano C, De Simone C, Cioni B, Meglio M, Bentivoglio AR. High frequency extradural motor cortex stimulation dramatically improves axial symptoms in a patient with Parkinson's disease. *Mov Disord* 2008; 23:1916-9.

Franzini A, Ferroli P, Dones I, Marras C, Broggi G. Chronic motor cortex stimulation for movement disorders: a promising perspective *Neurol Res* 2003; 25:123-126

Katayama Y, Fukaya C, Yamamoto T. Control of poststroke voluntary and involuntary movement disorders with deep brain or epidural cortical stimulation *Stereotact Funct Neurosurg* 1997; 69:73-79

Katayama Y, Oshima H, Fukaya C, Kawamata T, Yamamoto T. Control of post-stroke movement disorders using chronic motor cortex stimulation. *Acta Neurochir* 2002; 79(Suppl):89-92

Katayama Y, Kano T, Kobayashi K, Oshima H, Fukaia C, Yamamoto T. Feed-forward control of post-stroke movement disorders by on-demand type stimulation of the thalamus and motor cortex. *Acta Neurochir* Suppl 2006; 99:21-23

Kleiner-Fisman G, Fisman DN, Kahn FI, Sime E, Lozano A, Lang AE. Motor cortical stimulation for Parkinsonism in Multiple Systemic Atrophy. *Arch Neurol* 2003; 60:1554-1558

Lyons KE, Wilkinson SB, Pahwa R. Stimulation of the motor cortex for disabling essential tremor *Clin Neurol Neurosurg* 2006; 108:564-567

Munno D, Caporale S, Zullo G, et al. Neuropsychological assessment of patients with advanced Parkinson Disease submitted to extradural motor cortex stimulation. *Cog Behav Neurol* 2007; 20: 1-6

Nguyen JP, Pollin B, Fève A, Geny C, Cesaro P. Improvement of action tremor by chronic cortical stimulation. *Mov Disord* 1998; 13:84-88

Pagni CA, Zeme S, Zenga F. Further experience with extradural motor cortex stimulation for treatment of advanced Parkinson's diseases. Report of 3 new cases *J Neurosurg Sci* 2003; 47:189-193

Pagni CA, Zeme S, Zenga F, Maina R.Extradural motor cortex stimulation in advanced Parkinson's disease: the Turin experience. *Neurosurg* 2005; 57 (ONS Suppl 3): ONS-402

Pagni CA, Albanese A, Bentivoglio A, Broggi G, Canavero S, Cioni B, et al. Results by motor cortex stimulation in treatment of focal dystonia, Parkinson's disease and post-ictal spasticity. The experience of the Italian Study Group of the Italian Neurosurgical Society. *Acta Neurochir Suppl.* 2008;101:13-21.

Romito LM, Franzini A, Perani D, et al. Fixed dystonia unresponsive to pallidal stimulation improved by motor cortex stimulation. *Neurology* 2007; 68:875-876

Strafella AP, Lozano AM, Lang AE, Ko J-H, Poon Y-Y, Moro E. Subdural motor cortex stimulation in Parkinson's disease does not modify movement-related rCBF pattern. *Mov Disorders* 2007; 22: 2113-2116

Seijo FJ, Gutierrez JC, Alvarez Vega M, Fernandez Gonzalez F, Lozano Aragonese B, Blazquez M. Therapeutic extradural cortical stimulation for Parkinson's disease: report of six cases and review of the literature. (*submitted*)

Sidoti C, Agrillo U. Chronic cortical stimulation for amyotrophic lateral sclerosis: a report of four consecutive operated cases after a 2-year follow up *Neurosurg* 2006; 58:E384

Tani N, Saitoh Y, Kishima H, et al. Motor cortex stimulation for levodopa-resistant akinesia: case report. *Mov Disorder* 2007; 22:1645-1649

Tsubokawa T, Katayama Y, Yamamoto T, Hirayama T, Koyama S. Chronic motor cortex stimulation in patients with thalamic pain. *J Neurosurg* 1993; 78:393-401

Verhagen Metman L, Pahwa R, Lyons K, et al. Motor cortex stimulation in Parkinson's disease: a pilot study. *Neurology* 2006 (suppl 2) 66: A47

In: Encyclopedia of Neuroscience Research
Editors: Eileen J. Sampson and Donald R. Glevins
ISBN 978-1-61324-861-4

Chapter XV

Deep Brain Stimulation in Epilepsy: Experimental and Clinical Data*

M. Langlois[1], S. Saillet[1], B. Feddersen[5], L. Minotti[1, 2], L. Vercueil[1, 2], S Chabardès[1, 3], O. David[1], A. Depaulis[1], P. Kahane[1,2,4] and C. Deransart[† 1,2]

[1]Grenoble Institut des Neurosciences, INSERM U836-UJF-CEA,
[2]Neurology Department, and [3]Neurosurgery Department, Grenoble University Hospital,
[4]Institute for Children and adolescents with epilepsy – IDEE,
Hospices Civils de Lyon, France,
[5]Department of Neurology, Klinikum Grosshadern, University of Munich, Germany

1. Introduction

About 30% of epileptic patients do not respond to antiepileptic drugs (Kwan & Brodie, 2000), of whom only a minority can benefit from resective surgery. Such a therapeutic option is considered only in patients who suffer from focal seizures with an epileptogenic zone clearly identified and safely removable. Therefore, patients with seizures arising from eloquent cortices, or which are multifocal, bilateral, or generalized, represent a particular challenge to "new" or "alternative" therapies. For these patients, neurostimulation appears with a great potential (Polkey et al., 2003; Theodore and Fisher, 2004). Different approaches to neurostimulation in epileptic patients now exist and depend on (i) the brain region which is

---

* A version of this chapter was also published in *Recent Breakthroughs in Basal ganglia Research* , edited by Erwan Bezard published by Nova Science Publishers, Inc. It was submitted for appropriate modification in an effort to encourage wider dissemination of research.

† Corresponding author: Colin Deransart, Grenoble - Institut des Neurosciences, Centre de recherche Inserm U 836-UJF-CEA-CHU, Equipe 9: Dynamique des Réseaux Synchrones Epileptiques, Université Joseph Fourier - Faculté de Médecine, Domaine de la Merci, 38700 La Tronche, colin.deransart@ujf-grenoble.fr

targeted and (ii) the way the stimulation is applied (Oommen et al., 2005; Morrell, 2006; Theodore and Fisher, 2004; Vonck et al., 2007). The aim of neurostimulation in epilepsy is to reduce the probability of seizure occurrence and/or propagation, either by manipulating remote control systems (vagus nerve stimulation, deep brain stimulation), or by interfering with the epileptogenic zone itself (repetitive transcranial magnetic stimulation, cortical stimulation). In most cases, stimulation is delivered continuously or intermittently according to a scheduled protocol. In particular, new progress in biotechnology and EEG signal analysis now allows stimulation in response to detection of electrographic seizures (e.g., closed-loop stimulation). Here, we review the various experimental and clinical attempts that have been made to control epileptic seizures by the means of deep brain electrical stimulation.

2. Deep Brain Stimulation (DBS)

For more than two decades, stimulation of a number of deep brain targets has been shown to be feasible, safe, and effective in humans suffering from different forms of movement disorders. This has led to the development of deep brain stimulation (DBS) in an increasing number of neurological and non-neurological diseases, including epilepsy (Benabid et al., 2001). Although the cortex plays a crucial role in seizure generation, accumulating evidence has pointed to the role of subcortical structures in the clinical expression, propagation and control of epileptic seizures in humans (Semah, 2002; Vercueil and Hirsch, 2002). Based on experimental findings, DBS has been applied to a number of targets, including the cerebellum, different nuclei of the thalamus, and several structures of the basal ganglia system. Although encouraging, published results do not reach a definite conclusion and require further studies using animal models. Indeed, the study of the mechanisms of actions of such DBS on epileptic seizures is critical to understand the transitions between normal and paroxysmal activities of the epileptic networks.

2.1. Cerebellum

During the 1950s and 1960s, cortical cerebellar stimulation was shown to have antiepileptic properties on different animal models of seizures, mostly penicillin and cobalt foci in cats (Cooke and Snider, 1955; Dow et al., 1962; Mutani et al., 1969). Following this, and assuming cerebellar outflow is inhibitory in nearly all patients, Cooper and colleagues showed that seizures were modified or inhibited in 10 out of their 15 epileptic patients, without adverse effects (Cooper et al., 1973; 1976; 1978). These data raised the issue of distant modulation of cortical epileptogenicity by electrical currents. More especially, this study showed for the first time the feasibility and safety of a therapeutic stimulation technique in epileptic patients. Later, a large open study on 115 patients reported that 31 became seizure-free and 56 were significantly improved by stimulation of the cerebellum (Davis and Emmonds, 1992). Such promising results, however, were not confirmed in 3 controlled clinical trials involving 14 patients, of whom only 2 were improved (Krauss and Fisher, 1993; Van Buren et al., 1978; Wright et al., 1984). Additional animal studies conducted in monkeys with cortical focal seizures induced by alumina cream, or in kindled cats, did not confirm previous experimental findings (Ebner et al., 1980 ; Lockard et al., 1979 ; Majkowski et al., 1980) and the interest for cerebellar stimulation in epilepsy disappeared for many years. Recently, however, a double-blind, randomized controlled pilot study conducted in 5 patients suffering from intractable motor seizures has renewed the interest in such stimulation (Velasco et al., 2005). In this study, 10-Hz stimulations were applied to the upper medial surface of each cerebellar hemisphere, and parameters were adjusted to deliver a constant charge density of 2.0 microC/cm^2/phase. During the initial 3-month double-blind phase, seizures were significantly reduced when the patients were stimulated. Over the following 6-month open-label phase, where all the patients were stimulated, seizures were reduced by 41% (14-75%) and the difference was significant for tonic and tonic-clonic seizures. Effectiveness was maintained over 2 years and few complications occurred. Altogether, although cerebellar stimulation appears to possess antiepileptic effects in some patients and/or some forms of epilepsy, the rationale of such suppressive effects remains to be determined. No clinical studies targeting this structure are currently under progress.

2.2. Thalamus

Since the 1980s, different nuclei of the thalamus have been studied to understand the physiopathology of epilepsy because many interaction pathways exist between these nuclei and the cortex. Several thalamic targets have been stimulated to suppress seizures, mainly the anterior nucleus and the centromedian nucleus. There is limited proof from animal studies that stimulation of these structures can influence seizure threshold. However, there is clinical evidence that continuous stimulation of these targets in epileptic patients reduces seizure frequency and severity.

2.2.a. Anterior Thalamus (AN)

The anterior nucleus (AN) of the thalamus receives projections from the hippocampus via the fornix, the mamillary bodies and the mamillo-thalamic fascicle of Vicq d'Azir and has outputs to the cingulate cortex and, via the cingulum, to the entorhinal cortex and back to the hippocampus. It appears to be in close interaction with the circuit of Papez which is often involved in some forms of epilepsies (e.g., temporal lobe epilepsies). AN therefore is central in the network which underlies limbic seizures and, as such, represents an attractive target for DBS in epileptic patients. Cooper and his group, encouraged by their experience on cerebellar stimulation, were the first to direct their interest to this nucleus, based on the hypothesis that AN could act as a "pacemaker" for the cortex. They showed that bilateral chronic stimulation of AN in 6 epileptic patients resulted in 60% reduction of seizure frequency in 5 of them, as well as a decrease in EEG spikes (Cooper and Upton, 1985). Using an experimental approach, it was later shown that AN and mamillary bodies were involved in the genesis of pentylenetetrazol-induced seizures and were activated during ethosuximide-induced suppression of these seizures (Mirski and Ferrendelli, 1986a ; 1986b). In addition, the section of the mamillo-thalamic bundle prevented pentylenetetrazol-induced seizures in guinea pigs (Mirsky and Ferrendelli, 1984). Furthermore, it was reported that 100-Hz electrical stimulation of the mammilary nuclei and AN increased the seizure threshold of pentylenetetrazol in rats (Mirski and Fisher, 1994; Mirski et al., 1997). These anticonvulsant effects were dependent on the intensity of the stimulation rather than on its frequency. On the contrary, low-frequency AN stimulation tended to be proconvulsive (Mirski et al, 1997). More recently, high-frequency AN stimulation suppressed focal cortical and limbic seizures induced by intra-cortical or intra-amydaloid kainic acid injections, respectively (Takebayashi et al, 2007a and 2007b) and delayed both status epilepticus and seizures induced by pilocarpine although without complete suppression (Hamani et al., 2004, 2008). Finally, 100-Hz AN stimulation was found to *aggravate* recurrent seizures observed following status epilepticus produced by systemic kainic acid (Lado, 2006).

These experimental data gave weight to the need of re-assessing the effect of AN stimulation in epileptic patients, Four open-label trials were reported showing that seizure frequency was reduced from 20 to 92% and being statistically significant in 12 of the 18 patients (Hodaie et al., 2002 ; Kerrigan et al., 2004 ; Lim et al., 2007 ; Osorio et al., 2007). Two patients presented a complication (small frontal hemorrhage and extension erosion over the scalp), which did not result in major or permanent neurological deficit. A study showed that insertion of AN electrodes by itself could reduce seizures (Lim et al., 2007) and another one that the observed benefits did not differ between stimulation-on and stimulation-off periods (Hodaie et al., 2002), thus raising the issue of a lesional, placebo or carry-over effect. To address this question, a large multicenter prospective randomized trial of AN stimulation for partial and secondarily generalized seizures (Stimulation of the Anterior Nucleus of the Thalamus for Epilepsy or SANTE) is currently under investigation in North America. Whether AN stimulation could be more effective in temporal lobe epilepsy (Zumsteg et al., 2006) and whether other components of the circuit of Papez, namely the mamillary bodies and mamillo-thalamic tract (Duprez et al., 2005 ; van Rijckevorsel et al., 2005) are possible targets for DBS are important issues for clinical trials.

2.2.b. Centromedian Thalamus (CM)

In addition to the AN, attention was also directed towards one of the intralaminar nuclei of the thalamus, the centromedian nucleus (CM). This nucleus is part of the reticulothalamocortical system mediating cerebral cortex excitability (Jasper, 1991), and has been suggested to participate in the modulation of vigilance states (Velasco et al., 1979). Although experimental findings remain rare (Arduini and Lary Bounes, 1952), a first open-label study was conducted in 5 patients with bilateral CM stimulation at the end of the 1980s (Velasco et al., 1987). Initial results indicated an improvement of seizure frequency and EEG spiking over 3 months of chronic stimulation. Later, Velasco's group accumulated data in a cohort of 49 patients suffering from different forms of seizures and epilepsies (Velasco et al., 2001a ; 2002). Among these patients, 5 to 13 were followed over long-term follow-up studies (Velasco et al., 1993 ; 1995 ; 2000ab, 2006). Overall, the procedure was reported to be beneficial and generally well-tolerated, although a central nystagmus was induced in some cases (Taylor et al., 2000). A few patients were explanted because of repeated and multiple skin erosions (Velasco et al., 2006). It is interesting to note that a decrease of 80% of seizures were observed on average in patients with generalized tonic-clonic seizures and atypical absences of the Lennox-Gastaut syndrome, with a global improvement of patients in their ability scale scores (Velasco et al., 2006). By contrast, no improvements were found for either complex partial seizures or focal spikes in temporal regions. The best clinical results were seen when both electrodes contacts were located within the CM on both sides and when stimulation at 6-8 Hz and 60 Hz induced recruiting responses and regional DC shifts, respectively (Velasco et al., 2000a). Two hours of daily 130-Hz stimulation sessions (1-minute on, 4 minutes off), alternating the right and left CM were used. However, continuous bilateral stimulation led to faster and more significant results (Velasco et al. 2001b). As for AN stimulation, persistent antiepileptic effects were found 3 months or more after discontinuation of the stimulation (« off effect »), and possible plasticity which develops during the stimulation procedure was suggested (Velasco et al. 2001b). No such seizure suppression was found in a small placebo-controlled study conducted in 7 patients with mesial temporal lobe epilepsy. In this study, no statistically significant difference from the baseline in frequency of tonic-clonic seizures when the stimulator was on versus off (Fisher et al., 1992). In the open-label follow-up phase, however, 3 of 6 patients reported at least a 50% decrease in seizure frequency.

Up to now, very few animal studies have examined the role of the CM or of the parafascicular nucleus (PF) of the thalamus—which has similar connections—in the control of epileptic seizures. In a genetic model of absence epilepsy in the rat (GAERS), pharmacological activation of the PF was found to suppress spike-and-wave discharges (Nail-Boucherie et al., 2005). More recently, 130-Hz stimulation of this structure was reported to interrupt focal hippocampal seizures in a mouse model of mesiotemporal lobe epilepsy (Langlois et al., in preparation). Because of its unique location between cortical and limbic structures and the basal ganglia (see below), the CM/PF nuclei could well constitute an interesting target for DBS. More animal studies are clearly required to understand the role of this structure in the modulation of epileptic seizures.

2.3. Basal Ganglia

Since the beginning of the 1980s, experimental animal studies have suggested the existence of a "nigral control" of epileptic seizures (for review see Gale, 1995; Depaulis et al., 1994). Inhibition of the Substantia Nigra pars Reticulata (SNR) has potent anti-epileptic effects in different animal models of epilepsy (Deransart and Depaulis, 2002) and the GABAergic SNR output appears to be a critical relay in this control (Depaulis et al., 1990; Paz et al., 2005; 2007). Local manipulations of the basal ganglia that lead to an inhibition of the SNR neurons (e.g., activation of the striatum or pallidum, inhibition of the sub-thalamic nucleus) also had significant anti-epileptic effects (for review see Deransart and Depaulis, 2002), suggesting that different striato-nigral circuits are involved in the control of epileptic seizures. In humans, EEG, clinical and imaging data also support the involvement of the basal ganglia in the propagation and/or the control of epileptic discharges (Biraben et al., 2004; Bouilleret et al., 2008; Vercueil and Hirsch, 2002). Altogether, experimental and clinical data suggest a privileged role for the basal ganglia in the control of generation and/or spread of epileptic discharges in the cortex. Paradoxically, the therapeutic relevance of such findings was rarely considered until the 1990s.

2.3.a. Caudate Nucleus (CN)

Following experimental evidence that stimulation of the caudate nucleus (CN) has antiepileptic properties in different animal models of seizures (La Grutta et al., 1971, 1988; Mutani, 1969; Oakley and Ojemann, 1982 ; Psatta, 1983), Chkhenkeli and his group, as well as Sramka and colleagues, were the first to suggest the beneficial effect of striatal low-frequency stimulation (below 50 Hz) in epileptic patients (Chkhenkeli, 1978; Sramka et al., 1980). A decrease in focal and generalized discharges was observed in 57 patients bilaterally stimulated at low frequency (4–6Hz) in the CN (Chkhenkeli and Chkhenkeli, 1997). The study, however, was not controlled and the effects on seizures were not assessed. Interestingly, epileptic activity was worsened by stimulating the CN at higher frequency, a finding that was also reported in the aluminium-hydroxide monkey model of motor seizures (Oakley and Ojemann, 1982). Therefore, if one assumes that low-frequency stimulation is excitatory and high-frequency stimulation is inhibitory, these clinical data are in agreement with animal data (see Deransart and Depaulis, 2002). Indeed, activation of the striatum inhibits the SNR through GABAergic projections and therefore leads to seizure suppression (Deransart et al., 1998). Although further studies are needed, these results highlight the ability of the basal ganglia system to modulate cortical epileptogenicity.

2.3.b. Subthalamic Nucleus (STN)

In 1998, Vercueil et al. (1998) were the first to show that 130-Hz stimulation of the subthalamic nucleus (STN) could interrupt absence seizures in GAERS, a well-established genetic model of absence epilepsy (Danober et al., 1998; Marescaux et al. 1992). Since then, high-frequency stimulation of the subthalamic nucleus has been reported to protect against seizures induced by local kainate injection in the amygdala (Bressand et al., 1999; Loddenkemper et al., 2001; Usui et al., 2005) or by fluorothyl inhalation (Veliskova et al., 1996). This is in agreement with the antiepileptic effects reported after pharmacological inhibition of the STN on seizures induced by amygdala kindling (Deransart et al., 1998a), intravenous bicuculline or by its focal application into the anterior piriform cortex (Dybdal and Gale, 2000) and in GAERS (Deransart et al., 1996).

This led the group of Benabid at Grenoble University Hospital to perform the first STN stimulation in a 5-year-old girl with pharmacologically-resistant inoperable epilepsy caused by a focal centroparietal dysplasia (Benabid et al., 2002). Later, 11 additional patients suffering from different forms of epilepsy received high frequency STN stimulation at different institutions (Chabardès et al., 2002; Loddenkemper et al., 2001; Vesper et al., 2007). Overall, seizure occurrence was reduced by at least 50% in 7/12 cases, and the stimulation was well tolerated. Good responders suffered from very different epilepsy types including focal epilepsy, Dravet syndrome, Lennox-Gastaut syndrome and progressive myoclonic epilepsy. Surgical complications occurred in 2 patients, including infection of the generator in one, and a postimplantation subdural hematoma in another who later underwent surgical treatment, without sequelae (Chabardès et al., 2002). Bilateral stimulation appeared more effective than unilateral stimulation, in agreement with experimental data (Depaulis et al., 1994). However, whether it should be applied continuously or intermittently remains questionable (Chabardès et al., 2002). Furthermore, whether the optimal target in epileptic patients is the STN itself or, as is suggested in some patients, the SNR, remains an important issue (see below - Chabardès et al., 2002 ; Vesper et al., 2007). A double-blind cross-over multicentric study is in progress in France (STIMEP) and aims at evaluating the clinical effect of 130-Hz stimulation of the STN/SNR in patients with ring chromosome 20 epilepsy. These patients suffer from very long lasting epileptic seizures, evolving often into status epilepticus and are difficult to control with antiepileptic drugs. They exhibit a deficit of dopaminergic activity in the striatum as compared with normal subjects (Biraben et al., 2004), a finding which is in accordance with the critical role of striatal dopamine in the control of seizures (Deransart et al., 2000).

2.3.c. Substantia Nigra pars Reticulata (SNR)

In 1980, Gale and Iadarola were the first to correlate an increase of GABA in the SNR with antiepileptic effects (Gale and Iadarola, 1980). Later, they showed that the potentiation of the GABAergic neurotransmission within the SNR, by bilateral microinjections of GABAmimetic drugs, suppressed convulsions in various models of generalized seizures in the rat (Iadarola and Gale, 1982). The possibility that seizures are controlled by the SNR also emerged from pharmacological studies in GAERS showing that a bilateral inhibition of SNR suppresses cortical Spike-and-Waves discharges (SWDs) (Depaulis et al., 1988, 1989; Deransart et al., 1996, 1998, 2001). Since then, several studies have confirmed that inhibition of the SNR has a potent anti-epileptic effects in different animal models of epilepsy (Depaulis et al., 1994; Deransart and Depaulis, 2002; Paz et al.,2005; 2007).

In this context, it was shown that DBS applied to the SNR, also suppressed generalized convulsive seizures induced by fluorothyl inhalation (Velisek et al., 2002), amygdala-kindled seizures (Morimoto and Goddard,1987; Shi et al., 2006), absence seizures in GAERS (Feddersen et al., 2007) and also focal seizures in kainate treated mice (Deransart et al., 2004). In the model of generalized convulsive seizures induced by fluorothyl inhalation, bilateral and bipolar 130Hz SNR stimulation had anticonvulsivant effects in both adults and infant rats (Velisek et al., 2002). In amygdala-kindling, such stimulations were shown to induce a long lasting suppression of antiepileptogenesis (Shi et al., 2006). In GAERS, bilateral, bipolar, and monophasic SNR stimulations at a frequency of 60Hz and a pulse width of 60μs were defined as the optimal conditions to interrupt ongoing absence seizures without motor side effects (Feddersen et al., 2007). The threshold to interrupt epileptic seizures was lower using SNR stimulation compared to STN stimulation, using the same model and stimulation parameters. However, this last study showed that continuous stimulation fail to control the occurrence of seizures, in agreement with previous reports (Vercueil et al., 1998) and suggested that a refractory period of about 60 sec exists during which any stimulation is without effect. This study also showed that *continuous* stimulation of the SNR could even aggravate seizure occurrence. Adaptive stimulation may allow to alleviate this problem and to further specify the existence of a refractory period (see below).

3. Stimulation at Seizure Focus

Stimulating the epileptogenic cortex to interrupt epileptic seizures may appear paradoxical. Indeed, "stimulation" classically means "excitation" and the epilepsies are characterized by a pathological hyperexcitability and hypersynchrony of cortical neurons. The effects provoked by cortical stimulation, however, depend on the stimulation parameters used, the region which is stimulated, as well as the way that the stimulation is delivered (indirectly or directly). Furthermore, cortical stimulation is generally used to map functions in eloquent brain. It is known that cortical stimulation can evoke focal after-discharges that may evolve into clinical seizures. It has been shown that afterdischarges elicited by electrical stimulation via subdural electrodes can be interrupted by the application of brief bursts of 50-Hz electrical stimulation through subdural electrode contacts (Lesser et al. 1999). To date, a few studies have been conducted, including a limited number of patients, and therapeutic results are equivocal at best.

Several preclinical studies have found potential antiepileptic effects of brain stimulation in animal models. Notably, low-frequency (1 Hz) stimulation applied after kindling stimulation of the amygdala was found to inhibit the development of afterdischarges, an effect named *quenching* (Weiss et al., 1995). This quenching effect seems effective in adult as well as immature rats (Velisek et al., 2002). Interestingly, when applied immediately *before* the kindling stimulus, preemptive 1 Hz sine wave stimulation was also effective, thus suggesting some potential benefit for seizure prevention (Goodman et al., 2005). Other regions such as the hippocampus (Barbarosie and Avoli, 1997), the central piriform cortex (Yang et al., 2006; Zhu-ge et al., 2007) or the cerebral fastigial nucleus (Wang et al., 2008) may also appear as potentially effective targets for 1-Hz stimulation treatment of epilepsy. In general these data suggest that 1-Hz stimulation inhibits both acquisition and expression of kindling seizure by preventing afterdischarge generation and propagation in rat. Unexpectedly, such effects are also observed in the cerebral fastigial nucleus, suggesting that targets outside the limbic system may have a significant antiepileptic action.

In humans, both low- (1-Hz) and medium- (50 Hz) frequency stimulation have proven effective to reduce interictal epileptiform discharges (Kinoshita et al., 2005b; Yamamoto et al., 2002). Therapeutic stimulation, however, was applied at high frequency in almost all studies. The first attempt of therapeutic stimulation of temporal lobe structures was reported in 1980, in 3 patients, without clear benefit (Sramka et al., 1980). More recently, several investigators have tried continuous scheduled stimulation of epileptic foci, including hypothalamic hamartoma (Kahane et al., 2003), neocortical structures (Elisevich et al., 2006) and, mostly, mesio-temporal lobe (Tellez-Zenteno et al., 2006; Velasco et al., 2000c ; 2007; Vonck et al., 2002). The first pilot study of mesio-temporal lobe stimulation, conducted in 10 patients studied by intracranial electrodes before surgery, showed that stimulation stopped seizures and decreased the number of interictal EEG spikes in the 7 patients where the stimulated electrode was placed within the hippocampus or hippocampal gyrus (Velasco et al., 2000c). There were no side-effects on language and memory, and no histological damages were found in the stimulated tissue. Whether such an antiepileptic effect could be observed over a more prolonged stimulation procedure was later evaluated in a small open series conducted in 3 patients, all of whom exhibited more than 50% of seizure reduction after a mean follow-up of 5 months, without adverse events (Vonck et al., 2002).

Following this, 2 additional trials of hippocampal stimulation were conducted, leading to opposite results. In one double-blind study, the seizure outcome was significantly improved in all 9 patients over a long-term follow-up (Velasco et al., 2007), which showed more than 95% seizure reduction in the 5 patients with normal MRI, and 50-70% seizure reduction in the 4 patients who had hippocampal sclerosis. No adverse events were found but 3 patients were explanted after 2 years due to skin erosion in the trajectory system. It was suggested that beneficial effects of stimulation were associated with a high GABA tissue content and a low rate of cell loss (Cuellar-Herrera et al., 2004). By contrast, seizure frequency was reduced by only 15% in average in the 4 patients of the double-blind, multiple cross-over, randomized study of Tellez-Zenteno et al. (2006). Additionally, effects seemed to carry over into the off period, thus raising the issue of an implantation effect. Yet, no adverse events were found.

Overall, stimulation of hippocampal foci shows beneficial trends, but whether the effect is significant, and of clear clinical relevance, remains debatable.

Currently, a randomized controlled trial of hippocampal stimulation for temporal lobe epilepsy (METTLE) is recruiting patients to determine whether unilateral hippocampal electrical stimulation is safe and more effective than simply implanting an electrode in the hippocampus without electrical stimulation, or treating with medical therapy alone. A prospective randomized controlled study of neurostimulation in the medial temporal lobe for patients with medically refractory medial temporal lobe epilepsy is also currently recruiting patients for a controlled randomized stimulation versus resection (CoRaStiR) study (www.clinicaltrials.gov).

4. Adaptative Stimulation

Continuous scheduled brain stimulation, whatever the target (DBS, cortical stimulation), has appeared to be safe and of potential benefit in treating medically intractable epilepsies (see above). Limited, but growing data suggests that responsive (seizure-triggered) stimulation might also be effective (Morrel, 2006). Such a strategy is distinct from continuous scheduled stimulation as it aims at blocking seizures when they occur, rather than at decreasing cortical excitability chronically. It is motivated by the reduction of power consumption, the paroxysmal nature of the seizures and the possible behavioural side-effects induced by chronic stimulations. Also, it has been suggested that continuous stimulations may aggravate seizures in animals (Feddersen et al., 2007). Seizure-triggered stimulation requires an implanted stimulating device coupled with real-time signal analysis techniques. Usually, a seizure detection algorithm allows the delivery of a stimulation to interrupt seizure prior to, or concomitantly to, the onset of clinical symptoms. A number of algorithms to detect seizures do exist (see for instance Osorio et al., 2002; Grewal and Gotman, 2005). The main stumbling block, as for continuous stimulation, is to find, ideally following an automatic search, optimal stimulation parameters to abort seizures. To our knowledge, existing literature about automatic seizure-triggered stimulation in animal models *in vivo* is rather limited. Using similar techniques as VNS therapy, Fanselow and colleagues have shown a reduction of pentylenetetrazole-induced seizure activity in awake rats by seizure-triggered trigeminal nerve stimulation (Fanselow et al., 2000). Interestingly, seizure-triggered stimulation was more effective than the stimulation protocol involving a fixed duty cycle, in terms of the percent seizure reduction per second of stimulation (up to 78%). Currently, a preliminary study in Grenoble (France) is testing a new technology based on stimulation combined to seizure-detection to interrupt absence seizures in GAERS (Saillet et al., submitted). This should allow to better determine the optimal target and parameters of stimulation required by such technology.

In humans, responsive stimulation can shorten or terminate electrically-elicited afterdischarges using brief bursts of 50-Hz electrical stimulation (Lesser et al., 1999), the effect being greater at primary sites than at adjacent electrodes (Motamedi et al., 2002). Preliminary trials of responsive stimulation, however, did not consistently use a similar paradigm (Kossof et al., 2004 ; Fountas et al., 2005 ; Osorio et al., 2005). The effects of responsive stimulation were first evaluated in 4 patients using an external neurostimulator, which proved effective at automatically detecting electrographic seizures, delivering targeted electrical stimuli, and altering or suppressing ictal discharges (Kossoff et al., 2004). Another feasibility study confirmed these results using a cranially implantable device in 8 patients (Fountas et al., 2005). Detection and stimulation were performed using electrodes placed over the seizure focus, and 7 of the 8 patients exhibited more than a 45% decrease in their seizure frequency, with a mean follow-up time of 9.2 months. In the third pilot study, conducted in 8 patients, stimulation was delivered either directly to the epileptogenic zone (local closed-loop, n=4), or indirectly through the anterior thalami (remote closed-loop, n=4), depending on whether the epileptogenic zone was single, or multiple (Osorio et al., 2005). On average, a 55.5% and 40.8% decrease of seizure frequency was observed in the local closed-loop group and in the remote closed-loop group, respectively. Overall, none of the 20 patients enrolled in these 3 pilot studies had adverse events. Although promising, this new therapy needs further evaluation and a multi-institutional prospective clinical trial is underway in the USA. The Responsive Neurostimulation System (RNS), sponsored by NeuroPace Inc., is designed to continuously monitor brain electrical activity from the electrodes and, after identifying the "signature" of a seizure's onset, deliver brief and mild electrical stimulation with the intention of suppressing the seizure. The purpose of the RNS System Pivotal Clinical Investigation is to assess the safety and to demonstrate that the RNS System is effective as an add-on (adjunctive) therapy in reducing the frequency of seizures in individuals with partial onset seizures that are refractory to two or more AED medications. Whether, in the near future, closed-loop stimulators will be able to react using seizure-prediction algorithms represents a particularly challenging issue.

5. Conclusions

Neurostimulation in non-surgically remediable epileptic patients represents an emerging treatment. It has the advantage of reversibility and adjustability, but remains palliative so that surgical resection remains the gold standard treatment of drug-resistant epilepsies whenever this option is possible. VNS is the only approved stimulation therapy for epilepsy so far and, as such, it is licensed in many countries as an adjunctive therapy. Other stimulation techniques must be considered experimental although several controlled studies are currently under investigation. Notably, results of direct brain stimulation, although encouraging, are not conclusive and further investigations are required to evaluate the real benefit of this emerging therapy, in as much as the risks of haemorrhage and infection, although low (around 5%), do exist. However, pathological examination in post-mortem studies and temporal lobe resection, in Parkinson's disease or epilepsy, suggest that chronic stimulation does not induce neural injury and can be delivered safely (Haberler et al., 2000; Pilitsis et al., 2008; Velasco et al., 2000c). In any case, seizure types or epileptic syndromes which may respond to stimulation should be identified, as well as the type of stimulation that is likely to be of potential efficacy depending on the patient's characteristics. This requires to improve our knowledge on the neural circuits in which seizures start and propagate, to better understand the precise mechanisms of the supposed effect of neurostimulation, and to search for optimal stimulation parameters. The development of experimental research in this field, as well as rigorous clinical evaluation, is essential for further improvements in clinical efficacy.

6. References

Akamatsu N, Fueta Y, Endo Y, Matsunaga K, Uozumi T, Tsuji S. Decreased susceptibility to pentylenetetrazole-inducded seizures after low frequency transcranial magnetic stimulaiton in the rat. *Neurosci Lett* 2001; 310:153-156.

Andy OJ, Jurko MF. Seizure control by mesothalamic reticular stimulation. *Clin Electroencephalography* 1986 ; 17 : 52-60.

Arduini D, Lary Bounes GC. Action de la stimulation électrique de la formation réticulaire du bulbe et des stimulations sensorielles sur les ondes strychniques. *Electroencephalogr Clin Neurophysiol* 1952; 4: 502-12.

Barbarosie M, Avoli M. CA3-driven hippocampal-entorhinal loop controls rather than sustains in vitro limbic seizures. *J Neurosci* 1997; 17 : 9308-9314.

Benabid AL, Koudsié A, Benazzouz A, Vercueil L, Fraix V, Chabardès S, Lebas JF, Pollak P. Deep brain stimulation of the corpus luysi (subthalamic nucleus) and other targets in Parkinson's disease. Extension to new indications such as dystonia and epilepsy. *J Neurol* 2001; 248 (suppl 3): III37-47.

Benabid AL, Minotti L, Koudsie A, de Saint Martin A, Hirsch E. Antiepileptic effect of high-frequency stimulation of the subthalamic nucleus (corpus luysi) in a case of medically intractable epilepsy caused by focal dysplasia: a 30-month follow-up: technical case report. *Neurosurgery* 2002; 50: 1385–1391.

Benazzouz A, Piallat B, Pollak P, Benabid AL. Responses of substantia nigra pars reticulata and glomus pallidus complex to high frequency stimulation of the subthalamic nucleus in rat: electrophysiological data. *Neurosci Lett* 1995; 189 : 77-80.

Biraben A, Semah F, Ribeiro MJ, Douaud G, Remy P, Depaulis A. PET evidence for a role of the basal ganglia in patients with ring chromosome 20 epilepsy. *Neurology* 2004; 63: 73-77.

Bouilleret V, Semah F, Chassoux F, Mantzaridez M, Biraben A, Trebossen R, Ribeiro MJ. Basal ganglia involvement in temporal lobe epilepsy: a functional and morphologic study. *Neurology* 2008; 70:177-184.

Brasil-Neto JP, de Arauja DP, Teixeira WA, Araujo VP, Boechat-Barros R. *Arq Neuropsiquiatr* 2004 ; 62 : 21-25.

Bressand K, Dematteis M, Kahane P, Benazzouz A, Benabid AL. Involvement of the subthalamic nucleus in the control of temporal lobe epilepsy: study by hight frequency stimulation in rats. *Soc Neurosci* 1999 25:1656.

Cantello R, Rossi S, Varrasi C, Ulivelli M, Civardi C, Bartalini S, Vatti G, Cincotta M, Borghoresi A, Zaccara G, Quartarone A, Crupi D, Langana A, Inghilleri M, Giallonardo AT, Berardelli A, Pacifici L, Ferreri F, Tombini M, Gilio F, Quarato P, Conte A, Manganotti P, Bongiovanni LG, Monaco F, Ferrante D, Rossini PM. Slow repetitive TMS for drug-reistant epilepsy : clinical and EEG findings of a placebo-controlled trial. *Epilepsia* 2007; 48: 366-374.

Chabardès S, Kahane P, Minotti L, Koudsie A, Hirsch E, Benabid A-L. Deep brain stimulation in epilepsy with particular reference to the subthalamic nucleus. *Epileptic disord* 2002; 4 (suppl 3): 83–93.

Chen R, Classen J, Gerloff C, Celnik P, Wassermann EM, Hallett M, Cohen LG. Depression of motor cortex excitability by low-frequency transcranial magnetic stimulation. *Neurology* 1997; 48: 1398-1403.

Chkhenkeli SA. The inhibitory influence of the nucleus caudatus electrostimulation on the human's amygdalar and hippocampal activity at temporal lobe epilepsy. *Bull Georgian Acad Sci* 1978; 4/6: 406-11.

Chkhenkeli SA, Chkhenkeli IS. Effects of therapeutic stimulation of nucleus caudatus on epileptic electrical activity of brain in patients with intractable epilepsy. *Stereotact Funct Neurosurg* 1997; 69: 221-224.

Cooke PM, Snider RS. Some cerebellar influences on electrically-induced cerebral seizures. *Epilepsia* 1955; 4: 19-28.

Cooper I. *Cerebellar stimulation in man*. New York: Raven Press, 1978: 1–212.

Cooper IS, Upton ARM. The effect of chronic stimulation of cerebellum and thalamus upon neurophysiology and neurochemistry of cerebral cortex. In: Lazorthes Y, Upton ARM, eds. *Neurostimulation: an overview*. New York: Futura, 1985: 207-211.

Cooper IS, Amin I, Riklan M, Waltz JM, Poon TP. Chronic cerebellar stimulation in epilepsy. Clinical and anatomical studies. *Arch Neurol* 1976; 33: 559-70.

Cooper IS, Amin I, Gilman S. The effect of chronic cerebellar stimulation upon epilepsy in man. *Trans Am Neurol Assoc* 1973; 98: 192-6.

Cuellar-Herrera M, Velasco M, Velasco F, Velasco AL, Jimenez F, Orozco S, Briones M, Rocha L. Evaluation of GABA system and cell damage in parahippocampus of patients with temporal lobe epilepsy showing antiepileptic effects after subacute electrical stimulation. *Epilepsia* 2004; 45: 459-466.

Danober L, Deransart C, Depaulis A, Vergnes M, Marescaux C. Pathophysiological mechanisms of genetic absence epilepsy in rat. *Prog Neurobiol* 1998; 55: 27-57.

Davis R, Emmonds SE. Cerebellar stimulation for seizure control: 17-year study. *Stereotact Funct Neurosurg* 1992; 58: 200–208.

Depaulis A, Vergnes M, Depaulis A. Endogenous control of epilepsy: the nigral inhibitory system. *Prog Neurobiol* 1994; 42: 33-52.

Deransart C, Depaulis A. The control of seizures by the basal ganglia? A review of experimental data. *Epileptic Disord* 2002; 4 (suppl 3): S61-S72.

Deransart C, Marescaux C, Depaulis A. Involvement of nigral glutamatergic inputs in the control of seizures in a genetic model of absence epilepsy in the rat. *Neuroscience* 1996; 71: 721-728.

Deransart C, Lê BT, Marescaux C, Depaulis A. Role of the subthalamo-nigral input in the control of amygdale-kindled seizures in rat. *Brain Res* 1998a; 807: 78-83.

Deransart C, Riban V, Lê BT, Marescaux C, Depaulis A. Dopamine in the nucleus accumbens modulates seizures in a genetic model of absence epilepsy in the rat. *Neuroscience* 2000; 100: 335-344.

Deransart C, Depaulis A (2004) Le concept de contrôle nigral des épilepsies s'applique-t-il aux épilepsies partielles pharmacorésistantes? *Epilepsies* 16(2) 75-82.

Dow RS, Ferandez-Guardiola A, Manni E. The influence of the cerebellum on experimental epilepsy. *Electroencephalogr Clin Neurophysiol* 1962; 14: 383-398.

Duprez TP, Serieh BA, Raftopoulos C. Absence of memory dysfunction after bilateral mammillary body and mammillothalamic tract electrode implantation: preliminary experience in three patients. *AJNR Am J Neuroradiol* 2005; 26: 195-197.

Ebner TJ, Bantli H, Bloedel JR. Effects of cerebellar stimulation on unitary activity within a chronic epileptic focus in a primate. *Electroencephalogr Clin Neurophysiol* 1980; 49: 585-599.

Elisevich K, Jenrow K, Schuh L, Smith B. Long-term electrical stimulation-induced inhibition of partial epilepsy. Case report. *J Neurosurg* 2006; 105: 894-897.

Fanselow EE, Reid AP, Nicolelis MA. Reduction of pentylenetetrazole-induced seizure activity in awake rats by seizure-triggered trigeminal nerve stimulation. *J Neurosci.* 2000; 20(21):8160-8.

Feddersen B, Vercueil L, Noachtar S, David O, Depaulis A, Deransart C. Controlling seizures is not controlling epilepsy: a parametric study of deep brain stimulation for epilepsy. *Neurobiol Dis* 2007; 27: 292-300.

Feger J, Robledo P. The effects of activation or inhibition of the subthalamic nucleus on the metabolic and electrophysiological activities within the pallidal complex and the substantia nigra in the rat. *Eur J Neurosci* 1991; 3: 947-952.

Fisher RS, Uematsu S, Krauss GL, Cysyk B, McPherson R, Lesser RP, Gordon B, Schwerdt P, Rise M. Placebo-controlled pilot study of centromedian thalamic stimulation in treatment of intractable seizures. *Epilepsia* 1992; 33: 841-851.

Fountas KN, Smith JR, Murro AM, Politsky J, Park YD, Jenkins PD. Implantation of a closed-loop stimulation in the management of medically refractory focal epilepsy: a technical note. *Stereotact Funct Neurosurg* 2005; 83: 153-158.

Fregni F, Otachi PT, Do Valle A, Boggio PS, Thut G, Rigonatti SP, Pascual-Leone A, Valente KD. A randomized clinical trial of repetitive transcranial magnetic stimulation in patients with refractory epilepsy. *Ann Neurol* 2006; 60: 447-455.

Gale K, Iadarola MJ. Seizure protection and increased nerve-terminal GABA: delayed effects of GABA transaminase inhibition. *Science* 1980; 208: 288-291.

Goodman JH, Berger RE, Tcheng TK. Preemptive low-frequency stimulation decreases the incidence of amygdala-kindled seizures. *Epilepsia* 2005; 46: 1-7.

Grewal S, Gotman J. An automatic warning system for epileptic seizures recorded on intracerebral EEGs. *Clin Neurophysiol.* 2005;116(10):2460-72.

Haberler C, Alesch F, Mazal PR, Pilz P, Jellinger K, Pinter MM, Hainfellner JA, Budka H. No tissue damage by chronic deep brain stimulation in Parkinson's disease. *Ann Neurol* 2000; 48: 372-376.

Hamani C, Ewerton FI, Bonilha SM, Ballester G, Mello LE, Lozano AM. Bilateral anterior thalamic nucleus lesions and high-frequency stimulation are protective against pilocarpine-induced seizures and status epilepticus. *Neurosurgery* 2004; 54:191-195.

Hamani C, Hodaie M, Chiang J, del Campo M, Andrade DM, Sherman D, Mirski M, Mello LE, Lozano AM. Deep brain stimulation of the anterior nucleus of the thalamus: effects of electrical stimulation on pilocarpine-induced seizures and status epilepticus. *Epilepsy Res.* 2008 Feb;78(2-3):117-23

Hodaie M, Wennberg RA, Dostrovsky JO, Lozano AM. Chronic anterior thalamus stimulation for intractable epilepsy. *Epilepsia* 2002; 43: 603-608.

Iadarola MJ, Gale K. Substantia nigra: site of anticonvulsant activity mediated by gammaaminobutyric acid. *Science* 1982; 218 : 1237-1240.

Jasper H. Current evaluation of the concepts of centrencephalic and cortico-reticular seizures. *Electroencephalogr Clin Neurophysiol* 1991; 78: 2-11.

Jennum P, Klitgaard H. Repetitive transcranial magnetic stimulations of the rat. Effect of acute and chronic stimulations on pentylenetetrazole-induced clonic seizures. *Epilepsy Res* 1996; 23: 115-122.

Joo EY, Han SJ, Chung S-H, Cho J-W, Seo DW, Hong SB. Antiepileptic effects of low frequency repetitive transcranial magnetic stimulation by different stimulation durations and locations. *Clinical Neurophysiology* 2007; 118 : 702-708.

Kahane P, Ryvlin P, Hoffmann D, Minotti L, Benabid AL. From hypothalamic hamartoma to cortex: what can be learnt from depth recordings and stimulation? *Epileptic Disord* 2003; 5: 205-217.

Kerrigan JF, Litt B, Fisher RS, Cranstoun S, French JA, Blum DE, Dichter M, Shetter A, Baltuch G, Jaggi J, Krone S, Brodie MA, Rise M, Graves N. Electrical stimulation of the anterior nucleus of the thalamus for the treatment of intractable epilepsy. *Epilepsia* 2004; 45: 346-354.

Kinoshita M, Ikeda A, Begum T, Yamamoto J, Hitomi T, Shibasaki H. Low-frequency repetitive transcranial magnetic stimulation for seizure suppression in patients with extratemporal lobe epilepsy – a pilot study. *Seizure* 2005a; 14: 387-392.

Kinoshita M, Ikeda A, Matsuhashi M, Matsumoto R, Hitomi T, Begum T, Usui K, Takayama M, Mikuni N, Miyamoto S, Hashimoto N, Shibasaki H. Electric cortical stimulation suppresses epileptic and background activities in neocortical epilepsy and mesial temporal lobe epilepsy. *Clin Neurophysiol* 2005b; 116: 1291-1299.

Kobayashi M, Pascual-Leone A. Transcranial magnetic stimulation in neurology. *Lancet Neurol* 2003; 2: 145-156.

Kossoff EH, Ritzl EK, Politsky JM, Murro AM, Smith JR, Duckrow RB, Spencer DD, Bergey GK. Effect of an external responsive neurostimulator on seizures and electrographic discharges during subdural electrode monitoring. *Epilepsia* 2004; 45: 1560-1567.

Krauss GL, Fisher RS. Cerebellar and thalamic stimulation for epilepsy. *Adv Neurol* 1993; 63: 231–245.

Kwan P, Brodie M. Early identification of refractory epilepsy. *N Engl J Med* 2000; 342: 314-319.

Lado FA. Chronic bilateral stimulation of the anterior thalamus of kainate-treated rats increases seizure frequency. *Epilepsia* 2006; 47: 27-32.

La Grutta V, Amato G, Zagami MT. The importance of the caudate nucleus in the control of convulsive activity in the amygdaloid complex and the temporal cortex of the cat. *Electroencephalogr Clin Neurophysiol* 1971; 31: 57-69.

La Grutta V, Sabatino M, Gravante G, Morici G, Ferraro G, La Grutta G. A study of caudate inhibition on an epileptic focus in the cat hippocampus. *Arch Int Physiol Biochim* 1988; 96: 113–120.

Lesser RP, Kim SH, Beyderman L, Miglioretti DL, Webber WR, Bare M, Cysyk B, Krauss G, Gordon B. Brief bursts of pulse stimulation terminate afterdischarges caused by cortical stimulation. *Neurology* 1999; 53: 2073-2081.

Lim SN, Lee ST, Tsai YT, Chen IA, Tu PH, Chen JL, Chang HW, Su YC, Wu T. Electrical stimulation of the anterior nucleus of the thalamus for intractable epilepsy : a long term follow-up study. *Epilepsia* 2007; 48: 342-347.

Lockard JS, Ojemann GA, Congdon WC, DuCharme LL. Cerebellar stimulation in alumina-gel monkey model: inverse relationship between clinical seizures and EEG interictal bursts. *Epilepsia* 1979; 20: 223-234.

Loddenkemper T, Pan A, Neme S, Baker KB, Rezai AR, Dinner DS, Montgomery EB Jr, Lüders HO. Deep brain stimulation in epilepsy. *J Clin Neurophysiol* 2001; 18: 514-532.

Majkowski J, Karliński A, Klimowicz-Młodzik I. Effect of cerebellar stimulation of hippocampal epileptic discharges in kindling preparation. *Monogr Neural Sci* 1980; 5: 40-45.

Marescaux C, Vergnes M, Depaulis A. Genetic absence epilepsy in rats from strasbourg. *J Neural Transm Suppl* 1992; 35: 37-69.

Menkes DL, Gruenthal M. Slow-frequency repetitive transcranial magnetic stimulation in a patient withfocal cortical dysplasia. *Epilepsia* 2000; 4:240–42

Mirski MA, Ferrendelli JA. Interruption of the mammillothalamic tract prevents seizures in guinea pigs. *Science* 1984; 226: 72-74.

Mirski MA, Ferrendelli JA. Anterior thalamic mediation of generalized pentylenetetrazol seizures. *Brain Res* 1986a; 399: 212-223.

Mirski MA, Ferrendelli JA. Selective metabolic activation of the mammillary bodies and their connections during ethosuximide-induced suppression of pentylenetetrazol seizures. *Epilepsia* 1986b; 27: 194-203.

Mirski MA, Fisher RS. Electrical stimulation of the mammillary nuclei increases seizure threshold to pentylenetetrazol in rats. *Epilepsia* 1994; 35: 1309-1316.

Mirski MA, Rossell LA, Terry JB, Fisher RS. Anticonvulsant effect of anterior thalamic high frequency electrical stimulation in the rat. *Epilepsy Res* 1997; 28: 89-100.

Misawa S, Kuwabara S, Shibuya K, Mamada K, Hattori T. Low-frequency transcranial magnetic stimulation for epilepsia partialis continua due to cortical dysplasia. *J Neurol Sci* 2005; 234: 37-39.

Morrell M. Brain stimulation for epilepsy : can scheduled or responsive neurostimulation stop seizures? *Curr Opin Neurol* 2006; 19: 164-168.

Motamedi GK, Lesser RP, Miglioretti DL, Mizuno-Matsumoto Y, Gordon B, Webber WR, Jackson DC, Sepkuty JP, Crone NE. Optimizing parameters for terminating cortical afterdischarges with pulse stimulation. *Epilepsia* 2002; 43: 836-846.

Mutani R. Experimental evidence for the existence of an extrarhinencephalic control of the activity of the cobalt rhinencephalic epileptogenic focus, part 1: the role played by the caudate nucleus. *Epilepsia* 1969; 10: 337–350.

Mutani R, Bergamini L, Doriguzzi T. Experimental evidence for the existence of an extrarhinencephalic control of the activity of the cobalt rhinencephalic epileptogenic focus. Part 2. Effects of the paleocerebellar stimulation. *Epilepsia* 1969; 10: 351-362.

Nail-Boucherie K, Lê-Pham BT, Gobaille S, Maitre M, Aunis D, Depaulis A. Evidence for a role of the parafascicular nucleus of the thalamus in the control of epileptic seizures by the superior colliculus. *Epilepsia* 2005; 46: 141-145.

Nanobashvili Z, Chachua T, Nanobashvili A, Bilanishvili I, Lindvall O, Kokaia Z. Suppression of limbic motor seizures by electrical stimulation in thalamic reticular nucleus. *Exp Neurol* 2003; 181: 224-230.

Oakley JC, Ojemann GA. Effects of chronic stimulation of the caudate nucleus on a preexisting alumina seizure focus. Exp Neurol 1982; 75:360–67.

Oommen J, Morrell M, Fisher RS. Experimental electrical stimulation for epilepsy. *Curr Treat Options Neurol* 2005 ; 7 : 261-271.

Osorio I, Frei MG, Giftakis J, Peters T, Ingram J, Turnbull M, Herzog M, Rise MT, Schaffner S, Wennberg RA, Walczak TS, Risinger MW, Ajmone-Marsan C. Performance reassessment of a real-time seizure-detection algorithm on long ECoG series. *Epilepsia.* 2002;43(12):1522-35.

Osorio I, Frein MG, Sunderam S, Giftakis J, Bhavaraju NC, Schaffner SF, Wilkinson SB. Automated seizure abatement in humans using electrical stimulation. *Ann Neurol* 2005; 57: 258-268

Osorio I, Overman J, Giftakis J, Wilkinson SB. High frequency thalamic stimulation for inoperable mesial temporal epilepsy. *Epilepsia* 2007; 48: 1561-1571.

Paz JT, Deniau JM, Charpier S. Rhythmic Bursting in the Cortico-Subthalamo-Pallidal Network during Spontaneous Genetically Determined Spike and Wave Discharges. *J Neurosci* 2005; 25: 2092-2101.

Paz JT, Chavez M, Saillet S, Deniau JM, Charpier S. Activity of ventral medial thalamic neurons during absence seizures and modulation of cortical paroxysms by the nigrothalamic pathway. *J Neurosci* 2007; 27: 929-941.

Pilitsis JG, Chu Y, Kordower J, Bergen CD, Cochran EJ, Bakay RA. Postmortem study of deep brain stimulation of the anterior thalamus: case report. *Neurosurgery* 2008; 62: 530-532.

Polkey CE. Alternative surgical procedures ti help drug-resistant epilepsy – a review. *Epileptic Disord* 2003; 5 : 63-75.

Psatta DM. Control of chronic experimental focal epilepsy by feedback caudatum stimulations. *Epilepsia* 1983; 24: 444-454.

Rotenberg A, Muller P, Birnbaum D, Harrington M, Riviello JJ, Pascual-Leone A, Jensen FE. Seizure suppression by EEG-guided repetitive transcranial magnetic stimulation in the rat. *Clin Neurophysiol* 2008; 119:2697-2702.

Saillet S, Charvet G, Gharbi S, Depaulis A, Guillemaud R, David O. Closed loop control of seizures in a rat model of absence epilepsy using the BioMEATM system. *IEEE Neural Eng*. Submitted

Santiago-Rodriguez E, Cardenas-Morales L, Harmony T, Fernandez-Bouzas A, Porras-Kattz E, Hernandez A. Repetitive transcranial magnetic stimulation decreases the number of seizures in patients with focal neocortical epilepsy. *Seizure* 2008 ; may 19 [Epub ahead of print].

Semah F. PET imaging in epilepsy: basal ganglia and thalamic involvement. *Epileptic Disord* 2002; 4 (suppl 3): S55-S60.

Shi LH, Luo F, Woodward D, Chang JY. Deepbrain stimulation of the substantia nigra pars reticulata exerts long lasting suppression of amygdale-kindled seizures. *Brain Res* 2006; 1090 (1): 202-207.

Sramka M, Fritz G, Gajdosova D, Nadvornik P. Central stimulation treatment of epilepsy. *Acta Neurochir Suppl* 1980; 30: 183-187.

Takebayashi S, Hashizume K, Tanaka T, Hodozuka A. The effect of electrical stimulation and lesioning of the anterior thalamic nucleus on kainic acid-induced focal cortical seizure status in rats. *Epilepsia*. 2007**a** Feb;48(2):348-58.

Takebayashi S, Hashizume K, Tanaka T, Hodozuka A. Anti-convulsant effect of electrical stimulation and lesioning of the anterior thalamic nucleus on kainic acid-induced focal limbic seizure in rats. *Epilepsy Res*. 2007**b** May;74(2-3):163-70.

Tassinari CA, Cincotta M, Zaccara G, Michelucci R.Transcranial magnetic stimulation and epilepsy. *Clin Neurophysiol* 2003; 114: 777-798.

Taylor RB, Wennberg RA, Lozano AM, Sharpe JA. Central nystagmus induced by deep-brain stimulation for epilepsy. *Epilepsia* 2000; 41: 1637-1641.

Tellez-Zenteno JF, McLachlan RS, Parrent A, Kubu CS, Wiebe S. Hippocampal electrical stimulation in mesial temporal lobe epilepsy. *Neurology* 2006; 66: 1490-1494.

Tergau F, Naumann U, Paulus W, Steinhoff BJ. Low-frequency repetitive transcranial magnetic stimulation improves intractable epilepsy. *Lancet* 1999; 353: 2209.

Theodore WH, Fisher RS. Brain stimulation for epilepsy. *Lancet Neurol* 2004; 3: 111-118.

Theodore WH, Hunter K, Chen R, et al. Transcranial magnetic stimulation for the treatment of seizures: a controlled study. *Neurology* 2002; 59: 560-562.

Usui N, Maesawa S, Kajita Y, Endo O, Takebayashi S, Yoshida J. Suppression of secondary generalization of limbic seizures by stimulation of subthalamic nucleus in rat. *J Neurosurg* 2005; 102 (6): 1122-1129.

Van Buren JM, Wood JH, Oakley J, Hambrecht F. Preliminary evaluation of cerebellar stimulation by double blind stimulation and biological criteria in the treatment of epilepsy. *J Neurosurg* 1978; 48: 407-16.

van Rijckevorsel K, Abu Serieh B, de Tourtchaninoff M, Raftopoulos C. Deep EEG recordings of the mammillary body in epilepsy patients. *Epilepsia* 2005; 46: 781-785.

Velasco F, Velasco M, Cepeda C, Munoz H. Wakefulness-sleep modulation of thalamic multiple unit activity and EEG in man. *Electroencephalogr Clin Neurophysiol* 1979; 47: 597-606.

Velasco F, Velasco M, Ogarrio C, Fanghanel G. Electrical stimulation of the centromedian thalamic nucleus in the treatment of convulsives seizures: a preliminary report. *Epilepsia* 1987; 28: 421-430.

Velasco F, Velasco M, Velasco AL, Jimenez F. Effect of chronic electrical stimulation of the centromedian thalamic nuclei on various intractable seizure patterns, I: clinical seizures and paroxysmal EEG activity. *Epilepsia* 1993; 34: 1052-1064.

Velasco F, Velasco M, Velasco AL, Jimenez F, Marquez I, Rise M. Electrical stimulation of the centromedian thalamic nucleus in control of seizures : long term studies. *Epilepsia* 1995; 36: 63-71.

Velasco F, Velasco M, Jimenez F, Velasco AL, Brto F, Rise M, Carrillo-Ruiz JD. Predictors in the treatment of difficult to control seizures by electrical stimulation of the centromedian thalamic nucleus. *Neurosurgery* 2000a ; 47: 295-305.

Velasco M, Velasco F, Velasco AL, Jimenez F, Brito F, Marquez I. Acute and chronic electrical stimulation of the centromedian thalamic nucleus: modulation of reticulo-cortical systems and predictor factors for generalized seizure control. *Arch Med Res* 2000b; 31:304-315.

Velasco M, Velasco F, Velasco AL, Boleaga B, Jimenez F, Brito F, Marquez I. Subacute electrical stimulation of the hippocampus blocks intractable temporal lobe seizures and paroxysmal EEG activities. *Epilepsia* 2000c; 41: 158-169.

Velasco F, Velasco M, Jimenez F, Velasco AL, Marquez I. Stimulation of the central median thalamic nucleus for epilepsy. *Stereotact Funct Neurosurg* 2001a; 77: 228–232.

Velasco M, Velasco F, Velasco AL. Centromedian-thalamic and hippocampal electrical stimulation for the control of intractable epileptic seizures. *J Clin Neurophysiol* 2001b; 18: 495-513.

Velasco F, Velasco M, Jimenez F, Velasco AL, Rojas B. Centromedian nucleus stimulation for epilepsy. Clinical, electroencephalographic, and behavioral observations. *Thalamus & Related systems* 2002 ; 1: 387-398.

Velasco F, Carrillo-Ruiz JD, Brto F, Velasco M, Velasco A.L., Marquez I, Davis R. double-blind, randomized controlled pilot study of bilateral cerebellar stimulation for treatment of intractable motor seizures. *Epilepsia* 2005; 46: 1071-1081.

Velasco AL, Velasco F, Jimenez F, Velasco M, Casro G, Carrillo-Ruiz JD, Fanghanel G, Boleaga B. Neuromodulation of the centromedian thalamic nuclei in the treatment of generalized seizures and the improvement of the quality of life in patients with Lennox-Gastaut syndrome. *Epilepsia* 2006; 47: 1203-1212.

Velasco AL, Velasco F, Velasco M, Trejo D, Castro G, Carrillo-Ruiz JD. Electrical stimulation of the hippocampal epileptic foci for seizure control : a double-blind, long-term follow-up study. *Epilepsia* 2007; 48: 1895-1903.

Velisek L, Veliskova J, Moshe S L. Electrical stimulation of substantia nigra pars reticulate is anticonvulsant in adult and young male rats. *Exp Neurol* 2002; 173 (1): 145-152.

Velisek L, Velsikova J, Stanton PK. Low-frequency stimulation of the kindling focus delays basolateral amygdala kindling in immature rats. *Neurosci Lett* 2002 ; 326 ; 61-63.

Veliskova J, Claudio OI, Galanopoulou AS, Kyrozis A, Lado FA, Ravizza T, Velisek L, Moshe SL. Developmental espects of the basal ganglia and therapeutic perspectives. *Epileptic Disord* 4 2002; Suppl 3: S73-82.

Vercueil L, Hirsch E. Seizures and the basal ganglia: a review of thc clinical data. *Epileptic Disord* 2002; 4 (suppl 3): S47-S54.

Vercueil L., Benazzouz A., Deransart C., Bressand K., Marescaux C., Depaulis A., Benabid A.L. High-frequency stimulation of the sub-thalamic nucleus suppresses absence seizures in the rat: comparison with neurotoxic lesions. *Epilepsy Res* 1998; 31: 39-46.

Vesper J, Steinhoff B, Rona S, Wille C, Bilic S, Nikkhah G, Ostertag C. Chronic high-frequency deep brain stimulation of the STN/SNr for progressive myoclonic epilepsy. *Epilepsia* 2007; 48: 1984-1989.

Vonck K, Boon P, Achten E, De Reuck J, Caemaert J. Long-term amygdalohippocampal stimulation for refractory temporal lobe epilepsy. *Ann Neurol* 2002; 52: 556-565.

Vonck K, Boon P, Van Roost D. Anatomical and physiological basis and mechanism of action of neurostimulation for epilepsy. *Acta Neurochir* 2007; 97 (suppl): 321-328.

Wang S, Wu DC, Ding MP, Li Q, Zhuge ZB, Zhang SH, Chen Z. Low-frequency stimulation of cerebellar fastigial nucleus inhibits amygdaloid kindling acquisition in sprague-dawley rats. *Neurobiol Dis* 2008; 29 (1): 52_58.

Wassermann EM, Lisanby SH. Therapeutic application of repetitive transcranial magnetic stimulation: a review. *Clin Neurophysiol* 2001; 112: 1367-1377.

Weiss SRB, Li XL, Rosen JB, Li H, Heynen T, Post RM. Quenching : inhibition of the development and expression of amygdala kindled seizures with low frequency stimulation. *Neuroreport* 1995; 4: 2171-2176.

Wright GD, Mc Lellan DL, Brice JG. A double-blind trial of chronic cerebellar stimulation in twelve patients with severe epilepsy. *J Neurol Neurosurg Psychiatry* 1984; 47: 769-74.

Yamamoto J, Ikeda A, Satow T, Takeshita K, Takayama M, Matsuhashi M, Matsumoto R, Ohara S, Mikuni N, Takahashi J, Miyamoto S, Taki W, Hashimoto N, Rothwell JC, Shibasaki H. Low-frequency electric cortical stimulation has an inhibitory effect on cortical focus in mesial temporal lobe epilepsy. *Epilepsia* 2002 ; 43 : 491-495.

Yang LX, Jin CL, Zhu-Ge ZB, Wang S, Wei EQ, Bruce IC, Chen Z. Unilateral low-frequency stimulation of central piriform delays seizure development induced by amygdaloid-kindled in rats. *Neuroscience* 2006; 138: 1089-1096.

Zhu-Ge ZB, Zhu YY, Wu DC, Wang S, Liu LY, Hu WW, Chen Z. Unilateral low-frequency stimulation of central piriform cortex inhibits amygdaloid-kindled seizures in Sprague-Dawley rats. *Neuroscience* 2007; 146: 901-906.

Zumsteg D, Lozano AM, Wennberg RA. Mesial temporal inhibition in a patient with deep brain stimulation of the anterior thalamus for epilepsy. *Epilepsia* 2006; 47: 1958-1962.

In: Encyclopedia of Neuroscience Research
Editors: Eileen J. Sampson and Donald R. Glevins
ISBN 978-1-61324-861-4

Chapter XVI

Psychosurgery of Obsessive-Compulsive Disorder: A New Indication for Deep Brain Stimulation?*

Dominique Guehl, Abdelhamid Benazzouz, Bernard Bioulac and Pierre Burbaud
Service de Neurophysiologie Clinique, Université Victor Segalen Bordeaux 2, CNRS UMR 5543, Centre Hospitalier Pellegrin, Place Amélie-Raba Léon, 33076 Bordeaux, France
Emmanuel Cuny and Alain Rougier
Service de Neurochirurgie, Université Victor Segalen Bordeaux 2, Centre Hospitalier Pellegrin, Place Amélie-Raba Léon, 33076 Bordeaux, France
Jean Tignol
Service de Psychiatrie d'Adultes, Université Victor Segalen Bordeaux 2, Centre Hospitalier Charles Perrens, Centre Carreire, 121 rue de la Béchade, 33076 Bordeaux, France
Bruno Aouizerate
Service de Psychiatrie d'Adultes, Université Victor Segalen Bordeaux 2, Centre Hospitalier Charles Perrens, Centre Carreire, 121 rue de la Béchade, 33076 Bordeaux, France
Service de Neurophysiologie Clinique, Université Victor Segalen Bordeaux 2, CNRS UMR 5543, Centre Hospitalier Pellegrin, Place Amélie-Raba Léon, 33076 Bordeaux, France

* A version of this chapter was also published in *Recent Breakthroughs in Basal ganglia Research* , edited by Erwan Bezard published by Nova Science Publishers, Inc. It was submitted for appropriate modification in an effort to encourage wider dissemination of research

Abstract

Obsessive-compulsive disorder (OCD) is a relatively common psychiatric condition with an estimated lifetime prevalence of 2–3% of the general population. It is generally characterized by a chronic course leading to a profound impairment in psychosocial functioning and quality of life. Although its pathophysiology is still far from resolved, an ample body of literature suggests the role of an overactivity in the frontal-subcortical loops originating in the orbitofrontal cortex and the anterior cingulate cortex, respectively. Today, the well-established efficacy of antidepressants, acting preferentially by blocking serotonin reuptake, in addition to psychological treatments, have considerably changed the poor prognosis of the illness. However, both conventional therapeutic approaches failed to substantially alleviate obsessive-compulsive symptoms in 20–30% of cases. From these considerations, several surgical strategies have been successfully proposed for treating the resistant forms of OCD, including anterior capsulotomy, anterior cingulotomy, subcaudate tractotomy and limbic leucotomy. They have been performed in order to interrupt the orbitofrontal and anterior cingulate loops at either cortical or subcortical level, which have been found as disrupted in OCD. Despite these lesion-producing techniques have relatively low incidence of serious complications or undesirable effects, deep brain stimulation, as a reversible procedure, has recently been introduced in the field of OCD, primarily targeting the limb part of the internal capsule or the ventral caudate nucleus. It has been shown to be of promising benefit that should be considered with caution and confirmed by further research in this area.

Introduction

Obsessive-compulsive disorder (OCD) is a relatively common anxiety disorder that affects 2 to 3% of the general population (Robins et al., 1984; Karno et al., 1988). According to this estimated lifetime prevalence rate, OCD is the fourth most frequent psychiatric condition following the phobias, substance abuse and major depression. OCD, which is characterized by recurrent intrusive thoughts and repetitive, time-consuming behaviors, is a severely disabling mental illness owing to its intensity, the continuous and unchanging or deteriorative course of its symptoms and the disturbance in psychosocial functioning that they cause (Rasmussen and Tsuang, 1984; Koran et al., 1996; Antony et al., 1998; Skoog and Skoog, 1999).

Although the pathophysiology of OCD is still far from resolved, the existence of a biological basis for OCD is now clearly established and will be discussed from a phenomenological point of view, on the one hand, and in the light of our increasing knowledge of the physiology of prefrontal cortex-striatal-thalamic-cortical functional loops, on the other. The present chapter will then examine the therapeutic effects of various psychosurgical approaches used for the management of OCD forms, which were unresponsive to the prototypical treatments classically proposed in this psychiatric condition. All these procedures have consisted in performing lesions in order to produce an interruption of fibers connecting the thalamus to the prefrontal cortex through the anterior limb of the internal capsule. Although they have not been shown to generate serious complications and/or untoward effects in most cases, deep brain stimulation (DBS) has recently been considered as a potential alternative for treating resistant OCD.

1. Brief Overview of the Pathophysiology of OCD

1.1 Phenomenological Aspects

The heart of the obsessional process is the subject's underlying impression that "something is wrong" (Schwartz, 1998, 1999). In other words, obsessions may be thought of as the permanent perception of a mistake and/or error in certain behavioral situations. Compulsions occur as behavioral responses aiming to relieve the tensions or anxiety generated by the situation. If obtained, this relief may be felt to be a form of reward. Nevertheless, it is only transient, thereby creating a feeling of considerable anxiety. This leads to immediately reproduce the behavior in a cyclic manner on the basis of an internal motivational state through an expectation of the reward. Although it is undoubtedly simplistic, this phenomenological view is of interest because it suggests that several malfunctioning processes are altered within the OCD: 1) error recognition; and, 2) emotion and motivation and their activational aspects, e.g. activations for initiation and sustaining behavioral reactions and tendency to work for reward. Therefore, it can be expected in OCD that there is a dysfunction of the brain regions that mediate the processes of error detection and/or management of the reward and emotional systems, including orbitofrontal and anterior cingulate cortices and ventral striatum, respectively (Devinsky et al., 1995; Le Moal, 1995; Piazza and Le Moal, 1996; Koob and Le Moal, 1997; Piazza and Le Moal, 1997; Schwartz, 1998, 1999; Bush et al., 2000; Paus, 2001; Bush et al., 2002).

1.2 Anatomo-functional Data

Several studies have found a dysregulation in the frontal-subcortical circuits originating in the orbitofrontal and anterior cingulate cortices, respectively, in the pathogenesis of OCD. These loops concern discrete regions of the striatum and the pallidum and receive clearly delimited inputs from the cortex, which are subsequently integrated as far as the thalamus. Although they are part of a parallel model of organization, these circuits are permanently interrelated (Cummings, 1993).

The orbitofrontal cortex (OFC) and ventromedial areas are involved in appraisal in determining the emotional and motivational values of environmental information, and in integrating the subject's prior experience, which is crucial in decision-making (Charney and Bremner, 1999; Tremblay and Schultz, 1999, 2000a, 2000b; Krawczyk, 2002; Ramnani and Owen, 2004). The OFC also contributes to the selection, comparison and judgment of stimuli and error detection process (Rosenkilde et al., 1981; Thorpe et al., 1983; Ramnani and Owen, 2004).

The anterior cingulate cortex (ACC) is comprised of 1) a ventral or affective region which is intimately connected to the orbitofrontal cortex, ventral striatum, amygdala, hippocampus and hypothalamus. This subdivision could keep attention on the internal emotional and motivational status and participates to the regulation of autonomic responses; and, 2) a dorsal or cognitive region which is interconnected to the dorsolateral prefrontal cortex, parietal cortex, and premotor and supplementary motor areas. This cortical area serves a wide range of cognitive functions including attention, working memory, error detection, conflict monitoring and response selection with decisional and planning aspects in situations with high likelihood of making an error, and anticipation of incoming information (Niki and Watanabe, 1979; Devinsky et al., 1995; Shima and Tanji, 1998; Schwartz, 1999; Bush et al., 2000; Akkal et al., 2002; Bush et al., 2002; Shidara and Richmond, 2002).

Ventral striatum that is received important inputs from the OFC and ACC, is primarily implicated in the preparation, initiation and execution of behavioral responses oriented toward reward delivery after cognitive and emotional integration of behaviorally relevant information at the cortical level (Hollerman et al., 1998; Tremblay et al., 1998; Hassani et al., 2001).

The medial dorsal nucleus (MD) is the principal thalamic nucleus related to the prefrontal cortex. It is composed of several subdivisions, the main portion being centrally located and represented by the parvicellular part (MDpc), and its most medial portion corresponding to the magnocellular part (MDmc; Goldman-Rakic and Porrino, 1985). The MDpc receives inputs from the substantia nigra, which has a major role in cognition, and projects preferentially to the ACC (Barbas, 2000; McFarland and Haber, 2002). On the other hand, the MDmc, which receives inputs from the limbic portions of the basal ganglia (ventral striatum) and from the amygdala, projects mainly to the OFC (Barbas, 2000; McFarland and Haber, 2002).

1.3 Functional Neuroimagery

Functional imaging research has shown that several brain structures in OCD patients have abnormal functional activity. Most teams have reported an increased functional activity in the OFC, ACC, head of the caudate nucleus and thalamus (Saxena et al., 1998; Baxter, 1999; Aouizerate et al., 2004a). These functional abnormalities have been found in basal conditions and during provocation tests. Moreover, the therapeutic efficacy of antidepressants and cognitive-behavioral therapies seems to be associated with a progressive reduction in activity of the OFC, ACC and the caudate nucleus, thereby strengthening other data suggesting a functional deterioration in the orbitofrontal and anterior cingulate cortex system in the pathogenesis of OCD (Saxena et al., 1998; Baxter, 1999).

In conclusion, it seems in OCD that there is a dysfunction of the brain regions that belong to the orbitofrontal and anterior cingulate loops in view of evidences obtained from separate and complementary approaches

2. Psychosurgical Treatments

Over the last 30 years, distinct therapeutic approaches have been developed ranging from psychotropic drugs to psychotherapies and are now considered as effective for the management of OCD. In this respect, an ample body of literature has emerged and proven the efficacy of antidepressant agents for treating OCD. They preponderantly block the serotonin transporter located in the outer membrane of presynaptic axonal terminals of the brainstem raphe neurons, and thereby inhibit neurotransmitter reuptake (Flament and Bisserbe, 1997; Piggot and Seay, 1998; Goodman, 1999; McDougle, 1999; Pigott and Seay, 1999). Psychological treatments primarily based on cognitive-behavioral therapy with the current procedure termed “exposure with self-imposed response prevention” also appear to be effective in the treatment of OCD, alone or in combination with pharmacotherapy (Greist,

1994; Foa et al., 1998; 2002). However, at least 20-30 % of patients with OCD fail to respond adequately, with minimal to no change during these conventional treatments (Pigott and Seay, 1998; Pallanti et al., 2002). Therefore, psychosurgical approaches have been considered for the management of these chronic, severely distressing forms of OCD resistant to all currently available traditional treatments. The most frequently described procedures, which are anterior capsulotomy, anterior cingulotomy, subcaudate tractotomy and limbic leucotomy, have shown to be of significant benefit in the treatment of refractory OCD.

2.1 Anterior Capsulotomy

Developed in Sweden, anterior capsulotomy targets the fiber bundles in the anterior limb of the internal capsule connecting the medial dorsal nucleus of the thalamus and the frontal lobes. Two surgical techniques have been used, the radiofrequency thermolesion, or thermocapsulotomy and the newer radiosurgical, or gamma knife capsulotomy techniques (Mindus and Jenike, 1992; Jenike, 1998; Binder and Iskandar, 2000; Greenberg et al., 2000; Greenberg et al., 2003). The first report by Herner (1961) showed that at follow-up ranging from 24 to 80 months there were either good or fair results in 78 % of the 18 patients with OCD who underwent this surgical procedure. Bingley et al. (1977) found that all 35 patients with OCD responded favorably to this type of operation. Sixteen (46 %) were free of symptoms, 9 (26 %) were much improved, and 10 (28 %) were slightly improved at the mean 35-month follow-up. There was an improvement in the degree of social rehabilitation and working capacity. In the study by Kullberg (1977), general improvement, reflecting symptom reduction and social capacity, was considered as "excellent" in 1 (12.5 %), "good" in 2 (25 %), "moderate" in 2 (25 %), "slight" in 1 and "none" in 2 (25 %) of the 8 patients treated with capsulotomy at the end of the follow-up period (1 to 9 years). Fodstad et al. (1982) reported that two patients with OCD were much improved during the 24-month period following capsulotomy. In the review by Mindus et al. (1994), it has been mentioned that 5 (23 %) of the 22 patients with OCD exhibited a clinical aggravation at the 8-year follow-up after capsulotomy while 2 (9 %) were improved by 1 % to 25 %, 5 (23 %) improved by 26 % to 50 %, 3 (13 %) improved by 51 % to 75 %, and 7 (32 %) improved by 75 % to 100 %. Adverse effects of anterior capsulotomy in the initial series of 116 patients observed by Herner (1961) may be summarized as follows: three patients died 11, 28 and 33 months, respectively after the intervention; two from suicide and the third from a cardiac infarction. Operative complications were considerably few: stenosis of the aqueduct of the mid-brain in one case and epilepsy in four (3.4 %). The side effects on the emotional, the volitional and the intellectual status, as found in 55.5 % of the 18 patients with OCD, were mild and transient. Other untoward effects were urinary (and fecal) incontinence and weight gain but were not serious. In the study by Bingley et al. (1977), transient hemiparesis due to small subcortical hemorrhage was described in 2 (6 %) of the 35 OCD cases treated with capsulotomy. EEG showed abnormalities localized mainly to the frontal regions in 28 of these patients at 10-day follow-up before normalization after a more prolonged post-operative period. Post-operative mental confusion also occurred in some cases and lasted less than 1-3 weeks after the operation. Other mental changes consisted of diminution of inhibition, elevation of mood, and loss of initiative, but were not disturbing in most cases (Bingley et al., 1977; Kullberg, 1977; Chiocca and Martuza, 1990). At the 1-year follow-up, capsulotomy contributed to reverse personality changes observed pre-operatively (Mindus et al., 1988; Mindus and Nyman, 1991; Mindus et al., 1999).

2.2 Anterior Cingulotomy

Developed in the United States and Canada, anterior cingulotomy was the most currently neurosurgical procedure applied to the treatment of anxiety disorders including refractory OCD. Bilateral thermolesions are created by the radiofrequency in the anterior cingulate cortex (Broadman areas 24 and 32; Mindus and Jenike, 1992; Jenike, 1998; Binder and Iskandar, 2000; Greenberg et al., 2000; Greenberg et al., 2003). Whitty et al. (1952) first reported the satisfactory results of this surgical procedure in two patients with OCD. The therapeutic effects were much less favorable in the study by Kullberg (1977) showing that cingulotomy produced a slight reduction in OCD symptoms severity only in 1 (33 %) of the three OCD cases at the end of the follow-up period (1-8 years). In a larger sample, Ballantine et al. (1987) found that 8 (25 %) of the 32 patients with OCD were slightly improved, 10 (31.3 %) were markedly improved, 4 (12.5 %) were functionally normal on medication and/or psychotherapy maintenance, and 4 (12.5 %) were essentially well without any treatment. The remaining 6 patients (18.7 %) showed unchanged or worsened OCD symptoms at the mean follow-up of 8.6 years. Jenike et al. (1991), using Y-BOCS for assessment of OCD symptom severity, described that 6 (43 %) of the 14 patients with OCD treated with cingulotomy showed no improvement, while the remaining 8 (57 %) exhibited moderate to marked improvement at the mean 13.1-year follow-up, defined by a 50 % or greater reduction in Y-BOCS scores. However, 2 of these 8 patients considered their improvement related to the current treatment with SRI (fluoxetine) and CBT rather than cingulotomy. In the most recent prospective study, Dougherty et al. (2002) reported favorable effects of this surgical procedure in 44 patients with intractable OCD. At a mean follow-up of 32 months, 14 (32 %) were considered as responders according to the conservative criteria for treatment response and 6 others (14 %) were partial responders after one or more cingulotomies. Serious operative complications were relatively few including hemiplegias (0.03 %) and controllable seizure disorders (1 %) in the series of 696 bilateral cingulotomies (Ballantine et al., 1987). Death by suicide was observed in 18 patients over the mean 8.6-year follow-up, but all 18 of these patients had preoperative suicidal ideation and 13 had made suicide attempts before the operation. Transient mental confusion was reported in 2 (15.4 %) of the 13 patients treated with cingulotomy during the first post-operative days (Kullberg, 1977). Memory deficits were described in two (5 %) of the 44 patients who underwent this type of surgical intervention, but had resolved within 6 to 12 months. Apathy with loss of energy was noted in one patient (2 %) following cingulotomy. Urinary disturbances were reported in three (7 %) patients, which resolved within the first days postoperatively (Dougherty et al., 2002).

2.3 Subcaudate Tractotomy

Developed in the United Kingdom, subcaudate tractotomy is characterized by bilateral lesions made by radioactive yttrium-90 seeds placed in the substantia innominata, which is located close to the ventral part of the anterior capsule beneath the head of the caudate nucleus. The goal is to interrupt connections from orbitofrontal cortex to subcortical regions such as ventral caudate, nucleus accumbens, which then send projections to the ventromedial pallidum and thalamus (Mindus and Jenike, 1992; Jenike, 1998; Binder and Iskandar, 2000; Greenberg et al., 2000; 2003). There were relatively few reports of the efficacy of this surgical procedure in the treatment of refractory OCD. In the series of Strom-Olsen and Carlisle (1971), it has been found that at the follow-up ranging from 16 months to 8 years, 7 (35 %) of the 20 patients with OCD had no symptoms, 3 (15 %) were improved but with slight residual symptoms for which no treatment was needed, 3 (15 %) were improved but with persistent symptoms requiring treatment. The remaining 7 (35 %) patients exhibited unchanged or worsened symptoms. In most cases, the reducing effects on OCD symptom severity were delayed within 4-5 months after the operation. Bridges et al. (1973) also reported favorable effects of subcaudate tractotomy in 24 patients intractable OCD. Sixteen (67 %) of these patients showed a good outcome, as defined by no symptoms or mild residual symptoms after three years of follow-up. The remaining eight (33 %) patients had poor outcome with persistence of significant symptoms responsible for functional impairment in the patient's life. These results were confirmed by Goktepe et al. (1975) showing that nine (50 %) of the 18 patients with OCD had satisfactory response after a 2.5-year period of follow-up. In the most recent report by Hodgkiss et al. (1995), good outcome was only observed in 5 (33.3 %) of the 15 patients with OCD within one year after surgery. There was no post-operative mortality in the series of Strom-Olsen and Carlisle (1971), conducted on 150 patients with psychiatric conditions operated. However, there were three deaths by suicide in the series of 208 cases of Goktepe et al. (1975). The incidence of post-operative epilepsy in this series was 2.2 % (Goktepe et al., 1975). Weight gain was observed (Strom-Olsen and Carlisle, 1971). Psychological and behavioral changes occurred (Strom-Olsen and Carlisle, 1971; Goktepe et al., 1975). The commonly reported symptoms were tiredness and lethargy lasting for up to three months or irritability, outspokenness and volubility, which most often failed to have repercussions on family and social relationships. Neuropsychological testing revealed frontal lobe dysfunction 6 months after the operation (Kartsounis et al., 1991).

2.4 Limbic Leucotomy

Introduced in the United Kingdom, limbic leucotomy is based on bilateral lesions of cingulotomy, in addition to those of the original subcaudate tractotomy (Mindus and Jenike, 1992; Jenike, 1998; Binder and Iskandar, 2000; Greenberg et al., 2000; Greenberg et al., 2003). In the early report of Kelly et al. (Kelly and Mitchell-Heggs, 1973; Kelly et al., 1973), 13 (76 %) of the 17 patients with OCD were clinically improved within six weeks after the intervention. Of these 13 patients, 1 (8 %) was rated as "symptom free", 6 (46 %) as "much improved" and 6 (46 %) as "improved". The favorable effects even persisted after a mean 16- or 17-month period of follow-up (Kelly and Mitchell-Heggs, 1973; Kelly et al., 1973; Mitchell-Heggs et al., 1976). These data confirmed the efficacy of this surgical procedure in the treatment of OCD, as reported in the early controlled retrospective study by Tan et al. (1971) showing a more profound reduction in OCD symptom severity in 24 patients with OCD operated compared with their 13 controls at three months after the operation, and at the end of the five years of follow-up. Post-operative complications were as follows:

subarachnoid hemorrhage in one patient (4 %) and death in three patients (12.5 %; of whom one committed suicide; Tan et al., 1971). There also were transient confusion, headache, apathy, laziness, incontinence of urine, and rarely feces, which currently lasted from one to several days. Weight gain was observed. Poorer memory and concentration capacities, outspokenness or irritability were inconsistently reported (Tan et al., 1971; Kelly and Mitchell-Heggs, 1973; Kelly et al., 1973; Mitchell-Heggs et al., 1976).

Therefore, functional neurosurgical procedures, leading to an interruption of reciprocal connections between the frontal lobes and subcortical structures, have been demonstrated to be effective in the management of OCD unresponsive to both pharmacological and psychological strategies typically used in this condition.

2.5 Deep Brain Stimulation

Although the lesion techniques have been shown to cause relatively few serious undesirable effects, the reversible procedure deep brain stimulation (DBS) in the context of recent technological advances is becoming an important strategy in the treatment of severe forms of Parkinson's disease and other movement disorders (Malhi and Sachdev, 2002). This approach has recently been proposed as a potential therapeutic alternative in resistant OCD. When applied bilaterally in the anterior limb of internal capsule instead of bilateral anterior capsulotomy, DBS has been found to markedly improve OCD symptoms in three of the four OCD patients who underwent this procedure. The beneficial effects occurred almost immediately after the beginning of DBS (Nuttin et al., 1999). Headache was experienced during the first 10 seconds of stimulation and then disappeared. Long-term follow-up studies have confirmed the favorable effects of chronic anterior capsular stimulation. There was a sustained 35 % and greater reduction in Y-BOCS scores in two of the three cases of long-standing treatment-resistant OCD during the 33-month period of follow-up. At 1-year DBS, there was no deleterious effect or harmful consequence on neuropsychological functioning and personality traits (Gabriels et al., 2003). It has also been shown that a gradual decline in OCD symptom severity with recovery defined by scores in Y-BOCS below 8, which was obtained within the first 3 months of DBS application and remained stable over the seven following months. This was paralleled by a dramatic improvement in global psychosocial functioning (Anderson and Ahmed, 2003). Other brain regions have been proposed as target for DBS. In this context, a reduction in OCD manifestations by approximately 60 % was reported in two Parkinsonian patients with comorbid OCD after 6-month DBS of the subthalamic nucleus. However, symptomatic improvement was observed immediately after discharge from the neurosurgery unit (Mallet et al., 2002). The nucleus accumbens that represents the ventral and limbic part of the caudate nucleus has also been chosen as target for DBS. There was a clinical amelioration in three of the four patients with OCD and related anxiety disorders, which was achieved a few days to several weeks after beginning of DBS. No side effect was reported during the 24- to 30-month period of follow-up (Sturm et al., 2003). We also found favorable effects of DBS of the ventral caudate nucleus, including the nucleus accumbens and ventromedial head of the caudate nucleus in one patient suffering from resistant OCD with comorbid major depression (Aouizerate et al., 2004b). Reduction in depressive symptom severity occurred early during the first three months of DBS until remission obtained at 6 months while there was a delayed onset of the anti-obsessional effects, which was observed at 12-month DBS. This was parallel by a progressive improvement in global functioning. No adverse effect of any kind was noted. There was no deterioration of neuropsychological testing exploring specifically executive and mnemonic functions.

Conclusion

To summarize, a disturbance in frontal-subcortical loops originating in the OFC and ACC has extensively been documented to play a major role in the pathogenesis of OCD. Numerous neuroimaging studies have consistently shown abnormalities in functional activity in the OFC, ACC, ventral striatum and medial thalamus. These brain regions have been found implicated in cognitive and emotional processes that can be thought disrupted in OCD in the light of phenomenological considerations. Data collected from psychosurgery proposed in resistant OCD forms, which failed to successfully respond to both conventional pharmacological treatments and psychotherapies, confirm the involvement of such brain regions in the pathophysiology of OCD. The most widely used approaches (anterior capsulotomy, anterior cingulotomy, subcaudate tractotomy and limbic leucotomy) intended to perform lesions within the frontal-subcortical loops have been shown to produce an improvement in OCD symptoms in 40 to 60% of cases. They were relatively well-tolerated. However, chronic DBS has been proposed as reversible procedure for the management of refractory OCD. This surgical technique has benefited from methodological advances made in the field of Parkinson's disease and other movement disorders. Different target regions have been considered for DBS in OCD, such as the anterior limb of the internal capsule, the subthalamic nucleus or the nucleus accumbens. Although DBS seems to be the first-line choice since it is the least invasive, there are still many unknown factors. The most critical issue is to determine the optimal targets both in terms of their possible efficacy and their risks and benefits. In this perspective, further preclinical studies in laboratory animals would be helpful in order to provide more precise information about the anatomo-functional bases of OCD. The promising anti-obsessional effects and safety of DBS remain to be confirmed in larger populations of patients with intractable OCD.

References

Akkal D, Bioulac B, Audin J, Burbaud P (2002) Comparison of neuronal activity in the rostral supplementary and cingulate motor areas during a task with cognitive and motor demands. *Eur J Neurosci* **15**:887-904.

Anderson D, Ahmed A (2003) Treatment of patients with intractable obsessive-compulsive disorder with anterior capsular stimulation. Case report. *J Neurosurg* **98**:1104-1108.

Antony MM, Downie F, Swinson RP (1998) Diagnostic issues and epidemiology in obsessive-compulsive disorder. In: *Obsessive-compulsive disorder*. Theory, research, and treatment (Swinson RP, Antony MM, Rachman S, Richter MA, eds), pp 3-32. New-York: The Guilford Press.

Aouizerate B, Guehl D, Cuny E, Rougier A, Bioulac B, Tignol J, Burbaud P (2004a) Pathophysiology of obsessive-compulsive disorder: a necessary link between phenomenology, neuropsychology, imagery and physiology. *Prog Neurobiol* **72**:195-221.

Aouizerate B, Cuny E, Martin-Guehl C, Guehl D, Amieva H, Benazzouz A, Fabrigoule C, Allard M, Rougier A, Bioulac B, Tignol J, Burbaud P (2004b) Deep brain stimulation of the ventral caudate nucleus in the treatment of obsessive-compulsive disorder and major depression. Case report. *J Neurosurg* **101**:682-686.

Ballantine HT, Jr., Bouckoms AJ, Thomas EK, Giriunas IE (1987) Treatment of psychiatric illness by stereotactic cingulotomy. *Biol Psychiatry* **22**:807-819.

Barbas H (2000) Connections underlying the synthesis of cognition, memory, and emotion in primate prefrontal cortices. *Brain Res Bull* **52**:319-330.

Baxter LR (1999) Functional imaging of brain systems mediating obsessive-compulsive disorder: Clinical studies. In: *Neurobiology of mental illness* (Charney DS, Nestler EJ, Bunney BS, eds), pp 534-547. New York: Oxford University Press.

Binder DK, Iskandar BJ (2000) Modern neurosurgery for psychiatric disorders. *Neurosurgery* **47**:9-21; discussion 21-23.

Bingley T, Leksell L, Meyerson BA, Rylander G (1977) Long-term results of stereotactic anterior capsulotomy in chronic obsessive-compulsive neurosis. In: *Neurosurgical treatment in Psychiatry, Pain and Epilepsy* (Sweet WH, Obrador S, Martin-Rodriguez JG, eds), pp 287-299. Baltimore: University Park Press.

Bridges PK, Goktepe EO, Maratos J (1973) A comparative review of patients with obsessional neurosis and with depression treated by psychosurgery. *Br J Psychiatry* **123**:663-674.

Bush G, Luu P, Posner MI (2000) Cognitive and emotional influences in anterior cingulate cortex. *Trends Cogn Sci* **4**:215-222.

Bush G, Vogt BA, Holmes J, Dale AM, Greve D, Jenike MA, Rosen BR (2002) Dorsal anterior cingulate cortex: a role in reward-based decision making. *Proc Natl Acad Sci U S A* **99**:523-528.

Charney DS, Bremner JD (1999) The neurobiology of anxiety disorders: Clinical studies. In: *Neurobiology of mental illness* (Charney DS, Nestler EJ, Bunney BS, eds), pp 494-517. New York: Oxford University Press.

Chiocca EA, Martuza RL (1990) Neurosurgical therapy of obsessive-compulsive disorder. In: *Obsessive-compulsive disorders: Theory and management* (Jenike MA, Baer L, Minichiello WE, eds), pp 283-294. Chicago: Year Book Medical Publishers.

Cummings JL (1993) Frontal-subcortical circuits and human behavior. *Arch Neurol* **50**:873-880.

Devinsky O, Morrell MJ, Vogt BA (1995) Contributions of anterior cingulate cortex to behaviour. *Brain* **118** (Pt 1):279-306.

Dougherty DD, Baer L, Cosgrove GR, Cassem EH, Price BH, Nierenberg AA, Jenike MA, Rauch SL (2002) Prospective long-term follow-up of 44 patients who received cingulotomy for treatment-refractory obsessive-compulsive disorder. *Am J Psychiatry* **159**:269-275.

Flament MF, Bisserbe JC (1997) Pharmacologic treatment of obsessive-compulsive disorder: comparative studies. *J Clin Psychiatry* **58** Suppl 12:18-22.

Foa EB, Franklin ME, Kozac MJ (1998) Psychosocial treatments for obsessive-compulsive disorder. Literature review. In: *Obsessive-compulsive disorder. Theory, research, and treatment* (Swinson RP, Antony MM, Rachman S, Richter MA, eds), pp 258-276. New-York: The Guilford Press.

Foa EB, Franklin ME, Moser J (2002) Context in the clinic: how well do cognitive-behavioral therapies and medications work in combination? *Biol Psychiatry* **52**:987-997.

Fodstad H, Strandman E, Karlsson B, West KA (1982) Treatment of chronic obsessive compulsive states with stereotactic anterior capsulotomy or cingulotomy. *Acta Neurochir (Wien)* **62**:1-23.

Gabriels L, Cosyns P, Nuttin B, Demeulemeester H, Gybels J (2003) Deep brain stimulation for treatment-refractory obsessive-compulsive disorder: psychopathological and neuropsychological outcome in three cases. *Acta Psychiatr Scand* **107**:275-282.

Goktepe EO, Young LB, Bridges PK (1975) A further review of the results of sterotactic subcaudate tractotomy. *Br J Psychiatry* **126**:270-280.

Goldman-Rakic PS, Porrino LJ (1985) The primate mediodorsal (MD) nucleus and its projection to the frontal lobe. *J Comp Neurol* **242**:535-560.

Goodman WK (1999) Obsessive-compulsive disorder: diagnosis and treatment. *J Clin Psychiatry 60 Suppl* **18**:27-32.

Greenberg BD, Murphy DL, Rasmussen SA (2000) Neuroanatomically based approaches to obsessive-compulsive disorder. Neurosurgery and transcranial magnetic stimulation. *Psychiatr Clin North Am* **23**:671-686, xii.

Greenberg BD, Price LH, Rauch SL, Friehs G, Noren G, Malone D, Carpenter LL, Rezai AR, Rasmussen SA (2003) Neurosurgery for intractable obsessive-compulsive disorder and depression: critical issues. *Neurosurg Clin N Am* **14**:199-212.

Greist JH (1994) Behavior therapy for obsessive compulsive disorder. *J Clin Psychiatry* **55** Suppl:60-68.

Hassani OK, Cromwell HC, Schultz W (2001) Influence of expectation of different rewards on behavior-related neuronal activity in the striatum. *J Neurophysiol* **85**:2477-2489.

Herner T (1961) Treatment of mental disorders with frontal stereotactic thermo-lesions. A follow-up study of 116 cases. *Acta Psychiatr Scand Suppl.* **36**.

Hodgkiss AD, Malizia AL, Bartlett JR, Bridges PK (1995) Outcome after the psychosurgical operation of stereotactic subcaudate tractotomy, 1979-1991. *J Neuropsychiatry Clin Neurosci* **7**:230-234.

Hollerman JR, Tremblay L, Schultz W (1998) Influence of reward expectation on behavior-related neuronal activity in primate striatum. *J Neurophysiol* **80**:947-963.

Jenike MA (1998) Neurosurgical treatment of obsessive-compulsive disorder. Br J Psychiatry Suppl:79-90.

Jenike MA, Baer L, Ballantine T, Martuza RL, Tynes S, Giriunas I, Buttolph ML, Cassem NH (1991) Cingulotomy for refractory obsessive-compulsive disorder. A long-term follow-up of 33 patients. *Arch Gen Psychiatry* **48**:548-555.

Karno M, Golding JM, Sorenson SB, Burnam MA (1988) The epidemiology of obsessive-compulsive disorder in five US communities. *Arch Gen Psychiatry* **45**:1094-1099.

Kartsounis LD, Poynton A, Bridges PK, Bartlett JR (1991) Neuropsychological correlates of stereotactic subcaudate tractotomy. *A prospective study. Brain* **114** (Pt 6):2657-2673.

Kelly D, Mitchell-Heggs N (1973) Stereotactic limbic leucotomy--a follow-up study of thirty patients. *Postgrad Med J* **49**:865-882.

Kelly D, Richardson A, Mitchell-Heggs N, Greenup J, Chen C, Hafner RJ (1973) Stereotactic limbic leucotomy: a preliminary report on forty patients. *Br J Psychiatry* **123**:141-148.

Koob GF, Le Moal M (1997) Drug abuse: hedonic homeostatic dysregulation. *Science* **278**:52-58.

Koran LM, Thienemann ML, Davenport R (1996) Quality of life for patients with obsessive-compulsive disorder. *Am J Psychiatry* **153**:783-788.

Krawczyk DC (2002) Contributions of the prefrontal cortex to the neural basis of human decision making. *Neurosci Biobehav Rev* **26**:631-664.

Kullberg G (1977) Differences in effects of capsulotomy and cingulotomy. In: *Neurosurgical treatment in Psychiatry, Pain and Epilepsy* (Sweet WH, Obrador S, Martin-Rodriguez JG, eds), pp 301-308. Baltimore: University Park Press.

Le Moal M (1995) Mesocorticolimbic dopaminergic neurons. Functional and regulatory roles. In: *Psychopharmacology: The fourth generation of progress* (Bloom FE, Kupfer DJ, eds), pp 283-294. New York: Raven Press.

Malhi GS, Sachdev P (2002) Novel physical treatments for the management of neuropsychiatric disorders. *J Psychosom Res* **53**:709-719.

Mallet L, Mesnage V, Houeto JL, Pelissolo A, Yelnik J, Behar C, Gargiulo M, Welter ML, Bonnet AM, Pillon B, Cornu P, Dormont D, Pidoux B, Allilaire JF, Agid Y (2002) Compulsions, Parkinson's disease, and stimulation. *Lancet* **360**:1302-1304.

McDougle CJ (1999) The neurobiology and treatment of obsessive-compulsive disorder. In: *Neurobiology of mental illness* (Charney DS, Nestler EJ, Bunney BS, eds), pp 518-533. New York: Oxford University Press.

McFarland NR, Haber SN (2002) Thalamic relay nuclei of the basal ganglia form both reciprocal and nonreciprocal cortical connections, linking multiple frontal cortical areas. *J Neurosci* **22**:8117-8132.

Mindus P, Nyman H (1991) Normalization of personality characteristics in patients with incapacitating anxiety disorders after capsulotomy. Acta Psychiatr Scand 83:283-291.

Mindus P, Jenike MA (1992) Neurosurgical treatment of malignant obsessive compulsive disorder. *Psychiatr Clin North Am* **15**:921-938.

Mindus P, Rasmussen SA, Lindquist C (1994) Neurosurgical treatment for refractory obsessive-compulsive disorder: implications for understanding frontal lobe function. J Neuropsychiatry *Clin Neurosci* **6**:467-477.

Mindus P, Edman G, Andreewitch S (1999) A prospective, long-term study of personality traits in patients with intractable obsessional illness treated by capsulotomy. *Acta Psychiatr Scand* **99**:40-50.

Mindus P, Nyman H, Rosenquist A, Rydin E, Meyerson BA (1988) Aspects of personality in patients with anxiety disorders undergoing capsulotomy. *Acta Neurochir Suppl (Wien)* **44**:138-144.

Mitchell-Heggs N, Kelly D, Richardson A (1976) Stereotactic limbic leucotomy--a follow-up at 16 months. *Br J Psychiatry* **128**:226-240.

Niki H, Watanabe M (1979) Prefrontal and cingulate unit activity during timing behavior in the monkey. *Brain Res* **171**:213-224.

Nuttin B, Cosyns P, Demeulemeester H, Gybels J, Meyerson B (1999) Electrical stimulation in anterior limbs of internal capsules in patients with obsessive-compulsive disorder. *Lancet* **354**:1526.

Pallanti S, Quercioli L, Koran LM (2002) Citalopram intravenous infusion in resistant obsessive-compulsive disorder: an open trial. *J Clin Psychiatry* **63**:796-801.

Paus T (2001) Primate anterior cingulate cortex: where motor control, drive and cognition interface. *Nat Rev Neurosci* **2**:417-424.

Piazza PV, Le Moal ML (1996) Pathophysiological basis of vulnerability to drug abuse: role of an interaction between stress, glucocorticoids, and dopaminergic neurons. *Annu Rev Pharmacol Toxicol* **36**:359-378.

Piazza PV, Le Moal M (1997) Glucocorticoids as a biological substrate of reward: physiological and pathophysiological implications. *Brain Res Brain Res Rev* **25**:359-372.

Piggot TA, Seay SM (1998) Biological treatments for obsessive-compulsive disorder. Literature review. In: *Obsessive-compulsive disorder. Theory, research, and treatment* (Eds R. P. Swinson MMA, S. Rachman, M. A. Richter., ed), pp pp. 298-326.: The Guilford Press: New-York.

Pigott TA, Seay SM (1999) A review of the efficacy of selective serotonin reuptake inhibitors in obsessive-compulsive disorder. *J Clin Psychiatry* **60**:101-106.

Ramnani N, Owen AM (2004) Anterior prefrontal cortex: insights into function from anatomy and neuroimaging. *Nat Rev Neurosci* **5**:184-194.

Rasmussen SA, Tsuang MT (1984) The epidemiology of obsessive compulsive disorder. *J Clin Psychiatry* **45**:450-457.

Robins LN, Helzer JE, Weissman MM, Orvaschel H, Gruenberg E, Burke JD, Jr., Regier DA (1984) Lifetime prevalence of specific psychiatric disorders in three sites. *Arch Gen Psychiatry* **41**:949-958.

Rosenkilde CE, Bauer RH, Fuster JM (1981) Single cell activity in ventral prefrontal cortex of behaving monkeys. *Brain Res* **209**:375-394.

Saxena S, Brody AL, Schwartz JM, Baxter LR (1998) Neuroimaging and frontal-subcortical circuitry in obsessive-compulsive disorder. *Br J Psychiatry Suppl*:**26**-37.

Schwartz JM (1998) Neuroanatomical aspects of cognitive-behavioural therapy response in obsessive-compulsive disorder. An evolving perspective on brain and behaviour. *Br J Psychiatry Suppl*:**38**-44.

Schwartz JM (1999) A role of volition and attention in the generation of new brain circuitry. Toward a neurobiology of mental force. *J Consciousness studies* **6**:115-142.

Shidara M, Richmond BJ (2002) Anterior cingulate: single neuronal signals related to degree of reward expectancy. *Science* **296**:1709-1711.

Shima K, Tanji J (1998) Role for cingulate motor area cells in voluntary movement selection based on reward. *Science* **282**:1335-1338.

Skoog G, Skoog I (1999) A 40-year follow-up of patients with obsessive-compulsive disorder [see commetns]. *Arch Gen Psychiatry* **56**:121-127.

Strom-Olsen R, Carlisle S (1971) Bi-frontal stereotactic tractotomy. A follow-up study of its effects on 210 patients. *Br J Psychiatry* **118**:141-154.

Sturm V, Lenartz D, Koulousakis A, Treuer H, Herholz K, Klein JC, Klosterkotter J (2003) The nucleus accumbens: a target for deep brain stimulation in obsessive-compulsive- and anxiety-disorders. *J Chem Neuroanat* **26**:293-299.

Tan E, Marks IM, Marset P (1971) Bimedial leucotomy in obsessive-compulsive neurosis: a controlled serial enquiry. *Br J Psychiatry* **118**:155-164.

Thorpe SJ, Rolls ET, Maddison S (1983) The orbitofrontal cortex: neuronal activity in the behaving monkey. *Exp Brain Res* **49**:93-115.

Tremblay L, Schultz W (1999) Relative reward preference in primate orbitofrontal cortex. *Nature* **398**:704-708.

Tremblay L, Schultz W (2000a) Modifications of reward expectation-related neuronal activity during learning in primate orbitofrontal cortex. *J Neurophysiol* **83**:1877-1885.

Tremblay L, Schultz W (2000b) Reward-related neuronal activity during go-nogo task performance in primate orbitofrontal cortex. *J Neurophysiol* **83**:1864-1876.

Tremblay L, Hollerman JR, Schultz W (1998) Modifications of reward expectation-related neuronal activity during learning in primate striatum. *J Neurophysiol* **80**:964-977.

Whitty CW, Duffield JE, Tov PM, Cairns H (1952) Anterior cingulectomy in the treatment of mental disease. *Lancet* **1**:475-481.

In: Encyclopedia of Neuroscience Research
Editors: Eileen J. Sampson and Donald R. Glevins

ISBN 978-1-61324-861-4

Chapter XVII

Deep Brain Stimulation in Adult and Pediatric Dystonia*

Laura Cif, Simone Hemm, Nathalie Vayssiere and Philippe Coubes†
Research Group on Movement Disorders, Department of Neurosurgery
(Professor Philippe Coubes, Montpellier University Hospital,
34295 Montpellier, Cedex 5, France

Abstract

In numerous medical conditions, abnormal movements (dystonia, dyskinesia, myoclonus and tremor) are present, isolated or associated to other neurological deficits. Deep brain stimulation has been validated as a safe and efficient neurosurgical procedure in treating Parkinson's disease and essential tremor. Following the experience with ablative surgery for dystonia and high frequency stimulation for Parkinson's disease, chronic electrical stimulation of the internal globus pallidus (GPi) has been proposed to treat dystonic syndromes. Based on etiology and spread of symptoms, 82 patients were divided in subgroups separating, for each of them, children and adults: primary DYT1 dystonia, primary dystonia without DYT1 mutation, primary segmental dystonia, post anoxic cerebral palsy, pantothenate kinase associated neurodegeneration (PKAN) and mitochondrial diseases. Assessment was performed pre- and postoperatively by the Burke-Fahn's Marsden Dystonia Rating Scale, motor and disability sections. Bilateral electrode implantation was performed under general anaesthesia. High frequency (130 Hz), 450 microseconds pulse width and monopolar stimulation were used. Clinical improvement at three years was comparable for the groups of primary dystonia (82% for the motor scores) and for secondary dystonia and degenerative diseases, improvement was less important but still interesting (40%). The originality of the method consisted in the use of stereotactic MR imaging for the target selection without microelectrode

* A version of this chapter was also published in *Recent Reakthroughs in Basal ganglia Research* , edited by Erwan Bezard published by Nova Science Publishers, Inc. It was submitted for appropriate modification in an effort to encourage wider dissemination of research

† Correspondence should be addressed to Philippe Coubes: Tel/Fax: +33 (0) 4.67.33.74.64. E-mail address: p-coubes@chu-montpellier.fr or urmae@chu-montpellier.fr

recordings. This reduced the duration of the procedure and the risk of haemorrhage. Chronic high frequency stimulation of the GPi can be considered as an efficient and lasting treatment in primary dystonia and proposed in selected cases of secondary dystonia and degenerative disease.

Background and Purpose

In a wide range of medical conditions, a movement disorder can be the only symptom or associated with other neurological deficits. Most of them are of unknown origin. Dystonia, dyskinesia, myoclonus, tremor, atethosis are abnormal movements susceptible to be influenced by high frequency modulation of the basal ganglia. Nevertheless, the control of these symptoms is highly depending on the etiology. Movement disorders (primary dystonia, myoclonus dystonia, tardive dyskinesia), which are not associated to other neurologic deficits are more likely to be controlled by high frequency stimulation.

The first validated medical condition successfully treated by DBS was Parkinson's disease (Benabid et al., 1987). Gradually, the indications of this therapy were enlarged to include other diseases associating movement disorders as dystonia and dyskinesia of various etiologies. Based on the experience of our group (Gros et al., 1976) with brain lesioning surgery (pallidotomy) in a young patient with idiopathic generalized dystonia (Iacono et al., 1996) and deep brain stimulation in Parkinson's disease dyskinesias (Benabid et al., 1987; 1993), we reported, in 1996 the efficacy of DBS in a young lady presenting with severe life threatening primary generalized dystonia after chronic bilateral stimulation of the GPi (Coubes et al., 1999). Since then, 105 dystonic patients have been treated for bilateral stimulation of the Internal Globus Pallidus (GPi) in the department of Neurosurgery in Montpellier.

The originality of our method consists in the use of stereotactic MR imaging for the target determination, without microelectrode recordings. This considerably reduces the procedure's duration, which is important especially in children, as well as for the risk of hemorrhage (0 cases). Furthermore, the surgery, performed under general anesthesia is more adapted in patients with permanent involuntary movements.

We will present the population, the surgical procedure, the clinical management of the patients and the results obtained by DBS. In order to discuss the criteria for selecting the patients with dystonia-dyskinesia, we will briefly remember this medical condition. Whatever, others movement disorders being influenced by the DBS are sometimes associated to dystonia in patients to be operated.

Dystonia has been defined as a neurological syndrome characterized by involuntary, sustained muscle contractions, causing twisting and repetitive movements or abnormal postures (Fahn, 1988).

Its classification is a difficult task and can be done in several ways: according to the spread of the symptoms to different body parts, to the age of onset or to etiology.

According to distribution, dystonia is classified into one of the following categories: focal, segmental, multifocal, hemidystonia and generalized dystonia.

The age of onset is another important criterion for characterizing a patient. In case of early onset, the disease is more likely to generalize and severely worsen than for adult-onset dystonias.

Dystonia can also be classified by etiology. In primary dystonia, dystonia is the only symptom and can be sporadic or inherited (DYT1 dystonia; Ozelius et al., 1997). Dystonia-plus is a group of syndromes where dystonia is usually associated to another neurological condition such as parkinsonism or myoclonus (Dopa responsive dystonia, myoclonus-dystonia syndrome). The group of heredodegenerative dystonia includes numerous diseases and in this group, dystonia is typically not pure. Amino acid disorders, lipid disorders, Lesch-Nyhan disease, pantothenate kinase-associated neurodegeneration (PKAN), mitochondrial diseases, Wilson's disease, Huntington's disease, Juvenile parkinsonism-dystonia (Dwork et al., 1993; Ishikawa and Takahashi, 1998) and many others are heredodegenerative dystonias. Secondary dystonias are generated by insults such as drugs, strokes, tumors, infections (Fahn, 1988) and this subgroup includes also dystonia-dyskinesia secondary to cerebral palsy (CP).

In 40 % of the patients with early-onset dystonia a specific cause can be found. The etiologic diagnosis is established by clinical evaluation, neuroimaging and molecular analysis in primary dystonias. When the history of the disease, the clinical examination and the brain MRI suggest a heredogenerative dystonia, further investigations are performed such as blood work-up, urine sample analysis, CSF testing, electrophysiologic studies, muscle, skin, liver biopsies, PET, eye slit-lamp examination.

The identification of the etiology is very important for the prognosis of the disease and because of the existence of very few medical conditions having a specific treatment such as Dopa-Responsive dystonia, creatine deficiency or Wilson's disease.

Several drugs are available to treat dystonic symptoms but their efficacy is often limited, transient and difficult to assess (often because of the fluctuation of the symptoms). The most important drugs to be used are Levodopa (also as a diagnosis trial in Dopa responsive dystonia), anticholinergics, benzodiazepines, baclofen (Ford et al., 1998) and Botulinum toxin injections especially for the treatment of focal dystonias (Tsui et al., 1987)

Because of the poor efficacy of the pharmaceutical treatment in front of the very severe forms of movement disorders (especially generalized dystonias with onset in childhood) we were led to propose deep brain stimulation as another therapeutic strategy in order to control the symptoms generating life-threatening complications.

Patients (Population)

In the following report, we will present the results of 82 patients presenting with segmentary or generalized dystonia treated by bilateral chronic electrical stimulation of the GPi. The population was divided in subgroups separating, for each of them, children from adults. Group 1 included patients with primary generalized DYT1 dystonia, group 2 primary generalized dystonia without DYT1 mutation, group 3 primary segmentary dystonia (cervico-axial dystonia), group 4 generalized dystonia-dyskinesia due to postanoxic cerebral palsy, group 5 generalized dystonia secondary to PKAN and group 6 one generalized dystonia secondary to mitochondrial disease. For the reported patients, the follow-up was at least of 6 months.

Clinical Evaluation

Dystonic movements and abnormal postures were evaluated using the Burke-Fahn-Marsden-Dystonia-Rating scale (BFMDRS, motor and disability part; Burke et al., 1985) before the surgical procedure, several times during the post-operative hospital stay, every month during the first year and every three months afterwards.

Surgical Procedure

Bilateral electrode implantation was performed in a single surgical session under general anesthesia (Vayssiere et al., 2000; Coubes et al., 2002a; Vayssiere et al., 2002). The MR-compatible Leksell stereotactic frame was applied and a 3D-SPGR (spoiled gradient recall) acquisition was performed. The postero-ventral part of the GPi was located through axial, sagittal and coronal MRI studies (Figure 1A). The target coordinates (x, y, z) and the trajectory angles (α,β) were calculated using a dedicated software.

Two four contact electrodes (DBS 3389, Medtronic, Minneapolis) were implanted under strict profile radioscopic control. Immediate postoperative control MRI (Figure 1B) was obtained with the stereotactic frame on. Electrodes were connected to a pulse generator five days later (Itrel II or III, Kinetra and Soletra Medtronic, Minneapolis, USA), which was subcutaneously introduced in the abdominal area. In each patient final placement of the implanted devices and their connections were checked by a radiographic control during the postoperative hospital stay (Figure 2).

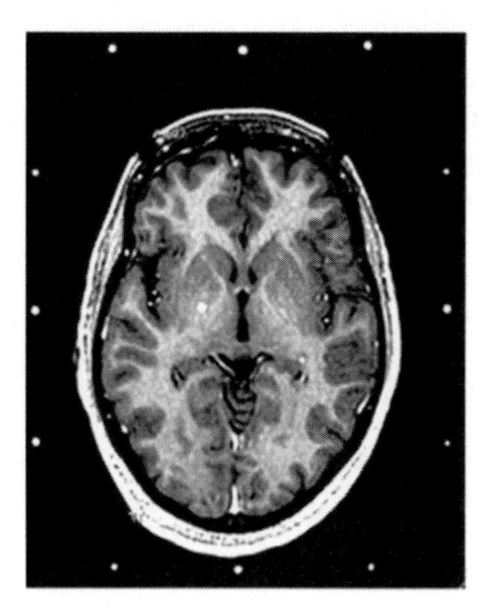
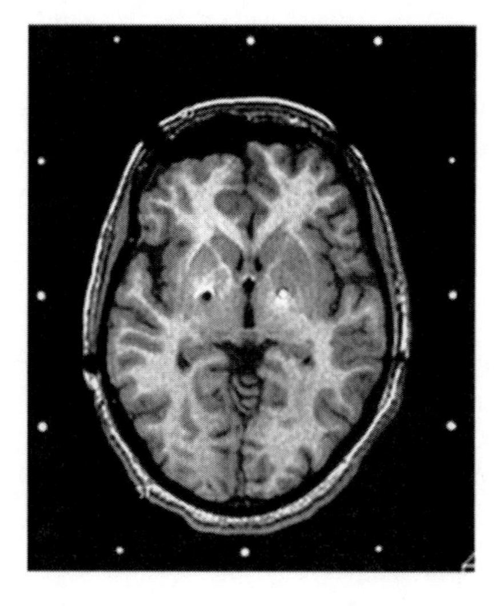

Figure 1. Stereotactic MRI. A. Pre-operative planning. B. Post-operative MRI to control the final electrode position (black points = electrode artefacts).

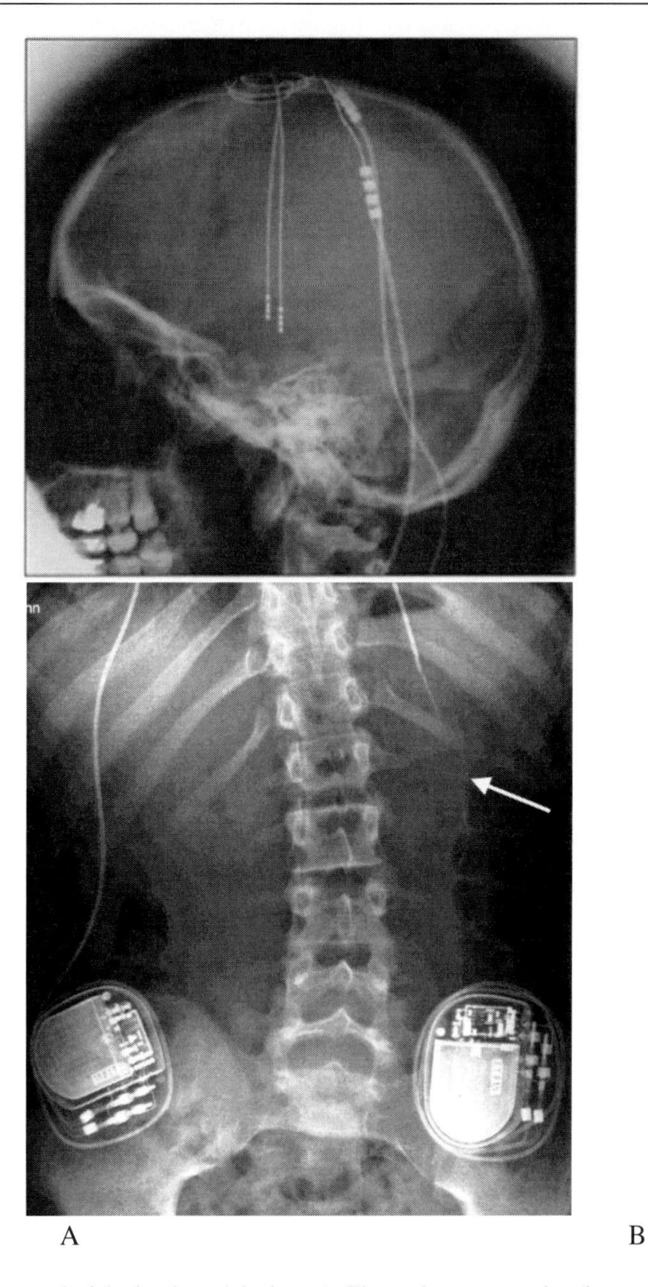

A B

Figure 2. Radiographic control of the implanted devices. A. Electrodes connected to the extensions. B. Neurostimulators with extension length reserve. Arrow indicates an extension fracture during sports.

Electric Parameter Settings

After implantation, stimulators were switched on. Electrical variables were set at high frequency (130 Hz), 450 μsec for the pulse width with one contact negatively activated. Intensity was progressively increased according to the needs of each patient (Hemm et al., 2004) and the clinical evolution. Usually the first levels for the voltage were between 0.5 V-

0.8 V. Over the time, we modified the electric parameters in several patients, by increasing the voltage or activating a second contact when requested. The mean steady state value of the voltage was 1.6 ± 0.3 V (Hemm et al., 2002).

Clinical Results

Motor and disability scores' evolution (BMFDRS) for primary and secondary dystonia is presented in tables I and II. Within each group, the improvement was progressive over time. With more than 3 years of follow-up, the clinical improvement was comparable for the two groups of primary dystonia (82% of improvement on the motor scale). The results obtained in the group of secondary dystonia and heredodegenerative diseases are less important but yet around 40 % with 3 years of follow-up. After three years, the disability score improvement was superior in the group of primary DYT1 dystonia (80 %) compared to non-DYT1 primary dystonia (56 %) and to secondary dystonia (19 %). The group of secondary dystonia and heredodegenerative diseases is a very heterogeneous group. This is why results for all etiologies should be presented separately.

Discussion

We report here our experience with bilateral chronic electrical stimulation of the GPi in the treatment of primary DYT1 positive and DYT1 negative generalized dystonia (Coubes et al., 1999; 2000; 2002b; Cif et al., 2004; Coubes et al., 2004), generalized dystonia–dyskinesia secondary to post-anoxic cerebral palsy, generalized dystonia secondary to PKAN syndrome and to mitochondrial diseases. We also report the results obtained by DBS in primary segmentary dystonia. Since the first child has been operated, 126 other patients underwent surgery for DBS in our department (Dystonia + Parkinson).

While at the beginning DBS was proposed in children with primary generalized dystonia (with or without DYT1 mutation), selection criteria were revisited and enlarged for including now other types of dystonia in children and adults. Several patients with generalized dystonia associating myoclonus underwent surgery for chronic electrical stimulation of the GPi and we could see an early and complete control of myoclonus. These findings led us to propose this treatment in a child with genetically proven myoclonus-dystonia syndrome (Cif et al., 2004) MDS, DYT11, mutation in the epsilon-amino sarcoglycan gene, (Gasser, 1998; Zimprich et al., 2001) and we obtained a very satisfactory improvement of his symptoms. We confirmed in a second patient with MDS the efficiency of GPi stimulation.

Being confronted with very severe clinical conditions in patients with secondary dystonia and heredodegenerative diseases, in which the efficacy of the medical treatment was poor and in which dystonic movements and postures were comparable with those met in primary dystonia, we proposed DBS in several selected patients of these groups.

The criteria for patient selection in this group were clinical, electrophysiological and based on brain imaging as well as on etiology.

Table I. Motor scores (BMFDRS) of dystonic children treated by DBS. Mean values ±SD. The values in brackets represent the mean improvement in percent. A reduction in the score indicates an improvement in function. BMFDRS denotes Burke-Marsden Fahn's Dystonia Rating Scale

Motor score		6 months after surgery/120 (%)[1]	1 year after surgery/120 (%)[1]	2 years after surgery/120 (%)[1]	>3 years after surgery/120 (%)[1]
PGD[2] DYT1+[3]	Children	80±20 n[5]=14	77±28 n[5]=13	76±25 n[5]=12	82±18 n[5]=10
	Adults	75±17 n[5]=10	75±21 n[5]=8	70±16 n[5]=5	73±13 n[5]=2
PGD[2] DYT1-	Children	72±15 n[5]=12	74±15 n[5]=12	66±26 n[5]=11	72±21 n[5]=9
	Adults	53±31 n[5]=19	64±33 n[5]=19	73±26 n[5]=16	71±26 n[5]=7
Cervico-axial dystonia	Children	77±0 n[5]=1	93±0 n[5]=1	93±12 n[5]=1	100±10 n[5]=1
	Adults	70±34 n[5]=7	88±20 n[5]=7	85±21 n[5]=7	84±12 n[5]=4
Post-anoxic cerebral palsy	Children	31±8 n[5]=5	37±6 n[5]=2	45±0 n[5]=1	51±0 n[5]=1
	Adults	41±25 n[5]=5	38±10 n[5]=5	45±15 n[5]=3	62±0 n[5]=1
PKAN[4]	Children	51±29 n[5]=4	59±25 n[5]=4	22±5 n[5]=2	33±13 n[5]=2
	Adults	71±12 n[5]=2	73±10 n[5]=2	74±1 n[5]=2	88±0 n[5]=1
Mitochondrial disease	Children	27±17 n[5]=3	46±30 n[5]=2	25±10 n[5]=2	42±6 n[5]=2

[1]The improvement in percent is calculated based on the maximal possible gain $(Score_{preop}-Score_{postop})/(Score_{preop})$.

[2]Primary generalized dystonia

[3]DYT1 mutation

[4]Pantothenate kinase-associated neurodegeneration

[5]Number of patients

Table II. Disability scores (BMFDRS) of dystonic children treated by DBS. Mean values ±SD. The values in brackets represent the mean improvement in percent. A reduction in the score indicates an improvement in function. BMFDRS denotes Burke-Marsden Fahn's Dystonia Rating Scale

Disability score		6 months after surgery/120 (%)[1]	1 year after surgery/120 (%)[1]	2 years after surgery/120 (%)[1]	>3 years after surgery/120 (%)[1]
PGD[2] DYT1+[3]	Children	58±39 n[5]=14	62±39 n[5]=13	63±27 n[5]=12	80±14 n[5]=10
	Adults	61±25 n[5]=10	64±29 n[5]=8	74±14 n[5]=52	58±3 n[5]=2
PGD[2] DYT1-	Children	42±23 n[5]=12	52±21 n[5]=12	54±24 n[5]=11	49±29 n[5]=9
	Adults	35±34 n[5]=19	52±31 n[5]=19	67±30 n[5]=16	58±36 n[5]=7
Cervico-axial dystonia	Children	100±0 n[5]=1	92±0 n[5]=1	92±0 n[5]=1	100±0 n[5]=1
	Adults	55±36 n[5]=7	74±22 n[5]=7	86±18 n[5]=7	74±12 n[5]=4
Cerebral palsy	Children	4±3 n[5]=5	11±4 n[5]=2	19±0 n[5]=1	19±0 n[5]=1
	Adults	21±22 n[5]=5	28±21 n[5]=5	17±12 n[5]=3	27±0 n[5]=1
PKAN[4]	Children	35±32 n[5]=4	28±28 n[5]=4	2±2 n[5]=2	7±14 n[5]=2
	Adults	31±20 n[5]=2	43±10 n[5]=2	62±7 n[5]=2	65±0 n[5]=1
Mitochondrial disease	Children	8±5 n[5]=3	10±3 n[5]=2	11±3 n[5]=2	8±1 n[5]=2

[1]The improvement in percent is calculated based on the maximal possible gain $(Score_{preop}-Score_{postop})/(Score_{preop})$.
[2]Primary generalized dystonia
[3]DYT1 mutation
[4]Pantothenate kinase-associated neurodegeneration
[5]Number of patients

We performed surgery for DBS in patients in whom dystonia and dyskinesia were prominent compared to other neurological deficits (especially motor deficit and spasticity), the motor pattern was preserved and in patients presenting with severe or life threatening symptoms due to dystonia-dyskinesia (swallowing difficulties, permanent opisthotonos, painful muscle spasms).

Electroencephalogram, electroretinogram, visual and brainstem, somatosensory evoked responses were obtained. Motor evoked potentials were performed in elder children to identify pyramidal tract impairment. In order to exclude brain abnormalities contraindicating surgery (major cortical atrophy, severe periventicular leucomalacia especially met in cerebral palsy, basal ganglia and thalamic lesions), brain MR under general anesthesia was performed in all patients.

As shown in tables I and II, best results were obtained within the group of patients with DYT1 mutation (Coubes et al., 1999; 2000; Roubertie et al., 2000; Coubes et al., 2004). The surgery of abnormal movements should intervene before the occurrence of skeletal deformities, which always diminish the outcome. Within the population of non-DYT1 dystonia and especially in secondary and heredodegenerative dystonias, results are not so predictable. The improvement of dysarthria is variable and often very few influenced by DBS.

Stimulation's switch-off systematically causes the recurrence of symptoms within some hours or days. Whatever, in several patients we could see a long lasting preservation of the clinical improvement for several weeks without stimulation.

In the secondary dystonia group, the efficacy of stimulation is far more limited. We were led to propose it for very handicapped patients for whom other therapeutic strategies failed to improve dystonia, as already mentioned. The clinical and etiological heterogeneity among this group almost prevents any global interpretation of these results. In this group, a frequently associated hypertonia of pyramidal origin influences the dystonic component.

An important negative prognostic factor under stimulation is the existence of a permanent hypertonia at rest, whatever may be its origin. Although we are not yet able to predict the long-term prognosis, we observed an interesting improvement with dyskinesia-dystonia secondary to a perinatal anoxia (dyskinetic forms of cerebral palsy accounting for less than 10 % of all forms CP; Lin, 2003), PKAN and mitochondrial diseases treated by GPi stimulation. A constant control of pain associated with muscle spasms was obtained in patients suffering from secondary dystonia.

The progression of the causal disease is of critical importance for patient's prognosis.

Using this surgical method based on MRI alone (Vayssiere et al., 2000; Coubes et al., 2002a; Vayssiere et al., 2002), the associated morbidity is low. We didn't observe hemorrhage due to the intracerebral tracts as reported before in other series (our experience reaches 236 electrodes; Starr et al., 1999). Secondary infection of the stimulation system remains the major complication of this technique and was observed in 4 patients. We summarize the complications observed in our population in table III.

Table III. Complications

Complications	Number
Hemorrage	0
Infection	4
Lead fracture	3
Extension fracture	3

The remarkable tolerance of the internal pulse generator must be also emphasized in children. We never observed any complication due to displacement with growth (loss of efficacy linked to the displacement of the electrode). As shown in figure 2 and 3, a residual length was enrolled around the battery and the electrode in order to compensate for growth in children and to provide some flexibility with movements in the system (Figure 2). Furthermore, it appears that growth does not interfere with stimulation, and the implantation of a single 90 cm extension compensates adequately for the growth of the child. We observed in two patients (1 child, 1 adult) an extension fracture due to adherences fixing the extension and limiting its mobility and flexibility.

The children's physical development (height, body weight) was followed as well as hormonal levels (Insuline-like Growth Factor-IGF1, Insuline–like Growth Factor Binding Protein 3-IGF-BP3, Estradiol, Testosterone, Folliculine Stimulating Hormone-FSH and Luteinizing Hormone-LH) in order to check puberty development.

Conclusion

Despite cost and complexity of the follow-up, bilateral chronic electrical stimulation can be proposed as first line treatment for early onset primary generalized dystonia when pharmacologically intractable and also early considered in segmentary dystonia in adults and in well-selected cases of secondary dystonia. It is conservative, adaptable, reversible and well tolerated by the whole population. It must be applied soon, especially in primary dystonia before neuro-ortopaedic sequels occur. The complication rate remains low.

For secondary dystonia, pallidal stimulation can partially improve dystonic syndromes with important control of pain and swallowing difficulties.

References

Benabid AL, Pollak P, Louveau A, Henry S, de Rougemont J (1987) Combined (thalamotomy and stimulation) stereotactic surgery of the VIM thalamic nucleus for bilateral Parkinson disease. *Appl Neurophysiol* **50**:344-346.

Benabid AL, Pollak P, Seigneuret E, Hoffmann D, Gay E, Perret J (1993) Chronic VIM thalamic stimulation in Parkinson's disease, essential tremor and extra-pyramidal dyskinesias. *Acta Neurochir Suppl (Wien)* **58**:39-44.

Burke RE, Fahn S, Marsden CD, Bressman SB, Moskowitz C, Friedman J (1985) Validity and reliability of a rating scale for the primary torsion dystonias. *Neurology* **35**:73-77.

Cif L, Valente EM, Hemm S, Coubes C, Vayssiere N, Serrat S, Di Giorgio A, Coubes P (2004) Deep brain stimulation in myoclonus-dystonia syndrome. *Mov Disord* **19**:724-727.

Coubes P, Roubertie A, Vayssiere N, Hemm S, Echenne B (2000) Treatment of DYT1-generalised dystonia by stimulation of the internal globus pallidus. *Lancet* **355**: 2220-2221.

Coubes P, Vayssiere N, El Fertit H, Hemm S, Cif L, Kienlen J, Bonafe A, Frerebeau P (2002a) Deep brain stimulation for dystonia. Surgical technique. *Stereotact Funct Neurosurg* **78**:183-191.

Coubes P, Echenne B, Roubertie A, Vayssiere N, Tuffery S, Humbertclaude V, Cambonie G, Claustres M, Frerebeau P (1999) [Treatment of early-onset generalized dystonia by chronic bilateral stimulation of the internal globus pallidus. Apropos of a case]. *Neurochirurgie* **45**:139-144.

Coubes P, Cif L, El Fertit H, Hemm S, Vayssiere N, Serrat S, Picot MC, Tuffery S, Claustres M, Echenne B, Frerebeau P (2004) Electrical stimulation of the globus pallidus internus in patients with primary generalized dystonia: long-term results. *J Neurosurg* **101**:189-194.

Coubes P, Cif L, Azais M, Roubertie A, Hemm S, Diakonoya N, Vayssiere N, Monnier C, Hardouin E, Ganau A, Tuffery S, Claustre M, Echenne B (2002b) [Treatment of dystonia syndrome by chronic electric stimulation of the internal globus pallidus]. *Arch Pediatr* **9** *Suppl* **2**:84s-86s.

Dwork AJ, Balmaceda C, Fazzini EA, MacCollin M, Cote L, Fahn S (1993) Dominantly inherited, early-onset parkinsonism: neuropathology of a new form. Neurology 43:69-74.

Fahn S (1988) Concept and classification of dystonia. *Adv Neurol* **50**:1-8.

Ford B, Greene PE, Louis ED, Bressman SB, Goodman RR, Brin MF, Sadiq S, Fahn S (1998) Intrathecal baclofen in the treatment of dystonia. *Adv Neurol* **78**:199-210.

Gasser T (1998) Inherited myoclonus-dystonia syndrome. Adv Neurol 78:325-334.

Gros C, Frerebeau P, Perez-Dominguez E, Bazin M, M PJ (1976) Long term results of stereotaxic surgery for infantile dystonia and dyskinesia. *Neurochirurgia (Stuttg)* **19**: 171-178.

Hemm S, N D, G M, N V, L C, P C (2002) Stimulated volume and energy consumption in improved dystonic patients treated by high frequency GPi stimulation. *Movement Disorders* **17**:S302.

Hemm S, Vayssiere N, Mennessier G, Cif L, Zanca M, Ravel P, Frerebeau P, P C (2004) Evolution of brain impedance in dystonic patients treated by GPi electrical stimulation. *Neuromodulation* **7**:67-75.

Iacono RP, Kuniyoshi SM, Lonser RR, Maeda G, Inae AM, Ashwal S (1996) Simultaneous bilateral pallidoansotomy for idiopathic dystonia musculorum deformans. *Pediatr Neurol* **14**:145-148.

Ishikawa A, Takahashi H (1998) Clinical and neuropathological aspects of autosomal recessive juvenile parkinsonism. *J Neurol* 245:P4-9.

Lin JP (2003) The cerebral palsies: a physiological approach. *J Neurol Neurosurg Psychiatry* **74** Suppl 1:i23-29.

Ozelius LJ, Hewett JW, Page CE, Bressman SB, Kramer PL, Shalish C, de Leon D, Brin MF, Raymond D, Corey DP, Fahn S, Risch NJ, Buckler AJ, Gusella JF, Breakefield XO (1997) The early-onset torsion dystonia gene (DYT1) encodes an ATP-binding protein. *Nat Genet* **17**:40-48.

Roubertie A, Echenne B, Cif L, Vayssiere N, Hemm S, Coubes P (2000) Treatment of early-onset dystonia: update and a new perspective. *Childs Nerv Syst* **16**:334-340.

Starr PA, Vitek JL, DeLong M, Bakay RA (1999) Magnetic resonance imaging-based stereotactic localization of the globus pallidus and subthalamic nucleus. *Neurosurgery* **44**:303-313; discussion 313-304.

Tsui JK, Fross RD, Calne S, Calne DB (1987) Local treatment of spasmodic torticollis with botulinum toxin. *Can J Neurol Sci* **14**:533-535.

Vayssiere N, Hemm S, Zanca M, Picot MC, Bonafe A, Cif L, Frerebeau P, Coubes P (2000) Magnetic resonance imaging stereotactic target localization for deep brain stimulation in dystonic children. *J Neurosurg* **93**:784-790.

Vayssiere N, Hemm S, Cif L, Picot MC, Diakonova N, El Fertit H, Frerebeau P, Coubes P (2002) Comparison of atlas- and magnetic resonance imaging-based stereotactic targeting of the globus pallidus internus in the performance of deep brain stimulation for treatment of dystonia. *J Neurosurg* **96**:673-679.

Zimprich A, Grabowski M, Asmus F, Naumann M, Berg D, Bertram M, Scheidtmann K, Kern P, Winkelmann J, Muller-Myhsok B, Riedel L, Bauer M, Muller T, Castro M, Meitinger T, Strom TM, Gasser T (2001) Mutations in the gene encoding epsilon-sarcoglycan cause myoclonus-dystonia syndrome. *Nat Genet* **29**:66-69.